T0306107

THE PREDICTIVE POWER OF COMPUTATIONAL
ASTROPHYSICS AS A DISCOVERY TOOL

IAU SYMPOSIUM 362

COVER ILLUSTRATION:

Supercomputer MareNostrum, Barcelona, Spain. Photo by D. Wiebe

IAU SYMPOSIUM PROCEEDINGS SERIES

Chief Editor
JOSÉ MIGUEL RODRÍGUEZ ESPINOSA, General Secretariat
Instituto de Astrofísica de Andalucía
Glorieta de la Astronomia s/n
18008 Granada
Spain
IAU-general.secretary@iap.fr

Editor
DIANA WORRALL, Assistant General Secretary
HH Wills Physics Laboratory
University of Bristol
Tyndall Avenue
Bristol
BS8 1TL
UK
IAU-assistant.general.secretary@iap.fr

INTERNATIONAL ASTRONOMICAL UNION

UNION ASTRONOMIQUE INTERNATIONALE

International Astronomical Union

THE PREDICTIVE POWER OF COMPUTATIONAL ASTROPHYSICS AS A DISCOVERY TOOL

PROCEEDINGS OF THE 362nd SYMPOSIUM OF THE INTERNATIONAL ASTRONOMICAL UNION VIRTUAL MEETING, ORIGINALLY PLANNED FOR CHAMONIX, FRANCE 8–12 NOVEMBER, 2021

Edited by

DMITRY BISIKALO
Institute of Astronomy of the Russian Academy of Sciences, Russia

DMITRI WIEBE
Institute of Astronomy of the Russian Academy of Sciences, Russia

and

CHRISTIAN BOILY
Observatoire astronomique, Université de Strasbourg, France

CAMBRIDGE
UNIVERSITY PRESS

Shaftesbury Road, Cambridge CB2 8EA, United Kingdom

One Liberty Plaza, 20th Floor, New York, NY 10006, USA

477 Williamstown Road, Port Melbourne, VIC 3207, Australia

314–321, 3rd Floor, Plot 3, Splendor Forum, Jasola District Centre, New Delhi – 110025, India

103 Penang Road, #05–06/07, Visioncrest Commercial, Singapore 238467

Cambridge University Press is part of Cambridge University Press & Assessment, a department of the University of Cambridge.

We share the University's mission to contribute to society through the pursuit of education, learning and research at the highest international levels of excellence.

www.cambridge.org
Information on this title: www.cambridge.org/9781108490665

© International Astronomical Union 2023

First published 2023

A catalogue record for this publication is available from the British Library

ISBN   978-1-108-49066-5   Hardback

Table of Contents

Preface x

Editors xi

List of Participants xii

Early structure formation in the THESAN radiation-magneto-hydrodynamics
simulations 1
    E. Garaldi, R. Kannan, A. Smith, V. Springel, R. Pakmor,
    M. Vogelsberger and L. Hernquist

Simulations of dark matter with frequent self-interactions 8
    Moritz S. Fischer

Simulations of the reionization of the clumpy intergalactic medium with a novel
particle-based two-moment radiative transfer scheme 15
    Tsang Keung Chan, Alejandro Benitez-Llambay, Tom Theuns and
    Carlos Frenk

Favoured Inflationary Models by SFC Baryogenesis 21
    Mariana Panayotova and Daniela Kirilova

The Effect of Non-Gaussian Primordial Perturbations on Large-Scale
Structure 26
    G. A. Peña and G. N. Candlish

The complex evolution of supermassive black holes in cosmological
simulations 33
    Peter H. Johansson, Matias Mannerkoski, Antti Rantala, Shihong Liao,
    Alexander Rawlings, Dimitrios Irodotou and Francesco Rizzuto

Nature of star formation in first galaxies 39
    Mahavir Sharma

The formation of Supersonically Induced Gas Objects (SIGOs)
with H_2 cooling 45
    Yurina Nakazato, Gen Chiaki, Naoki Yoshida, Smadar Naoz,
    William Lake and Chiou Yeou

An idealized setup for cosmological evolution of baryonic gas
in isolated halos 51
    S. Dattathri and P. Sharma

Detecting cosmic filamentary network with stochastic Bisous model 54
    Moorits Mihkel Muru

Absorption spectra from galactic wind models: a framework to link PLUTO
simulations to TRIDENT 56
    Benedetta Casavecchia, Wladimir E. Banda-Barragán, Marcus Brüggen
    and Fabrizio Brighenti

Hydrodynamic Simulations and Time-dependent Photoionization Modeling of
Starburst-driven Superwinds 64
    A. Danehkar, M. S. Oey and W. J. Gray

Numerical modelling of X-shaped radio galaxies using back-flow model 70
    Gourab Giri and Bhargav Vaidya

Radiation Magnetohydrodynamic Simulations of Soft X-ray Emitting Regions in
Active Galactic Nuclei 76
    Taichi Igarashi, Yoshiaki Kato, Hiroyuki R. Takahashi, Ken Ohsuga,
    Yosuke Matsumoto and Ryoji Matsumoto

Cold Gas in Outflow: Evidence for Delayed Positive AGN Feedback 82
    Yu Qiu

A New Code for Relativistic Hydrodynamics and its Application to
FR II Radio Jets 87
    Jeongbhin Seo, Hyesung Kang and Dongsu Ryu

Studying Magnetic Field Amplification in Interacting Galaxies Using Numerical
Simulations 94
    Simon Selg and Wolfram Schmidt

X-ray spectral and image spatial models of NGC 3081 with *Chandra* data 100
    O.V. Kompaniiets, Iu.V. Babyk, A.A. Vasylenko, I.O. Izviekova and
    I.B. Vavilova

Simulating star formation in spiral galaxies 105
    Steven Rieder, Clare Dobbs, Thomas Bending, Kong You Liow and
    James Wurster

The CNN classification of galaxies by their image morphological
peculiarities 111
    D. Dobrycheva, V. Khramtsov, M. Vasylenko and I. Vavilova

Velocity-space substructures and bar resonances in an *N*-body Milky Way 116
    Tetsuro Asano, Michiko S. Fujii, Junich Baba, Jeroen Bédorf,
    Elena Sellentin and Simon Portegies Zwart

Fate of escaping orbits in barred galaxies 122
    Debasish Mondal and Tanuka Chattopadhyay

Dynamical evolution modeling of the Collinder 135 & UBC 7 binary
star cluster 128
    Marina Ishchenko, Peter Berczik and Nina Kharchenko

Multiparticle collision simulations of dense stellar systems and plasmas 134
    P. Di Cintio, M. Pasquato, L. Barbieri, H. Bufferand, L. Casetti,
    G. Ciraolo, U. N. di Carlo, P. Ghendrih, J. P. Gunn, S. Gupta, H. Kim,
    S. Lepri, R. Livi, A. Simon-Petit, A. A. Trani and S.-J. Yoon

A novel generative method for star clusters from hydro-dynamical
simulations 141
    Stefano Torniamenti

Stellar Population Photometric Synthesis with AI of S-PLUS galaxies 148
Vitor Cernic, for the S-PLUS collaboration

Clustering stellar pairs to detect extended stellar structures 150
Sergey Sapozhnikov and Dana Kovaleva

Kinetic modeling of auroral events at solar and extrasolar planets 152
Valery I. Shematovich and Dmitry V. Bisikalo

Stellar activity effects on the atmospheric escape of hot Jupiters 158
Hiroto Mitani, Riouhei Nakatani and Naoki Yoshida

Study of the non-thermal atmospheric loss for exoplanet π Men c 164
Anastasia A. Avtaeva and Valery I. Shematovich

Multi-component MHD model for hydrogen-helium extended envelope
of hot Jupiter 167
Y.G. Gladysheva, A.G. Zhilkin and D.V. Bisikalo

Modeling Stellar Jitter for the Detection of Earth-Mass Exoplanets via Precision
Radial Velocity Measurements 169
Samuel Granovsky, Irina N. Kitiashvili and Alan A. Wray

Thermal atmospheric escape of close-in exoplanets 173
Eugenia S. Kalinicheva and Valery I. Shematovich

Inference of Magnetic Fields and Space Weather Hazards of Rocky Extrasolar
Planets From a Dynamical Geophysical Model 175
*Varnana M. Kumar, Thara N. Sathyan, T.E. Girish, P.E. Eapen,
Biju Longhinos and J. Binoy*

Hydrodynamical Simulations of Misaligned Accretion Discs in Binary Systems:
Companions tear discs 177
S. Doğan, C. J. Nixon, A. R. King, J. E. Pringle and D. Price

Hybrid magnetic structures around spinning black holes connected to a
surrounding accretion disk 184
I. El Mellah, B. Cerutti, B. Crinquand and K. Parfrey

Mass ejection from neutron-star mergers 190
*Masaru Shibata, Sho Fujibayashi, Kota Hayashi, Kenta Kiuchi and
Shinya Wanajo*

Merging of spinning binary black holes in globular clusters 203
*Margarita Sobolenko, Peter Berczik, Manuel Arca Sedda,
Konrad Maliszewski, Mirek Giersz and Rainer Spurzem*

Predicting the Expansion of Supernova Shells Using Deep Learning toward
Highly Resolved Galaxy Simulations 209
*Keiya Hirashima, Kana Moriwaki, Michiko Fujii, Yutaka Hirai,
Takayuki Saitoh and Junichiro Makino*

Toward Realistic Models of Core Collapse Supernovae: A Brief Review 215
Anthony Mezzacappa

Hot spot drift in synchronous and asynchronous polars: synthesis of light curves   228
     Andrey Sobolev, Dmitry Bisikalo and Andrey Zhilkin

The predictive power of numerical simulations to study accretion and outflow in
T Tauri Stars   234
     Ana I. Gómez de Castro

Feedback from the Vicinity of Massive Protostars in the First
Star Formation   246
     Kazutaka Kimura

A cloud-cloud collision in Sgr B2? 3D simulations meet SiO observations   250
     Wladimir E. Banda-Barragán, Jairo Armijos-Abendaño and Helga Dénes

Formation process of the Orion Nebula Cluster   258
     Michiko S. Fujii, Long Wang, Takayuki R.Saitoh, Yutaka Hirai and
     Yoshito Shimajiri

PION: Simulations of Wind-Blown Nebulae   262
     Jonathan Mackey, Samuel Green, Maria Moutzouri, Thomas J. Haworth,
     Robert D. Kavanagh, Maggie Celeste, Robert Brose, Davit Zargaryan and
     Ciarán O'Rourke

Infrared appearance of wind-blown bubbles around young massive stars   268
     Maria S. Kirsanova and Yaroslav N. Pavlyuchenkov

Numerical 2D MHD simulations of the collapse of magnetic rotating protostellar
clouds with the Enlil code   273
     Sergey Khaibrakhmanov, Sergey Zamozdra, Natalya Kargaltseva,
     Andrey Zhilkin and Alexander Dudorov

Two-dimensional MHD model of gas flow dynamics near a young star with a jet
and a protoplanetary disk   279
     V. V. Grigoryev, D. V. Dmitriev and T. V. Demidova

PRESTALINE: A package for analysis and simulation of star forming regions . .   282
     G. Van Looveren, O. V. Kochina and D. S. Wiebe

Episodic accretion onto a protostar   288
     Tomoyuki Hanawa, Nami Sakai and Satoshi Yamamoto

Photoevaporation of Protoplanetary Disks   294
     Ayano Komaki, Riouhei Nakatani and Naoki Yoshida

Modeling protoplanetary disk evolution in young star forming regions   300
     Martijn J. C. Wilhelm, Simon Portegies Zwart, Claude
     Cournoyer-Cloutier, Sean Lewis, Brooke Polak, Aaron Tran,
     Mordecai-Mark Mac Low and Stephen L. W. McMillan

Long-Term Evolution of Convectively Unstable Disk   306
     Lomara Maksimova and Yaroslav Pavlyuchenkov

Heliosphere in the Local Interstellar Medium   309
     Nikolai V. Pogorelov

3D Realistic Modeling of Solar-Type Stars to Characterize the Stellar Jitter . . . 324
Irina Kitiashvili, Samuel Granovsky and Alan Wray

Dynamic complexity based analysis on the relationship between solar activity and
cosmic ray intensity 330
Vipindas V., Sumesh Gopinath, Vinod Kumar R. and Girish T.E.

Advances and Challenges in Observations and Modeling of the Global-Sun
Dynamics and Dynamo 333
*Alexander Kosovichev, Gustavo Guerrero, Andrey Stejko, Valery Pipin
and Alexander Getling*

From evolved Long-Period-Variable stars to the evolution of M31 353
*Maryam Torki, Mahdieh Navabi, Atefeh Javadi, Elham Saremi,
Jacco Th. van Loon and Sepideh Ghaziasgar*

Applicability of the Bulirsch-Stoer algorithm in the circular restricted
three-body problem 356
Tatiana Demidova

A radiation hydrodynamics scheme on adaptive meshes using the Variable
Eddington Tensor (VET) closure 358
*Shyam H. Menon, Christoph Federrath, Mark R. Krumholz, Rolf Kuiper,
Benjamin D. Wibking and Manuel Jung*

Interplay of various particle acceleration processes in
astrophysical environment 365
Sayan Kundu and Bhargav Vaidya

GP-MOOD: a positivity-preserving high-order finite volume method for
hyperbolic conservation laws 373
Dongwook Lee and Rémi Bourgeois

Converting sink particles to stars in hydrodynamical simulations 380
Kong You Liow, Steven Rieder, Clare Dobbs and Sarah Jaffa

Modelling astrophysical fluids with particles 382
Stephan Rosswog

Detecting vortices in fluid dynamics simulations using computer vision 398
Thomas Rometsch

Modeling gravitational few-body problems with TSUNAMI and OKINAMI 404
Alessandro A. Trani and Mario Spera

Plasmoid Dominated Magnetic Reconnection and Particle Acceleration 410
Arghyadeep Paul, Sirsha Nandy and Bhargav Vaidya

Author Index 417

Preface

This special volume includes contributions from the IAUS 362 Symposium, The Predictive Power of Computational Astrophysics as a Discovery Tool. Computational astrophysics indeed rapidly becomes an indispensable tool for data-handling and making scientific discoveries in astronomy. A spectacular example is the precise calculation of gravitational wave forms coupled with sophisticated algorithms for signal analysis, together enabling a reliable gravitational wave detection. The main objective of this Symposium was to capitalize on these and other exciting advances. Our intention was to bring together both top scientists and students in a broad variety of research fields to summarize major achievements and outstanding challenges from theory and observations.

The initial plan was to hold a meeting in France, in June 2020. For obvious reasons we first had to postpone the Symposium until 2021, and then we decided to make it a fully online event in November 2021. Still, this has not prevented us from having a diverse and fruitful meeting. We have considered various options for organizing this event, including available commercial solutions, and finally decided that a professional Zoom account (kindly provided by the IAU) in combination with a dedicated Slack working space would fit all our needs. That proved to be a viable solution. We also offered a WonderMe space for private discussions.

The list of registered participants is just over 200 and consists of representatives from 35 countries, with two most significant delegations from the USA and Russia. While online meetings do have some disadvantages, the total participant number far exceeds the number of participants that had registered for the initial dates. Online format has allowed the Symposium to be attended by people from underrepresented countries, which would otherwise have not been able to participate.

The scientific program of the Symposium was quite extended with 17 invited talks, 95 contributed talks, and 23 posters, and consisted of whole-day sessions. As we had to take time zones into account, it was impossible to organize truly topical sessions, but we still succeeded in keeping some subject organization, while respecting speakers' comfort. Two technical support teams from Japan and Russia provided Zoom and Slack functioning, distribution of links, time keeping etc.

Overall, the symposium was very inspiring and, hopefully, useful. All the presented talks were quite informative. Of course, some of them raised lots of questions, but it is a normal situation in science, and it does not mean that they are not interesting. The symposium participants were of very different levels, from students to a Nobel prize winner. The range of topics was also very broad, therefore the SOC spent a lot of time trying to combine wideness and deepness. Finally, we managed to reach a good combination of reviews and contributed talks that gave us both an extensive introduction to main topics and the highest level of specific studies.

We wish to express our gratitude to Edouard Audit and his team at CEA for setting up the registration desk and handling the bookings. The Symposium would have been impossible without technical and financial assistance of the University of Tokyo and the Institute of Astronomy of the Russian Academy of Sciences.

Dmitry Bisikalo
Dmitri Wiebe
Christian Boily

Editors

Dmitry Bisikalo
Institute of Astronomy of the Russian Academy of Sciences, Russia

Dmitri Wiebe
Institute of Astronomy of the Russian Academy of Sciences, Russia

Christian Boily
Observatoire astronomique, Université de Strasbourg, France

SOC:

Dmitry Bisikalo (Russia), chair
Christian Boily (France), co-chair
Tomoyuki Hanawa (Japan), co-chair
James Stone (USA), co-chair
Edouard Audit (France)
Barbara Ercolano (Germany)
Michiko Fujii (Japan)
Erik Katsavounidis (USA)
Irina Kitiashvili (USA)
Michela Mapelli (Italy)
Garrelt Mellema (Sweden)
Shazrene Mohamed (South Africa)
Elisabete M. de Gouveia Dal Pino (Brazil)
Dongsu Ryu (South Korea)
Dmitri Wiebe (Russia)
Feng Yuan (China)
Simon Portegies Zwart (The Netherlands)

LOC:

Edouard Audit (France), chair
Christian Boily (France)
Michiko Fujii (Japan)
Dmitri Wiebe (Russia)

List of Participants

Abdellaoui, S.
Acharya, A.
Achikanath Chirakkara, R.
Arminjon, M.
Arthur, L.
Asano, T.
Asher, A.
Audit, E.
Avtaeva, A.
Babyk, Iu
Banda-Barragan, W.
Barnes, L.
Barnes, R.
Berczik, P.
Betranhandy, A.
Bhattacharjee, A.
Birky, J.
Bisikalo, D.
Bisnovatyi-Kogan, G.
Bland-Hawthorn, J.
Boily, Ch.
Bromm, V.
Brown, A.
Buslaeva, A.
Calderón, D.
Canameras, R.
Casanueva, C.
Casavecchia, B.
Caux, E.
Cernic, V.
Chan, T.K.
Cho, H.
Dall'Amico, M.
Danehkar, A.
Dattathri, Sh
Demidova, T.
d'Emilio, V.
Dencs, Z.
Dharmawardena, Th.
Di Cintio, P.
Dobrycheva, D.
Dogan, S.
Dupont, J.
Eggenberger Andersen, O.
El Mellah, I.
Elbakyan, V.
Faria, S.
Fischer, M.

Fujii, M.
Garaldi, E.
Gaspari, M.
Giri, G.
Gladysheva, Yu.
Glines, F.
Gomez de Castro, A.I.
Granovsky, S.
Grete, Ph.
Grigoryev, V.
Grudić, M.
Guszejnov, D.
Hanawa, T.
Hennebelle, P.
Hirashima, K.
Hu, Ch.-Y.
Huijuan, W.
Hutter, A.
Igarashi, T.
Ilhe, Sh.
Illarionov, E.
Ishchenko, M.
Ishizaki, R.
Izzo, L.
Jack, D.
Johansson, P.
Kalinicheva, E.
Katsavounidis, E.
Katsianis, A.
Kaygorodov, P.
Keppens, R.
Kewley, L.
Khaibrakhmanov, S.
Kimura, K.
Kirsanova, M.S.
Kitiashvili, I.
Klessen, R.
Kochina, O.
Komaki, A.
Kondratyev, I.
Kosovichev, A.
Koudmani, S.
Kundu, S.
Lamberts, A.
Lane, H.
Lecavelier des Etangs, A.
Lee, D.
Li, H.

Liao, Sh.
Liow, K.Y.
Mackey, J.
Maksimova, L.
Malhan, Kh.
Mapelli, M.
Massaro Acha, A.
Mathew, S.S.
Mayank, P.
Mellema, G.
Mendez, O.
Menon, Sh.
Mezzacappa, A.
Mignon-Risse, R.
Mitani, H.
Mohamed, Sh.S.
Moiseenko, S.
Mondal, D.
Muru, M.M.
Nakazato, Yu.
Nebot, A.
Nobels, F.
O'Connor, E.
Ocvirk, P.
Okazaki, A.
Panayotova, M.
Paul, A.
Peña, G.
Pillepich, A.
Pogorelov, N.
Portegies Zwart, S.
Prozesky, A.
Qiu, Y.
Ramakrishnan, V.
Rawlings, A.
Rieder, S.
Riggs, J.
Ripperda, B.
Rodrigues, Th.
Rometsch, Th.
Rosswog, S.
Ryu, D.
Sakurai, Yu.
Sapozhnikov, S.
Savanov, I.
Selg, S.

Seo, J.
Shang, H.
Sharma, M.
Shaw, Ch.
Shematovich, V.
Shibata, M.
Sobolenko, M.
Sobolev, A.
Sotillo Ramos, D.
Souami, D.
Soudais, A.
Sow Mondal, S.
Spurzem, R.
Starkenburg, T.
Taani, A.
Takiwaki, T.
Tomisaka, K.
Topchieva, A.
Torki, M.
Torniamenti, S.
Trani, A.
Vipin Das, M.
Väisälä, M.
van der Merwe, Ch.
Van Looveren, G.
Varnana, M K.
Vasylenko, M.
Vavilova, I.
Vogelsberger, M.
Volvach, A.
Vorobyov, E.
Wagner, A.
Wang, W.
Wang, J.
Wang, l.
Weiss, R.
Wiebe, D.
Wilhelm, M.
Wu, Y.
Yao, Zh.
Yesilirmak, B.
Yuan, F.
Zapartas, M.
Zhi, H.
Zhou, Y.
Zolotarev, R.

The Predictive Power of Computational Astrophysics as a Discovery Tool
Proceedings IAU Symposium No. 362, 2023
D. Bisikalo, D. Wiebe & C. Boily, eds.
doi:10.1017/S1743921322001636

Early structure formation in the THESAN radiation-magneto-hydrodynamics simulations

E. Garaldi[1] ⓘ, R. Kannan[2], A. Smith[3]†, V. Springel[1], R. Pakmor[1], M. Vogelsberger[3] and L. Hernquist[2]

[1]Max-Planck Institute for Astrophysics, Karl-Schwarzschild-Str. 1, D-85741 Garching, Germany

[2]Center for Astrophysics | Harvard & Smithsonian, 60 Garden Street, Cambridge, MA 02138, USA

[3]Department of Physics, Massachusetts Institute of Technology, Cambridge, MA 02139, USA

Abstract. The formation of the first galaxies in the Universe is the new frontier of both galaxy formation and reionization studies. This creates a fierce new challenge, i.e. to simultaneously understand in a unique and coherent picture the processes of galaxy formation and reionization, and – crucially – their connection. To this end, we present the THESAN suite of cosmological radiation-magneto-hydrodynamical simulations. They are unique since they: (i) cover a very broad range of spatial and temporal scales; (ii) include an unprecedentedly-broad range of physical processes for simulations of such scales and resolution; (iii) exploit knowledge accumulated at low redshift to minimize the number of free parameters in the physical model; (iv) use a variance-suppression technique in the production of initial conditions to increase their statistical fidelity. Finally, the THESAN suite includes multiple runs of the same initial conditions, exploring current unknowns in the physics of dark matter and ionizing sources.

Keywords. galaxies: high-redshift – cosmology: large-scale structure of universe – radiative transfer – methods: numerical

1. Introduction

The Epoch of Reionization (EoR) is a landmark in the evolution of the Universe. During this time, the hydrogen residing in the intergalactic medium (IGM) between galaxies was ionised by the energetic radiation produced by stars within primeval galaxies. As such, the EoR links the high-redshift cosmic structures with the Universe today, providing ground for assessing our understanding of galaxy formation (which comes mostly from low-redshift observations) and playing a key role in shaping the formation and evolution of galaxies. For these reasons, the formation of galaxies at redshift $z \gtrsim 6$ is the new frontier in the study of both galaxy formation and cosmic reionization.

Despite its importance, the EoR is among the least understood epochs in galaxy evolution. The reasons are twofold. On the one hand, observationally probing such a remote epoch is remarkably challenging, and only recently have surveys (e.g. REBELS and ALPINE) and instruments (e.g. `ALMA`, `JWST`, `CCAT-p`, `SPHEREx`, `LOFAR`, `MWA`, `HERA` and `SKA`) started to provide detailed information on the $z \gtrsim 6$ Universe. On the other hand, theoretical investigation is hindered by the multi-scale nature of the physical processes involved. These difficulties have not, however, prevented us from developing a relatively

† NHFP Einstein Fellow.

coherent picture of the EoR in the last decades. In particular, its timeline is constrained by a number of measurements to the redshift range $5.5 \lesssim z \lesssim 10$ and its tail-end (where individual ionized regions coalesce) to occur at $5 \lesssim z \lesssim 6$. These measurements come from the evolution of Lymanα – hereafter Lyα – emitters and the number density of Lyman break galaxies (Ota et al. 2008; Ono et al. 2012; Pentericci et al. 2014; Choudhury et al. 2015; Tilvi et al. 2014; Mesinger et al. 2015), from the dark pixel statistics (McGreer et al. 2011, McGreer et al. 2015, Lu et al. 2020), from GRB afterglow (Totani et al. 2006; Chornock et al. 2006), from the Doppler widths of Lyα absorption lines in the quasar near zones (Bolton et al. 2012), from the quasar damping wing (Mortlock et al. 2011, Schroeder et al. 2013, Greig et al. 2017, Greig et al. 2019, Wang et al. 2020), from CMB modeling (Robertson et al. 2013), from the Gunn-Peterson optical depth (Fan et al. 2006; Davies et al. 2018; Yang et al. 2020; Bosman et al. 2021), from the angular correlation function of Lyα emitters (Sobacchi & Mesinger 2015), from the rest-frame UV continuum of galaxies (Schenker et al. 2014), from the detection of Lyα emission in Lyman break galaxies (Mason et al. 2018, Mason et al. 2019, Hoag et al. 2019, Jung et al. 2020), from a combination of Lyα luminosity function, clustering and line profile (Ouchi et al. 2010) and from the Lyα visibility (Dijkstra et al. 2011). This process was most likely powered by a large number of star-forming galaxies with stellar mass $M_{\rm star} \lesssim 10^8 \, {\rm M_\odot}$ and relatively-high escape fractions of ionizing photons (see e.g. Haardt & Madau 2012).

Numerical studies of the EoR are among the most challenging in astrophysics, as they simultaneously necessitate large volumes, adequate resolution to capture sub-galactic processes, and an accurate treatment of Radiation Transport (RT). While a handful of simulations with at least some of these characteristics have become available in the last few years (Aurora, SPHINX, Renaissance, Obelisk, CoDa, CoDaII, CROC), they often compromise on one of them. Most importantly, these simulations are only calibrated and tested against observations at $z \gtrsim 5$, which are not only scarce and of lower quality compared to $z = 0$ ones, but also are degenerate with the astrophysical processes these simulations aim at studying.

In this proceedings we will introduce THESAN, a new suite of simulations designed to bypass the aforementioned problems and provide a comprehensive view of a large volume of the high-z Universe. We intend that this suite will provide a firm ground upon which future investigations can be built. For this reason, we will make the data public in the near future and welcome proposals for collaborative work.

2. Design and calibration

The THESAN simulations are run using the AREPO code (Springel 2010), which solves the equations of magneto-hydrodynamics on a moving mesh, generated as the Voronoi tessellation of a set of mesh-generating points that approximately follow the gas flow. Gravity is computed using a hybrid Tree-PM approach, which separately computes long-range (via a particle mesh approach) and short-range (through a hierarchical oct-tree) forces. The time integration of these equations is carried out in a hierarchical fashion.

One of the main characteristics of THESAN is the fact that it features a suite of simulations exploring the (quite uncertain) high-redshift physics. For computational reasons, such exploration is carried out using a slightly lower resolution compared to the main run (named THESAN-1). These complementary runs explore the effects of numerical resolution (THESAN-2), of the galaxy mass-dependence of photon escape (THESAN-HIGH-2 and THESAN-LOW-2), and of the nature of dark matter (THESAN-SDAO-2). In addition, we performed simulations using the original IllustrisTNG model (THESAN-TNG-2 and THESAN-TNG-SDAO-2), without any radiation (THESAN-NORT-2), as well as dark-matter only runs that extend to $z = 0$ (THESAN-DARK-1 and THESAN-DARK-2). The main features of these runs are reported in Table 1. All simulations share

Table 1. Overview of the THESAN simulation suite. From left to right the columns indicate the name of the simulation, initial particle number, the mass of the dark matter and gas particles, the minimum softening length of gas, star and dark matter particles, the minimum cell size at $z = 5.5$, the final redshift of the simulation, and the escape fraction of ionizing photons from the birth cloud (if applicable).

Name	$N_{\rm part}$	$m_{\rm DM}$ [$M_\odot$]	$m_{\rm gas}$ [ckpc]	ϵ [pc]	$r_{\rm cell}^{\rm min}$ [pc]	$z_{\rm end}$	$f_{\rm esc}$
THESAN-1	2×2100^3	3.12×10^6	5.82×10^5	2.2	~ 10	5.5	0.37
THESAN-2	2×1050^3	2.49×10^7	4.66×10^6	4.1	~ 35	5.5	0.37
THESAN-WC-2	2×1050^3	2.49×10^7	4.66×10^6	4.1	~ 35	5.5	0.43
THESAN-HIGH-2	2×1050^3	2.49×10^7	4.66×10^6	4.1	~ 35	5.5	0.8
THESAN-LOW-2	2×1050^3	2.49×10^7	4.66×10^6	4.1	~ 35	5.5	0.95
THESAN-SDAO-2	2×1050^3	2.49×10^7	4.66×10^6	4.1	~ 35	5.5	0.55
THESAN-TNG-2	2×1050^3	2.49×10^7	4.66×10^6	4.1	~ 35	5.5	–
THESAN-TNG-SDAO-2	2×1050^3	2.49×10^7	4.66×10^6	4.1	~ 35	5.5	–
THESAN-NORT-2	2×1050^3	2.49×10^7	4.66×10^6	4.1	~ 35	5.5	–
THESAN-DARK-1	2×2100^3	3.70×10^6	–	2.2	–	0.0	–
THESAN-DARK-2	2×1050^3	2.96×10^7	–	4.1	–	0.0	–

the same initial conditions, and follow the evolution of a patch of the Universe of size $L_{\rm box} = 95.5\,h^{-1}\,{\rm Mpc}$.

2.1. *The galaxy formation model*

To maximise the physical fidelity of THESAN across the entire history of the Universe, we employ as sub-resolution physics prescriptions the IllustrisTNG model (Weinberger et al. 2017; Pillepich et al. 2018), which has been extensively investigated at low redshift and proved to match many of the observed galaxy properties. This choice ensures that (i) the THESAN simulations have a galaxy formation model that is plausible throughout the entire history of the Universe, and (ii) the simulations have a single free parameter to be tuned (the unresolved stellar escape fraction of ionising photons, $f_{\rm esc}$), since all other parameters are kept fixed to the original IllustrisTNG values. The IllustrisTNG model includes, among others, the following: stochastic star formation, mass, energy and metal return from supernovae and AGB stars, explicit tracking of 9 metal species, stellar winds, black holes accretion and (bimodal) feedback, and magnetic fields originating from a uniform cosmological seed field.

The only modification to the original IllustrisTNG model is the replacement of its spatially-uniform UV background with a self-consistent RT scheme, which allows us to properly capture the EoR and that we describe in the following section. Finally, we also include in THESAN a model for dust evolution (following McKinnon et al. 2017), which includes production from stellar evolution and ISM growth, and destruction from supernovae, astration and thermal sputtering.

2.2. *Self-consistent radiation transport*

The propagation of radiation is followed using a moment-based approach, where the first two moments of the RT equation are solved employing the M1 closure relation (Levermore 1984), thanks to the AREPO-RT (Kannan et al. 2019) extension of AREPO. To reduce the computational load, only the UV part of the source spectrum is tracked employing three energy bins defined by the following thresholds: $[13.6, 24.6, 54.4, \infty)\,{\rm eV}$, in order to bracket the ionization energies of hydrogen and helium. Each resolution element tracks, for each bin, the comoving photon number density and flux. Within each of these bins, radiation is assumed to follow the spectral *shape* of a 2 Myr old, quarter-solar metallicity stellar population, as predicted by the BPASS library

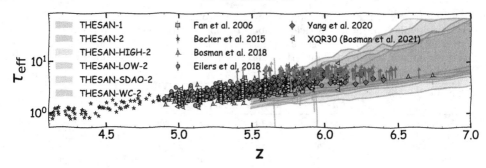

Figure 1. Evolution of the effective Lyα optical depth (τ_{eff}). The shaded regions show the central 95% of the data computed from synthetic spectra, while symbols report individual observations as indicated in the legend.

(Eldridge et al. 2017), which we use to compute the source spectra including the contribution from binary stellar systems. The predicted photon flux from the BPASS library is attenuated by a constant stellar escape fraction f_{esc} that captures the effect of the (unresolved) birth cloud of the star particles. The value of f_{esc} is approximately tuned to obtain the desired reionization history, which we match to the so-called 'late' reionization (i.e. an EoR ending at $z \lesssim 5.5$).

3. First results

We begin by showing the evolution of the Lyα effective optical depth (τ_{eff}) in a selection of THESAN runs in Fig. 1. Almost all THESAN runs agree well with the measurements available in the redshift range where they overlap ($5.5 \leq z \lesssim 6.5$). The only exception is THESAN-LOW-2, where only galaxies hosted by haloes of mass $M_{\mathrm{halo}} \leq 10^{10}\ \mathrm{M_\odot}$ are allowed to have non-zero escape fraction. This run completes reionization much earlier than the others (since small galaxies are mostly dominant at higher redshift), and therefore has a significantly more transparent IGM than the other THESAN runs. When investigating the distribution of τ_{eff} in THESAN-1 at a given redshift (not shown), we find that it is slightly shifted to more transparent values with respect to observations, suggesting a (slightly) too early reionization.

3.1. *The IGM – galaxy connection*

The combination of large volume, high resolution and realistic galaxy formation model renders THESAN a unique tool to study the connection between the IGM properties and the high-z galaxy population. In this Section, we do so by investigating how the Lyα flux in quasars (QSOs) depends on the distance between the line of sight and nearby galaxies. In particular, we follow the approach originally suggested in Kakiichi et al. (2018) and study the quantity $\langle T(r) \rangle / \bar{T} - 1$, where $T(r)$ is the transmissivity (i.e. the transmitted flux divided by the continuum value at the same location) at a given distance r between a pixel in the QSO spectrum and a galaxy, the angular brackets indicate an average over all galaxy – pixel pairs at distance r, and $\bar{T}$ indicates the average over all pairs, regardless of their distance.

In Fig. 2 we show how the (normalised) transmissivity depends on distance for different galaxy selections at $z = 5.5$. In particular, lines in different shades of grey correspond to selections based on the mass of the dark matter halo hosting the galaxy, the purple line displays the effect of selecting galaxies with stellar mass $M_{\mathrm{star}} \geq 10^{10}\ \mathrm{M_\odot}$ and the orange line shows the result of identifying galaxies based on their CIV absorption along the same QSO sightline. We compare our results to two observational campaigns. Unlike previous

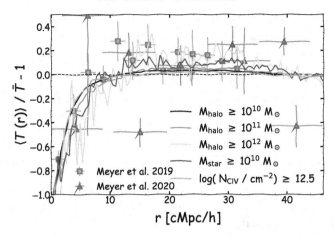

Figure 2. Average (normalised) transmitted flux $\langle T(r) \rangle$ as a function of the galaxy-spectrum distance r, normlised by its global average $\bar{T}$, for different galaxy-selection criteria at $z = 5.5$.

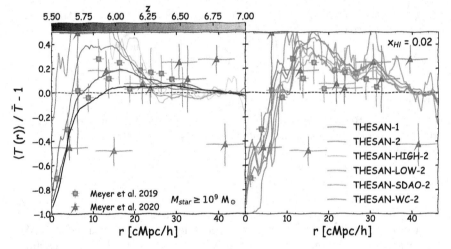

Figure 3. Same as Fig. 2, but only showing the $M_{star} \geq 10^9\,M_{\odot}$ selection as a function of redshift (left panel) and of photon escape and dark matter model (right). Notice that in the right panel, we matched the volume-averaged neutral fraction $x_{HI} = 0.02$ rather than the redshift in the different THESAN runs, in order to remove dependency on the reionization history.

studies, THESAN is able to recover for the first time the flux modulation as a function of r. However, the flux enhancement at intermediate scales $10 \lesssim r/(h^{-1}\,\mathrm{Mpc}) \lesssim 30$ is reproduced only by the most extreme of our galaxy selections (i.e. $M_{halo} \geq 10^{12}\,M_{\odot}$). This is, however, in tension with the estimated host mass of the observed galaxy population by Meyer et al. (2019), which is estimate to be $M_{halo} \gtrsim 10^{10}\,M_{\odot}$. Additionally, trying to mimic the selection procedure of Meyer et al. (2019) by selecting galaxies based on their CIV absorption results in a much smaller flux enhancement than observed.

In order to investigate the origin of this discrepancy between THESAN and the observations, we show in Fig. 3 how $\langle T(r) \rangle / \bar{T}$ depends on the redshift (left) and on some of the high-z physics (right), for the $M_{star} \geq 10^9\,M_{\odot}$ selection. The left panel shows that this measure is very sensitive to the redshift, with the flux enhancement moving to smaller r and larger values with increasing redshift. This is fully compatible with its interpretation as a proximity effect. The ionised bubbles around galaxies grow (on average) with time, producing a more extended region of flux enhancement (thanks to the suppressed neutral fraction). Simultaneously, $\bar{T}$ strongly decreases with increasing redshift outside of these regions, following the evolution of the average neutral fraction in the universe. Hence, for a given flux within the ionised bubbles the ratio $\langle T(r)\rangle/\bar{T}$ is much higher at higher redshift.

The left panel of Fig. 3 also shows the sensitivity of $\langle T(r)\rangle/\bar{T}$ to the timing of reionization. Hence, it can be used as promising tool to constrain the latter, provided it is not (too) sensitive on other properties of the EoR. We test this in the right panel of Fig. 3, where we report the flux modulation in the different THESAN runs, hence exploring different photon escape and dark matter models. Notice that we have factored out the different reionization histories of these runs by matching the volume-averaged neutral fraction ($x_{\mathrm{HI}} = 0.02$) in the runs instead of their redshift. We note here that there will be some residual effect due to the simulations being at slightly different stages of the structure formation process, but the reionization histories are similar enough to render this residual difference negligible for all but the THESAN-LOW-2 run, which completes reionization significantly earlier than the others. The panel shows that $\langle T(r)\rangle/\bar{T}$ is quite insensitive to the specific radiation escape and dark matter models employed, making it possible to use the flux excess as a way to constrain the timing of reionization. Hence, comparing the values obtained with available observations, we conclude that reionization occurs slightly too early in THESAN, despite it being designed to simulate a 'late' reionization history ending at $z \lesssim 5.5$. This is consistent with the conclusion obtained from the analysis of the effective optical depth distribution.

Discussion

Christian Boily: Regarding the optical depth, you said your modelling matches the data points. It seems to me that your models were actually significantly below the lower [observational] values.

Enrico Garaldi: Of course it depends on the source model. What I am referring to is the red band in the plot, which corresponds to our fiducial run. It goes through the data fairly well. I did not have the time to show it, but what you can do is to take thin redshift slices and check the distribution of optical depth at any given redshift. We match the observations well, although we are slightly too transparent at every redshift, which indicates that an even-later reionization better matches current data, but the difference is not large. This green band instead is THESAN-LOW-2, the model where only low-mass galaxies contribute to reionization, and it is instead in significant tension with the data. There is of course the caveat that all these models have an escape fraction included.

Mahavir Sharma: I wonder if you have a prediction on the 21cm power spectrum and a comparison to the MWA or LOFAR data.

Enrico Garaldi: There is definitely in the papers, but I have here only one plot. These are the predictions for different redshifts, and the straight lines are the predicted sensitivities for LOFAR, HERA and SKA-LOW.

References

Ota K., et al., 2008, ApJ, 677, 12
Ono Y., et al., 2012, ApJ, 744, 83
Pentericci L., et al., 2014, ApJ, 793, 113
Choudhury T. R., Puchwein E., Haehnelt M. G., Bolton J. S., 2015, MNRAS, 452, 261
Tilvi V., et al., 2014, ApJ, 794, 5

Mesinger A., Aykutalp A., Vanzella E., Pentericci L., Ferrara A., Dijkstra M., 2014, MNRAS, 446, 566
McGreer I. D., Mesinger A., Fan X., 2011, MNRAS, 415, 3237
McGreer I. D., Mesinger A., D'Odorico V., 2015, MNRAS, 447, 499
Lu T.-Y., et al., 2020, ApJ, 893, 69
Totani T., Kawai N., Kosugi G., Aoki K., Yamada T., Iye M., Ohta K., Hattori T., 2006, PASJ, 58, 485
Chornock R., Berger E., Fox D. B., Lunnan R., Drout M. R., Fong W.-f., Laskar T., Roth K. C., 2013, ApJ, 774, 26
Bolton J. S., Becker G. D., Raskutti S., Wyithe J. S. B., Haehnelt M. G., Sargent W. L. W., 2012, MNRAS, 419, 2880
Mortlock D. J., et al., 2011, Nature, 474, 616
Schroeder J., Mesinger A., Haiman Z., 2013, MNRAS, 428, 3058
Greig B., Mesinger A., Haiman Z., Simcoe R. A., 2017, MNRAS, 466, 4239
Greig B., Mesinger A., Bañados E., 2019, MNRAS, 484, 5094
Wang F., et al., 2020, ApJ, 896, 23
Robertson B. E., et al., 2013, ApJ, 768, 71
Fan X., et al., 2006, AJ, 132, 117
Davies F. B., et al., 2018, ApJ, 864, 142
Yang J., et al., 2020, ApJ, 904, 26
Bosman S. E. I., et al., 2021, arXiv e-prints, p. arXiv:2108.03699
Sobacchi E., Mesinger A., 2015, MNRAS, 453, 1843
Schenker M. A., Ellis R. S., Konidaris N. P., Stark D. P., 2014, ApJ, 795, 20
Mason C. A., Treu T., Dijkstra M., Mesinger A., Trenti M., Pentericci L., de Barros S., Vanzella E., 2018, ApJ, 856, 2
Mason C. A., et al., 2019, MNRAS, 485, 3947
Hoag A., et al., 2019, ApJ, 878, 12
Jung I., et al., 2020, ApJ, 904, 144
Ouchi M., et al., 2010, ApJ, 723, 869
Dijkstra M., Mesinger A., Wyithe J. S. B., 2011, MNRAS, 414, 2139
Haardt F., Madau P., 2012, ApJ, 746, 125
Springel V., 2010, MNRAS, 401, 791
Weinberger R., Springel V., Hernquist L., Pillepich A. et al., 2017, MNRAS, 465, 3291
Pillepich A., Nelson D., Hernquist L., Springel V., Pakmor R., et al., 2018, MNRAS, 473, 4077
McKinnon R., Torrey P., Vogelsberger M., Hayward C. C., Marinacci F., 2017, MNRAS, 468, 1505
Levermore C. D., 1984, J. Quant. Spectrosc. Radiative Transfer, 31, 149
Kannan R., Vogelsberger M., Marinacci F., McKinnon R., et al. 2019, MNRAS, 485, 117
Eldridge J. J., Stanway E. R., Xiao L.,et al. 2017, Publ. Astron. Soc. Australia, 34, e058
Fan X., Strauss M. A., Becker R. H. White R. L. et al., 2006, AJ, 132, 117
Becker G. D., Bolton J. S., Madau P., Pettini M. et al., 2015, MNRAS, 447, 3402
Bosman S. E. I., Fan X., Jiang L., Reed S., Matsuoka Y. et al., 2018, MNRAS, 479, 1055
Bosman S. E. I., Davies F. B., Becker G., Keating L. et al., 2021, arXiv e-prints (2108.03699)
Eilers A.-C., Davies F. B., Hennawi J. F., 2018, ApJ, 864, 53
Yang J., Wang F., Fan X., Hennawi J., Davies F. et al., 2020, ApJ, 904, 26
Kakiichi K., Ellis R. S., Laporte N., Zitrin A., Eilers A.-C. et al., 2018, MNRAS, 479, 43
Meyer R. A., Bosman S. E. I., Kakiichi K., Ellis R. S., 2019, MNRAS, 483,19
Meyer R. A., Kakiichi K., Bosman S. E. I., Ellis, R. S. et al., 2020, MNRAS, 494, 1560

The Predictive Power of Computational Astrophysics as a Discovery Tool
Proceedings IAU Symposium No. 362, 2023
D. Bisikalo, D. Wiebe & C. Boily, eds.
doi:10.1017/S1743921322001247

Simulations of dark matter with frequent self-interactions

Moritz S. Fischer

Hamburger Sternwarte, Universität Hamburg, Gojenbergsweg 112, D-21029 Hamburg,
Germany

Abstract. Self-interacting dark matter (SIDM) is promising to solve or at least mitigate small-scale problems of cold collisionless dark matter. N-body simulations have proven to be a powerful tool to study SIDM within the astrophysical context. However, it turned out to be difficult to simulate dark matter (DM) models that typically scatter about a small angle, for example, light mediator models. We developed a novel numerical scheme for this regime of frequent self-interactions that allows for N-body simulations of systems like galaxy cluster mergers or even cosmological simulations. We have studied equal and unequal mass mergers of galaxies and galaxy clusters and found significant differences between the phenomenology of frequent self-interactions and the commonly studied large-angle scattering (rare self-interactions). For example, frequent self-interactions tend to produce larger offsets between galaxies and DM than rare self-interactions.

Keywords. astroparticle physics, methods: numerical, galaxies: haloes, dark matter

1. Introduction to SIDM

The cosmological standard model ΛCDM has been quite successful in explaining the observed large scale structure. Large cosmological N-body simulations have been used, e.g. the Millenium simulations (Springel et al. 2005), to make predictions from ΛCDM. The first such simulations were DM-only and did not take baryonic physics into account. For about two decades, it is known that these simulations deviate on small, i.e. galactic, scales from the observed matter distribution. But at the same time, they are remarkably successful in explaining the distribution of matter on large scales.

There exist several problems or maybe better curiosities on small scales, together they form the small-scale crisis of ΛCDM. One of these problems is the core-cusp problem. DM-only simulations predict cuspy haloes that are fairly well described by an NFW profile. However, cored haloes with a lower central density are observed. Besides, there exists the too-big-to-fail problem, the diversity problem and the plane of satellite problem, and maybe more. For a review of the small-scale challenges see Bullock & Boylan-Kolchin (2017).

Many potential solutions have been proposed to the small-scale problems. Several of them rely on an alternative DM model, e.g. warm DM (Dodelson & Widrow 1994), self-interacting DM (Spergel & Steinhardt 2000) or fuzzy DM (Hu et al. 2000). Another branch of solutions relies on the inclusion of baryonic physics in cosmological simulations. Unfortunately, modelling the baryonic processes comes with high uncertainty, but it has been shown that a proper inclusion of the processes, like star formation, supernovae and AGNs can at least contribute to a solution of the small-scale crisis. Besides, there are also other attempts to solve the problems on small scale by introducing an alternative theory of gravity. In addition, improvements in modelling the internal structure of observed galaxies may contribute to a solution too (Oman et al. 2019).

From now on, we will focus on one particular potential solution, which is DM with self-interactions. SIDM refers to a class of particle physics models that assume that DM particles interact with each other through an additional force beyond gravity. But this affects only DM and no interaction with standard model particles is assumed. SIDM has been studied for about two decades and it has been shown that self-interactions can resolve or at least mitigate several small-scale problems. For instance, SIDM can create density cores in DM haloes, by transferring heat inward. For a review of SIDM, see Tulin & Yu (2018).

There exist several methods to model SIDM. The gravothermal fluid model and the Jeans approach make simplifying assumptions such that they can only be applied to relaxed haloes. In contrast, N-body simulations do not simplify the problem but are computationally much more expensive. In the following, we focus on N-body simulation, i.e. describe how to faithfully model self-interactions even in complicated systems.

To model SIDM, one needs to solve the Vlasov-Poisson equation with an additional collision term, see Eq. (1). Thus, SIDM is neither collisionless like CDM nor fully collisional like a fluid. Hence, the 6d phase-space information is required.

$$\frac{\partial f}{\partial t} + \vec{v} \cdot \nabla_x f - \nabla_x \Phi \cdot \nabla_v f = \left(\frac{\partial f}{\partial t}\right)_{\text{coll}}. \tag{1}$$

The self-interactions are described by the collision term. This term follows from the differential cross-section of the particle physics model. Here we distinguish two regimes, the large-angle and the small-angle scattering. If the typical scattering angle is large, a significant amount of momentum is transferred per scattering event. Thus not many scattering events are necessary to alter the DM distribution. Therefore, the self-interactions are called rare. On the other hand, if particles scatter at small angles, only a little amount of momentum is transferred between the particles per interaction. Hence, this type of interaction must be frequent to have a significant effect on the distribution of DM. It has turned out that this regime is more difficult to model within N-body simulations. For about two decades researchers have been performing N-body simulations of SIDM. However, almost all of them fall into the regime of rare self-interactions, mostly an isotropic cross-section has been studied. However, in Fischer et al. (2021a) we introduced a novel scheme that allows, for the first time, to model fSIDM within N-body simulations from first principles. In contrast to the work of Kummer et al. (2019), we can simulate more complicated setups like mergers or do cosmological simulations. So far we have focused on idealized simulations of equal and unequal mass mergers (Fischer et al. 2021a,b).

2. N-body Simulation with Self-Interactions

In this section, we describe the basic principles of the numerical formulation used to model self-interactions within N-body codes. Before we explain the new scheme for frequent self-interactions, we describe the state-of-the-art scheme for rare self-interactions, and lastly, we show a test problem to validate our numerical scheme and the implementation.

The today widely spread Monte Carlo scheme for rSIDM goes back to Burkert (2000) and has been highly improved by Rocha et al. (2013). The idea is to treat the interaction between N-body particles like physical particles. If two numerical particles are close to each other, a random number is drawn to decide whether they interact with each other or not. In the case of interaction, another random number is needed to decide which angle they scatter. All numerical particle interactions are treated pairwise, this allows conserving momentum and energy explicitly. Note that SIDM physics could be described deterministically at the scales of interest as long as one studies the limit of many physical particles. Thus, it is not affected by the stochastic nature of the interaction of individual particles. However, when modelling self-interactions within the N-body method difficulties arise. To overcome them, random numbers and the formulation analogous to the interaction of physical particles have been introduced.

In principle, the scheme for rSIDM could describe the limit of fSIDM, but time-step limitations make it impractical to use it. The interaction probability of two numerical particles per simulation time step must always be smaller than unity, which gives a time-step constraint. In the fSIDM limit, this time-step constraint implies a time-step of zero. Thus, the simulation can no longer be advanced in time. To overcome these problems, a different formulation of the collision term is required.

For the fSIDM scheme, we no longer describe the interaction of numerical particles like physical particles. As for the rSIDM scheme, we use a stochastic process that converges in the limit of many particles against the deterministic collective behaviour of many physical SIDM particles. At least there exists another stochastical process in the limit of fSIDM, besides the particle physics one, that fulfills this condition. We use this process to overcome the problems with the rSIDM scheme described above. In our fSIDM scheme, physical particles interact with each other if they are close. There is no interaction probability anymore. For close pairs of particles, we use an effective description based on a drag force to describe the self-interactions.

The drag force description we use goes back to Kahlhoefer et al. (2014). The idea is the following: A DM particle is travelling through a constant background density at rest. While doing so, it scatters frequently with the background particles; every interaction alters the direction of motion by a tiny angle. The corresponding velocity changes perpendicular to the direction of motion average out, but the one parallel to the direction of motion sum up and decelerate the DM particle. This is the drag force we use for the numerical scheme. The energy taken from the forward motion goes into the perpendicular component. To understand this, it might be helpful to think of a phase-space patch travelling through the background density. The energy taken from the forward motion does not dissipate but increases the velocity dispersion perpendicular to the direction of motion. The phase-space patch is heated and its particles no longer have the same velocity. But their mean direction of motion does not change.

In order to apply this to the numerical particles, we treat close pairs in two steps. First, we compute the momentum change due to the drag force and decelerate the particles. Here, the numerical particles represent phase-space patches that overlap in configuration space and thus can interact and decelerate each other. Secondly, we re-add the energy lost in the first step into a random direction but within a plane perpendicular to the direction of motion.

We have implemented the novel scheme for fSIDM into the cosmological N-body code GADGET-3, which is a successor of GADGET-2 (Springel 2005). Our implementation conserves momentum and energy explicitly. This makes the parallelisation more complicated than, for example, in SPH. But reasonable large simulations with fSIDM can be run. For DM-only simulations, the self-interactions slow down the simulation by a factor of 4 or more. The exact number depends largely on the specific simulation setup.

To demonstrate that the implementation in GADGET-3 works as expected, we use a test problem similar to Rutherford's experiment. A beam of DM particles scatters on a target consisting of DM particles and the distribution of the deflection angles is measured. In contrast to Rutherford, we do not use a thin but a thick target. An analytical solution for this problem is given by Moliere (1948). In Fig. 1, we compare our simulation results (black) to the exact solution (blue). The distributions of the deflection angle agree well

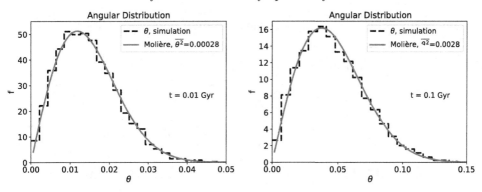

Figure 1. The angular deflection test problem is shown. The distribution of the deflection angle is given for two different times that the particle is travelling within the target. This corresponds to different target thicknesses, i.e. the left-hand panel gives a target that is thinner than the one of the right-hand panel. The black curve shows the simulation result and the blue one gives the exact solution. This figure is a reproduction of Fig. 3 of Fischer et al. (2021a).

with each other. From this, we can conclude that the implementation of our scheme works as expected. Hence, we turn our focus to an astrophysical motivated problem in the next section.

3. Merging Galaxy Clusters and SIDM

In this section, we study mergers of galaxy clusters since they are observationally and theoretically well-studied systems. The most famous system in this context is probably the Bullet Cluster, which has also been studied in the context of SIDM (Randall et al. 2008; Robertson et al. 2017a,b). If DM undergoes self-interactions, one would expect that the DM haloes of the clusters behave differently than their galaxies. In particular, if self-interactions are frequent, a drag force that decelerates the DM component may arise. The galaxies are not affected by the drag force and thus an offset between the DM and the galactic component can occur. As claimed by Kim et al. (2017) small offsets can also arise in the case of isotropic scattering if the DM density is large enough. Similar to Kim et al. (2017) we study equal mass mergers with head-on collisions. The haloes follow an NFW profile and have a virial mass of $10^{15} M_\odot$. In Fig. 2, we show the position of the DM density peaks along the merger axis of our simulations with various DM models. In particular, we simulate collisionless DM, rSIDM, and fSIDM employing several cross-sections. The scattering of the SIDM models is elastic and velocity-independent. Note that to match rSIDM and fSIDM, we use the momentum transfer cross-section. For more details see Fischer et al. (2021a). If the cross-section is increasing, the merger time becomes smaller and for a very large cross-section, the DM haloes coalesce on contact. During the infall phase the peak position is mostly unaffected by the self-interactions, but at about the first pericentre passage, differences arise. If one compares rSIDM to fSIDM, differences appear to be small, i.e. the DM models behave similarly. For a cross-section of $1.5 \, \text{cm}^2/\text{g}$, the plot shows the largest difference between the models. However, we are interested in the DM-galaxy offsets. The simulation contains collisionless particles that follow the same NFW profile as the DM component to mimic the galaxies of the cluster. At the centre of each halo, we placed a more massive particle to mimic the brightest cluster galaxy (BCG). In Fig. 3, we show the offset between the DM component and the BCG position as a function of time relative to the first pericentre passage. For a given cross-section, the offset is much larger for the fSIDM runs. First, the DM peaks are in between the galaxy peaks, and at a later time, it is vice versa (the sign of the offset changes). We measure the maximum offset for the latter case and show it for even more

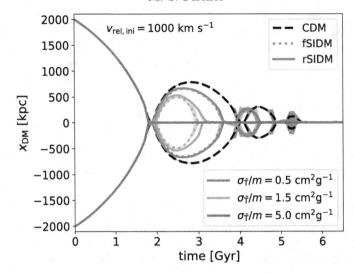

Figure 2. The peak position of the DM haloes as a function of time for several cross-sections is shown. This figure is a reproduction of the lower panel of Fig. 8 of Fischer et al. (2021a).

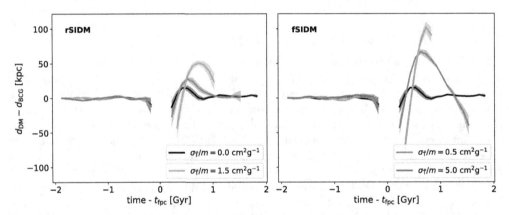

Figure 3. DM-galaxy offset as a function of time relative to the first pericentre passage (t_{fpc}). The offset is positive when the galaxy peaks are in between the DM peaks and negative otherwise. When the DM peaks are close, peak finding becomes inaccurate and we do not display the offsets. This figure is a reproduction of the lower panels of Fig. 9 of Fischer et al. (2021a).

cross-sections in Fig. 4. For large cross-sections, the offsets become zero as the DM haloes coalesce on contact, thus the type of offset shown here can no longer occur. For small cross-sections, the offset increases with cross-section and the maximum offset is much larger for fSIDM than for rSIDM. It becomes clear that very large offsets can only be explained by fSIDM but not by rSIDM.

4. Conclusions

From the work presented here, we can draw two main conclusions.

First, it is possible to model DM with frequent self-interactions within N-body simulations. The presented numerical scheme relies on an effective drag force arising from self-interactions. Moreover, we have demonstrated through a test problem similar to Rutherford's experiment that the numerical scheme allows to accurately simulate small-angle scattering.

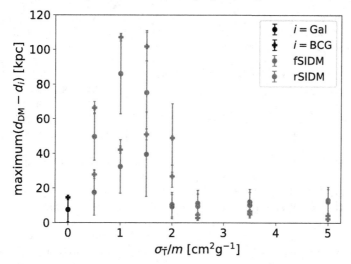

Figure 4. Maximum DM-galaxy offset as a function of the cross-section. Only offsets where the galaxy peaks are in between the DM peaks are considered. We display offsets for rSIDM (red) and fSIDM (green) measured relative to the peak of the galaxy distribution and the BCG position. This figure is a reproduction of Fig. 10 in Fischer et al. (2021a).

Secondly, we found that rSIDM and fSIDM show different phenomenologies. In particular, we have demonstrated that the size of the offset between galaxies and DM in galaxy cluster mergers depends crucially on the shape of the differential cross-section. It is worth mentioning that this is not only the case for a fairly large cross-section, but also true for a strength of self-interactions that is within the current observational bounds, e.g. from dark matter density cores. Although there have been observational claims of fairly large offsets in the literature (e.g. Harvey et al. 2015), this can not be taken as evidence for fSIDM as a more thorough analysis of observations does not confirm these large offsets (Wittman et al. 2018). In the future, a combination of multiple measurements, e.g. offsets and dark matter cores, could possibly allow to constrain the shape of the differential cross-section, i.e. discriminate between rSIDM and fSIDM.

Acknowledgements

It is a pleasure to acknowledge all of my collaborators who contributed to the work I have described here. They are Marcus Brüggen, Kai Schmidt-Hoberg, Klaus Dolag, Felix Kahlhoefer, Antonio Ragagnin and Andrew Robertson. This work is funded by the Deutsche Forschungsgemeinschaft (DFG, German Research Foundation) under Germany's Excellence Strategy – EXC 2121 "Quantum Universe" – 390833306.

References

Bullock, J. S. & Boylan-Kolchin, M. 2017, Small-Scale Challenges to the ΛCDM Paradigm. *ARAA*, 55(1), 343–387.

Burkert, A. 2000, The Structure and Evolution of Weakly Self-interacting Cold Dark Matter Halos. *APJL*, 534(2), L143–L146.

Dodelson, S. & Widrow, L. M. 1994, Sterile neutrinos as dark matter. *PRL*, 72(1), 17–20.

Fischer, M. S., Brüggen, M., Schmidt-Hoberg, K., Dolag, K., Kahlhoefer, F., Ragagnin, A., & Robertson, A. 2021,a N-body simulations of dark matter with frequent self-interactions*. *Monthly Notices of the Royal Astronomical Society*, stab1198.

Fischer, M. S., Brüggen, M., Schmidt-Hoberg, K., Dolag, K., Ragagnin, A., & Robertson, A. 2021,b Unequal-mass mergers of dark matter haloes with rare and frequent self-interactions. *Monthly Notices of the Royal Astronomical Society*, 510b(3), 4080–4099.

Harvey, D., Massey, R., Kitching, T., Taylor, A., & Tittley, E. 2015, The nongravitational interactions of dark matter in colliding galaxy clusters. *Science*, 347(6229), 1462–1465.

Hu, W., Barkana, R., & Gruzinov, A. 2000, Fuzzy Cold Dark Matter: The Wave Properties of Ultralight Particles. *PRL*, 85(6), 1158–1161.

Kahlhoefer, F., Schmidt-Hoberg, K., Frandsen, M. T., & Sarkar, S. 2014, Colliding clusters and dark matter self-interactions. *MNRAS*, 437(3), 2865–2881.

Kim, S. Y., Peter, A. H. G., & Wittman, D. 2017, In the wake of dark giants: new signatures of dark matter self-interactions in equal-mass mergers of galaxy clusters. *MNRAS*, 469(2), 1414–1444.

Kummer, J., Brüggen, M., Dolag, K., Kahlhoefer, F., & Schmidt-Hoberg, K. 2019, Simulations of core formation for frequent dark matter self-interactions. *MNRAS*, 487(1), 354–363.

Moliere, G. 1948, Theorie der streuung schneller geladener teilchen ii mehrfach-und vielfach-streuung. *Zeitschrift für Naturforschung A*, 3(2), 78–97.

Oman, K. A., Marasco, A., Navarro, J. F., Frenk, C. S., Schaye, J., & Benítez-Llambay, A. r. 2019, Non-circular motions and the diversity of dwarf galaxy rotation curves. *MNRAS*, 482(1), 821–847.

Randall, S. W., Markevitch, M., Clowe, D., Gonzalez, A. H., & Bradač, M. 2008, Constraints on the Self-Interaction Cross Section of Dark Matter from Numerical Simulations of the Merging Galaxy Cluster 1E 0657-56. *ApJ*, 679(2), 1173–1180.

Robertson, A., Massey, R., & Eke, V. 2017,a Cosmic particle colliders: simulations of self-interacting dark matter with anisotropic scattering. *MNRAS*, 467a(4), 4719–4730.

Robertson, A., Massey, R., & Eke, V. 2017,b What does the Bullet Cluster tell us about self-interacting dark matter? *MNRAS*, 465b(1), 569–587.

Rocha, M., Peter, A. H. G., Bullock, J. S., Kaplinghat, M., Garrison-Kimmel, S., Oñorbe, J., & Moustakas, L. A. 2013, Cosmological simulations with self-interacting dark matter – i. constant-density cores and substructure. *Monthly Notices of the Royal Astronomical Society*, 430(1), 81–104.

Spergel, D. N. & Steinhardt, P. J. 2000, Observational evidence for self-interacting cold dark matter. *Physical Review Letters*, 84(17), 3760–3763.

Springel, V. 2005, The cosmological simulation code GADGET-2. *MNRAS*, 364(4), 1105–1134.

Springel, V., White, S. D. M., Jenkins, A., Frenk, C. S., Yoshida, N., Gao, L., Navarro, J., Thacker, R., Croton, D., Helly, J., & et al. 2005, Simulations of the formation, evolution and clustering of galaxies and quasars. *Nature*, 435(7042), 629–636.

Tulin, S. & Yu, H.-B. 2018, Dark matter self-interactions and small scale structure. *PHYSREP*, 730, 1–57.

Wittman, D., Golovich, N., & Dawson, W. A. 2018, The Mismeasure of Mergers: Revised Limits on Self-interacting Dark Matter in Merging Galaxy Clusters. *ApJ*, 869(2), 104.

The Predictive Power of Computational Astrophysics as a Discovery Tool
Proceedings IAU Symposium No. 362, 2023
D. Bisikalo, D. Wiebe & C. Boily, eds.
doi:10.1017/S1743921322001235

Simulations of the reionization of the clumpy intergalactic medium with a novel particle-based two-moment radiative transfer scheme

Tsang Keung Chan[1,2] (ORCID), **Alejandro Benitez-Llambay[3], Tom Theuns[1,2]**
and Carlos Frenk[1,2]

[1]Institute for Computational Cosmology, Durham University, South Road,
Durham DH1 3LE, UK
email: `tsang.k.chan@durham.ac.uk`

[2]Department of Physics, Durham University, South Road, Durham DH1 3LE, UK

[3] Dipartimento di Fisica G. Occhialini, Università degli Studi di Milano Bicocca, Piazza della
Scienza, 3 I-20126 Milano MI, Italy

Abstract. The progress of cosmic reionization depends on the presence of over-dense regions that act as photon sinks. Such sinks may slow down ionization fronts as compared to a uniform intergalactic medium (IGM) by increasing the clumping factor. We present simulations of reionization in a clumpy IGM resolving even the smallest sinks. The simulations use a novel, spatially adaptive and efficient radiative transfer implementation in the SWIFT SPH code, based on the two-moment method. We find that photon sinks can increase the clumping factor by a factor of $\sim$10 during the first ~ 100 Myrs after the passage of an ionization front. After this time, the clumping factor decreases as the smaller sinks photoevaporate. Altogether, photon sinks increase the number of photons required to reionize the Universe by a factor of $\eta \sim 2$, as compared to the homogeneous case. The value of η also depends on the emissivity of the ionizing sources.

Keywords. radiative transfer, cosmology:early universe, galaxies:intergalactic medium

---

1. Introduction

During the epoch of reionization (EoR) at redshifts $z \gtrsim 6$, the universe underwent a global phase transition from mostly neutral to mostly ionized due to photoionization of the intergalactic medium (IGM), most likely by hot stars in the first galaxies. This change in neutral fraction strongly affects the transmission properties of the IGM in Lyman lines and hence the detectability and observability of galaxies (see e.g. Robertson 2021 for a recent review). Photoionization also increases the IGM's temperature from ~ 10 K to $\gtrsim 10^4$ K, which affects the growth of small density perturbations and suppresses the formation and growth of dwarf galaxies (Efstathiou 1992; Bullock, Kravtsov & Weinberg 2000). Such suppression helps reconciling the low amplitude of the observed luminosity function of the Milky Way with the high amplitude of the subhalo mass function as inferred from simulations (Moore *et al.* 2000; Benson *et al.* 2002). It is the first time ever that galaxies impact so dramatically the global gas properties of the universe.

Several current and upcoming observations will explore the EoR with unprecedented levels of detail. The ongoing and upcoming observations of the 21-cm signal from neutral hydrogen by the Low-Frequency Array (LOFAR) and in future the Square Kilometre

Array (SKA), will explore the geometry and timing of reionization. The James Webb Space Telescope (JWST) will directly observe galaxies in the EoR to an unprecedented depth and will likely be able to constrain the sources of reionization.

Currently, there are two major outstanding questions regarding the nature of reionization. Firstly, which sources dominate the emissivity of ionizing radiation? During the EoR, active galactic nuclei and galaxies more luminous than the Milky Way are likely too rare to provide the required photon budget. Fainter galaxies are likely much more numerous, however observationally it is challenging to determine the luminosity function at such faint levels (see e.g. Sharma *et al.* 2016). In either case, the fraction of photons that escape from their natal galaxy is uncertain.

Secondly, what are the sinks of ionizing photons and how do they evolve? Each ionizing photon can ionize one neutral hydrogen atom, however the hydrogen ion can recombine again. Therefore, all sources combined need to emit more than one ionizing photon per hydrogen atom to complete reionization. The volume-averaged recombination rate per unit volume can be written as

$$\langle \frac{\mathrm{d}n_{\mathrm{HI}}}{\mathrm{d}t}|_{\mathrm{rec}} \rangle = \langle \alpha_r \, n_e \, n_{\mathrm{HII}} \rangle = c_l \, \alpha_r \, \langle n_{\mathrm{H}} \rangle^2 \equiv c_l \frac{\langle n_{\mathrm{H}} \rangle}{\tau_r} \,. \tag{1.1}$$

Here, $n_e \approx 1.08 n_{\mathrm{HII}}$ is the electron density, with the factor 1.08 applicable if Helium is singly ionized, α_r is the recombination coefficient, and τ_r a characteristic recombination time. The factor $c_l > 1$ is usually called the 'clumping factor': it quantifies how much the recombination rate in a clumpy IGM is higher compared to that in a uniform IGM. Using the value of the case-B recombination rate at a temperature of $T = 2.2 \times 10^4 \mathrm{K}$ for the recombination coefficient, the product $\tau_r(z) \, H(z) \approx ((1 + 6.2)/(1 + z))^{3/2}$ for the values of the cosmological parameters from Planck Collaboration (2014), where $H(z)$ is the Hubble constant. This shows that for $c_l \approx 1$, recombinations cease to be important below a redshift $z \sim 6$ in the uniform IGM.

However, recombinations are enhanced in regions of higher density, such as the filaments of the cosmic web, minihalos† and more massive halos. They increase the value of the clumping factor c_l and hence also the number of photons per baryon required to reionize the universe. In this work, we investigate the evolution of c_l due to photoevaporation of minihalos, using high resolution radiation-hydrodynamic simulations that resolve even the smallest minihalos. We provide an overview of our simulations and methodology in the next section. We then present our results, discussions, and a future outlook.

2. Overview of our method

The IGM's temperature is coupled to that of the CMB by residual electrons up to a redshift of ≈ 137, after which the gas cools adiabatically (see e.g. Peebles 1993). The value of the Jeans mass,

$$M_J = \left(\frac{5 k_{\mathrm{B}} T}{G \mu m_{\mathrm{H}}} \right)^{3/2} \left(\frac{3}{4 \pi \rho} \right)^{1/2} , \tag{2.1}$$

is then $\sim 4 \times 10^3 (\frac{1+z}{9})^{3/2} \mathrm{M}_\odot$ before reionization, where z is the reionization redshift. Here, ρ is the total density (baryons plus dark matter), and the numerical values assume a temperature $T = 1.6$ K, and mean molecular weight, $\mu = 1.23$, as is appropriate for

---

† Minihalos are the halos that can retain baryons but cannot undergo atomic cooling due to their weak gravitational potential.

a neutral, primordial gas. Minihalos with mass larger than M_J will contribute to the clumping factor c_l and need to be resolved numerically ‡.

Photoheating during reionization increases the temperature rapidly to a value of $T_r \approx \epsilon_\gamma/(3k_B) = 2.2 \times 10^4 \, (\epsilon_\gamma/6.33 \text{ eV})$K, where ϵ_γ is the mean energy of the ionizing photons; the numerical value assumes that the ionizing photons are emitted by hot stars with effective temperature 10^5K (see e.g. Chan *et al.* 2021). Halos with virial temperature below T_r or, equivalently, virial mass below

$$M_{\text{vir}} \lesssim 10^9 M_\odot \left(\frac{9}{1+z} \right)^{3/2}, \qquad (2.2)$$

will lose a significant fraction of their baryons through photoevaporation (Okamoto Gao & Theuns 2008), which decreases c_l. We aim to simulate the photoevaporation of these minihalos and other small-scale structures numerically.

Accurately modelling the evolution of c_l requires high mass resolution to resolve halos with mass M_J. Emberson, Thomas & Alvarez (2013) found that the dark matter particle mass, m_{dm}, should not exceed 100 $M_\odot$, for the simulations to converge. Photoevaporation significantly reduces c_l, so radiation hydrodynamics is necessary (Park *et al.* 2016; D'Aloisio *et al.* 2020). Finally, Emberson, Thomas & Alvarez (2013) claims that the simulation volume needs to be at least 1 cMpc (the "c" stands for comoving) on a side to account for sample variance.

To fulfil these requirements, we set up a high-resolution radiation-hydrodynamics cosmological simulation at high redshift. We adopt the following cosmological parameters: $h = 0.678$, $\Omega_m = 0.307$, $\Omega_\Lambda = 0.693$, and $\Omega_b = 0.0455$. We generate initial conditions using MUSIC (Hahn & Abel 2011). Our cubic simulation volume has a side-length, $l_{\text{box}} = 800$ ckpc, and it is filled with 512^3 gas and dark matter particles ($m_{\text{DM}} = 100 \, M_\odot$). Runs are performed with SWIFT (Schaller *et al.* 2016) to which we added the SPH-M1RT radiation hydrodynamics module (Chan *et al.* 2021). We use the grey approximation for the ionization cross-section, adopt the on-the-spot approximation for recombination, and use the optically-thin approximation in the calculation of the photoheating. We reduce the speed of light for computational efficiency, decreasing c by a factor of 10 at the mean density, verifying that this does not impact our results. The detailed methodology and tests will be presented elsewhere (Chan *et al.*, in prep.).

We set the redshift of reionization to $z_r = 8$. For redshifts $\leq z_r$, we inject plane-parallel radiation from two opposite sides of the cubic volume with a blackbody spectrum of temperature 10^5K. The simulation follows the propagation of the two ionization fronts through the simulated volume, calculating in detail shadowing and photoevaporation of gas in halos.

3. Preliminary Results

We performed simulations with two different values of the imposed photoionization rate at the edges, $\Gamma_{-12} = 0.03$ and $\Gamma_{-12} = 0.3$, where Γ_{-12} is the photo-ionization rate in units of 10^{-12}s^{-1}; these values bracket those used by D'Aloisio *et al.* (2018). The left panel of Fig. 1 shows the dark matter and neutral gas structures at $z = 8$ for the higher value of Γ_{-12}. At the left and right edges, the gas is highly ionized, in the middle of the volume, the gas is still neutral. Island of neutral gas in the highly ionized IGM correspond to dense structures with short recombination times. These include the filamentary structures of the cosmic web as well as gas in halos. The latter are visible as small yellow dots in the

---

‡ Pre-reionization X-ray sources can heat the IGM, boost the Jeans mass above Eq. 2.1, and reduce the IGM clumping. However, Eq. 2.1 still represents the minimum possible Jeans mass of the IGM, especially given that pre-reionization X-ray sources are highly uncertain.

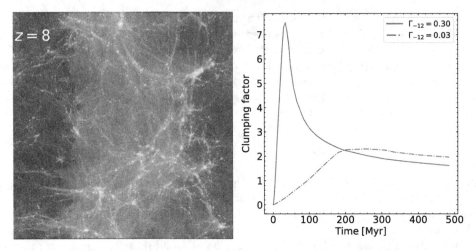

Figure 1. *Left panel:* A projection of neutral gas (yellow) blended with dark matter (background) in a simulation with $\Gamma = 0.3 \times 10^{-12} \mathrm{s}^{-1}$; *Right panel:* Evolution of the clumping factor, c_l, in our simulations since the onset of reionization with different values of Γ_{-12}, as indicated in the legend.

panel. The intervening density structures deform the initially flat ionization front into a corrugated sheet.

The right panel of Fig. 1 shows the evolution of c_l for the two choices of photo-ionization rate. Initially, the ionized gas density, n_{HII}, and the clumping factor, c_l, both increase as the ionization front traverses the volume. The clumping factor is larger for higher values of Γ_{-12}. As the IGM is ironed out by photoheating, c_l decreases again. Our results agree qualitatively with those from Park *et al.* (2016) and D'Aloisio *et al.* (2020).

4. Discussions and Conclusion

The evolution of the clumping factor is a crucial ingredient of the physics of reionization. We find that the inhomogeneous IGM can hinder the progress of reionization by increasing the clumping factor, c_l. For the first ~ 100 Myrs after the passage of an ionization front, c_l may be increased by up to a factor of 10 compared to a homogeneous IGM†. The clumping factor is even larger for higher values of the photoionization rate and/or lower redshift of reionization (Chan *et al.* in prep.). These results are consistent with previous findings, e.g. Park *et al.* (2016) and D'Aloisio *et al.* (2020). However, our values for c_l are smaller than found in studies that neglect radiation-hydrodynamic effects, such as those of e.g. Emberson, Thomas & Alvarez (2013). The hydrodynamic response of ionizing photons can reduce c_l in the IGM within a few tens of Myrs. We find that it takes only ~ 1 extra photon to ionize the majority of the clumpy IGM, as compared to a homogeneous universe.

The small-scale structure of the IGM also affects the IGM opacity. Becker *et al.* (2021) present evidence for a rapid evolution of the mean free path of ionizing photons, λ_{MF}, between $z = 6$ and $z = 5$. Cain *et al.* (2021) and Davies *et al.* (2021) argued that this implies a significant evolution of the mean value of Γ_{-12}. It would be interesting to study the correlation between c_l, λ_{MF} and $\bar{\Gamma}_{-12}$. We are currently performing a suite of high-resolution radiation-hydrodynamic cosmological simulation that span different box sizes, redshifts of reionization and imposed photoionization rates (Chan *et al.* in prep.).

---

† See, e.g. Shapiro & Giroux (1987); Ciardi *et al.* (2006), for the effect of clumping on cosmological ionization front.

We will use this suite to investigate the impact of minihalos on reionization and the evolution of λ_{MF}.

5. Acknowledgement

This work was supported by the Science and Technology Facilities Council (STFC) astronomy consolidated grant ST/P000541/1 and ST/T000244/1. We acknowledge support from the European Research Council through ERC Advanced Investigator grant, DMIDAS [GA 786910] to CSF. ABL acknowledges support by the European Research Council (ERC) under the European Union's Horizon 2020 research and innovation program (GA 757535) and UNIMIB's Fondo di Ateneo Quota Competitiva (project 2020-ATESP-0133). This work used the DiRAC@Durham facility managed by the Institute for Computational Cosmology on behalf of the STFC DiRAC HPC Facility (www.dirac.ac.uk). The equipment was funded by BEIS capital funding via STFC capital grants ST/K00042X/1, ST/P002293/1, ST/R002371/1 and ST/S002502/1, Durham University and STFC operations grant ST/R000832/1. DiRAC is part of the National e-Infrastructure. This work made use of SPHViewer (Benitez-Llambay 2015).

References

Becker G. D., D'Aloisio A., Christenson H. M., Zhu Y.,Worseck G. & Bolton J. S. 2021, *MNRAS*, 508, 1853
Benitez-Llambay A. 2015, py-sphviewer: Py-SPHViewerv1.0.0, doi:10.5281/zenodo.21703
A. J. Benson, C. S. Frenk, C. G. Lacey, C. M. Baugh & S. Cole 2002, *MNRAS*, 333, 177
Bullock J. S., Kravtsov A. V. & Weinberg D. H. 2000, *ApJ*,539, 517
Cain C., D'Aloisio A., Gangolli N. & Becker G. D. 2021, *ApJL*, 917, L37
Chan T. K., Theuns T., Bower R. & Frenk C. 2021, *MNRAS*, 505, 5784
Ciardi B., Scannapieco E., Stoehr F., Ferrara A., Iliev I. T., & Shapiro P. R. 2006, *MNRAS*, 366, 689
D'Aloisio A., McQuinn M., Davies F. B. & Furlanetto S. R.,2018, *MNRAS*, 473, 560
D'Aloisio A., McQuinn M., Trac H., Cain C. & Mesinger A. 2020, *ApJ*, 898, 149
Davies F. B., Bosman S. E. I., Furlanetto S. R., BeckerG. D. & D'Aloisio A. 2021, *ApJL*, 918, L35
Efstathiou, G. 1992, *MNRAS*, 256, 43P
Emberson J. D., Thomas R. M. & Alvarez M. A., 2013, *ApJ*, 763, 146
Moore B., Ghigna S., Governato F., Lake G., Quinn T., Stadel J. & Tozzi P. 2000, *ApJL*, 524, L19
Hahn O. & Abel T. 2011, *MNRAS*, 415, 2101
Haiman Z., Abel T. & Madau P. 2001, *ApJ*, 551, 599
Okamoto T., Gao L. & Theuns T., 2008, *MNRAS*, 390, 920
Park H., Shapiro P. R., Choi J.-h., Yoshida N., Hirano S. & Ahn K. 2016, *ApJ*, 831, 86
Peebles, P. J. E. 1993, Principles of Physical Cosmology by P.J.E. Peebles. Princeton University Press, 1993. ISBN: 978-0-691-01933-8
Planck Collaboration, Ade, P. A. R., Aghanim, N., et al. 2014, *A&A*, 571, A16
Robertson, B. E. 2021, *arXiv*, 2110.13160
Schaller M., Gonnet P., Chalk A. B. G. & Draper P. W. 2016, *arXiv*, 1606.02738
Shapiro P. R. & Giroux M. L. 1987, *ApJL*, 321, L107
Sharma M., Theuns T., Frenk C., Bower R., Crain R., Schaller M., Schaye J., 2016, *MNRAS*, 458, L94
Theuns T. 2021, *MNRAS*, 500, 2741

Discussion

HUI LI: Hi TK. Nice to see you again. I have a question about the clumping factor. I totally agree with you that with high resolution you can resolve more clumpiness for the simulation and then change the recombination rate. But my question is at which resolution we can fully resolve this. This is really a desperate problem, because the higher the resolution the more clumpiness you will observe in the simulation.

TSANG KEUNG CHAN: That is a very good question. There were some studies a few years ago that did a very high resolution and found that convergence of the clumpiness is around the mass resolution we use here, so we take this mass resolution. But we have radiation hydrodynamics. They post-processed the simulations with radiative transfer, so they can run faster. However, they do not consider, e.g., radiative heating, so they overestimated the clumpiness. But even in that case, they also found convergence at this mass scale. Thus, we should be able to resolve the clumping factor at this resolution.

CHIA YU HU: My question is also related to the clumping factor. Is the clumping factor in your clumping factor defined globally? Or do you use a scale to define that when you do the averaging?

TSANG KEUNG CHAN: At least in our study and also other paper, we define the clumping factor in the whole box. This will depend on the box size you choose because they have different structures.

CHIA YU HU: But it could be that the small scale clumping doesn't play a role in a large scale clumping.

TSANG KEUNG CHAN: It depends on how large the clumping factor is. We find the clumping factor can be up to 10 or 20. If there are also many minihalos in large boxes, then supposedly they should also contribute significantly to the clumping factor.

The Predictive Power of Computational Astrophysics as a Discovery Tool
Proceedings IAU Symposium No. 362, 2023
D. Bisikalo, D. Wiebe & C. Boily, eds.
doi:10.1017/S174392132200151X

Favoured Inflationary Models by SFC Baryogenesis

Mariana Panayotova[1] and Daniela Kirilova[2]

[1]Institute of Astronomy with National Astronomical Observatory,
Bulgarian Academy of Sciences
email: `mariana@astro.bas.bg`

[2]Institute of Astronomy with National Astronomical Observatory,
Bulgarian Academy of Sciences
email: `dani@astro.bas.bg`

Abstract. We provide analysis of the baryon asymmetry generated in the Scalar Field Condensate (SFC) baryogenesis model obtained in new inflation, chaotic inflation, Starobinsky inflation, MSSM inflation, quintessential inflation, considering both cases of efficient thermalization after inflation and also delayed thermalization. We have found that baryon asymmetry generated in SFC baryogenesis model is considerably bigger than the observed one for the new inflation, new inflation model by Shafi and Vilenkin, MSSM inflation, chaotic inflation with high reheating temperature and the simplest Shafi-Vilenkin chaotic inflationary model. Therefore, strong diluting mechanisms are needed to reduce the baryon excess to its observational value today for these models. We have shown that for the SFC baryogenesis model a successful generation of the observed baryon asymmetry is possible in Modified Starobinsky inflation, chaotic inflation with low reheating temperature, chaotic inflation in SUGRA and quintessential inflationary model.

Keywords. early univese

---

1. Introduction

Here we present shortly the results of our study of SFC baryogenesis models in different inflationary scenarios, published in Kirilova&Panayotova (2021).

Cosmic and gamma-ray data indicate that there is no significant antimatter quantity up to galaxy cluster scales of 10-20 Mpc Steigman (1976), Steigman (2008), Stecker (1985), Ballmoos (2014), Dolgov (2015)†. Hence, a generation of the observed baryon asymmetry from initially matter-antimatter symmetric state of the very early Universe must have happened in the period after inflation, but before Big Bang Nucleosynthesis (BBN) epoch.

Baryon asymmetry is usually described by the baryon density or the baryon to photon ratio:

$$\beta = (N_b - N_{\bar{b}})/N_\gamma \sim N_b/N_\gamma = \eta, \qquad (1.1)$$

† However, small quantities of antimatter, even in our Galaxy have been observed, see refs. Dolgov (2021), Dolgov (2022), where indications about the presence of 14 anti-stars in our Galaxy are presented. The fractional density of compact anti-stars in the universe up to 10% does not contradict the existing observational bounds, see for instance ref. Blinnikov, Dolgov&Postnov (2015) and references there in. The observational limits on anti-stars are less constraining than on gas clouds of antimatter because surface annihilation on the stars is not as efficient as the volume annihilation ref. Steigman (1976).

and is known with high precision from BBN and Cosmic Microwave Background (CMB) measurements. Namely, $\eta \sim 6 \times 10^{-10}$.

1.1. *SFC baryogenesis model short description*

There exist various baryogenesis models which generate successfully this number at different epochs before BBN - GUT baryogenesis, SUSSY baryogenesis, baryogenesis through leptogenesis, warm baryogenesis, etc. Here we discuss the SFC baryogenesis Dolgov&Kirilova (1990), Dolgov&Kirilova (1991) based on the Afleck and Dine baryogenesis scenario Affleck&Dine (1985).

According to SFC baryogenesis model at the end of inflation besides the inflaton ψ, there existed a complex scalar field φ, carrying baryon charge. B is not conserved at large φ due to the presence of B non-conserving (BV) self-interaction terms in the potential $V(\varphi)$, while at small φ BV is negligible. During inflation because of the rise of quantum fluctuations of φ, a condensate $<\varphi> \neq 0$ with a nonzero baryon charge B was formed Vilenkin&Ford (1982), Bunch&Davies (1978), Starobinsky (1982).

At the end of inflation φ starts to oscillate around its equilibrium and its amplitude decreases due to the universe expansion and particle creation processes of scalar field to fermions Dolgov&Kirilova (1990), Kirilova&Panayotova (2007). B which survives until B-conservation epoch t_B, is transferred to fermions. For more details about SFC baryogenesis model see Kirilova&Panayotova (2015) and Kirilova&Panayotova (2021).

2. Baryon asymmetry production in different inflationary models and different reheating scenarios

2.1. *Description of the numerical analysis*

We have provided numerical analysis of the SFC baryogenesis model. We have used the following equation of motion describing the evolution of φ:

$$\ddot{\varphi} + 3H\dot{\varphi} + \frac{1}{4}\Gamma_\varphi \dot{\varphi} + U'_\varphi = 0, \qquad (2.1)$$

where $a(t)$ is the scale factor, H is the Hubble parameter $H = \dot{a}/a$. $\Gamma_\varphi = \alpha\Omega$ is the rate of particle creation, $\Omega = 2\pi/T$, where T is the period of the field oscillations. The analytically estimated value: $\Omega_0 = \lambda^{1/2}\varphi_0$, is used as an initial condition of the frequency in the numerical analysis.

The field potential was chosen of the form:

$$U(\varphi) = m^2\varphi^2 + \frac{\lambda_1}{2}|\varphi|^4 + \frac{\lambda_2}{4}(\varphi^4 + \varphi^{*4}) + \frac{\lambda_3}{4}|\varphi|^2(\varphi^2 + \varphi^{*2}). \qquad (2.2)$$

The following assumptions were made: the mass is $m \ll H_I$ and $m = 10^2 - 10^4$ GeV, the self-coupling constants λ_i are of the order of the gauge coupling constant α. The energy density of φ at the inflationary stage is of the order H_I^4, hence

$$\varphi_o^{max} \sim H_I \lambda^{-1/4}, \quad \dot{\varphi}_o = (H_I)^2, \quad B_0 = H_I^3. \qquad (2.3)$$

We have developed a program in fortran 77 using 4th order Runge-Kutta method to solve the system of ordinary differential equations, corresponding to the equation of motion for the real and imaginary part of φ and B. We have provided a numerical analysis Kirilova&Panayotova (2007), Kirilova&Panayotova (2012), Kirilova&Panayotova (2014), Kirilova&Panayotova (2015) of the evolution of φ,

$$\varphi(t) = x + iy \quad and \quad B(t) = -i(\dot{\varphi}^*\varphi - \dot{\varphi}\varphi^*) \qquad (2.4)$$

from the inflationary stage until B conservation epoch for about 100 sets of parameters of SFC baryogenesis model.

The parameters ranges studied are: $\alpha = 10^{-3} - 5 \times 10^{-2}$, $H_I = 10^7 - 10^{12}$ GeV, $m = 100 - 1000$ GeV, $\lambda_1 = 10^{-3} - 5 \times 10^{-2}$, $\lambda_{2,3} = 10^{-4} - 5 \times 10^{-2}$.

The produced baryon asymmetry β in SFC baryogenesis model depends on the generated baryon excess B at the epoch t_B, the reheating temperature of the Universe T_R and the value of the Hubble parameter at the end of inflation H_I. Namely:

$$\beta \sim N_B / T_R^3 \sim B T_R / H_I. \tag{2.5}$$

T_R and H_I values depend on the type of inflation and reheating. First consideration of the SFC baryogenesis model in different inflationary scenarios and preliminary results were reported in refs. Kirilova&Panayotova (2019), Kirilova&Panayotova (2020). In the work Kirilova&Panayotova (2021) we considered all B-excess values in the whole range of the studied parameter sets of the SFC baryogenesis model and different values of T_R and H_I corresponding to the studied inflationary models and reheating scenarios.

We have considered the following inflationary models: the new inflation Linde (1982), Albrecht&Steinhardt (1982), Shafi-Vilenkin model of new inflation, chaotic inflation Linde (1985), Linde (1990), Shafi-Vilenkin model of chaotic inflation, chaotic inflation in SUGRA, Starobinsky inflation Kofman, Linde&Starobinsky (1985), MSSM inflation and quintessential inflation.

We have discussed different possibilities for reheating. Reheating temperature T_R depends on the model of reheating, perturbative or non-perturbative decay of the inflaton, inflaton decay rate, spectrum of the decay particles, thermalization after inflation (efficient or delayed) Marko et.al. (2020).

2.2. *Production of the baryon asymmetry – results and discussion*

Our calculations show that for the new inflation model Linde (1982), Albrecht&Steinhardt (1982) for $H_I = 10^{10}$ GeV and $T_R = 10^{14}$ GeV, the obtained baryon asymmetry β is by order of magnitude bigger than the observed for all sets of parameters. We calculated baryon asymmetry also for new inflation by Shafi and Vilenkin for $H_I = 3 \times 10^9$ GeV and $T_R = 3 \times 10^7$ GeV, Chaotic inflation for $H_I \in [10^{11}, 10^{12}]$ GeV and $T_R \in [10^{12}, 10^{14}]$ GeV, Shafi and Vilenkin Chaotic inflation for $T_R = 10^{12} - 10^{13}$ GeV and $H_I \in [5 \times 10^9, 10^{12}]$ GeV and MSSM inflation for $H_I = 1$ GeV, $T_R = 2 \times 10^8$ GeV. In these models, values orders of magnitude bigger than the observed baryon asymmetry are generated. Hence, for these inflationary models strong diluting mechanisms are necessary to reduce the resultant baryon excess to the value observed today.

The numerical analysis showed that baryon asymmetry equal to the observed one can be produced by the SFC baryogenesis model in the following inflationary models: Modified Starobinsky inflation with $T_R = 0.1(\Gamma M_{Pl})^{1/2} = 10^9$ GeV, $H_I = 10^{11}$ GeV; chaotic inflation with efficient thermalization for $H_I = 10^{12}$ GeV, $T_R = 6.2 \times 10^9$ GeV and for $H_I = 10^{11}$ GeV and $T_R = 1.9 \times 10^9$ GeV; chaotic inflation with delayed thermalization for $H_I = 10^{12}$ GeV, $T_R = 4.5 \times 10^8$ GeV; chaotic inflation with monomial potential with $p = 2/3$ and $T_R = 10^9$ GeV and $H_I \sim 10^{11}$ GeV; Chaotic inflation in SUGRA with $T_R > 10^9$ GeV; and Quintessential inflation with $T_R = 2 \times 10^5$ GeV and decay into massless particles, $H_I = 10^{12}$ GeV. The particular parameters sets of the SFC baryogenesis model are listed in Table 1. So, choosing the inflationary model, it is possible to fix the SFC baryogenesis model parameters.

In Figure 1 we present the successful inflationary models, corresponding to different reheating temperatures in the plane of α - $\lambda_{2,3}$ with fixed parameters for the SFC baryogenesis model, namely: $\lambda_1 = 5 \times 10^{-2}, m = 350$ GeV and $H_I = 10^{12}$ GeV. The models correspond to different reheating temperatures, see Table 1.

Table 1. Successful production of the observed baryon asymmetry β for particular sets of SFC model parameters in different inflationary scenarios. Table from ref. Kirilova&Panayotova (2021).

Starobinsky Inflation	$H_I = 10^{11}$ GeV; $T_R = 10^9$ GeV	$\lambda_1 = \alpha = 5 \times 10^{-2}$, $\lambda_2 = \lambda_3 = 10^{-2}$, $m = 100$ GeV, $\beta = 9.3 \times 10^{-10}$	
	$H_I = 10^{12}$ GeV; $T_R = 10^9$ GeV	$\lambda_1 = 5 \times 10^{-2}$, $\alpha = 3 \times 10^{-2}$, $\lambda_2 = \lambda_3 = 10^{-3}$, $m = 350$ GeV, $\beta = 6.6 \times 10^{-10}$	$\lambda_1 = \alpha = 5 \times 10^{-2}$, $\lambda_2 = \lambda_3 = 10^{-3}$, $m = 350$ GeV, $\beta = 8.0 \times 10^{-10}$
Quintessential Inflation	$H_I = 10^{12}$ GeV; $T_R = 2 \times 10^5$ GeV	$\lambda_1 = 5 \times 10^{-3}$, $\alpha = 10^{-3}$, $\lambda_2 = \lambda_3 = 10^{-4}$, $m = 350$ GeV, $\beta = 4.6 \times 10^{-10}$	$\lambda_1 = 10^{-2}$, $\alpha = 10^{-3}$, $\lambda_2 = \lambda_3 = 10^{-4}$, $m = 350$ GeV, $\beta = 7.8 \times 10^{-10}$
Chaotic Inflation, Efficient Thermalization	$H_I = 10^{12}$ GeV; $T_R = 6.2 \times 10^9$ GeV	$\lambda_1 = \alpha = 5 \times 10^{-2}$, $\lambda_2 = \lambda_3 = 10^{-2}$, $m = 350$ GeV, $\beta = 7.4 \times 10^{-10}$	
Chaotic Inflation, Delayed Thermalization	$H_I = 10^{12}$ GeV; $T_R = 4.5 \times 10^8$ GeV	$\lambda_1 = \alpha = 10^{-2}$, $\lambda_2 = \lambda_3 = 10^{-3}$, $m = 350$ GeV, $\beta = 9.5 \times 10^{-10}$	$\lambda_1 = \alpha = 5 \times 10^{-2}$, $\lambda_2 = \lambda_3 = 10^{-3}$, $m = 350$ GeV, $\beta = 3.6 \times 10^{-10}$

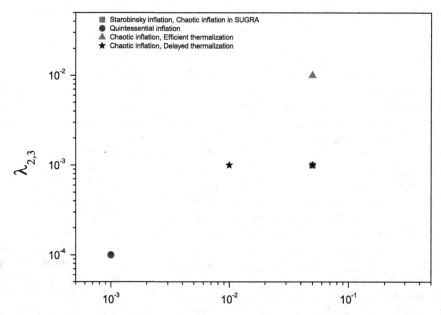

Figure 1. The figure presents different inflationary models in the α - $\lambda_{2,3}$ plane for which successful SFC baryogenesis is achieved for the following parameters: $\lambda_1 = 5 \times 10^{-2}, m = 350$ GeV and $H_I = 10^{12}$ GeV.

For successful production of β, the parameters α and λ have the same values in Starobynsky and Chaotic inflation. In case of chaotic inflation in SUGRA, our analysis has shown that for $T_R = 10^9$ GeV, the results coincide with these for Starobynsky inflation.

Quintessential inflationary model, however, needs an order of magnitude smaller parameters values to produce the observed baryon asymmetry.

3. Conclusion

The numerical analysis of the SFC baryogenesis model with different reheating temperatures of several inflationary scenarios has shown that:

(i) SFC baryogenesis model overproduces baryon asymmetry for the following inflationary models: new inflation, new inflation model by Shafi and Vilenkin, chaotic inflation with high reheating temperature, the simplest Shafi-Vilenkin chaotic inflationary model and MSSM inflation. For these models strong diluting mechanisms are needed to reduce the baryon excess to its observational value.

(ii) SFC baryogenesis model produces close to the observed baryon asymmetry value in the following inflationary models: Modified Starobinsky inflation, chaotic inflation with lower reheating temperature, chaotic inflation in SUGRA and Quintessential inflation. In case of delayed thermalization, a successful SFC baryogenesis may be achieved in the chaotic inflationary models.

(iii) However, choosing for the value of α a value close to α_{GUT}, SFC baryogenesis cannot be realized in Quintessential inflation. Thus, in this case, SFC baryogenesis favors Starobynsky and Chaotic inflation.

4. Acknowledgements

We acknowledge the partial financial support by project DN18/13-12.12.2017 of the Bulgarian National Science Fund of the Bulgarian Ministry of Education and Science.

References

Kirilova, D., Panayotova, M. 2021, *Galaxies*, 9, 49
Steigman, G. 1976 *Ann. Rev. Astron. Astrophys.*, 14, 339
Steigman, G. 2008, *J. Cosmol. Astropart. Phys.*, 0910, 001
Stecker, F. 1985, *Nucl. Phys. B*, 252, 25
Ballmoos, P. 2014, *Hyperfine Interact.*, 228, 91
Dolgov, A. 2015, *EPJ Web of Conferences*, 95, 03007
Dolgov, A. 2021, *Proceedings of 20th Lomonosov Conference on Elementary Particle Physics, ICNFP 2021* e-Print:2112.15255
Dolgov, A. 2021, *20th Lomonosov Conference on Elementary Particle Physics* e-Print:2201.04529
Blinnikov S., Dolgov A., Postnov K. 2015, *Phys.Rev.D* 92 2, 023516
Pettini, M., Cooke, R. 2012, *Mon. Not. Roy. Astron. Soc.*, 425, 2477
Ade, P., et.al. [Planck Collaboration] 2016, *Astron. Astrophys.*, 594, A13
Dolgov, A., Kirilova, D. 1990, *Sov. J. Nucl. Phys.*, 51, 172
Dolgov, A., Kirilova, D. 1991, *J. Moscow Phys. Soc.*, 1, 217
Affleck, I., Dine, M. 1985, *Nucl. Phys. B*, 249, 361
Vilenkin, A., Ford, L. 1982, *Phys. Rev. D*, 26, 1231
Bunch, T., Davies, P. 1978, *Proc. R. Soc. London, Ser. A*, 360, 117
Starobinsky, A. 1982, *Phys. Lett. B*,117, 175
Kirilova, D., Panayotova, M. 2007, *Bulg. J. Phys.*, 34 s2, 330
Kirilova, D., Panayotova, M. 2012, *Proc. 8th Serbian-Bulgarian Astronomical Conference (VIII SBGAC), Leskovac, Serbia 8-12 May*
Kirilova, D., Panayotova, M. 2014, *BAJ*, 20, 45
Kirilova, D., Panayotova, M. 2015, *Advances in Astronomy*, 465
Linde, A. 1982, *Phys. Lett. B*, 108, 389
Albrecht, A., Steinhardt, P. 1982, *Phys. Rev. Lett.*, 48, 1220
Linde, A. 1985, *Phys. Lett. B*, 129, 177, (1983); *Phys. Lett. B*, 162, 281
Linde, A. 1990, *Particle Physics and Inflationary Cosmology, Harwood, Chur, Switzerland*
Kofman, L., Linde, A., Starobinsky, A. 1985, *Phys. Lett. B*, 157, 36
Marko, A., Gasperis, G., Paradis, G., Cabella, P. 2020, *arXiv:1907.06084*
Kirilova, D., Panayotova, M. 2019, *AIP Conf. Proc.*, 2075, 090017
Kirilova, D., Panayotova, M. 2020, *Publ. Astron. Soc. "Rudjer Bockovic", Proc. XII SB Astronomical Conference, 25–29 September 2020, Sokobanja, Serbia*, 20, 39

The Predictive Power of Computational Astrophysics as a Discovery Tool
Proceedings IAU Symposium No. 362, 2023
D. Bisikalo, D. Wiebe & C. Boily, eds.
doi:10.1017/S1743921322001788

The Effect of Non-Gaussian Primordial Perturbations on Large-Scale Structure

G. A. Peña and **G. N. Candlish**

Instituto de Física y Astronomía, Universidad de Valparaíso, Gran Bretaña 1111,
Valparaíso, Chile
email: greco.pena@postgrado.uv.cl

Abstract. The late-time effect of primordial non-Gaussianity offers a window into the physics of inflation and the very early Universe. In this work we study the consequences of a particular class of primordial non-Gaussianity that is fully characterized by initial density fluctuations drawn from a non-Gaussian probability density function, rather than by construction of a particular form for the primordial bispectrum. We numerically generate multiple realisations of cosmological structure and use the late-time matter polyspectra to determine the effect of these modified initial conditions. In this non-Gaussianity has only a small imprint on the first polyspectra, when compared to a standard Gaussian cosmology. Furthermore, some of our models present an interesting scale-dependent deviation from the Gaussian case in the bispectrum and trispectrum, although the signal is at most at the percent level.

Keywords. Large-scale structure of Universe – inflation – methods: numerical

1. Introduction

One of the current challenges in cosmology is to understand the physical processes that gave rise to the primordial inhomogeneities of the Universe. These inhomogeneities are generated during cosmic inflation in the standard model of cosmology. There is a wide variety of inflationary models that spontaneously give rise to the initial conditions of the Universe. While the simplest models, assuming a single inflationary scalar field, lead to a nearly Gaussian distribution of primordial perturbations, more complex models with larger numbers of degrees of freedom can produce measureable levels of non-Gaussianity (here a comprehnsive review: Celoria & Matarrese (2018)).

If primordial perturbations are the seeds that gave rise to the structures of the Universe, we can assume that there is an underlying signal corresponding to the primordial non-Gaussianities. In this work we consider a simple approach to test this signal on the formation of large-scale structure. We use a recently-proposed characterization of this primordial non-Gaussianity in terms of a non-Gaussian PDF describing the primordial curvature perturbations Chen et al. (2018a,b)). Using this approach will allow us to generate initial conditions for our simulations by simply changing the PDF from which we draw our initial sample of density perturbations. A significant difference of this approach compared to previous approaches to studying non-Gaussianity in cosmological simulations is that we do not limit ourselves to specifying a particular form of the primordial bispectrum, as we have the full PDF available to us.

In this work we aim to perform a preliminary exploration of the parameter space of this novel characterization of the primordial non-Gaussianity. In order to extract the signal from that induced through the non-linear process of gravitational collapse we must average over several realisations of cosmological evolutions. Thus we would, in principle,

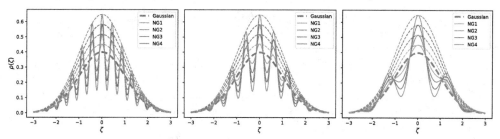

Figure 1. The non-Gaussian probability density distributions for the primordial curvature perturbations considered in this work (see Eq. (108) in Chen et al. (2018a)). The thick blue dashed line is the fiducial Gaussian distribution. All of our PDFs have unit variance. The PDF cut in the horizontal range between $\zeta_{min} = -4$, and $\zeta_{max} = 4$.

require multiple N-body simulations where we vary the initial sample of random numbers as well as the parameters of the PDF. To facilitate the generation of these realisations we have used the L-PICOLA code (Howlett et al. (2015)), which has been verified to be sufficiently accurate for our purposes by comparing with a smaller number of full N-body simulations run using the RAMSES code (Teyssier (2002)).

2. Simulations

2.1. *Initial conditions*

The inflationary scenario that gives rise to the generation of the non-Gaussian initial conditions comprises perturbations in an axion-like isocurvature field (with sinusoidal potential) which are coupled to the primordial curvature field. From this scenario rise non-Gaussian initial conditions characterized by the probability density distribution of the primordial curvature perturbation (see Eq. (104) of Chen et al. (2018a,b)). As PDF is characterized by an oscillatory modulation, we generate 4 different amplitudes (referred to as non-Gaussianity levels), and 3 different frequencies, as shown in Fig. 1. The thick blue dashed line indicates a Gaussian PDF with a standard deviation of unity. The orange, green, red, and purple lines correspond to the four levels of non-Gaussianity we consider. The dashed lines correspond to Gaussian envelopes that touch the upper peaks of the oscillations, and each panel represents a different modulated frequency. Note that we are considering large deviations from the amplitude of Gaussianity to explore the feasibility of detecting this type of NG in n-point late-time statistics; observational constraints such as those of the CMB could probably exclude our non-Gaussian primordial PDF.

We generate a sample of each of our non-Gaussian PDFs by using a simple accept-reject technique, whereby we generate uniformly distributed random values within a bounding box that includes the highest peak of the oscillatory PDF and is cut in the horizontal range between $\zeta_{min} = -4$, and $\zeta_{max} = 4$. This technique consists in accepting the values that belong to the distribution while rejecting those points that do not. The procedure ends when we generate N values for all distributions, where N is the total number of particles in our simulations, chosen to be equal to the number of points used in our discretized density field mesh. The final step is to sort the non-Gaussian distributions according to the ordering of the Gaussian one. In other words, we generate a "white noise field" using the PDFs at all points.

This procedure was repeated for all 4 levels of non-Gaussianity, all 3 frequencies, and 5 different values of the initial seed of the uniform random number generator, to generate 5 different statistical realizations. Therefore, in total, we produced initial conditions for 65 models, 5 of them being fiducial Gaussian models. To provide clarity when indicating a model, we have adopted the following naming:

- Initial condition type: G for Gaussian, NG for non-Gaussian.
- Level of non-Gaussianity: 1, 2, 3, or 4. This only applies for NG models. The meaning of these levels will be clarified shortly.
- Frequency of non-Gaussianity: f1, f2, or f3. This only applies for NG models. The meaning of these levels will be clarified shortly.
- Random realisations: r1, r2, r3, r4 or r5. This will only be necessary when we refer to one specific realization. For most results, we average over all realizations.

For our simulations, we have used the N-body code, RAMSES, and the mock catalog, L-PICOLA. To generate the initial conditions for RAMSES we have passed the white noise to MUSIC (Hahn & Abel (2011)), where it is rearranged in a discretized grid. Then it is transformed to Fourier space and multiplied with a k-space transfer function generated by CAMB:

$$\delta(\vec{k}) = c k^{n_s/2} T(k) \mu(\vec{k}), \tag{2.1}$$

where $\mu(\vec{k})$ is the Fourier-transformed set of white noise, the transfer function $T(k)$ corresponds to a standard LCDM cosmology, and c is a normlisation constant. The real space over-density field $\delta(r)$ is then obtained by inverse Fourier transformation. Finally the 2LPT method is applied to generate the initial conditions.

In the case of L-PICOLA, we have modified the initial condition generation to read and use our random numbers, where the 2LPT method is also used. Therefore, the generation of the initial conditions of L-PICOLA is the same as MUSIC. We use RAMSES for just one realization with frequency f1 (5 models). All other models are generated with L-PICOLA.

For all our simulations, we have assumed a standard ΛCDM cosmology with the following parameters: $\Omega_m = 0.3$, $\Omega_b = 0.04$, $\Omega_\Lambda = 0.7$, $\sigma_8 = 0.88$, and $n_s = 0.96$. In addition, for all realisations we have used a 500 Mpc box, and a particle number of 256^3. In the case of RAMSES, which employs the AMR method, we have set a coarser grid resolution of 256^3 points, with 6 levels of refinement. In the case of L-PICOLA we have a fixed grid resolution of 256^3 points.

2.2. Polyspectra Analysis

An important feature of the nature of the non-Gaussian PDF is that only the even n-point functions are non-zero, while the odd n-point functions are zero just like a Gaussian PDF. On the other hand, possible evidence of primordial non-Gaussianities is mixed with non-Gaussianities coming from the gravitational collapse of the large-scale structure at all orders of the n-point function making it difficult to decode its signature. Therefore, as preliminary work, we decided to focus on the search for the possible signature of this primordial Gaussianity by considering the power spectrum, bispectrum, and trispectrum. As we did not find a major significance in the power spectrum, in this paper we only report the results for the bispectra and trispectra.

As a first step, we focused our analyses on symmetric configurations. For the bispectrum we choose $k_1 = k_2 = k_3$, and for the trispectrum we choose $k_1 = k_2 = k_3 = k_4$. For the latter case, we have considered all (possibly folded) quadrilaterals with equal side lengths.

We calculated the polyspectra using the Pylians code (Villaescusa-Navarro 2018). We have modified the code and added the trispectrum calculation for our purposes. We use 35 linearly-spaced bins in the range $2.2 k_F \leq k \leq k_F N_k/3$, where the fundamental frequency is $k_F = 2\pi/L$, with $L = 500$ Mpc (box length), and N_k is the number of points used in our discretised Fourier space in each dimension, i.e, $N_k = 256$. We avoided small values of k corresponding to scales close to the box size as they are highly affected by the sample variance due to the small number of k configurations. At the other extreme we set the upper limit on k to $k_F N_k/3$ to avoid very high values of k where the estimator is expected to perform poorly (Sefusatti et al. (2016)). This also ensures that we are below the fundamental limit set by the Nyquist frequency $k_{Nyq} = k_F N_k/2$.

3. Results

3.1. *Variance due to differing realizations*

This section shows only our main results. We performed a normalized polyspectra analysis of variance on the results of the L-PICOLA simulations. We chose to work mainly with L-PICOLA because of its accuracy and speed. Our comparative analysis with RAMSES showed an excellent agreement of less than 1% at all scales. Therefore, we can have high confidence that the non-Gaussian polyspectra normalized by the Gaussian polyspectra are very well represented by L-PICOLA.

To remove the variance arising from differing realizations of the large-scale structure, we determine the average normalized polyspectra by averaging over all 5 realizations for each model. As discussed earlier, we consider only symmetrical configurations of the wavenumbers in this work, so we may treat all polyspectra as depending upon a single value of k. In this way, we can attempt to cancel out the contribution to these polyspectra arising from the non-linear structure formation, which will vary in each realization. We then use the various realizations to estimate the minimum and maximum values for the normalized polyspectra, which we will refer to as the variance around our averaged normalized polyspectra.

In the following plots (Fig. 2 and Fig. 3) of this section, we show the results of the analysis of variance in the polyspectra for models NG2 (left column), and NG4 (right column) for our frequencies f1, f2, and f3 (first, second and third row, respectively) at $z = 0$. In all plots, the red solid line represents the NG2 model, and the blue solid line represents the NG4 model, while the light grey strip will show the level of variance around these lines.

3.1.1. *Bispectra*

The results for the averaged normalized bispectra, at $z = 0$, are given in Fig. 2. The clearest thing to be seen in these results is that the variance increases as amplitude and frequency increase (see Fig. 1). The larger variance is given by the NG4f3 model, which shows a variance of $\sim 16\%$ (without considering the larger scales), and the rest of the models are below $\sim 10\%$. On the other hand, we can see a little scale dependence in the NG4f2 model, with a larger deviation from the equality line (black dashed line in the figure). We also can see a similar scale dependence, but smaller, in the NG2f2 model. These models presents a feature that we refer to as a "dip" between $k \sim 0.2$ h/Mpc and $k \sim 0.8$ h/Mpc. Apart from the presence of this "dip" in the models using frequency f2, there is little evidence of deviation from the Gaussian case in the other models.

This leads to an interesting conclusion regarding the sensitivity of the bispectrum to this form of non-Gaussianity, which is by construction a symmetric oscillatory correction to an underlying Gaussian. This does not violate the property of Gaussian distributions that the odd-n moments are identically zero. Thus the 3-point correlation function (whose k-space analog is the bispectrum under consideration here) in perfect Gaussian conditions would vanish. Normally the presence of non-linear structure induces non-Gaussianities such that the bispectrum is not zero. What we apparently find, however, is that structure formation driven from this primordial non-Gaussianity imprints a marginally statistically significant signal in the bispectrum at the limit of detectability above the "noise" resulting from non-linearities, at least for this choice of parameters.

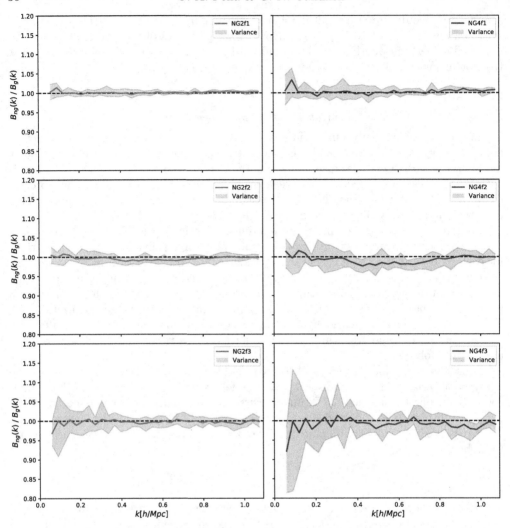

Figure 2. Non-Gaussian bispectra normalised with respect to the Gaussian bispectra, at $z = 0$. *Left column*: NG2 models, *right column*: NG4 models. *Top row*: frequency f1, *middle row*: frequency f2, *bottom row*: frequency f3. The red and blue lines indicate the average normalized bispectra (for the NG2 and NG4 models respectively) while the light grey strip shows the degree of variance around this average arising from the individual realizations.

3.1.2. Trispectra

In the case of the trispectrum, we have considered working with quadrilaterals in equilateral configurations, which are generally folded in 3-dimensional space. Then we take $k_1 = k_2 = k_3 = k_4$, where we have not set a restriction on the other two additional degrees of freedom. On the other hand, it is worth remembering that the trispectrum is the analog in Fourier space of the connected 4-point correlation function, where the disconnected parts are given by-products of the power spectra, which correspond to the disconnected 2-point correlation functions (Verde & Heavens (2001)). Thus, including for the equilateral configuration considered here, the trispectrum is an independent statistical measure that goes beyond the power spectrum.

The results for the averaged normalized trispectra, at $z = 0$, are shown in Fig. 3. Here, we must take into consideration that we have a smaller range of values shown on the

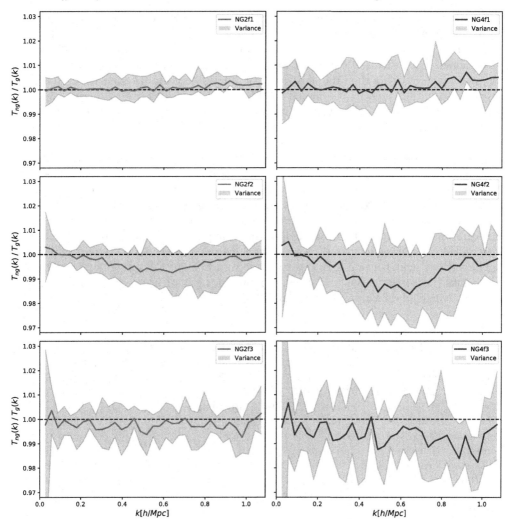

Figure 3. Non-Gaussian trispectra normalised with respect to the Gaussian trispectra, at $z = 0$. The panels are as for Fig. 2.

vertical axes of the bispectra of Fig. 2. Then, we see a lower sensitivity in all models if we compare it to the bispectra.

For models of frequency f2, interestingly, we have again a "dip" in a similar range ($0.3 \lesssim k \lesssim 0.9$ h/Mpc) to that shown in the bispectra of frequency f2. In the case of the other frequencies, there is very little obvious scale dependence, as seen for the bispectra. The f3 models show some indications of suppression of the trispectrum across all accessible scales, with the smaller scales more suppressed in the NG4f3 model. It is also worth pointing out that the variance over realizations in the trispectra is lower than seen for the bispectra, being at most 4% for the NG4f3 model. However, this level of variance implies that these results are not statistically significant.

4. Conclusions

Our study has focussed on analyzing the low-n correlation functions (specifically their Fourier space analogs: the bispectrum and trispectrum) to search for possible signatures of this type of primordial non-Gaussianity at low redshift. We have found that using the averaged normalized polyspectra, there are some scale-dependent deviations from the Gaussian model, which may be at the limit of detectability. Interestingly, the most significant signal is seen in the bispectrum for the models with frequency f2. For all other models, even the most extreme case NG4f3 we see only sub-percent deviations from Gaussianity, well within the sample variance of differing non-linear realizations. Thus our best-case scenario, given by the model NG4f2 shows deviations from Gaussianity in the bispectrum at the level of 2%. The statistical significance of this deviation by considering sample variance is rather marginal but may be detectable if sufficient precision can be obtained.

It is also noteworthy that the signal present in the frequency f2 models is a scale-dependent suppression of the symmetric bispectrum and the symmetric trispectrum. Consideration of non-symmetric configurations in k-space may well lead to further insights, and a more extensive exploration of the parameter space may uncover models whose late-time n-point correlation functions show more significant deviations of a similar form. Such an exploration may help to disentangle the precise relationship between the non-Gaussian frequency and the scale dependence of the deviations from Gaussianity.

The authors acknowledge financial support from FONDECYT Regular No. 1181708. GP thanks the Postgrado en Astrofísica program of the Instituto de Física y Astronomía of the Universidad de Valparaíso for funding.

References

Celoria M., & Matarrese S. 2018, *J. Cosmology Astropart. Phus.*, 039

Chen X., Palma G. A., Scheihing H. B., & Sypsas S. 2018a, *Phys. Rev. D.*, 98, 083528

Chen X., Palma G. A., Scheihing H. B., & Sypsas S. 2018, *Phys. Rev. Lett.*, 121, 161302

Hahn, O. & Abel, T. 2011, *MNRAS*, 415, 3

Howlett C., Manera M., & Percival W. J. 2015, *Astronomy and Computing*, 12, 109

Sefusatti E., Crocce M., Scoccimarro R., & Couchman H. M. P. 2016, *MNRAS*, 460, 3624

Teyssier R. 2002, *A&A*, 382, 412

Verde L., & Heavens A. F. 2001, *ApJ*, 553, 14

Villaescusa-Navarro F. 2018, Pylians: Python libraries for the analysis of numerical simulations (ascl:1811.008)

Discussion

ANONYMOUS: The use of L-PICOLA seems to be quite useful in the field of cosmological simulation, can you tell us a little more about it?

GRECO: L-PICOLA is a is a dark matter halo catalog generator known as mocks catalogues. The main advantage of using such codes is that they can evolve a dark matter distribution from early times to the present day much faster than a full non-linear N-body simulation.

The Predictive Power of Computational Astrophysics as a Discovery Tool
Proceedings IAU Symposium No. 362, 2023
D. Bisikalo, D. Wiebe & C. Boily, eds.
doi:10.1017/S1743921322001673

The complex evolution of supermassive black holes in cosmological simulations

Peter H. Johansson[1] , Matias Mannerkoski[1], Antti Rantala[2],
Shihong Liao[1], Alexander Rawlings[1], Dimitrios Irodotou[1],
Francesco Rizzuto[1]

[1]Department of Physics, Gustaf Hällströmin katu 2, FI-00014, University of Helsinki, Finland
email: `Peter.Johansson@helsinki.fi`

[2]Max-Planck Institut für Astrophysik, Karl-Schwarzschild-Str 1, D-85748, Garching, Germany

Abstract. We present here self-consistent zoom-in simulations of massive galaxies forming in a full cosmological setting. The simulations are run with an updated version of the KETJU code, which is able to resolve the gravitational dynamics of their supermassive black holes, while simultaneously modelling the large-scale astrophysical processes in the surrounding galaxies, such as gas cooling, star formation and stellar and AGN feedback. The KETJU code is able to accurately model the complex behaviour of multiple SMBHs, including dynamical friction, stellar scattering and gravitational wave emission, and also to resolve Lidov–Kozai oscillations that naturally occur in hierarchical triplet SMBH systems. In general most of the SMBH binaries form at moderately high eccentricities, with typical values in the range of $e = 0.6 - 0.95$, meaning that the circular binary models that are commonly used in the literature are insufficient for capturing the typical binary evolution.

Keywords. galaxies: supermassive black holes, galaxies: formation, galaxies: elliptical and lenticular, methods: numerical

1. Introduction

In the ΛCDM model galaxies grow hierarchically through mergers and gas accretion (e.g. Naab & Ostriker 2017). As all massive galaxies contain supermassive black holes (SMBHs) in their centres, the hierarchical growth of galaxies will invariably lead to SMBH mergers, which typically proceed through a three-stage process (Begelman et al. 1980). At large separations the evolution of the SMBHs are driven by dynamical friction until a binary forms. In the next phase the SMBH binary hardens through three-body scattering with individual stars (Hills & Fullerton 1980) and then finally at subparsec scales the binary coalesces due to the emission of gravitational waves (Peters 1964).

Modelling this entire SMBH coalescence process in a full cosmological simulation has been very challenging due to the inability of simultaneously modelling the small-scale SMBH dynamics and global galactic-scale astrophysical processes in simulations that include gravitational softening (e.g. Ryu et al. 2018). Instead, the parsec-scale dynamics has typically been modelled by postprocessing the simulations using semi-analytic methods based on orbit-averaged equations (Kelley et al. 2017) or by resimulating selected regions of galaxies by separate stand-alone N-body codes (Khan et al. 2016).

Here we present self-consistent cosmological zoom-in simulations run with our updated KETJU code (Rantala et al. 2017, Rantala et al. 2018, Mannerkoski et al. 2021), which is able to resolve the dynamics of merging SMBHs down to tens of Schwarzschild radii, while simultaneously modelling astrophysical processes in the surrounding galaxies, such as gas cooling, star formation and stellar and AGN feedback.

2. Simulations

In the KETJU code the dynamics of SMBHs and their surrounding stellar particles is integrated with the high-accuracy regularised integrator MSTAR (Rantala et al. 2020), whereas the dynamics of the remaining particles is computed with the standard GADGET-3 leapfrog method (Springel 2005). The gravitational interactions of SMBHs with other SMBHs and stellar particles are computed without softening while the interactions between stellar particles are softened in order to avoid energy errors when particles enter and exit the regularised KETJU region. The effects of general relativity, such as binary precession and gravitational wave (GW) emission are modelled by including post-Newtonian correction terms up to order 3.5 between each pair of SMBHs (Mora & Will 2004). In addition, we also now include the 1PN corrections for general N-body systems, which could potentially affect the long-term evolution of triple and multiple SMBH systems (e.g. Will 2014).

The gas component is modelled using the SPHGAL smoothed particle hydrodynamics implementation (Hu et al. 2014). We include metal-dependent gas cooling that tracks 11 individual elements and use a stochastic star formation model with a critical hydrogen number density threshold of $n_{\rm H} = 0.1$ cm^{-3}. The model also includes feedback from supernovae (both type II and Ia) and massive stars, as well as the production of metals through chemical evolution (Aumer et al. 2013). Galaxies with dark matter halo masses of $M_{\rm DM} = 10^{10} h^{-1} M_\odot$ are seeded with SMBHs with masses of $M_{\rm BH} = 10^5 h^{-1} M_\odot$, which first grow through standard Bondi–Hoyle–Lyttleton accretion and BH merging, with the maximum accretion rate capped at the Eddington limit, assuming a fixed radiative efficiency of $\epsilon_r = 0.1$. A total of 0.5% of the rest mass energy of the accreted gas is coupled to the surrounding gas as thermal feedback (Johansson et al. 2009a).

We run two cosmological zoom-in simulations starting at a redshift of $z = 50$, with the initial conditions generated using the MUSIC software package (Hahn & Abel 2011). The first simulation (simulation 1) targets a dark matter halo with a virial mass of $M_{200} \sim 7.5 \times 10^{12} M_\odot$, whereas in the second simulation (simulation 2) we target a more massive system of $M_{200} \sim 2.5 \times 10^{13} M_\odot$, covering a larger initial comoving volume of $(10h^{-1}$ Mpc$)^3$. The high-resolution zoom-in regions are initially populated with both gas and dark matter particles, with masses of $m_{\rm gas} = 3 \times 10^5\ M_\odot$ and $m_{\rm DM} = 1.6 \times 10^6\ M_\odot$, respectively. The baryonic particles have gravitational softenings of $\epsilon_{\rm bar} = 40h^{-1}$ pc for stars and gas and $\epsilon_{\rm DM} = 93h^{-1}$ pc for the dark matter particles. The simulations are run initially with standard GADGET-3, until the SMBHs have grown to be sufficiently massive ($M_{\rm BH} \sim 7.5 \times 10^7 M_\odot$) to allow for detailed dynamical modelling using the KETJU code, as the algorithmically regularised integrator requires a BH to stellar particle mass ratio of $\sim 500 - 1000$ in order to provide accurate results (Mannerkoski et al. 2019).

3. Resolving SMBH triplet systems

Simulation 1 was run with standard GADGET-3 until redshift $z \approx 0.62$. At this point the target halo hosted three massive galaxies (A, B and C), all containing their individual central SMBHs with masses in excess of $10^8 M_\odot$ ($M_{\rm BH,A} = 8.4 \times 10^8 M_\odot$, $M_{\rm BH,B} = 1.1 \times 10^8 M_\odot$ and $M_{\rm BH,C} = 2.1 \times 10^8 M_\odot$, see Mannerkoski et al. 2021). At this stage we turned on the KETJU integration as the mass ratio between the SMBHs and the stellar particles was now sufficiently large. The radii of the regularised KETJU regions were set to $120h^{-1}$ pc, corresponding to three times the baryonic softening length.

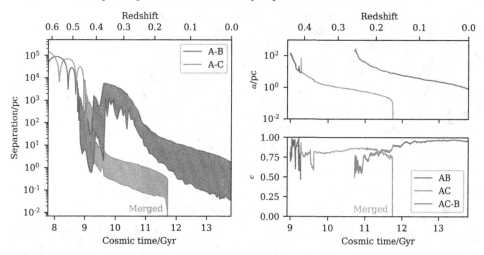

Figure 1. Left: The separations of the A-B and A-C SMBHs over the duration of the KETJU simulation, with the shaded regions showing the range of rapid oscillations. Right: The evolution of the semimajor axis a (top) and the eccentricity e (bottom) for the SMBHs in the system (figure adapted from Mannerkoski et al. 2021).

Galaxy B merges with galaxy A at a redshift of $z \approx 0.48$ and during the merger the two SMBHs sink towards the centre of the merger remnant forming a binary (AB-binary) with a semi-major axis of $a_{AB} \approx 100$ pc. This binary hardens through stellar scattering over the next ~ 250 Myr reaching a semi-major axis of $a_{AB} \approx 10$ pc (see Fig. 1). However, before this binary enters into the gravitational wave dominated regime, galaxy C merges with the AB galaxy remnant bringing in SMBH-C in the process, which results in a three-body interaction between the three SMBHs. Initially, the three-body interaction causes rapid changes in the eccentricity of the AB-binary and finally SMBH-B is ejected from the centre, with SMBH-C instead replacing it in the new AC-binary.

After a few hundred Myr, SMBH-B falls back towards the AC-binary resulting in an interaction with the AC-binary, which can be seen from the small SMBH separations and the dip in the AC eccentricity in Fig. 1. This interaction ejects SMBH-B to an even wider orbit, and it takes it around one Gyr to sink back into the centre. In the meantime, the AC-binary hardens due to stellar scattering and finally merges driven by gravitational wave emission, roughly ~ 3 Gyr after the galaxies merged. The remaining AB-binary also hardens due to stellar scattering, but does not have time to merge before the simulation ends at $z = 0$.

The eccentricity of the AC-binary also exhibits small oscillations after SMBH-B enters into a sub ~ 100 pc hierarchical configuration. At this stage the inner binary has a semi-major axis of $a_{AC} \approx 0.4$ pc, while SMBH-B is on a much wider orbit with $a_{AC-B} \approx 20$ pc and an eccentricity of $e_{AC-B} \approx 0.79$ at an inclination of $i_{AC-B} \approx 90.8°$. Here we are in fact witnessing Lidov–Kozai oscillations (Lidov 1962) suppressed by the relativistic precession of the inner orbit, due to the fact that the binary precession period ($\sim 6 \times 10^5$ yr) is much shorter than the Lidov–Kozai oscillation period ($\sim 4 \times 10^7$ yr) for this particular system (e.g. Blaes et al. 2002).

4. Simulating systems with multiple SMBHs

In simulation 2 a larger comoving volume of $(10h^{-1} \text{ Mpc})^3$ was run initially with GADGET-3 until redshift $z \approx 0.815$, after which the integration was continued with KETJU turned on (Mannerkoski et al. 2022). At the start of the KETJU simulation

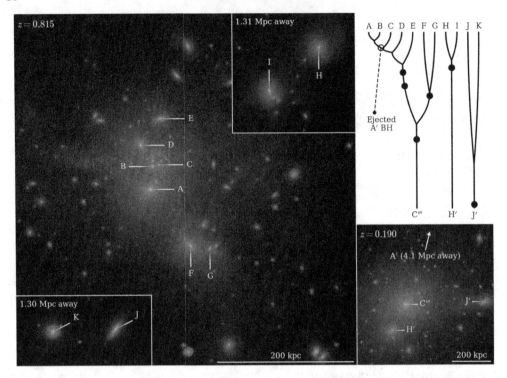

Figure 2. Left: The initial state of the KETJU run, with the galaxies and SMBHs indicated. The main panel shows the central group of galaxies, with two more distant galaxy pairs shown as insets in the corners. Top right: A schematic merger tree of the galaxies and their SMBHs, with time proceeding from top to bottom. The lines depict galaxy mergers, while the circles indicate SMBH binary mergers. The final state of the KETJU simulation at $z = 0.190$ is shown in the bottom right corner, with the remaining SMBHs labelled (Mannerkoski et al. 2022).

the volume contained 11 massive galaxies, with SMBHs that are resolved with their individual regularised regions. The galaxies are shown in the left panel of Fig. 2 with seven galaxies (A-G), located in a central group that is collapsing within a halo with a total virial mass of $M_{200} \approx 2 \times 10^{13} M_\odot$. In addition, there are two more distant galaxy pairs, with H and I located in a halo with a virial mass of $M_{200} \approx 2.5 \times 10^{12} M_\odot$, and K and J found in a halo with a virial mass of $M_{200} \approx 1.3 \times 10^{12} M_\odot$. Due to the high number of massive black holes in this simulation, we lowered the gravitational softening to $\epsilon_\star = 20h^{-1}$ pc for the KETJU simulation, which allowed us to resolve regularised regions around each SMBH with a radius of $60h^{-1}$ pc.

The galaxies and their constituent SMBHs undergo multiple mergers during the KETJU simulation, which is depicted schematically in the top right panel of Fig. 2. In this simulation we also include a description for SMBH spins and model their gravitational wave driven merger kicks using an analytic model based on numerical relativity fitting functions from Zlochower & Lousto (2015). Typically the SMBH merger remnants experience rather modest kicks of $v_{\rm kick} \lesssim 500$ km/s, the exception being the AB-SMBH remnant, which receives a very large kick of $v_{\rm kick} = 2257$ km/s, which is sufficient to eject the SMBH from its host galaxy. Thus, galaxy A is temporarily lacking a SMBH, however this situation is rapidly remedied with the subsequent mergers of galaxies C, D and E, which bring in their central SMBHs replacing the ejected SMBH. The fact that the original SMBH was ejected from this galaxy has important consequences for the evolution of the galaxy on the $M_{\rm BH} - \sigma$ plane, as the galaxy will have an undermassive SMBH with

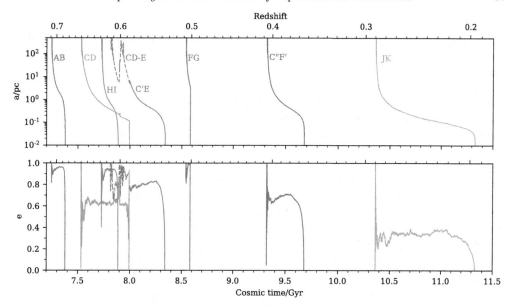

Figure 3. The evolution of the semimajor axis a (top) and the eccentricity e (bottom) of the SMBH binaries in simulation 2. The dashed line shows the parameters of the outer orbit in the hierarchical CD-E triplet system. The binaries form with eccentricities in a broad range of $e \sim 0.3 - 0.9$ (figure adapted from Mannerkoski et al. 2022).

respect to the observed relation (Johansson et al. 2009b, Kormendy & Ho 2013, see also Mannerkoski et al. 2022 for details).

In Fig. 3 we show the evolution of the semi-major axis and eccentricity for all the resolved massive SMBH mergers in the simulation as a function of redshift. In general most of the SMBH binaries form at moderately high eccentricities, with typical values in the range of $e = 0.6 - 0.95$ and limited eccentricity evolution during the hardening process. The relatively high eccentricities result in short binary lifetimes with the SMBH coalescense typically occurring within $\sim 200 - 500$ Myr. However, there are some notable exceptions, for example the FG-binary has an extremely high eccentricity of $e = 0.998$, which results in a very rapid gravitational wave driven merger within just a few tens of Myr. For this binary most of the eccentricity growth occurs when the binary semimajor axis is still above ~ 10 pc, and the mass ratio of the binary is large ($\sim 7:1$), implying that resonant dynamical friction (Rauch & Tremaine 1996) might also be operational, in addition to simple stellar scattering (Quinlan 1996).

The JK-binary on the other hand has a low eccentricity of only $e = 0.35$, and is formed after a nearly circular orbit galaxy merger. The low eccentricity results in a slow merger process and it takes nearly a Gyr for the black holes to merge after forming a hard binary. Finally, similarly to simulation 1, a SMBH triplet (CD-E) is also occurring in this simulation (Fig. 3). After a strong gravitational interaction with the CD-binary, SMBH-E settles into a hierarchical triplet configuration around the inner binary. However, contrary to SMBH triplet in simulation 1, the outer period is in this case shorter than the relativistic period of the inner binary. This results in Lidov–Kozai oscillations (Lidov 1962) that eventually excite the CD-binary eccentricity from $e \approx 0.55$ to a very high value of $e \approx 0.9$, and the increased eccentricity is sufficient to drive the CD-binary to a near instant merger through the increased emission of gravitational waves.

5. Conclusions

We have demonstrated here that the KETJU code can be used to resolve the detailed small-scale dynamics of tens of SMBHs evolving in a complex cosmological environment over extended periods of time. All SMBH binary systems found in our simulations were driven to merger by stellar interactions without any signs of stalling. Our simulated binaries typically formed on highly eccentric orbits, indicating that the circular binary models that are commonly used in the literature are insufficient for capturing the typical binary evolution. In addition, we found that systems with multiple interacting SMBHs naturally occur in a ΛCDM setting and it is important to capture their dynamics accurately, which can only be done with direct integrations of the type presented here. Finally, we stress the importance of simultaneously modelling the accurate small-scale SMBH dynamics and gas dynamics, which will be in particular important when making gravitational wave predictions for LISA (Amaro-Seoane et al. 2022), as it will be mostly sensitive to somewhat lower-mass SMBHs in the mass range of $M_{BH} \sim 10^5 - 10^7 M_\odot$, which are expected to reside in late-type gas-rich galaxies.

Acknowledgments

The authors acknowledge the support by the ERC via Consolidator Grant KETJU (no. 818930) and the support of the Academy of Finland grant 339127.

References

Amaro-Seoane, P., et al., 2022, *Living Reviews in Relativity* submitted, ArXiv:2203.06016
Aumer, M., White, S. D. M., Naab, T., & Scannapieco, C. 2013, *MNRAS*, 434, 3142
Begelman, M. C., Blandford, R. D., & Rees, M. J. 1980, *Nature*, 287, 307
Blaes, O., Lee, M. H., & Socrates, A. 2002, *ApJ*, 578, 775
Hahn, O., & Abel, T. 2011, *MNRAS*, 415, 2101
Hills, J. G., & Fullerton, L. W. 1980, *AJ*, 85, 1281
Hu, C.-Y., Naab, T., Walch, S., Moster, B. P., & Oser, L. 2014, *MNRAS*, 443, 1173
Johansson, P. H., Naab, T., & Burkert, A. 2009a, *ApJ*, 690, 802
Johansson, P. H., Burkert, A., & Naab, T. 2009b, *ApJL*, 707, L184
Kelley, L. Z., Blecha, L., & Hernquist, L. 2017, *MNRAS*, 464, 3131
Khan, F. M., Fiacconi, D., Mayer, L., Berczik, P., & Just, A. 2016, *ApJ*, 828, 73
Kormendy, J., & Ho, L. C. 2013, *ARA&A*, 51, 511
Lidov, M. L. 1962, *P&SS*, 9, 719
Mannerkoski, M., Johansson, P. H., Pihajoki, P., Rantala, A., & Naab, T. 2019, *ApJ*, 887, 35
Mannerkoski, M., Johansson, P. H., Rantala, A., Naab, T., & Liao, S. 2021, *ApJL*, 912, L20
Mannerkoski, M., Johansson, P. H., Rantala, A., Naab, T., Liao, S., & Rawlings, A. 2022, *ApJ* in press, ArXiv:2112.03576
Mora, T., & Will, C. M. 2004, *PhRvD*, 69, 104021
Naab, T. & Ostriker, J.P. 2017, *ARAA*, 55, 59
Quinlan, G. D. 1996, *NewA*, 1, 35
Peters, P. C. 1964, *Physical Review*, 136, 1224
Rantala, A., Pihajoki, P., Johansson, P. H., et al. 2017, *ApJ*, 840, 53
Rantala, A., Johansson, P. H., Naab, T., Thomas, J., & Frigo, M. 2018, *ApJ*, 864, 113
Rantala, A., Pihajoki, P., Mannerkoski, M., Johansson, P. H., et al. 2020, *MNRAS*, 492, 4131
Rauch, K. P., & Tremaine, S. 1996, *NewA*, 1, 149,
Ryu, T., Perna, R., Haiman, Z., Ostriker, J. P., & Stone, N. C. 2018, *MNRAS*, 473, 3410
Springel, V. 2005, *MNRAS*, 364, 1105
Will, C. M. 2014, *PhRvD*, 89, 044043
Zlochower, Y., & Lousto, C. O. 2015, *PhRvD*, 92, 024022

The Predictive Power of Computational Astrophysics as a Discovery Tool
Proceedings IAU Symposium No. 362, 2023
D. Bisikalo, D. Wiebe & C. Boily, eds.
doi:10.1017/S174392132200120X

Nature of star formation in first galaxies

Mahavir Sharma ⓘ

Indian Institute of Technology (IIT) Bhilai, GEC Campus, Sejbahar, Raipur, 492015, India
email: `mahavir@iitbhilai.ac.in`

Abstract. One of the primary foci of research in astrophysics is on developing a rigorous understanding of the first galaxies. This entails studying the physical processes such as accretion, cooling and star formation in first galaxies, and also investigating the consequences of these processes in the present day Universe. We investigate the star formation in the early galaxies and its subsequent evolution using the EAGLE simulation and find that the star formation has a smooth evolutionary behaviour at low redshifts leading to a main sequence of star formation that can be explained by deterministic models using accretion history as an input. In contrast, at high redshift (> 6), most of the galaxies are bursty. At high redshift, instead of exhibiting a main sequence in SFR $-$ M_h plane, they bunch-up around a halo mass of $\approx 10^9$ $M_\odot$ and SFR ≈ 0.1 $M_\odot$ yr^{-1}. As a consequence, the reionization of the Universe is led by low mass haloes hosting brighter galaxies that are undergoing intense bursts. Furthermore, the bursts in the infant galaxies lead to a poorly mixed interstellar medium in which the stars can form from gas enriched predominantly by a single nucleosynthetic channel. The lower mass subset of the stars formed in first galaxies resemble the carbon enhanced metal poor stars in our Galaxy while the higher mass ones reionized the Universe.

Keywords. galaxies: high-redshift, galaxies: formation, galaxies: evolution, stars: abundances, ISM: structure, ISM: evolution

---

1. Introduction

The Universe plunged into the dark ages after the electrons recombined with protons at redshift $z \approx 1100$, that is also known as the surface of last scattering when the cosmic microwave background photons were emitted. During the dark ages the primordial perturbations in the neutral hydrogen in the Universe kept on evolving (e.g. Dodelson 2003). How exactly the first stars and galaxies emerged out of the growing density perturbations is a challenge to the 21st century astrophysics (e.g. Bromm 2013; Naab & Ostriker 2017).

At the observational front, there has been a quest to detect the first galaxies using state of the art instruments such as the Hubble Space Telescope (Beckwith et al. 2006). The James Webb Space Telescope (JWST) (Gardner et al. 2006) is also expected to provide vital information on the first galaxies, such as their star formation rates (SFRs) and the nebular emission lines from them. Furthermore, the Universe was not completely dark during the dark ages, since the neutral hydrogen can be seen through its hyperfine spin-flip 21 cm transition (e.g. Pritchard & Loeb 2012). Detecting the 21 cm signal from the early Universe is a primary focus of the mega radio astronomy projects such as the Low Frequency Array (van Haarlem et al. 2013), the Murchison Widefield Array (e.g. Beardsley et al. 2019) and the upcoming Square Kilometre Array (Weltman et al. 2020). A significant effort in this direction is devoted to detect the progress of the Universe during the epoch when it is undergoing a transition from the dark neutral to the bright ionized state. This transition known as the 'cosmic reionization' was brought about by the star formation in first galaxies.

Theoretical effort has focused on simulating the formation of the first galaxies (Schaye et al. 2015; Crain et al. 2015; Vogelsberger et al. 2014; Naab & Ostriker 2017). Another step in this direction was to deploy the outcomes of galaxy formation into a radiative transfer calculation in the evolving Universe to reproduce the progress of reionization and to predict the evolving power spectrum of the 21-cm signal (e.g. Sokasian et al. 2001; Santos et al. 2010; Wise et al. 2014; Mesinger et al. 2011).

Therefore, an accurate modelling of the formation of galaxies and stars lies at the heart of the current problems in cosmology. Understanding their evolution to the present day has also been a major theme of research, and an objective of the milestone computational efforts such as the MILLENNIUM simulation (Springel et al. 2005) and more recently the EAGLE (Schaye et al. 2015; Crain et al. 2015; McAlpine et al. 2016) and ILLUSTRIS simulation (Vogelsberger et al. 2014). These computational studies do an excellent job in reproducing the galaxy luminosity functions and the stellar mass to halo mass relations. They can reproduce the evolutionary history of cosmic star formation, and also do very well in reproducing the quenching of star formation and thus in quantifying the active and passive galaxies.

A rigorous understanding of the formation of first stars and galaxies from the primordial metal poor gas at the end of cosmic dawn has been a challenge (e.g. Greif et al. 2010). The behaviour of star formation at high redshift is markedly different from that at low redshift. The high redshift galaxies are compact and the SFR for a particular mass, or equivalently the specific SFR increases with redshift (e.g. Shibuya et al. 2015; Sharma et al. 2016). Furthermore, the simulations often find that the star formation at the beginning is abrupt and features episodes of high activity followed by periods of quiescence (e.g. Sharma et al. 2016; Ceverino et al. 2017).

In this work, we investigate the star formation in galaxies in the EAGLE simulation (Schaye et al. 2015; McAlpine et al. 2016). In particular, we draw a comparison between the behaviour of star formation in high redshift haloes and in low redshift haloes. We also discuss the implications of the nature of star formation in high-redshift galaxies for the epoch of reionization and for the population of metal poor stars in the Milky Way.

2. The evolution of the star formation in galaxies

The models of galaxy formation aim to derive the star formation and its evolution over the cosmic ages. Most of these models rely on an interplay of gas accretion, formation of stars and mass ejection by outflow i.e. feedback (e.g. Bouché et al. 2010; Davé et al. 2012; Lilly et al. 2013; Neistein & Dekel 2008; Dekel et al. 2013; Dekel & Mandelker 2014). In these models, coupled equations for the rate of change of stellar mass due to star-formation and rate of change of gas mass due to accretion, star-formation and outflows are solved to model the evolution of galaxies with redshift.

Recently we proposed the $I\kappa\epsilon\alpha$ model that introduces a new method of deriving the salient features of galaxy evolution such as the evolution of the SFR, existence of the main sequence of star formation and the stellar mass function of galaxies, purely from the energy arguments (Sharma & Theuns 2020). The galaxies are virialized systems, therefore the addition of matter through cosmological accretion makes them even more bound as their total energy decreases. However, they compensate for it by forming stars that raise the internal energy. This mechanism is analogous to the behaviour of main sequence stars which also are in virial equilibrium and any decrease in total energy via cooling is compensated by the nuclear energy generation. In galaxies the accretion plays a similar role as the cooling plays in stars so as to make the system even more bound, and the star formation plays the role of nuclear fusion.

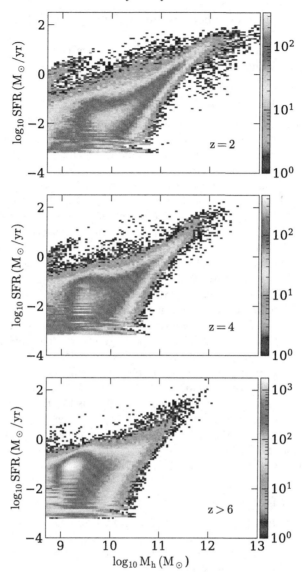

Figure 1. The distribution of star formation rates (SFR) and halo masses (M_h) of galaxies in the EAGLE simulation (Schaye et al. 2015; McAlpine et al. 2016) at redshift $z > 6$ (bottom panel), $z = 4$ (middle panel) and at $z = 2$ (top panel). In the top panel corresponding to $z = 2$, most of the galaxies are distributed along a diagonal ridge indicating the main sequence of galaxies. However at high redshift (bottom panel), most of the galaxies are concentrated in a narrow range of M_h and SFR due to the bursty nature of star formation.

In the $I\kappa\epsilon\alpha$ model the SFR $\propto M_h^{5/3}$ and the proportionality constant and M_h are redshift dependent (Sharma & Theuns 2020). The redshift dependence is almost the same for different halo masses (e.g. Correa et al. 2015). Given the relation between the SFR and halo mass, and further considering the existence of a relation between halo mass and stellar mass (e.g. Behroozi et al. 2013), it is easy to explain the existence of the main sequence of galaxies at low redshifts (see the top and the middle panel in Fig. 1).

At what redshift the main sequence emerges? The same question can also be stated as whether the evolution and behaviour of star formation at high redshift is the same as at low redshift. When looking at the bottom panel in Fig. 1 for redshift $z > 6$, we notice that there is a minor hint of the main sequence trend, but it is eclipsed by an exceptionally high SFR in galaxies of a specific halo mass. This indicates that the infant galaxies at high redshift have an abrupt behaviour of star formation.

3. The bursty star formation in first galaxies

The nature of star formation is strikingly different at high redshift, that cannot and should not be described by the smooth evolutionary models, which are more suited at low redshifts to explain the trends such as the main sequence. The high redshift galaxies have higher SFRs (Shibuya et al. 2015; Sharma et al. 2016), and more importantly the star formation is very bursty. This is also apparent in Fig. 1. At low redshifts (top and middle panel) the galaxies occupy a ridge in the SFR $- M_\mathrm{h}$ plane that can be identified as the main sequence. Clearly the main sequence is quite well established at redshift 2 (top panel) while it is still developing when looking at redshift 4 (middle panel). At redshift > 6 (bottom panel) there is only a hint of the main sequence at best. Instead a high concentration of galaxies around a halo mass of $\approx 10^9$ $M_\odot$ and SFR ≈ 0.1 $M_\odot$ yr^{-1} is clearly evident. This is an indication of a characteristic halo mass in which the star formation history begins with a (most-probable) SFR of roughly 0.1 $M_\odot$ yr^{-1} that can be understood as the characteristic SFR of the initial burst.

A newly formed galaxy acquires primordial gas, and it begins its journey at the onset of the first burst of stars. The first burst has a huge effect on a tiny infant galaxy as it clears almost all of its gas (see Fig. 2 in Sharma et al. 2018 and Fig. 4 in Sharma et al. 2019). As a result, the SFR is drastically reduced to almost zero. Then after a period of no activity the gas builds up again and the next burst begins, however it is likely that the second burst is not as powerful and damaging as the first one, because the dark matter halo mass and the stellar mass of the galaxy and hence the specific SFR would be lower during the second and subsequent bursts. The cycle is repeated, and with each subsequent burst the galaxy keeps on growing. Slowly and steadily its SFR transforms from being dominated by bursts to a smooth evolution.

4. Implications of bursty star formation

The smooth evolution of star formation driven by cosmological accretion leads to the emergence of main sequence of star forming galaxies in the present day Universe. Then what could be the consequences of the bursty nature of star formation? In the following, we discuss two of the major consequences.

Most of the stars that we see in the present day galaxies (e.g. the Sun), exhibit signatures of all the elements from helium to iron in their spectra. However, finding stars that lack these elements has been a challenge (e.g. Beers & Christlieb 2005). The old stars which lack heavy elements supposedly formed in first galaxies when the Universe was metal (heavy elements) free. The metals are produced in subsequent cycles of star formation.

A subset of metal poor stars detected in our Galaxy shows unusual enhancement of carbon, characterised by carbon enhanced metal poor (CEMP) stars can be further classified into CEMP-no (that lack slow or rapid neutron capture process elements) and CEMP-s (those with abundance of slow neutron capture process elements such as Ba) (Aoki et al. 2007; Frebel et al. 2006; Frebel & Norris 2015).

Most of the normal stars acquire their metals through either the SNe or through the AGB nucleosynthesis channel, and presumably in a well mixed interstellar medium that was replenished from multiple cycles of star formation. However, the CEMP stars likely formed in interstellar medium of first galaxies, because those galaxies had a poorly mixed interstellar medium due to bursts. Therefore stars could form from gas enriched predominantly by a single enrichment channel, either via the metal-poor SNe-type-II channel or via the AGB channel, the former can produce the CEMP-no stars and the latter can result into the formation of the CEMP-s stars (Sharma et al. 2018).

In a burst, stars are formed with a range of masses and lifetimes. The low-mass ones have long lifetimes while the high-mass ones are short-lived. Therefore the CEMP stars we see today are the low mass stars formed during the initial bursts.

Their high-mass counterparts had short lifetimes. They provided most of the ionizing photons that reionized the Universe. As we see in the bottom panel of Fig. 1, bursts were more likely to occur in the low-mass haloes ($\approx 10^9$ M$_\odot$), but the corresponding galaxies were not the faintest. Hence at high redshift, the bursty brighter galaxies that were hosted by low-mass haloes were the major contributors of the ionizing photons (Sharma et al. 2016, 2017). These bursty galaxies can be investigated in detail by the JWST (Gardner et al. 2006).

5. Acknowledgments

We acknowledge the Virgo Consortium for making their simulation data available. The EAGLE simulations were performed using the DiRAC-2 facility at Durham, managed by the ICC, and the PRACE facility Curie based in France at TGCC, CEA, Bruyères-le-Châtel. We thank IIT Bhilai and the department of science and technology (DST), India, for providing the support to carry out this research. We also thank the IAU for supporting the presentation of this work. We thank an anonymous referee for constructive comments.

References

Aoki W., Beers T. C., Christlieb N., Norris J. E., Ryan S. G., Tsangarides S., 2007, The Astrophysical Journal, 655, 492
Beardsley A. P., et al., 2019, PASA, 36, e050
Beckwith S. V. W., et al., 2006, AJ, 132, 1729
Beers T. C., Christlieb N., 2005, ARA&A , 43, 531
Behroozi P. S., Wechsler R. H., Conroy C., 2013, ApJ, 770, 57
Bouché N., et al., 2010, ApJ, 718, 1001
Bromm V., 2013, Reports on Progress in Physics, 76, 112901
Ceverino D., Glover S. C. O., Klessen R. S., 2017, MNRAS, 470, 2791
Correa C. A., Wyithe J. S. B., Schaye J., Duffy A. R., 2015, MNRAS, 450, 1514
Crain R. A., et al., 2015, MNRAS, 450, 1937
Davé R., Finlator K., Oppenheimer B. D., 2012, MNRAS, 421, 98
Dekel A., Mandelker N., 2014, MNRAS, 444, 2071
Dekel A., Zolotov A., Tweed D., Cacciato M., Ceverino D., Primack J. R., 2013, MNRAS, 435, 999
Dodelson S., 2003, Modern cosmology
Frebel A., Norris J. E., 2015, ARA&A , 53, 631
Frebel A., et al., 2006, The Astrophysical Journal, 652, 1585
Gardner J. P., et al., 2006, Space Sci. Rev., 123, 485
Greif T. H., Glover S. C., Bromm V., Klessen R. S., 2010, The Astrophysical Journal, 716, 510
Lilly S. J., Carollo C. M., Pipino A., Renzini A., Peng Y., 2013, ApJ, 772, 119
McAlpine S., et al., 2016, Astronomy and Computing, 15, 72
Mesinger A., Furlanetto S., Cen R., 2011, MNRAS, 411, 955
Naab T., Ostriker J. P., 2017, ARA&A , 55, 59
Neistein E., Dekel A., 2008, MNRAS, 383, 615
Pritchard J. R., Loeb A., 2012, Reports on Progress in Physics, 75, 086901
Santos M. G., Ferramacho L., Silva M. B., Amblard A., Cooray A., 2010, MNRAS, 406, 2421

Schaye J., et al., 2015, MNRAS, 446, 521

Sharma M., Theuns T., 2020, MNRAS, 492, 2418

Sharma M., Theuns T., Frenk C., Bower R., Crain R., Schaller M., Schaye J., 2016, MNRAS, 458, L94

Sharma M., Theuns T., Frenk C., Bower R. G., Crain R. A., Schaller M., Schaye J., 2017, MNRAS, 468, 2176

Sharma M., Theuns T., Frenk C. S., Cooke R. J., 2018, MNRAS, 473, 984

Sharma M., Theuns T., Frenk C., 2019, MNRAS, 482, L145

Shibuya T., Ouchi M., Harikane Y., 2015, ApJS, 219, 15

Sokasian A., Abel T., Hernquist L. E., 2001, New A, 6, 359

Springel V., et al., 2005, Nature, 435, 629

Vogelsberger M., et al., 2014, Nature, 509, 177

Weltman A., et al., 2020, PASA, 37, e002

Wise J. H., Demchenko V. G., Halicek M. T., Norman M. L., Turk M. J., Abel T., Smith B. D., 2014, Monthly Notices of the Royal Astronomical Society, 442, 2560

Yoon J., et al., 2018, ApJ, 861, 146

van Haarlem M. P., et al., 2013, A&A, 556, A2

Discussion

KATSIANIS: Hello, I enjoyed your talk. Very nice. It has been found that cosmological simulations like EAGLE and IllustrisTNG suffer from resolution limitations (one may run the model at another resolution and get different results if feedback prescriptions are not changed). This raises the question if the models are physical or they are just very well tuned. For example the gas phase metallicity relation even at $z \approx 0$ from EAGLE is different between the high resolution and average (reference) resolution runs. The SFRs of TNG are different between the TNG300 and TNG100 runs besides that the model was tuned to reproduce the CSFRD at the reference resolution. How can we trust that $I\kappa\epsilon\alpha$ is reproducing realistic results since it was tuned to reproduce properties from a simulation that has these limitations. Are there any other observables that the model can reproduce and do you have in mind to check against something specific?

SHARMA: Hi, yes the model does reproduce the SFRs in haloes e.g. the Milky Way that we compared in Sharma & Theuns (2020). We compared the mass-metallicity relations as well with observations. Further, we also predicted the low mass end of the SHMF. Because the model is based entirely on an energy/virial argument, the parameters in the model are not the same as that in the simulation. In fact, one would notice that the outflow feedback is an output of our model, not an input. As for the main sequence, we compared with the main sequence in reference simulation (or automatically with observations as both agree).

The parameters of the model are purely motivated by virial equilibrium argument. Those parameters wouldn't have a unique relation with the implementation of feedback in simulations. So, even if the feedback in the simulation has to be changed with resolution, that would mean that the interpreted relations of that feedback to our $I\kappa\epsilon\alpha$ parameters would be different, but our $I\kappa\epsilon\alpha$ parameters would still be the same.

The Predictive Power of Computational Astrophysics as a Discovery Tool
Proceedings IAU Symposium No. 362, 2023
D. Bisikalo, D. Wiebe & C. Boily, eds.
doi:10.1017/S1743921322001284

The formation of Supersonically Induced Gas Objects (SIGOs) with H_2 cooling

Yurina Nakazato[1], Gen Chiaki[2], Naoki Yoshida[3,4], Smadar Naoz[5,6], William Lake[5,6] and Chiou Yeou[5,6]

[1]Department of Physics, The University of Tokyo, 7-3-1 Hongo, Bunkyo,
Tokyo 113-0033, Japan,
email: yurina.nakazato@phys.s.u-tokyo.ac.jp

[2]Astronomical Institute, Tohoku University, 6-3, Aramaki, Aoba-ku, Sendai,
Miyagi 980-8578, Japan

[3]Kavli Institute for the Physics and Mathematics of the Universe (WPI), UT Institute for
Advanced Study, The University of Tokyo, Kashiwa, Chiba 277-8583, Japan

[4]Research Center for the Early Universe, School of Science, The University of Tokyo,
7-3-1 Hongo, Bunkyo, Tokyo 113-0033, Japan

[5]Department of Physics and Astronomy, UCLA, Los Angeles, CA 90095

[6]Mani L. Bhaumik Institute for Theoretical Physics, Department of Physics and Astronomy,
UCLA, Los Angeles, CA 90095, USA

Abstract. During the recombination of the universe, supersonic relative motion between baryons and dark matter (DM) generally existed. In the presence of such streaming motions, gas clumps can collapse outside of virial radii of their closest dark matter halos. Such baryon dominant objects are thought to be self-gravitating and are called supersonically induced gas objects; SIGOs. We perform three-dimensional hydrodynamical simulations by including H_2 chemical reactions and stream velocity and follow SIGO's formation from $z = 200$ to $z = 25$. SIGOs can be formed under the influence of stream velocity, and H_2 cooling is effective in contracting gas clouds. We follow its further evolution with higher resolution. We find that there are SIGOs which become Jeans unstable outside of the virial radius of the closest DM halos. Those SIGOs are gravitationally unstable and trigger star formation.

Keywords. cosmology: theory, methods: numerical, galaxies: high-redshift, stars: Population III

---

1. Introduction

Recent observations have revealed the history of the universe. At the recombination epoch, the universe became neutral and transparent. At that time, Hydrogen and Helium atoms formed. After that, the first star formation started at around $z \sim 30$. The first stars consisted of only Hydrogen and Helium atoms. The first stars emitted UV radiation, which led to reionization and ended the dark age. An important effect to be considered in star formation in the early universe is the so-called stream velocity (SV). It is a relative velocity between baryons and DM, which originates from baryon acoustic oscillations. The typical value of the relative velocity is 30 km/s at the recombination period, which is about five times greater than the sound speed, so it causes supersonic gas flow. Furthermore, Tseliakhovich & Hirata (2010) argue that the velocity field is coherent over a few mega-parsec scales, thus stream velocity causes non-trivial effects on the first bound objects. Naoz & Narayan (2014) show analytically that SV can form baryon density peaks outside of the virial radius of its closest DM halo. Such objects are called Supersonically Induced Gas Objects (SIGOs) and are thought to be progenitors of star clusters in the early universe. The existence of SIGOs has been identified in several simulations. Popa et al. (2016) perform simulations with adiabatic gas and show that SIGOs are able to form in the early universe. Recently, Chiou et al. (2019) and Chiou et al. (2021) incorporate atomic hydrogen cooling in their hydrodynamical simulations. They show that atomic hydrogen cooling alone does not help SIGOs to condense to runaway collapse.

We argue that the reason why those SIGOs in previous research could not collapse is that gas clumps did not cool efficiently. In this study, we incorporate H_2 cooling in our simulations. Radiative cooling of H_2 plays a vital role in the early universe since the primordial gas consists of only H and He. H_2 cooling can lower the temperature of primordial gas clouds to ~ 200K (Yoshida et al. (2006)). We expect H_2 cooling enables SIGOs to collapse.

2. Method

We use the moving mesh code AREPO (Springel (2010)) and perform cosmological hydrodynamical simulations. In the parent simulations, we adopt the simulation boxsize of 1.4cMpc/h with 512^3 DM particles and 512^3 gas cells, and follow the gas evolution from $z = 200$ to $z = 25$. Regarding to the chemistry, we use cooling library GRACKLE (Smith et al. (2017); Chiaki & Wise (2019)) and calculate 49 reactions of 15 chemical species; $e, H, H^+, He, He^+, He^{++}, H^-, H_2, H_2^+, D, D^+, HD, HeH^+, D^-$ and HD^+. We run four simulations with/without SV (2v/0v) and with/without H_2 cooling (H2/H). Those runs are labeled as "2vH2", "2vH", "0vH2" and "0vH" respectively.

SIGOs are identified by the following two conditions. The first one is that gas clouds are outside of the virial radii of their closest DM halos. The second one is that the local baryon fraction around the gas halos is over 60%. We select one of the SIGOs detected in Run-2vH2 case and call it "S1". To follow the evolution of S1, we cut out 10 kpc of the region around it and restart the simulation from $z = 25$. The second simulation can refine the gas cell automatically and follow the SIGO's evolution much more precisely than the parent one. The detailed settings are described in Nakazato et al. (2021).

3. Result

3.1. SIGO's morphology in parent simulation

Figure 1 shows the gas clouds formation from $z = 31$ to $z = 25$. The colormap shows the gas number density around S1.

We consider S1 which is identified as a SIGO in Run-2vH2. S1 is located at the center of Figure 1 at $z = 25$ with a physical size of ~ 1 kpc. The distance between S1 and its closest DM halo is four times greater than the virial radius of the DM halo. The large gas filament that contains S1 is displaced by ~ 5 comoving kpc to the right from the underlying filamentary structure of DM. This is owing to the stream velocity which flows from left to right (which corresponds to the direction of $+x$).

By definition, S1 is not hosted by DM halos and thus its baryon density is lower than ordinary primordial gas clouds hosted by DM halos (non-SIGOs). For instance, the halo located at the bottom of Figure 1 is identified as a non-SIGO hosted by a DM halo. We refer to it as S2. In comparison with number density, S1 has a much smaller maximum density of ~ 7.9 /cc at $z = 25$ than S2 of 6.89×10^3 /cc.

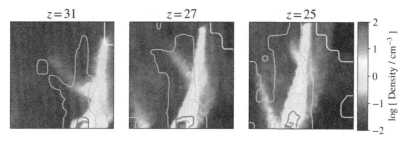

Figure 1. The projected number density of baryons in Run-2vH2. Contours show the DM density. White, pink and red contour lines show 2, 20, and 200 times of the critical density respectively. The side of the length is 40 ckpc.

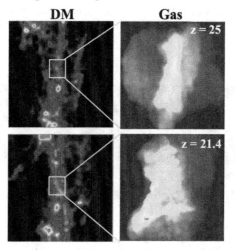

Figure 2. The upper and bottom panels show the S1 at $z = 25$ and $z = 21.4$, when S1 reached Jeans instability. Left: projected DM density distribution around S1 normalized by the critical density of DM. The side of the length is 5 kpc. Right: projected gas number density of S1. The side of the length is 1 kpc.

3.2. *High resolution simulation of S1*

We conduct a high-resolution simulation around S1. S1 become Jeans unstable at $z = 21.4$ (see Figure 3). Figure 2 shows the evolution of S1 from $z = 25$ to $z = 21.4$. We see that S1 reaches high density without being hosted by DM halos.

Figure 3 shows the thermal evolution of S1 at $z = 21.4$. H_2 cools S1 to ~ 200 K and contracts S1 to ~ 100 /cc. We calculate the ratio of the enclosed mass to the Jeans mass $M_J = (\pi/6) \, c_s^3/(G^{3/2}\rho^{1/2})$. We use the mass-weighted average temperature and gas density at all radii. We define the center of the radial profile as the highest density position and calculate the enclosed mass. From the right panel of Figure 3, it is clearly seen that M_{enc}/M_J is over 1 at $z = 21.4$, which suggests that S1 is Jeans unstable. The Jeans mass at $z = 21.4$ is $M_J = 5 \times 10^4 \, M_\odot$, which is over 50 times larger than normal (without SV) Jeans mass $M_J \sim 10^3 \, M_\odot$ (Abel et al. (2002)).

3.3. *Comparison of the same region in each run*

In order to confirm whether a SIGO is also formed in the other runs, we have checked the same region as Figure 1 in Run-0vH2 and 2vH. Figure 4 shows three projected gas density colormaps. From left to right, the colormaps are for Run-0vH2, 2vH2 and 2vH. In the case of 2vH (the most right), there are no gas clumps formed. This is due to

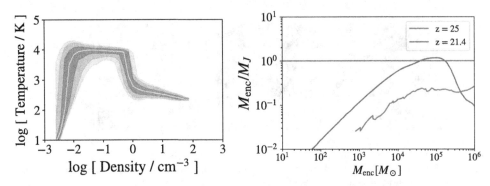

Figure 3. The left figure is a density-temperature phase diagram of S1 at $z = 21.4$. The right figure is radial profiles of the ratio of the enclosed gas mass to the Jeans mass. The red line indicates that S1 reaches Jeans instability.

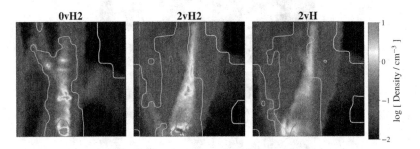

Figure 4. Colormaps of the gas density in the region with the same coordinates as the region of SIGO S1 identified by Run-2vH2 ($z = 25$). Note that 2vH2 corresponds to Figure 1, and the contour lines represent the DM density. Gas clumps S1 and S3 are located in the center of the Run-2vH2 and 0vH2 figures, respectively. The length of one side of each color map is 40 ckpc.

the combined effects of SV and cooling inefficiencies. The stream velocity smoothes the baryon density peak (Park et al. (2020)), and causes an offset between the density peak of baryon and that of DM. The offset effectively delays the gas contraction. Atomic hydrogen cooling is efficient for gas with 8000 K (Barkana & Loeb (2001)). On the other hand, molecular hydrogen cooling can lower gas to 200 K, where the gas contracts to the density of $n_{\mathrm{gas}} \sim 10^3$ /cc, reaches Jeans instability, and finally starts runaway collapse (Yoshida et al. (2006)). In Run-2vH, which does not include the H_2 chemical reactions, the gas cooling is inefficient and the clouds contraction to higher density is delayed. Due to these two effects, there was no gas clouds corresponding to S1 in Run-2vH case.

Run-0vH2 is performed without SV but with H_2 cooling. Since there is no SV, the position of the DM density peak traces the position of the gas density peak. H_2 cools the gas efficiently, and we find a gas clump corresponding to S1 at the center of figure 4. We name it S3. Figure 5 shows the radial profiles of the baryon fractions of S1 and S3 at $z = 25$, respectively. The green line is the average baryon fraction in the universe, $\Omega_b / \Omega_m = 0.044/0.27 = 0.16$. For S1, the baryon fraction F_{bar} is always larger than the cosmic mean fraction in the region $r \lesssim 10^3$ pc. In S3, F_{bar} is always smaller than the cosmic mean baryon fraction. This means that S3 is hosted by a DM halo. We conclude that SIGOs are formed via the combined effects of SV and H_2 cooling; SV causes the offset of the density peak between gas and DM, whereas H_2 cooling enables the gas to condense to reach high densities.

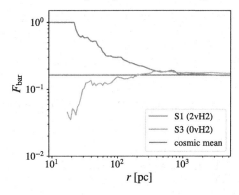

Figure 5. Radial profiles of baryon fraction at $z = 25$. Blue and orange lines shows the fraction for S1 and S3 respectively. The green line is a cosmological baryon density. We calculate the baryon fraction as (encolosed gas mass)/(enclosed (DM+ gas) mass).

4. Conclusion

We have performed cosmological hydrodynamical simulations and followed the formation of a SIGO. We find, for the first time, that a SIGO forms in the environment with stream velocity and condenses via H_2 cooling enough to become Jeans unstable. The SIGO is expected to experience runaway collapse and become a star-forming cloud.

Previous studies have conducted cosmological simulations and showed the characteristics of SIGOs (Popa et al. (2016); Chiou et al. (2018, 2019, 2021); Lake et al. (2021)). A recent study by Schauer et al. (2021) also incorporates non-equilibrium chemical reactions into their simulations and follows the formation of SIGOs. They conclude that H_2 cools such DM-deficient gas clumps to the density of $n_{gas} \sim 10$ /cc, which is not enough to gravitational collapse. They calculate the metallicity required to make the gas clump gravitationally unstable and find that if the gas contains the metal of $Z \sim 10^{-3} Z_\odot$, it can be cooled and condensed enough to runaway collapse.

In our study, we conduct high-resolution simulation after S1 reaches $n_{gas} \sim 10$/cc, and follow the SIGO's contraction with H_2 cooling. It is important to follow the long-term chemo-thermal evolution of individual gas clouds to identify SIGOs that finally collapse gravitationally. We show that a SIGO contracts slowly with its free-fall time, which is also the same as H_2 cooling time scale. The SIGO eventually cools down to 200 K and reaches high-density without being hosted by DM halos.

In our future study, we will investigate physical properties of the other 50 SIGOs which are located in Run-2vH2 to study statistics (Nakazato in prep.). We run additional 50 high-resolution simulations and examine the rate of SIGOs which collapse without being hosted or swallowed by the nearby halos. Moreover, we will follow further evolution of the SIGO to see if it fragments and forms a star cluster by introducing sink particles in our simulations. The protostellar evolution in star-forming regions of the SIGO will unveil the properties of the first star cluster. Since the SIGO we have studied here is a DM poor object, it can be a candidate progenitor of a globular cluster. Combing these future studies will reveal the fate of star-forming SIGOs and their typical properties such as mass, baryon fraction and so on, and clarify the relationship between SIGOs and globular clusters.

5. Q & A

In this section, we attach questions and answers that followed our talk at the IAU Symposium 362.

1. *Does the high-resolution simulation also include DM background?*

We also include DM particles in our high-resolution simulation. As mentioned in Section 2, we cut off a region of 10 kpc on a side, center on S1. Notice that we refine only gas cells but DM particles. The zoom-in simulations, in which the cut region is reverted to the initial redshift $z = 200$ and follow SIGOs' evolution, will be performed in the next study.

2. *What is exactly the stream velocity physically doing that it's allowing for SIGOs' formation?*

Stream velocity produces the spatial offset between the density peak of baryons and one of DM (Naoz & Narayan (2014)). This offset allows for clouds to exist outside of the corresponding DM halo. The following gas contraction is triggered by H_2 cooling. (The detail is in the next question and Section 3.3.)

3. *What do you mean by hydrogen cooling? Is H_2 cooling the recombination of* H^+ *and* e^- *or the molecular bond between two hydrogen atoms?*

For primordial gas, H_2 is generated via the following "H^- process"

$$H + e^- \rightarrow H^- + \gamma, \quad H + H^- \rightarrow H_2 + e^-, \tag{5.1}$$

where the electrons act as a catalyst (Yoshida et al. (2006)). The main cooling process for metal-free gas is the emission from the vibration and rotation transition of H_2. This radiation cooling enables primordial clouds to contract effectively.

References

Abel, T., Bryan, G. L., & Norman, M. L. 2002, Science, 295, 93.
Barkana, R. & Loeb, A. 2001, PhR, 349, 125.
Chiaki, G. & Wise, J. H. 2019, MNRAS, 482, 3933.
Chiou, Y. S., Naoz, S., Marinacci, F., et al. 2018, MNRAS, 481, 3108.
Chiou, Y. S., Naoz, S., Burkhart, B., Marinacci, F., & Vogelsberger, M. 2019, ApJL, 878, L23
Chiou, Y. S., Naoz, S., Burkhart, B., Marinacci, F., & Vogelsberger, M. 2021, ApJ, 906, 25
Lake, W., Naoz, S., Chiou, Y. S., et al. 2021, ApJ, 922, 86.
Naoz, S., & Narayan, R. 2014, ApJL, 791, L8
Nakazato, Y., Chiaki, G., Yoshida, N., et al. 2021, arXiv:2111.10089
Park, H., Ahn, K., Yoshida, N., et al. 2020, ApJ, 900, 30.
Popa, C., Naoz, S., Marinacci, F., & Vogelsberger, M. 2016, MNRAS, 460, 1625
Schauer, A. T. P., Bromm, V., Boylan-Kolchin, M., et al. 2021, ApJ, 922, 193.
Smith, B. D., Bryan, G. L., Glover, S. C. O., et al. 2017, MNRAS, 466, 2217.
Springel, V. 2010, MNRAS, 401, 791.
Tseliakhovich, D., & Hirata, C. 2010, PhRvD, 82
Yoshida, N., Omukai, K., Hernquist, L., et al. 2006, ApJ, 652, 6.

The Predictive Power of Computational Astrophysics as a Discovery Tool
Proceedings IAU Symposium No. 362, 2023
D. Bisikalo, D. Wiebe & C. Boily, eds.
doi:10.1017/S1743921322001764

An idealized setup for cosmological evolution of baryonic gas in isolated halos

S. Dattathri[1,2] and P. Sharma[1]

[1]Department of Physics, Indian Institute of Science, Bangalore-560012, India

[2]Department of Astronomy, University of Michigan, 1085 S. University Avenue, Ann Arbor, MI, 48109, USA

Abstract. We use hydrodynamical simulations to study the evolution of baryonic gas in a cosmologically evolving dark matter halo. We model both the inner and outer regions of the halo using a density profile that transitions from an inner NFW profile to a flat profile far from the halo. Metallicity-dependent radiative cooling and AGN jet feedback are implemented, which lead to heating and cooling cycles in the core. We analyze the evolution of gas and the central supermassive black hole (SMBH) across cosmological time. We find that the properties of the gas and the SMBH are correlated across halo masses and feedback efficiencies.

Keywords. Galaxy: active-Galaxies: intergalactic medium-Galaxies: halos

1. Introduction

According to the standard ΛCDM (Λ cold dark matter) paradigm of structure formation in the Universe, the evolution of galaxy clusters is primarily governed by dark matter halo dynamics. Whereas, in addition to following the dark matter gravity, the gas (which constitutes $\sim 80\%$ of baryons within massive halos) is strongly affected by radiative cooling and feedback heating powered by accretion onto a central supermassive black hole (SMBH). Modern cosmological galaxy formation simulations evolve baryons including these processes, and with sub-grid models for star formation and SMBH growth.

Another class of numerical simulations consists of idealized halo simulations that focus on various aspects of baryonic physics. While these simulations provide insight into gas evolution, they typically lack cosmological evolution. Most of the idealized simulations of isolated halos assume a static dark matter halo (e.g. Prasad et al (2015)), and various important parameters (such as the metallicity of the IGM, e.g. Choudhury et al (2019)) are not evolved cosmologically. While modern galaxy formation simulations reach very high resolutions, they cannot achieve the resolution achievable in single-halo simulations (e.g., zoom-in simulations). Therefore, to focus on the most basic physical processes governing the halo gas, we carry out simulations of the halo gas in cosmologically growing halos. This approach can provide a useful middle ground between fully cosmological and isolated, cosmologically non-evolving halo simulations.

2. Setup

We solve the standard hydrodynamical equations in spherical coordinates in 2D with external gravity due to the dark matter halo, radiative cooling, and mass and momentum injection due to AGN jet feedback, using the astrophysical MDH code PLUTO (Mignone et al (2007)). The dark matter halo density profile is modelled according to Diemer & Kravtsov (2014), who proposed a new density profile that accurately models

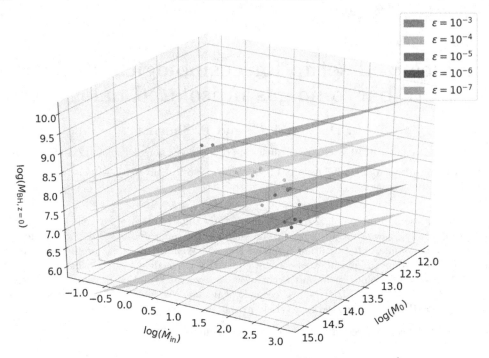

Figure 1. Properties of the halo gas and central SMBH in the $\log(M_0)$ - $\log(\dot{M}_{\rm in})$ - $\log(M_{\rm BH, z=0})$ space for different feedback efficiencies.

the dark matter density beyond the virial radius. Metallicity-dependent radiative cooling and AGN jet feedback are implemented, which strongly affects the dynamics of the gas in the halo core. The feedback is modelled by the accretion of cold gas onto a central SMBH, and is characterized by the feedback efficiency ϵ. Our simulations start at $z = 6$, with an initial gas density profile that follows the dark matter as $\rho_{\rm g} = 0.2\rho_{\rm DM}$.

3. Results and Discussion

Radiative cooling and AGN feedback lead to a self-regulating cycle of heating and cooling in the halo core is formed. Cooling initially dominates and cold gas flows toward the center, leading to a large accretion rate onto the SMBH. This launches powerful jets that heat up the gas, forming low density bubbles and cavities. The hot gas mixes with the surrounding medium, which eventually leads to decrease in the accretion rate and jet power. Cooling starts to dominate again, and the cycle is repeated.During the heating portion of the cycle, the jet power, accretion rate onto the black hole, and the value of $\min(t_{\rm cool}/t_{\rm ff})$ increase, while the cold gas mass within 5 kpc decreases. The opposite is true for the cooling portion of the cycle.

The effect of the AGN feedback on the halo gas can be quantified by the suppression of the cold gas inflow rate relative to the respective cooling flow. Runs with greater feedback efficiency lead to greater suppression because the gas is heated and ejected from the core to a greater degree. This also leads to longer feedback cycles, as the gas takes longer to cool again. Massive halos have deeper potential wells, so a greater amount of energy is required to disrupt the core to the same degree.

The accretion rate onto the central SMBH is initially limited by the Eddington rate, leading to exponential growth of the black hole. This is the quasar phase of black hole evolution. The duration of the quasar phase is longer for lower black hole seed masses.

After the quasar phase, the accretion rate is no longer limited by the Eddington rate. This slower, self-regulated growth is the radio phase.

We find that the present-day black hole mass, the halo mass, the cold gas inflow rate, and the feedback efficiency are strongly correlated, and the values lie on a fundamental plane, as shown in Figure 1.

References

Diemer B. & Kravtsov A. V., 2014, *ApJ*, 789, 1
Mignone A., et al., 2007, *ApJS*, 170, 228
Prasad D., Sharma P., & Babul A., 2015, *ApJ*, 811, 108
Choudhury P. P., Kauffmann G., & Sharma P., 2019, *MNRAS*, 485, 3430

The Predictive Power of Computational Astrophysics as a Discovery Tool
Proceedings IAU Symposium No. 362, 2023
D. Bisikalo, D. Wiebe & C. Boily, eds.
doi:10.1017/S1743921322001879

Detecting cosmic filamentary network with stochastic Bisous model

Moorits Mihkel Muru ⓘ

Tartu Observatory, University of Tartu, Observatooriumi 1, 61602 Tõravere, Estonia
email: `moorits.mihkel.muru@ut.ee`

Abstract. The Bisous model is a tool that uses stochastic methods to detect the network of galactic filaments. This model is explicitly developed to detect the structure from observational data, using only galaxy positions as input. This paper shows that the Bisous model gives reliable results and including photometric data improves the resulting filamentary network. We used MULTIDARK-GALAXIES catalogue to create a mock with photometric redshifts and samples with different galaxy number densities. We found that the filaments detected with the Bisous model are reliable; 85% of the detected filaments are unchanged compared to results with more complete input data. Adding photometric data improves the fraction of galaxies in filaments. Using the confusion matrix technique, we found the false discovery rate to always be below 5% when using photometric data.

Keywords. methods: data analysis, methods: statistical, galaxies: statistics, large-scale structure of the Universe

---

1. The Bisous filament finder

The Bisous model is a stochastic tool used to identify the spines of the filaments using the spatial distribution of galaxies or haloes (Tempel *et al.* 2014, Tempel *et al.* 2016).

First, the Bisous randomly populates the volume with cylinders. Each configuration of cylinders in the volume has a defined energy value, which depends on the position of the cylinders in relation to the underlying data of haloes and the interconnectedness of the cylinders (see Tempel *et al.* 2016 for mathematical definitions). Using the Metropolis-Hastings algorithm and the simulated annealing procedure, the Bisous model minimises the system's energy by randomly adding, removing, or changing the cylinders†. The results are averaged over a hundred Bisous runs to suppress Poisson noise from the random process.

2. Data

We used the MULTIDARK-GALAXIES galaxy catalogue, which uses a dark matter-only simulation with a semi-analytical model for galaxies (Knebe *et al.* 2018, Klypin *et al.* 2016).

For mock photometric data, we added a random shift to galaxies' distances. We created three types of samples. Photometric-only samples with gaussian uncertainties with standard deviations of $\sigma = 1$ to $10\,\text{Mpc}$ and mixed samples with both photometric data ($\sigma = 5\,\text{Mpc}$ and $10\,\text{Mpc}$), the standard deviation is constant within a single sample. Spectroscopic data with $s = 10$ to $50\,\%$ of the brightest galaxies that have exact distances. All of these samples have the same amount of galaxies but different amounts of spectroscopic galaxies. For example, $\sigma 5 s 40$ means 40% of the brightest galaxies have

---

† For an animation, see: `https://www.aai.ee/~elmo/sdss-filaments/sdss_filaments.mp4`

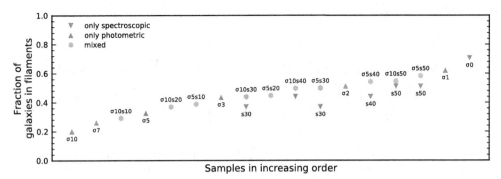

Figure 1. Fraction of galaxies in filaments with different samples. The sample $\sigma 0$ indicates the ideal case where all the galaxies have spectroscopic redshifts, we use it as our ground truth. Other spectroscopic-only samples are for reference to show how adding photometric data (in mixed samples) improves the fraction of galaxies in filaments.

exact positions, other 60% have distances with random shifts with a gaussian ($\sigma = 5$ Mpc) profile. For reference, we also included the spectroscopic-only samples; for example, $s30$ has only 30% of the brightest galaxies. These spectroscopic-only samples have fewer galaxies than photometric-only and mixed samples.

3. Results

Bisous model with different galaxy number densities. The results are presented in Muru & Tempel (2021). In summary, the higher the galaxy input density, the more complete the detected filamentary network compared to the results from highest density data, which we use as a ground truth. We also showed that the filaments detected by the Bisous model are reliable. That means 85% of the filaments do not change even if higher density input data is used.

Usefulness of photometric data. In SDSS DR12, there are about 100 times more photometric galaxies than spectroscopic ones and using those would help to increase input data density. The distance for photometric galaxies has high uncertainty but Kruuse *et al.* (2019) show that photometric galaxies are correlated with the filamentary pattern. The fraction of galaxies in filaments with photometric and mixed samples in Figure 1 show that adding photometric galaxies improves the number of galaxies in filaments. This result indicates that improving the input data galaxy number density with galaxies that only have photometric redshifts improves the overall detected filamentary network in contrast to only using galaxies with spectroscopic redshifts. We analysed the results with the confusion matrix technique using $\sigma 0$ (all the galaxies have spectroscopic redshifts) as ground truth. We found that the false discovery rate is below 5% for every sample, both mixed and photometric-only.

References

Klypin A., Yepes G., Gottlöber S., Prada F., Heß S. 2016, *MNRAS*, 457, 4340

Knebe A., Stoppacher D., Prada F., Behrens C., Benson A., Cora S.A., Croton D.J., et al. 2018, *MNRAS*, 474, 5206

Kruuse M., Tempel E., Kipper R., Stoica R.S. 2019, *A&A*, 625, A130

Muru, M.M., & Tempel, E. 2021, *A&A*, 649, A108

Tempel E., Stoica R. S., Martínez V. J., Liivamägi L. J., Castellan G., Saar E. 2014, *MNRAS*, 438, 3465

Tempel E., Stoica R. S., Kipper R., Saar E. 2016, *Astronomy and Computing*, 16, 17

The Predictive Power of Computational Astrophysics as a Discovery Tool
Proceedings IAU Symposium No. 362, 2023
D. Bisikalo, D. Wiebe & C. Boily, eds.
doi:10.1017/S1743921322001211

Absorption spectra from galactic wind models: a framework to link PLUTO simulations to TRIDENT

Benedetta Casavecchia[1,2], **Wladimir E. Banda-Barragán**[2], **Marcus Brüggen**[2] **and Fabrizio Brighenti**[1]

[1]Dipartimento di Fisica e Astronomia, Università di Bologna, Via Gobetti 93/2, 40122, Bologna, Italy
email: `benedett.casavecchia@studio.unibo.it`

[2]Hamburger Sternwarte, Universität Hamburg, Gojenbergsweg 112, D-21029 Hamburg, Germany

Abstract. Galactic winds probe how feedback regulates the mass and metallicity of galaxies. Galactic winds have cold gas, which is mainly observable with absorption and emission lines. Theoretically studying how absorption lines are produced requires numerical simulations and realistic starburst UV backgrounds. We use outputs from a suite of 3D PLUTO simulations of wind-cloud interactions to first estimate column densities and temperatures. Then, to create synthetic spectra, we developed a python interface to link our PLUTO simulations to TRIDENT via the YT-package infrastructure. We produce UV backgrounds accounting for the star formation rate of starbursts. For this purpose, we use fluxes generated by STARBURST99, which are then processed through CLOUDY to create customised ion tables. Such tables are subsequently read into TRIDENT to generate absorption spectra. We explain how the various packages and tools communicate with each other to create ion spectra consistent with spectral energy distributions of starburst systems.

Keywords. hydrodynamics, methods: numerical, Galaxy: halo, ISM: evolution

---

1. Introduction

It is now well known how important it is to no longer consider galaxies as isolated systems. Galaxies are dynamic systems that acquire gas from the surrounding media. They can grow via merger events or by accreting material from the intergalactic medium, which then provides the fuel for star formation. Each galaxy is immediately surrounded by a galactic corona known as the circumgalactic medium (CGM), which contains gas, dust, and cosmic rays (Tumlinson et al. 2017). Observations of such material reveal that the CGM is complex because of feedback processes that allow for the exchange of matter and energy. Feedback can be due to star formation or active galactic nuclei (AGN), and it often has a strong impact on a galaxy's evolution, its star formation rate (SFR) and its metallicity.

In this project we solely focus on stellar feedback, which is observed in starburst galaxies, where most of the released energy comes from the SNe explosions and stellar winds from young massive OB-type stars, and manifests as galactic winds (see section 8.7 of Cimatti et al. 2019; Veilleux et al. 2020; Danehkar et al. 2021). For nearby galaxies, outflow patterns have been studied in emission and absorption lines, but for distant systems, observations are generally limited to the study of absorption lines due to cold gas

Table 1. Initial conditions for single cloud models interacting with galactic winds. Column 1 indicates the name of the model, column 2 shows the value to which the cooling floor is set, and column 3 indicates the orientation of the magnetic field with respect to the wind. The computational domain of the simulation is $384 \times 768 \times 384$, which corresponds to a physical domain of $(120 \times 240 \times 120)$ pc; thus, 32 cells cover the initial radius of the cloud. Initially, the cloud has a spherical shape with uniform density distribution, it is at rest with a temperature of 8.06×10^2 K and has a polytropic index $\gamma = 5/3$. The wind-cloud density contrast $\chi = \rho_{\text{cloud}}/\rho_{\text{wind}} \approx 700$, the wind Mach number $\mathcal{M}_{\text{wind}} = 4$ and plasma beta parameter $\beta = P_{\text{th}}/P_{\text{mag}} \approx 100$; where β is defined as the ratio of the thermal pressure to the magnetic pressure of the plasma.

(1) Model	(2) Cooling Floor [K]	(3) Magnetic Field Orientation
rwc-r32-al-f1	50	Aligned
rwc-r32-tr-f1	50	Transverse
rwc-r32-al-f2	10^4	Aligned
rwc-r32-tr-f2	10^4	Transverse

around galaxies (e.g. Tumlinson et al. 2017 and Chisholm et al. 2018). These are detected either down-the-barrel with the stellar continuum of the galaxy in the background or transversely, i.e., along the lines of sight of distant quasars (QSOs). From the study of absorption lines it is possible to extract relations for the estimation of the SFR and the stellar mass of star-forming galaxies, as for example was done in Tchernyshyov et al. (2021), where the authors used the O VI column density for their estimates.

Simulations, as is often the case in astrophysics, play a key role in better understanding the physics behind the observed phenomena. Indeed, there are many examples of simulations that attempt to emulate the behaviour of atomic/molecular clouds interacting with galactic winds (e.g. Scannapieco and Brüggen 2015; Goldsmith and Pittard 2018; Banda-Barragán et al. 2019; de la Cruz et al. 2021). Our project ultimately aims at finding a link between numerical models and observations. For the first part of our project, which we present in this paper, we have created a tool that is able to produce column density maps and absorption spectra of different ions from grid-based simulations, specifically from the PLUTO code (Mignone et al. 2007), in order to compare simulations of galactic winds with observations.

2. Simulations

To develop the framework presented in this paper, we have utilised data from existing simulations of wind-cloud models based on Banda-Barragán et al. (2016) and Banda-Barragán et al. (2018). All the simulations presented here use a 3D Cartesian coordinate system (X_1, X_2, X_3) with uniformly-spaced grid cells, and resort to standard methods for ideal, grid-based hydrodynamics, available in the PLUTO code (Mignone et al. 2007), to solve the mass, momentum, and energy conservation laws. Further details on the numerics of these models will be presented in upcoming publications.

Our models account for interactions between isolated spherical clouds and supersonic winds in the presence of weak magnetic fields. The gas is subjected to radiative heating and cooling, so it can lose or gain thermal energy depending on its temperature and density. During the computation, the cooling and heating rates are read from a table (preliminary produced by the CLOUDY code by Ferland et al. 2017) and then they are removed or added to the energy conservation equation (in a similar fashion to what is described in Banda-Barragán et al. 2021 for shock-multicloud models, see also Banda-Barragán et al. 2020). The initial conditions and domain sizes of our models are displayed in Table 1.

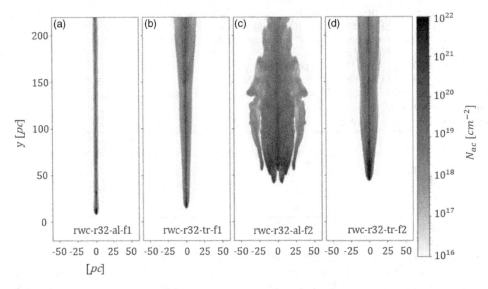

Figure 1. Column number density of the clouds integrated along the X_3 axis. The 4 panels aim to demonstrate that the choice of the cooling floor and magnetic field orientation can drastically change the evolution of wind-swept atomic clouds. The images are taken at the same time and represent the 4 models shown in Table 1. They show the generation of filaments as the clouds begin to lose material when interacting with the wind. Some differences between the models persist; in panel a) and b) the clouds are not completely destroyed by the end of the simulation and their centre of mass does not move excessively from the starting point. On the other hand, in the models with the cooling floor set to 10^4 K, shown in panels c) and d), the clouds do not have the chance to cool down beyond that value, so they expand and are therefore displaced along the wind direction. Finally, in the latter two models, the orientation of the magnetic field has a stronger influence on the cloud evolution, leading to evaporation if the magnetic field is aligned with the wind, and a slower disintegration when it is transverse.

From a physical viewpoint, the simulations we analysed are wind-tunnel simulations representing 3D sections of a global galactic wind. In particular, they emulate the behaviour of atomic clouds interacting with supersonic flows that represent galactic winds. The main features that our four simulations have in common are mentioned in the caption of Table 1. All of them have been implemented taking into account the same atomic radiative processes, the same magnetic field configurations, and a uniform cloud density distribution with the same density contrast with respect to the wind. On the other hand, the differences that have been implemented in the initial conditions of our four panels include: a different cooling floor (50 K or 10^4 K) and a different direction of the magnetic field (parallel to the wind direction or transverse to it). As can be seen in Figure 1, these variations in the initial conditions lead to very different dynamical evolutions.

3. Generation of Synthetic Spectra

Here we discuss the implementation of a framework able to produce synthetic spectra and column density maps of atomic/molecular clouds interacting with galactic winds from the simulation dataset described in the previous section. The flowchart shown in Figure 2 provides a summary of the framework. In the next subsections we will explain in more detail how each component of the code and the various software packages interact with each other, in particular how spectra are produced by combining two Python tool-kits, YT (Turk et al. 2011) and TRIDENT (Hummels et al. 2017), together with

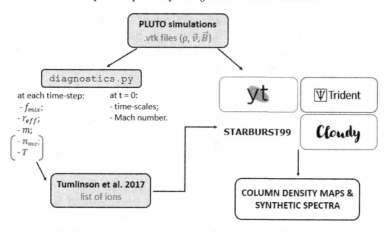

Figure 2. Flow chart summarising the main components of the framework that produces column density maps and synthetic spectra from PLUTO simulations. The starting point is the datasets provided by PLUTO, which we initially analysed using the script `diagnostics.py` in order to find which ions would be included in the synthetic spectra. The right-hand side of the diagram shows which software we used to produce the ion density maps and spectra.

the STARBURST99 (Vázquez and Leitherer 2017) software, and the CLOUDY Code (Ferland et al. 2017).

3.1. *Preliminary analysis and data preparation*

The first part of the framework involves reading the `.vtk` files produced by the PLUTO simulations. Then, we analyse the data in order to study thermodynamical aspects of these systems. For this purpose, we have created a script in Python called `diagnostics.py` that receives as input the `.vtk` files with the gas density, velocity, magnetic field of each cell and normalisation factors in c.g.s units (ρ_0, v_0, L_0, plus the mean particle mass, μ, and the polytropic index γ).

Our analysis in Python provides two types of outputs: a set of diagnostics calculated at $t = 0$ such as: magnetic field strength, initial temperature, Mach number and time-scales; and a set of time-dependent quantities averaged over the whole simulation volume and calculated at each simulation time step, such as: number density, cloud mass, coordinates of the cloud's centre of mass, its effective radius, temperature and mixing fraction. Our scripts also produce column density maps of cloud gas (both edge-on and down-the-barrel) and 2D histograms which allow us to determine which ions should exist, given the number density-temperature ranges characteristic of these interactions. Using these results, and the inputs from Figure 6 in Tumlinson et al. (2017) we can readily identify which ions should be typically observed in our simulated galactic winds, and can therefore be targeted in our research.

3.2. *Reading PLUTO data into the YT framework*

A more detailed picture of how the synthetic spectra are generated can be deduced from the flow chart in Figure 2, which shows the key steps of the process. The first package we used was YT. YT is a toolkit for visualisation and analysis of datasets produced by 3D simulations and it is also the first dependency of TRIDENT. YT has loaders for several commonly-used hydrodynamic codes, but does not have a native interpreter for PLUTO data, so it was necessary to create a script to fill this gap. The Python script in matter is called `Pluto_YT_interface.py`, which converts the content of each .vtk file into a series of numpy arrays. Thanks to the `yt.load_uniform_grid` function, it is possible to load the data and display temperature and density maps. Subsequently, this function creates a dictionary with all the data fields and their units in c.g.s. and defines a computational domain that coincides with the simulation one.

3.3. *Feeding TRIDENT and producing spectra*

Once the grid-based data have been read into YT as a "StreamHandler" and once the ions to be analysed have been chosen, we produce column density maps and temperature-number density 2D histograms to compare with those produced earlier by the `diagnostics.py` script. If everything matches, the following step is to produce synthetic spectra exploiting the TRIDENT package. The first function used is `trident.add_ion_fields` which generates projected density maps for each ion in order to better understand their spatial distribution within the cloud.

The second step is to use the `trident.make_simple_ray` function to save all the field values of the ions intercepted by the beam and save them in an .h5 file (see Hummels et al. 2017). Finally, the last part of the code involves the `trident.SpectrumGenerator` class which receives as input the ray and the range of wavelengths in Angstrom, the resolution $\Delta\lambda$ and it generates as output the spectrum. In this final stage it is possible to decide the type and intensity of noise in order to produce a more realistic spectrum, and also to produce it in velocity space.

3.4. *Customised UV Backgrounds*

TRIDENT uses by default the Haardt and Madau (2012) UV background, which only takes into account the cosmic UV background radiation. However, since the atomic clouds investigated in this study are located close to regions of high star formation, it is necessary to consider the extra UV radiation produced by the evolution of e.g. OB stars (see also Cottle et al. 2018). The creation of a more realistic background is briefly summarised in the top right-hand part of Fig. 2. We used the STARBURST99 software and the CLOUDY code, in particular the library `Cloudy_cooling_tools` by Smith et al. (2008); Smith et al. (2017) to produce new ion tables to replace "hm2012_hr.h5" in the Trident configuration file. The steps followed to produce the UV background are:

• Choosing a Spectral Energy Distribution (SED) from those produced by STARBURST99 that best represents the stellar population near the simulated atomic cloud. We chose a SED assuming solar metallicity and 3-Myr-old starburst, in agreement with McClure-Griffiths et al. (2013) concerning the observation of atomic hydrogen in the Galactic centre of our Galaxy. SEDs have wavelengths in $\mathring{A}$ on the X-axis and luminosity in erg s^{-1} $\mathring{A}^{-1}$ on the Y-axis.

• We converted the brightnesses to specific intensities and produced an .out file with a customised format, so that it can be read by CLOUDY. The Python script that handles this procedure is `sb99_cloudy_interface.py`, and the wavelengths with relative luminosities given as input have been converted to E [Ryd] and $log(J_\mu)$. We also converted the SEDs representing the Haardt and Madau (2012) profile provided by CLOUDY and TRIDENT into the same units to evaluate more clearly how important it is to consider a UV background which takes into account star formation. The SED produced with STARBURST99 has a flux at least one order of magnitude higher than the others, in the wavelength range that we are interested in. Our results also appear to be in agreement with Werk et al. (2014).

• Then, we ran a CLOUDY simulation in parallel using 16 cores with the `CIAOLoop` code, located inside `Cloudy_cooling_tools`, on SUPERMUC-NG in order to obtain ion tables that take into account radiation from stellar feedback. The simulation receives as input the .out file produced in the previous step and was run with a temperature range

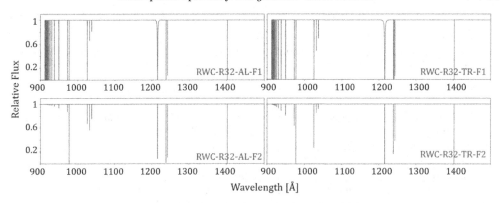

Figure 3. Synthetic spectra of the single cloud models shown in Figure 1. The 4 spectra were obtained at the same time as the images in Figure 1, where the top two panels correspond to models with cooling floor at 50 K, while the bottom two panels correspond to models with 10^4 K as cooling floor, and are all derived imposing zero noise. The lines in absorption represent the following ions: H I, O VI, C III, C IV, Si IV, N V, Ne VIII, and Mg II.

from 1 to 10^9 K, hydrogen number density of 10^{-9} cm^{-3} $- 10^4$ cm^{-3} with step size of 0.0125 dex and at redshift $= 0$ (as advised in Hummels et al. 2017 and Cottle et al. 2018).

- Finally, by exploiting the `convert_ion_balance_tables` function imported from `cloudy_grids` inside `Cloudy_cooling_tools` we generated the ion field output file in `.h5` format, which we then placed inside the `.trident` folder as the new default background.

4. Preliminary Results and Conclusions

The resulting down-the-barrel spectra obtained by following the framework described above can be viewed in Fig. 3. The top two panels correspond to models with a cooling floor at 50 K, while the bottom two panels correspond to models with 10^4 K as the cooling floor. By comparing different models representing the interaction of a single atomic cloud with galactic winds, each of them with a different cooling floor and different orientation of the magnetic field, it is possible to see how both phenomena influence the morphology of the atomic cloud and potentially the spectra. We find that the morphological variations of the clouds, which we mentioned earlier, are also reflected in the resulting spectra.

While we are currently working on quantifying how much the cooling floor and magnetic fields affect the column density of certain ions such as O VI, some qualitative features of our spectra already draw our attention. For instance, Figure 3 reveals that changing the orientation of the magnetic field in the top two models does not cause large variations in the spectra; while in the bottom two panels the depth of some lines varies depending on the orientation of the magnetic field. Moreover, quenching radiative cooling at higher temperatures has the main effect of reducing the width of the Lyα absorption line.

As briefly presented here, the overall aim of our project is to create a framework able to produce synthetic spectra from galactic wind simulations. For the code development part, the PLUTO-TRIDENT interface is now ready to use. The advantage of this tool is that it is quite flexible and can be used to study the physics and chemistry of different astrophysical systems on different scales. In addition, it makes it possible to produce column density maps for individual ions and spectra with customised UV backgrounds that take into account the high star formation that occurs in a starburst galaxy. A further future goal is to produce synthetic spectra simulating observations with different instruments.

The authors gratefully acknowledge the Gauss Centre for Supercomputing e.V. (www.gauss-centre.eu) for funding this project by providing computing time (via grant pn34qu) on the GCS Supercomputer SuperMUC-NG at the Leibniz Supercomputing

Centre (www.lrz.de). WEBB is supported by the Deutsche Forschungsgemeinschaft (DFG) via grant BR2026/25, and by the National Secretariat of Higher Education, Science, Technology, and Innovation of Ecuador, SENESCYT. We also thank the referee for their helpful feedback on our paper.

References

Banda-Barragán, W. E., Brüggen, M., Federrath, C., Wagner, A. Y., Scannapieco, E., & Cottle, J. 2020, Shock-multicloud interactions in galactic outflows - I. Cloud layers with lognormal density distributions. *MNRAS*, 499(2), 2173–2195.

Banda-Barragán, W. E., Brüggen, M., Heesen, V., Scannapieco, E., Cottle, J., Federrath, C., & Wagner, A. Y. 2021, Shock-multicloud interactions in galactic outflows - II. Radiative fractal clouds and cold gas thermodynamics. *MNRAS*, 506(4), 5658–5680.

Banda-Barragán, W. E., Federrath, C., Crocker, R. M., & Bicknell, G. V. 2018, Filament formation in wind-cloud interactions- II. Clouds with turbulent density, velocity, and magnetic fields. *MNRAS*, 473(3), 3454–3489.

Banda-Barragán, W. E., Parkin, E. R., Federrath, C., Crocker, R. M., & Bicknell, G. V. 2016, Filament formation in wind-cloud interactions - I. Spherical clouds in uniform magnetic fields. *MNRAS*, 455(2), 1309–1333.

Banda-Barragán, W. E., Zertuche, F. J., Federrath, C., García Del Valle, J., Brüggen, M., & Wagner, A. Y. 2019, On the dynamics and survival of fractal clouds in galactic winds. *MNRAS*, 486(4), 4526–4544.

Chisholm, J., Bordoloi, R., Rigby, J. R., & Bayliss, M. 2018, Feeding the fire: tracing the mass-loading of 10^7 K galactic outflows with O VI absorption. *MNRAS*, 474(2), 1688–1704.

Cimatti, A., Fraternali, F., & Nipoti, C. 2019, Introduction to Galaxy Formation and Evolution. From Primordial Gas to Present-Day Galaxies. *arXiv e-prints*, arXiv:1912.06216.

Cottle, J. N., Scannapieco, E., & Brüggen, M. 2018, Column Density Profiles of Cold Clouds Driven by Galactic Outflows. *ApJ*, 864(1), 96.

Danehkar, A., Oey, M. S., & Gray, W. J. 2021, Catastrophic Cooling in Superwinds. II. Exploring the Parameter Space. *ApJ*, 921(1), 91.

de la Cruz, L. M., Schneider, E. E., & Ostriker, E. C. 2021, Synthetic Absorption Lines from Simulations of Multiphase Gas in Galactic Winds. *ApJ*, 919(2), 112.

Ferland, G. J., Chatzikos, M., Guzmán, F., Lykins, M. L., van Hoof, P. A. M., Williams, R. J. R., Abel, N. P., Badnell, N. R., Keenan, F. P., Porter, R. L., & Stancil, P. C. 2017, The 2017 Release Cloudy. *Rev. Mex. Astron. Astrofis.*, 53, 385–438.

Goldsmith, K. J. A. & Pittard, J. M. 2018, A comparison of shock-cloud and wind-cloud interactions: effect of increased cloud density contrast on cloud evolution. *MNRAS*, 476(2), 2209–2219.

Haardt, F. & Madau, P. 2012, Radiative Transfer in a Clumpy Universe. IV. New Synthesis Models of the Cosmic UV/X-Ray Background. *ApJ*, 746(2), 125.

Hummels, C. B., Smith, B. D., & Silvia, D. W. 2017, Trident: A Universal Tool for Generating Synthetic Absorption Spectra from Astrophysical Simulations. *ApJ*, 847(1), 59.

McClure-Griffiths, N. M., Green, J. A., Hill, A. S., Lockman, F. J., Dickey, J. M., Gaensler, B. M., & Green, A. J. 2013, Atomic Hydrogen in a Galactic Center Outflow. *ApJ*, 770(1), L4.

Mignone, A., Bodo, G., Massaglia, S., Matsakos, T., Tesileanu, O., Zanni, C., & Ferrari, A. 2007, PLUTO: A Numerical Code for Computational Astrophysics. *ApJS*, 170(1), 228–242.

Richter, P., Nuza, S. E., Fox, A. J., Wakker, B. P., Lehner, N., Ben Bekhti, N., Fechner, C., Wendt, M., Howk, J. C., Muzahid, S., Ganguly, R., & Charlton, J. C. 2017, An HST/COS legacy survey of high-velocity ultraviolet absorption in the Milky Way's circumgalactic medium and the Local Group. *A&A*, 607, A48.

Scannapieco, E. & Brüggen, M. 2015, The Launching of Cold Clouds by Galaxy Outflows. I. Hydrodynamic Interactions with Radiative Cooling. *ApJ*, 805(2), 158.

Smith, B., Sigurdsson, S., & Abel, T. 2008, Metal cooling in simulations of cosmic structure formation. *MNRAS*, 385(3), 1443–1454.

Smith, B. D., Bryan, G. L., Glover, S. C. O., Goldbaum, N. J., Turk, M. J., Regan, J., Wise, J. H., Schive, H.-Y., Abel, T., Emerick, A., O'Shea, B. W., Anninos, P., Hummels, C. B., & Khochfar, S. 2017, GRACKLE: a chemistry and cooling library for astrophysics. *MNRAS*, 466(2), 2217–2234.

Tchernyshyov, K., Werk, J. K., Wilde, M. C., Prochaska, J. X., Tripp, T. M., Burchett, J. N., Bordoloi, R., Howk, J. C., Lehner, N., O'Meara, J. M., Tejos, N., & Tumlinson, J. 2021, The CGM^2 Survey: Circumgalactic O VI from dwarf to massive star-forming galaxies. *arXiv e-prints*, arXiv:2110.13167.

Tumlinson, J., Peeples, M. S., & Werk, J. K. 2017, The Circumgalactic Medium. *ARA&A*, 55(1), 389–432.

Turk, M. J., Smith, B. D., Oishi, J. S., Skory, S., Skillman, S. W., Abel, T., & Norman, M. L. 2011, yt: A Multi-code Analysis Toolkit for Astrophysical Simulation Data. *The Astrophysical Journal Supplement Series*, 192, 9.

Vázquez, G. A. & Leitherer, C. Modeling Small Stellar Populations Using Starburst99. In Charbonnel, C. & Nota, A., editors, *Formation, Evolution, and Survival of Massive Star Clusters* 2017, volume 316, pp. 359–360.

Veilleux, S., Maiolino, R., Bolatto, A. D., & Aalto, S. 2020, Cool outflows in galaxies and their implications. *A&ARv*, 28(1), 2.

Werk, J. K., Prochaska, J. X., Tumlinson, J., Peeples, M. S., Tripp, T. M., Fox, A. J., Lehner, N., Thom, C., O'Meara, J. M., Ford, A. B., Bordoloi, R., Katz, N., Tejos, N., Oppenheimer, B. D., Davé, R., & Weinberg, D. H. 2014, The COS-Halos Survey: Physical Conditions and Baryonic Mass in the Low-redshift Circumgalactic Medium. *ApJ*, 792(1), 8.

Discussion

Q: In the density-temperature plot, the majority of the gas is placed in a line which reflects the adiabatic expansion of the cloud. However, there is some hot and low-density gas that does not follow the adiabatic relation; why?

A: Shock heating causes the adiabatic expansion of the cloud and produces the hot gas we see on the n_H vs. T maps. There is also radiative heating in these models, but this plays a smaller role in heating up hot gas.

Q: Do you think relative motions can affect the evolution of the clouds?

A: Adding relative motion to these models would not be needed because we can always change the reference frame of the simulation (via a Galilean transformation) into one where the cloud is not moving at t=0. To study the interaction of moving/evolving clouds with winds, wind-filament and shock-multicloud models are more adequate than single-cloud models.

Q: Have you already compared your theoretical spectra with some observed within our Galaxy, in particular the absorption lines of some ions such as O VI?

A: Studying O VI in our Galaxy is definitely something we can do in future work, there are also other ions in the Milky Way (e.g. Richter et al. 2017).

The Predictive Power of Computational Astrophysics as a Discovery Tool
Proceedings IAU Symposium No. 362, 2023
D. Bisikalo, D. Wiebe & C. Boily, eds.
doi:10.1017/S1743921322001570

Hydrodynamic Simulations and Time-dependent Photoionization Modeling of Starburst-driven Superwinds

A. Danehkar⬡, M. S. Oey and W. J. Gray

Department of Astronomy, University of Michigan, Ann Arbor, MI 48109, USA
email: danehkar@eurekasci.com

Abstract. Thermal energies deposited by OB stellar clusters in starburst galaxies lead to the formation of galactic superwinds. Multi-wavelength observations of starburst-driven superwinds pointed at complex thermal and ionization structures which cannot adequately be explained by simple adiabatic assumptions. In this study, we perform hydrodynamic simulations of a fluid model coupled to radiative cooling functions, and generate time-dependent non-equilibrium photoionization models to predict physical conditions and ionization structures of superwinds using the MAIHEM atomic and cooling package built on the program FLASH. Time-dependent ionization states and physical conditions produced by our simulations are used to calculate the emission lines of superwinds for various parameters, which allow us to explore implications of non-equilibrium ionization for starburst regions with potential radiative cooling.

Keywords. Stars: winds, outflows – galaxies: starburst – hydrodynamics – ISM: bubbles – radiation mechanisms: general – galaxies: star clusters – intergalactic medium

---

1. Introduction

Thermal and mechanical feedback from OB stars in stellar clusters displaces the surrounding medium in starburst regions on a large scale and forms a galactic-scale outflow named superwind (Heckman et al. 1990), accompanied by a narrow shell and sometime by a hot bubble called superbubble (Weaver et al. 1977). The physical properties of the expanding wind region prior to the bubble and shell have been obtained by Chevalier & Clegg (1985) using adiabatic fluid equations that yield the outflow density $n \propto r^{-2}$ and temperature $T \propto r^{-4/3}$. However, the fluid equations coupled to radiative cooling functions studied by Silich et al. (2004) depict a deviation from the adiabatic temperature, which may explain strong cooling and suppressed superwinds seen in observations of some star-forming galaxies (Oey et al. 2017; Turner et al. 2017; Jaskot et al. 2017). In particular, semi-analytic studies and hydrodynamic simulations demonstrated that radiative cooling is heavily dependent on the metallicity, mass-loss rate, and wind velocity (Silich et al. 2004; Tenorio-Tagle et al. 2005; Gray et al. 2019a; Danehkar et al. 2021). In the case of starburst galaxies where metallicity is low, high mass-loss rates and low outflow velocities contribute to substantial radiative cooling (Danehkar et al. 2021).

Photoionization calculations were performed to identify the superwind models with strong radiative cooling (Gray et al. 2019a; Danehkar et al. 2021). However, emission lines in photoionization (PIE) and collisional ionization equilibrium (CIE) calculated by Danehkar et al. (2021) did not make a clear distinction between those with and without substantial radiative cooling. Previously, photoionization models built with time-dependent non-equilibrium ionization (NEI) states by Gray et al. (2019a) and Gray et al.

(2019b) also indicate that ions such as O VI and C IV could behave differently where plasma is in the NEI case. Non-equilibrium conditions occur when the radiative cooling timescale τ_{cool} is shorter than the CIE timescale τ_{CIE} (see e.g. Gnat & Sternberg 2007). In the expanding wind region where plasma is in transition from CIE to PIE at temperatures below 10^6 K, NEI conditions may emerge (Vasiliev 2011) and NEI states could have substantial deviations from CIE states (Gnat & Sternberg 2007; Vasiliev 2011; Oppenheimer & Schaye 2013), which can be identified using the C IV, N V, and O VI lines.

Recently, Danehkar et al. (2021) reported emission line fluxes calculated based on CIE+PIE assumptions using the physical conditions obtained from hydrodynamic simulations with the MAIHEM module in the FLASH program. Similarly, we also computed emission line fluxes following Gray et al. (2019a) for NEI conditions using the time-dependent NEI states and physical properties predicted by our hydrodynamic simulations (Danehkar et al. 2022), showing that the C IV and O VI lines have different behaviors in non-equilibrium photoionized expanding wind regions.

2. Hydrodynamic Simulations

To study starburst-driven superwinds, we consider starburst feedback from a spherically symmetric stellar cluster parameterized by the cluster radius R_{sc}, the mass-loss rate $\dot{M}$, and the stellar wind velocity V_∞. The radiation field is characterized by the total stellar luminosity and spectral energy distribution (SED). The surrounding medium has a density n_{amb}, while its temperature T_{amb} is dependent on the radiation field and determined by a CLOUDY model.

Our hydrodynamic simulations are performed using the MAIHEM atomic and cooling package (Gray et al. 2015; Gray & Scannapieco 2016; Gray et al. 2019b) in the frame work of the FLASH program (Fryxell et al. 2000), which obtains the solutions for the following one-dimensional spherically symmetric fluid equations coupled to the radiative cooling and photo-heating functions:

$$\frac{d\rho}{dt} + \frac{1}{r^2}\frac{d}{dr}\left(\rho u r^2\right) = q_m, \tag{2.1}$$

$$\frac{d\rho u}{dt} + \rho u \frac{du}{dr} + \frac{dP}{dr} = -q_m u, \tag{2.2}$$

$$\frac{d\rho E}{dt} + \frac{1}{r^2}\frac{d}{dr}\left[\rho u r^2\left(\frac{u^2}{2} + \frac{\gamma}{\gamma-1}\frac{P}{\rho}\right)\right] = \sum_i n_i \Gamma_i - \sum_i n_i n_e \Lambda_i + q_e, \tag{2.3}$$

where r is the radius, ρ the density, u the velocity, P the thermal pressure, E the total energy per unit mass, $\gamma = 5/3$ the specific heat ratio, $q_m = \dot{M}/(\frac{4}{3}\pi R_{sc}^3)$ and $q_e = (\frac{1}{2}\dot{M}V_\infty^2)/(\frac{4}{3}\pi R_{sc}^3)$ the mass and energy deposition rate per unit volume, respectively, n_i the number densities of ions, n_e the electron number density, Λ_i the radiative cooling rates for a specified temperature from Gnat & Ferland (2012), $\Gamma_i = \int_{\nu_{0,i}}^\infty (4\pi J_\nu/h\nu)h(\nu - \nu_{0,i})\sigma_i(\nu)d\nu$ the photo-heating rates obtained from the given radiation field J_ν and the photoionization cross-section $\sigma_i(\nu)$ (Verner & Yakovlev 1995; Verner et al. 1996), ν the frequency, $\nu_{0,i}$ the ionization frequency, and h the Planck constant.

To set the boundary conditions, we employ the semi-analytic radiative assumptions adopted by Silich et al. 2004, which are based on the adiabatic solutions obtained by Chevalier & Clegg (1985). Accordingly, the density, temperature, and velocity at $r = R_{sc}$ are set to $\rho = \dot{M}/(2\pi R_{sc}^2 V_\infty)$, $T = (\frac{1}{2}V_\infty)^2\mu/(\gamma k_B)$, and $u = \frac{1}{2}V_\infty$, respectively (μ the mean mass per particle, and k_B the Boltzmann constant). For the initial conditions, we set the ambient density specified by an input parameter and the ambient temperature determined by our CLOUDY model, while the medium outside the cluster radius is in stationary states ($u = 0$) at $t = 0$.

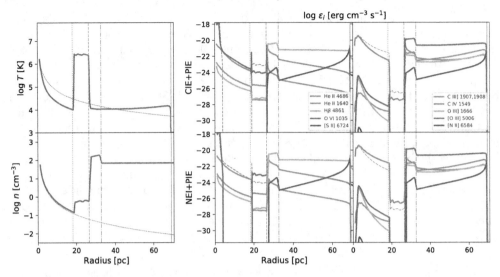

Figure 1. *Left Panels*: The temperature T and density n profiles (solid red lines) predicted by our MAIHEM simulations with the adiabatic solutions (red dashed lines). *Right Panels*: The line emissivities ε_i of the emission lines He II $\lambda4686,\lambda1640$, Hβ $\lambda4861$, O VI $\lambda1035$, [S II] $\lambda6724$ (left), C III] $\lambda\lambda1907,1908$, C IV $\lambda\lambda1549$, O III] $\lambda1666$, [O III] $\lambda5006$, and [N II] $\lambda6584$ (right) calculated by the model in collisional ionization and photoionization equilibrium (CIE+PIE; Danehkar et al. 2021), and in non-equilibrium photoionization (NEI with PIE in the ambient medium; Danehkar et al. 2022). The bubble boundaries, the shell end, and Strömgren sphere are depicted by dotted, dashed, dash-dotted, and solid vertical gray lines, respectively. The wind parameters are: $V_\infty = 457\,\mathrm{km\,s^{-1}}$, $\dot{M} = 0.607 \times 10^{-2}\,\mathrm{M_\odot\,yr^{-1}}$, $t = 1\,\mathrm{Myr}$, the stellar cluster $R_{sc} = 1\,\mathrm{pc}$ and $M_\star = 2 \times 10^6\,\mathrm{M_\odot}$, and the medium $n_{amb} = 100\,\mathrm{cm^{-3}}$ and $Z/Z_\odot = 0.5$. The O VI and C IV lines predicted by NEI are overplotted by dashed lines in the CIE+PIE panel, and vice versa.

The radiation field J_ν included in our hydrodynamic simulations is generated by the stellar population synthesis program Starburst99 (Levesque et al. 2012; Leitherer et al. 2014) for the rotational stellar models (Ekström et al. 2012; Georgy et al. 2012) with an initial mass function with slope $\alpha = 2.35$ within the range 0.5–150 $\mathrm{M_\odot}$ and the total stellar mass $M_\star = 2 \times 10^6\,\mathrm{M_\odot}$, which are associated with the mass-loss rate $\dot{M} = 10^{-2}\,\mathrm{M_\odot\,yr^{-1}}$ at 1 Myr. The Starburst99 radiation field is applied to the photo-heating rates in MAIHEM to perform non-equilibrium calculations, as well as CLOUDY photoionization models.

Figure 1 (left panels) presents the temperature T and density n radial profiles (solid red lines) generated by our MAIHEM simulation for a model with substantial radiative cooling, along with the expected adiabatic solutions without radiative cooling (dashed lines). The four distinctive regions of a typical superwind defined by Weaver et al. (1977) are also separated by dotted, dashed, and dash-dotted vertical gray color lines, namely expanding wind region (before dotted), bubble (between dotted and dashed), shell (between dashed and dash-dotted), and ambient medium (after dash-dotted vertical lines). The Strömgren sphere (solid vertical gray color line) are determined by a CLOUDY photoionization run on the density profile (pure PIE) following Danehkar et al. (2021).

Figure 2 shows the mean radiative temperature over the mean adiabatic temperature, $f_T \equiv T_{wind}/T_{adi}$, of the expanding wind region predicted by our MAIHEM hydrodynamic simulations for different wind parameters (V_∞ and $\dot{M}$), ambient densities (n_{amb}), metallicity ($Z/Z_\odot$), and a stellar cluster with $R_{sc} = 1\,\mathrm{pc}$ and $M_\star = 2 \times 10^6\,\mathrm{M_\odot}$, and current age $t = 1\,\mathrm{Myr}$. The catastrophic cooling (CC) and catastrophic cooling bubble (CB) wind modes, which are with and without bubbles, have $f_T < 0.75$, while the adiabatic bubble (AB) and pressure-confined (AP) mode have $0.75 < f_T < 1.25$. Moreover, the adiabatic

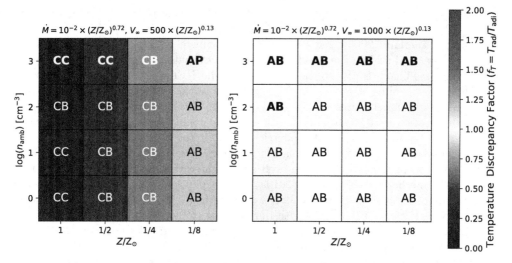

Figure 2. The mean radiative temperature $T_{\rm rad}$ with respect to the mean adiabatic temperature $T_{\rm adi}$ of the expanding wind region for various wind parameters $V_{\infty} = 500$, and $1000\ {\rm km\,s^{-1}}$ and $\dot{M} = 10^{-2} \times (Z/Z_{\odot})^{0.72}\ {\rm M_{\odot}\,yr^{-1}}$, metallicity $Z/Z_{\odot} = 0.125$, 0.25, 0.5, and 1, and ambient media with $\log n_{\rm amb} = 0$, 1, 2, and 3 ${\rm cm^{-3}}$, surrounding a stellar cluster characterized by $R_{\rm sc} = 1$ pc, $M_{\star} = 2 \times 10^{6}\ {\rm M_{\odot}}$, and age $t = 1$ Myr. The wind models are identified as adiabatic bubble (AB), catastrophic cooling (CC), catastrophic cooling bubble (CB), and pressure-confined (AP), with optically-thick status (bold font), based on criteria defined by Danehkar et al. (2021).

pressure-confined (AP) mode is assigned to those models where the bubble expansion is confined by the ambient thermal pressure (see Danehkar et al. 2021 for more detail). Optically-thick models having neutral ambient medium are also displayed with the bold font. It can be seen that increasing mass-loss rates and decreasing wind velocities result in enhanced radiative cooling in the expanding wind region.

3. Time-dependent Photoionization Models

The physical conditions and NEI states of the expanding wind region made by our hydrodynamic simulations with MAIHEM, along with the radiation field generated by Starburst99, are used to construct non-equilibrium photoionization models with the CLOUDY program (Ferland et al. 2013, 2017). Excluding the NEI states, Danehkar et al. (2021) incorporated the physical properties into a grid of CLOUDY models, which describe the CIE+PIE cases. Inclusion of the NEI states predicted by MAIHEM allows us to emulate non-equilibrium photoionization in the expanding wind region (Danehkar et al. 2022) for which pure PIE is still present in the ambient medium.

In the non-equilibrium conditions, gas kinematic and ionization structures are closely interconnected. The time-dependent NEI states are linked to collisional and dielectronic recombination, collisional ionization, and photoionization rates. The number density of ions n_i for each chemical element in the non-equilibrium cases can be described as

$$\frac{1}{n_{\rm e}}\frac{dn_i}{dt} = n_{i+1}\alpha_{i+1} - n_i\alpha_i + n_{i-1}S_{i-1} - n_iS_i + \frac{1}{n_{\rm e}}n_{i-1}\zeta_{i-1} - \frac{1}{n_{\rm e}}n_i\zeta_i, \qquad (3.1)$$

where α_i includes the radiative recombination rate (Badnell 2006) and dielectronic recombination rates (references given in Table 1 of Gray et al. 2015) for the ionic species i, S_i is the collisional ionization rates (Voronov 1997), $\zeta_i = \int_{\nu_{0,i}}^{\infty}(4\pi J_{\nu}/h\nu)\sigma_i(\nu)d\nu$ is the photoionization rates calculated using the specified background radiation field J_{ν} made by Starburst99 and the photoionization cross-section $\sigma_i(\nu)$.

The non-equilibrium cases appear when the CIE timescale $\tau_{\rm CIE} \approx 1/(n_e\alpha_i + n_eS_i)$ (Mewe 1999) is longer than the cooling timescale $\tau_{\rm cool} = 3(n_i + n_e)k_BT/(2n_i^2\Lambda_i)$ (Dopita & Sutherland 2003). For less dense environment ($\lesssim 1\,{\rm cm}^{-3}$) that is typical of the expanding wind region, the ions C IV, N V, and O VI satisfy the condition $\tau_{\rm CIE} \geq \tau_{\rm cool}$ at temperatures below 10^6 K, so they could be in the NEI situations in the presence of strong radiative cooling. For $\tau_{\rm CIE} \ll \tau_{\rm cool}$, plasma is in the CIE conditions.

Figure 1 (right panels) presents the emissivities of Hβ, low-excitation [S II] and [N II], and high-excitation He II, [O III], and C III], as well as highly-ionized C IV and O VI calculated by CLOUDY for the physical properties (without the NEI states) produced by our MAIHEM simulation associated with plasma in photoionization and collisional ionization equilibrium (top panel; CIE+PIE), as well as the emissivities computed using CLOUDY following the method of Gray et al. (2019a) for the physical conditions and NEI states generated by our MAIHEM simulation corresponding to the non-equilibrium photoionization situations (bottom panel; NEI with pure PIE in the ambient medium). Large grids for various model parameters are provided as interactive figures by Danehkar et al. (2021) for combined CIE+PIE conditions and Danehkar et al. (2022) for combined NEI+PIE situations, and hosted on this website https://galacticwinds.github.io/superwinds/. As seen in Figure 1, the O VI and C IV emissivity profiles in NEI are not the same as those with CIE (see green and orange dashed lines), particularly in the expanding wind region affected by radiative cooling. This behavior can be explained by the time-dependent ionization states of the ions O VI and C IV at temperatures below 10^6 K when rapid cooling faster than ionization processes occurs ($\tau_{\rm cool} < \tau_{\rm CIE}$).

4. Implications for Starburst Galaxies

Our NEI calculations indicate that radiative cooling could enhance the C IV 1550 Å in metal-rich and the O VI 1035 Å doublet in metal-poor environments. The C IV emission line were found in some metal-poor starburst galaxies that are good candidates for suppressed or minimal wind signatures (Senchyna et al. 2017; Berg et al. 2019a,b). As discussed by Gray et al. (2019a), these observations might be associated with kinematic features of suppressed bipolar superwinds rather than resonant scattering mentioned by Berg et al. (2019a). An O VI absorption line associated with a weak outflow was found in Haro 11, while the O VI emission luminosity suggests some cooling loss (Grimes et al. 2007). The O VI $\lambda 1035$ doublet absorption identified in a gravitationally lensed, galaxy has the features of weak low-ionization winds (Chisholm et al. 2018), which may also be explained by the mass-loss effect under non-equilibrium conditions. Moreover, observations of a star-forming galaxy depict an extended halo in the O VI image, and a weak O VI absorbing outflow in the spectrum (Hayes et al. 2016), which might be an indication of suppressed winds.

The enhancements of the C IV and O VI lines could be related to substantial radiative cooling as suggested by Gray et al. (2019a) and Gray et al. (2019b). Time-dependent NEI states calculated by de Avillez & Breitschwerdt (2012) also imply that O VI can be produced in NEI at $10^{4.2-5}$ K below the temperatures that produce O VI in CIE. Similarly, our NEI calculations (before the shell; see Figure 1) also show that the C IV and O VI emission lines do not behave the same in CIE and NEI, especially in the outflow region strongly impacted by radiative cooling.

Time-dependent ionization processes could also be sensitive to time-evolving ionizing sources. Our radiation field was made by Starburst99 for a typical age of 1 Myr. The radiation field calculated by Starburst99 can evolve with the age, which can affect the formation of radiative cooling in starburst-driven superwinds. Our future hydrodynamic simulations with time-evolving radiation fields will help us to understand better the implication of time-dependent non-equilibrium ionization for star-forming regions.

References

Badnell N. R., 2006, *ApJS*, 167, 334

Berg D. A., Chisholm J., Erb D. K., Pogge R., et al., 2019a, *ApJ*, 878, L3

Berg D. A., Erb D. K., Henry R. B. C., et al., 2019b, *ApJ*, 874, 93

Chevalier R. A., Clegg A. W., 1985, *Nature*, 317, 44

Chisholm J., Bordoloi R., Rigby J. R., Bayliss M., 2018, *MNRAS*, 474, 1688

Danehkar A., Oey M. S., Gray W. J., 2021, *ApJ*, 921, 91

Danehkar A., Oey M. S., Gray W. J., 2022, *ApJ*, 937, 68

de Avillez M. A., Breitschwerdt D., 2012, *ApJ*, 761, L19

Dopita M. A., Sutherland R. S., 2003, Astrophysics of the Diffuse Universe. Springer: Berlin

Ekström S. et al., 2012, *A&A*, 537, A146

Ferland G. J. et al., 2017, *RMxAA*, 53, 385

Ferland G. J. et al., 2013, *RMxAA*, 49, 137

Fryxell B. et al., 2000, *ApJS*, 131, 273

Georgy C., Ekström S., Meynet G., Massey P., et al., 2012, *A&A*, 542, A29

Gnat O., Ferland G. J., 2012, *ApJS*, 199, 20

Gnat O., Sternberg A., 2007, *ApJS*, 168, 213

Gray W. J., Oey M. S., Silich S., Scannapieco E., 2019a, *ApJ*, 887, 161

Gray W. J., Scannapieco E., 2016, *ApJ*, 818, 198

Gray W. J., Scannapieco E., Kasen D., 2015, *ApJ*, 801, 107

Gray W. J., Scannapieco E., Lehnert M. D., 2019b, *ApJ*, 875, 110

Grimes J. P. et al., 2007, *ApJ*, 668, 891

Hayes M., Melinder J., Östlin G., Scarlata C., et al., 2016, *ApJ*, 828, 49

Heckman T. M., Armus L., Miley G. K., 1990, *ApJS*, 74, 833

Jaskot A. E., Oey M. S., Scarlata C., Dowd T., 2017, *ApJ*, 851, L9

Leitherer C., Ekström S., Meynet G., Schaerer D., et al., 2014, *ApJS*, 212, 14

Levesque E. M., Leitherer C., Ekstrom S., Meynet G., Schaerer D., 2012, *ApJ*, 751, 67

Mewe R., 1999, in X-ray spectroscopy in Astrophysics, van Paradijs J., Bleeker J. A., eds.,
    Lectrure Notes in Physics, Springer: Berlin

Oey M. S., Herrera C. N., Silich S., Reiter M., et al., 2017, *ApJ*, 849, L1

Oppenheimer B. D., Schaye J., 2013, *MNRAS*, 434, 1043

Senchyna P. et al., 2017, *MNRAS*, 472, 2608

Silich S., Tenorio-Tagle G., Rodríguez-González A., 2004, *ApJ*, 610, 226

Tenorio-Tagle G., Silich S., Rodríguez-González A., Muñoz-Tuñón C., 2005, *ApJ*, 620, 217

Turner J. L., Consiglio S. M., Beck S. C., Goss W. M., et al., 2017, *ApJ*, 846, 73

Vasiliev E. O., 2011, *MNRAS*, 414, 3145

Verner D. A., Ferland G. J., Korista K. T., Yakovlev D. G., 1996, *ApJ*, 465, 487

Verner D. A., Yakovlev D. G., 1995, *A&AS*, 109, 125

Voronov G. S., 1997, Atom. Data Nucl. Data Tabl., 65, 1

Weaver R., McCray R., Castor J., Shapiro P., Moore R., 1977, *ApJ*, 218, 377

The Predictive Power of Computational Astrophysics as a Discovery Tool
Proceedings IAU Symposium No. 362, 2023
D. Bisikalo, D. Wiebe & C. Boily, eds.
doi:10.1017/S1743921322001569

Numerical modelling of X-shaped radio galaxies using back-flow model

Gourab Giri[⬡] and Bhargav Vaidya

Department of Astronomy, Astrophysics and Space Engineering, Indian Institute
of Technology Indore, Simrol 453552, Madhya Pradesh, India
emails: `gourab@iiti.ac.in`, `bvaidya@iiti.ac.in`

Abstract. The focus of this work is to comprehensively understand hydro-dynamical back-flows and their role in dynamics and non-thermal spectral signatures particularly during the initial phase of X-shaped radio galaxies. In this regard, we have performed axisymmetric (2D) and three dimensional (3D) simulations of relativistic magneto-hydrodynamic jet propagation from tri-axial galaxies. High-resolution dynamical modelling of axisymmetric jets has demonstrated the effect of magnetic field strengths on lobe and wing formation. Distinct X-shape formation due to back-flow and pressure gradient of ambient is also observed in our 3D dynamical run. Furthermore, the effect of radiative losses and diffusive shock acceleration on the particle spectral evolution is demonstrated, which particularly highlights how crucial their contributions are in the emission signature of these galaxies. This imparts a significant effect on the galaxy's equipartition condition, indicating that one must be careful in extending its use in estimating other parameters, as the criterion evolves with time.

Keywords. Galaxies: jets, Acceleration of particles, Magnetohydrodynamics (MHD), Methods: numerical

---

1. Introduction

Extended radio galaxies have radio jets that emanate from the central Active Galactic Nuclei (AGN) and travel along a route specified by the central black hole's spin axis. Some of these radio galaxies show significant distortion in their jet forming winged radio sources. They can be identified by observing the presence of two double lobe structures aligned at an angle to each other (forming X-shape) (Kraft et al. 2005). Due to this peculiarity, several formation mechanisms have been attributed to describe its formation process along with their plausible support from low frequency radio observations (see Gopal-Krishna et al. 2012). However, to date there exists no general agreement between the proposed models. Here, we will study the Back-flow model (Leahy Williams 1984; Capetti et al. 2002) as the possible origin of X-morphology using numerical simulations. As per the Back-flow model, back flow of jet material formed as a result of the pressure imbalance at the jet head are strongly deflected by the ambient medium itself, forming an X-shaped source. Several simulation studies have already been conducted in this regard, addressing numerous dynamical characteristics of the synthetic morphology having X-shape (Hodges-Kluck & Reynolds 2012; Rossi et al. 2017). However, in terms of their emission modelling, these studies have used rather proxies (i.e. pressure as emissivity) or a simplistic treatment of it ignoring the crucial effect of several micro-physical processes (e.g. cooling and re-energization processes) on particle spectral upgradation and hence on the emission.

In this work, we are intending to address some of the fundamental aspects of the Back-flow model, and explore its associate signatures on the appeared morphology. Then, we plan to describe the spectral aspect of these radio galaxies by accurately incorporating the effects of ongoing micro-physical processes like radiative and adiabatic losses, as well as the particle re-energization process due to diffusive shocks. This allows us to track the spectral history of emitting particles throughout the galaxy, and investigate how much crucial re-energization processes are in keeping the structure active, allowing us to distinguish between distinct formation mechanisms.

2. Simulation Setup

Our numerical simulations were performed using the relativistic magneto-hydrodynamic (RMHD) module of PLUTO code (Mignone et al. 2007).

• **Dynamical modelling:**

We have initialized a tri-axial ellipsoidal galaxy medium using the King's density profile (Cavaliere & Fusco-Femiano 1976) defined (in Cartesian system) as

$$\rho = \frac{\rho_0}{(1 + x^2/a^2 + y^2/b^2 + z^2/c^2)^{\frac{3}{4}}} \tag{2.1}$$

where ρ_0 is the central density of the galaxy (1 amu/cc) and a, b and c are the effective core radii (4/3, 8/3 and 1 kpc respectively). The galaxy posses an initial isothermal atmosphere, static equilibrium of which is maintained by a gravity profile that is obtained using the hydro-static equilibrium equation. At the centre of this distribution, we have introduced a jet nozzle which will continuously inject bi-directional relativistic jet (radius 200 pc) having Lorentz factor $\Gamma = 5$ and which is under-densed as $10^{-6}\rho_0$. A toroidal B-field in the jet injection region is incorporated which is defined as $B_x = -B_t r sin\vartheta$ and $B_y = B_t r cos\vartheta$ where r and ϑ are the polar coordinates in the perpendicular plane to the jet ejection axis and B_x, B_y are the components of the field defined in that plane. The value of B_t is constant and is determined through the jet magnetization parameter σ (set to 0.01) which is the ratio of Poynting flux to the matter energy flux ($\frac{B_t^2}{\Gamma^2 \rho h}$; h is the specific enthalpy). To be specific, we have followed the initial configuration of Rossi et al. (2017) in modelling the dynamics of X-shaped radio galaxies (XRGs), focusing on their early formation phase in a galactic medium (i.e. evolution up to 4 Myr).

• **Spectral modelling:**

Additionally, we have used the hybrid framework of the PLUTO code, which helps us inject Lagrangian macro-particles into the domain, to model the spectral properties of these galaxies (Vaidya et al. 2018; Mukherjee et al. 2021). We continuously inject these particles with the jet in order to fill the computational domain with $\sim 10^6$ particles by the end of the simulation. Each macro-particle represents an ensemble of non-thermal electrons which follows a power law pattern (initially) defined as $N(\gamma) = N_0 \gamma^{-p}$ with index $p = 6$. The value of N_0 can be obtained by $\int_{\gamma_{min}}^{\gamma_{max}} N(\gamma)d\gamma = n_{micro}$ where n_{micro} is the non-thermal electron number density. Its value is set by assuming that the energy density of the injected electrons (U_e) is a fraction of the magnetic energy density $\left(\frac{B_{dyn}^2}{8\pi}\right)$ at the injection region, i.e. (in cgs units)

$$U_e = m_e c^2 \int_{\gamma_{min}}^{\gamma_{max}} \gamma N(\gamma)d\gamma \left(= \frac{B_{eq}^2}{8\pi}\right) = \epsilon \frac{B_{dyn}^2}{8\pi} \tag{2.2}$$

The equipartition magnetic energy density is represented by the term in the bracket in Eq. (2.2) (Hardcastle et al. 2002). In reality, B_{eq} (the equipartition magnetic field) is rather a proxy for magnetic field defined such that magnetic energy associated is equivalent to the radiating electrons energy and therefore cannot be considered as physical

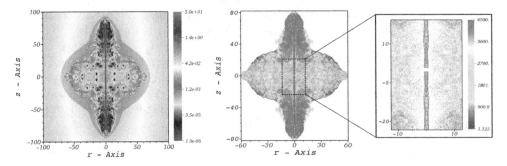

Figure 1. *Left:* density distribution of the galaxy, obtained from our 2D-run, is shown at time 3.26 Myr. *Right:* distribution of the Lagrangian macro-particles injected into the domain where the associate color bar represents the time of their injection. A higher value of it indicates a freshly injected plasma which suggests that the accumulation of older particles occurs in the wings. A zoomed version of this map represents the particle injection process from the central injection region i.e. the jet nozzle. Here length, density and time are defined with respect to 200 pc, 1 amu/cc and 651.8 yrs respectively.

magnetic field. Here, ϵ represents the fraction which is set to the value of 2.7×10^{-4} under the limiting Lorentz factors 10^2 ($\gamma_{\min}$) and 10^{10} ($\gamma_{\max}$) of the electrons. The choice of ϵ is rather arbitrary, resulting in a sub-equipartition strength of emission with $B_{\mathrm{eq}} = 0.01 B_{\mathrm{dyn}}$. Hence, we obtain the value of n_{micro} as $10^{-3}\rho_j$ where ρ_j is the jet density. The initial spectral distribution of the electrons will be modified over time depending on the micro-physical processes it undergoes like radiative losses, adiabatic cooling or diffusive shock acceleration which we considered in our simulations. For this reason, the particles are insensitive to the initial conditions when they travel some distance away from the injection region. The radiative losses considered here are the synchrotron and the inverse Compton (IC-CMB) emission.

We simulate a 3D Cartesian domain having size $(16 \times 32 \times 16)$ kpc, distributed in $192 \times 384 \times 192$ grids. We note here that a simpler version of this model in 2D is also performed by us in sufficiently high resolution (1536×1536 grids for (20×20) kpc domain) to understand the formation of X-morphology and effect of Back-flow on the morphological signature (in high resolution). The simulation was conducted in cylindrical geometry in $r - z$ plane (axisymmetric run) where z-axis represents the major axis of the galaxy (in 3D setup, it is the y-axis).

3. Formation of X-shaped morphology

The result of our simulation has been highlighted in Fig. 1, where in the left panel, the density distribution of the galaxy at time 3.26 Myr is represented (obtained from the 2D run). The strong bow shock and the contact discontinuity of the cocoon formed due to the jet-ambient interaction are highlighting a distinct X-shape of the galaxy where the jet is travelling along the major axis of the ambient. In this case, the over-pressured cocoon, formed due to the the continuous inflation of back-flowing material from jet, expands rapidly along the maximum pressure gradient path of the galaxy i.e. along the minor axis of the tri-axial medium. The high resolution map also shows the formation of several turbulent features in the cocoon, especially also in the wings, which play a crucial role in influencing the emission aspect of these structures via re-energization (discussed later). In the right part of Fig. 1, we show the particle distribution map of the galaxy where the associate colorbar represents the time of their injection into the domain. A higher value of injection time indicates a freshly injected particle. From figure, it is evident that the accumulation of older particles occurs in the wings, whereas the active lobes are filled with fresh plasma, satisfying the basic prediction of the Back-flow model. The zoomed version of this map represents the continuous particle injection process from the jet nozzle where the recollimations of jets are visible. We observed that the prominence of the formed X-shape reduces with the increasing angle between the major axis of the galaxy and the jet ejection axis (as typically observed), indicating the crucial role of ambient medium in shaping these morphologies.

The observed shape of the structure is however sensitive to the initial value of the magnetic field in the jet. A higher value of B-field is capable of suppressing the extent of the wing which in turn make the lobe expansion faster leading it to form a classical radio galaxy, even though it travels along the major axis of a highly elliptical ambient medium. Whereas, a weaker value of it will increase the prominence of the formed X-shape. The average B-field value at the end of the simulation for 2D-case is found to be several time higher than the values obtained in our 3D runs. This is due to the dimensional restriction of the cocoon expansion as the simulation has been carried out in 2D. As a result of which, the cocoon becomes incapable of reducing the B-field values that in turn affect the emission signature of the galaxy as well (as the particles cool down faster).

The basic characteristics described above, also hold true in our 3D run, however due to no dimensional restrictions, the structure evolve self consistently there (Fig. 2). For our further quantitative analysis, we will now focus on the results obtained from our 3D simulation.

4. Diffusive shock and equipartition condition

While determining the strength of the magnetic field from observation of synchrotron radiation of radio galaxies, it is mostly assumed that the radiating electrons are in equipartition with the magnetic energy of the cocoon. The equipartition condition is written in Eq. 2.2. However, several studies based on the IC-CMB measurement or the analytical modeling of radio galaxies have shown an overestimated value of B_{eq} compared to its actual value (B_{dyn}) (Croston et al. 2005; Kraft et al. 2005; Mahatma et al. 2020).

Using the Lagrangian macro-particles injected into the domain, we have intended here to verify the evolution of equipartition condition of the simulated XRG (in 3D). The particles injected into the domain have started with a sub-equipartition strength as discussed in Section 2. Subsequently, we see that the condition is getting updated for the particles as they undergo through the micro-physical processes considered in the simulation. In this regard, we have tracked four particles injected into the domain at different times and plotted the ratio of the equipartition B-field strength (B_{eq}) to the dynamical B-field strength (B_{dyn}), where $B_{eq} = B_{dyn}$ indicates true equipartition. The trajectories of these particles are plotted in the top panel of Fig. 2 over the 2D slice of the formed X-shaped structure at 3.78 Myr (for visualisation and) to show where the particles end up at 3.78 Myr. In the bottom panel, the corresponding evolution of magnetic field ratios (i.e. B_{eq}/B_{dyn}) are shown for these particles. From figure, we can observe that all the particles are showing an overall increasing nature of the B-field ratio with time from their initial phase marked by the dashed grey line. The primary causes of this growth are as follows: (a) the particles are subjected to diffusive shocks at the shock locations that they encounter in their trajectory, and (b) the adiabatic expansion of the cocoon leading to a subsequent decrease of B_{dyn} field strength. We observed that the strength of these shocks varies between compression ratio values of 1.2 to 4 that correspond to weakest to strongest shocks, which further hints that shock re-acceleration plays a crucial role in updating the particle spectra and hence the emission of these galaxies. These shocks put a temporary hold on the drastic cooling of particles, and as a consequence the spectra will be flatter than expected from spectral ageing. An imprint of this on the emission

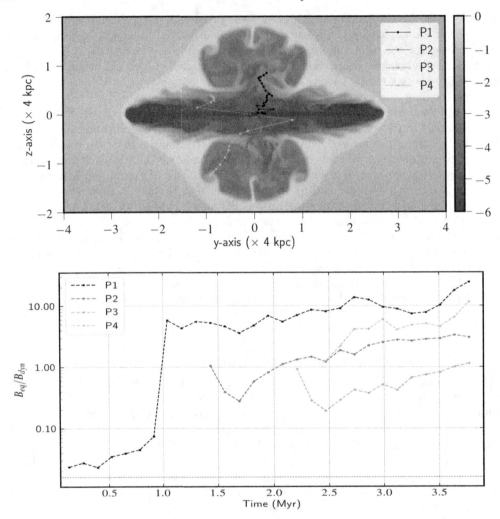

Figure 2. *Top:* *y-z* slice (at $x = 0$) of our simulated XRG (in 3D) at time 3.78 Myr, over-plotted with the trajectories of four Lagrangian particles injected into the domain through the jet nozzle at different times. The corresponding colorbar represents the log variation of density of the galaxy. *Bottom:* evolution of B_{eq}/B_{dyn} with time for the same particles where the dashed grey line represents the initial condition for the injected particles. A value of $B_{eq} = B_{dyn}$ indicates the true equipartition between the radiating electrons and the magnetic field of the galaxy.

signatures of wing and active lobe of XRGs have elaborately discussed by us in the study Giri et al. (2022).

One point to note here is that the trajectories of particles P1 and P2 do not contradict the ideal back-flow scenario. These particles were injected into the domain at a very early stage of the simulation when the wing-lobe structure was not prominent enough to distinguish. Moreover, the frequency of data saving of the simulation is also quite low for tracking the proper trajectories of these particles at their initial phase. Despite the fact, the changes in particle spectra concurrent to the micro-physical processes encountered by them are adequately captured, which is evident in Fig. 2 (bottom), where they show higher $\frac{B_{eq}}{B_{dyn}}$ values at their first detection than the initial injection condition.

The above discussion indicates that the equipartition condition of these radio galaxies varies with time as the particles evolve continuously under the cooling or re-energization processes in the turbulent cocoon and wing. An in-depth examination of the role of diffusive shocks and cooling in wing and lobe is discussed further in Giri et al. (2022). Here, in particular, we show that several particles alter their initial equipartition condition by shifting towards higher magnetic field ratios governed by the combined effect of the above mentioned micro-scaled processes. This indicates that the equipartition criterion of radio galaxies is not a ready to take assumption, and one must be careful in making its use in evaluating other physical parameters, for example, the spectral age of a radio galaxy (Mahatma et al. 2020; Giri et al. 2022).

5. Summary

We have performed high resolution 2D axi-symmetric and full scale 3D relativistic magneto-hydrodynamic simulation of jet propagation from triaxial galaxy to understand the formation of X-shaped radio galaxies based on the back-flow model. We affirm the role of pressure gradient of the ambient and hence the initial jet ejection direction in forming the X-shaped morphology as suggested by Capetti et al. (2002). In this case, the wing length is also governed by the magnetic field strength as strong fields tend to suppress the wing length due to stronger tension force. Further for our 3D simulations, we have adopted the model from Rossi et al. (2017) and have demonstrated the formation of X-shaped radio galaxies. We also show that older particles tend to accumulate in the wings, whereas lobes are mostly filled with freshly injected particles from the AGN.

Along with dynamical studies, we have also investigated the effect of back flow on the non-thermal particle spectral signatures of these galaxies which we have modelled using a hybrid approach considering the role of the diffusive shock acceleration (DSA) and the radiative and adiabatic cooling. We show that the role of DSA cannot be ignored in these galaxies as they keep the radio structure active during its evolution by re-accelerating particles against their drastic cooling. We have quantified the ratio of equipartition magnetic field B_{eq} with dynamical magnetic fields B_{dyn}. During the course of evolution, the injected particles show a systematic shift from an initial sub-equipartition population $B_{\mathrm{eq}} < B_{\mathrm{dyn}}$ to approximately equipartition due to consistent evolution of particle distribution in presence of radiative losses from local magnetic fields and diffusive shock acceleration. This analysis indicates that equipartition between the radiating and magnetic energy is rather a dynamic process which evolves with time. Such an evolution has significant impact on the estimation of spectral age of radio galaxies as was demonstrated by Mahatma et al. (2020); Giri et al. (2022).

References

Capetti, A., Zamfir, S., Rossi, P., et al. 2002, A&A, 394, 39
Cavaliere, A. & Fusco-Femiano, R. 1976, A&A, 500, 95
Croston J. H., Hardcastle M. J., Harris D. E., et al. 2005, ApJ, 626, 733
Giri, G., Vaidya, B., Rossi, P., et al. 2022, A&A, DOI:10.1051/0004-6361/202142546
Gopal-Krishna Biermann P. L., Gergely L. Á., Wiita P. J., 2012, Research in Astronomy and Astrophysics, 12, 127
Hardcastle, M. J., Birkinshaw, M., Cameron, R. A., et al. 2002, ApJ, 581, 948
Hodges-Kluck E. J., Reynolds C. S., 2011, ApJ, 733, 58
Kraft R. P., Hardcastle M. J., Worrall D. M., Murray S. S., 2005, ApJ, 622, 149
Leahy, J. P. & Williams, A. G. 1984, MNRAS, 210, 929
Mahatma, V. H., Hardcastle, M. J., Croston, J. H., et al. 2020, MNRAS, 491, 5015
Mignone, A., Bodo, G., Massaglia, S., et al. 2007, ApJS, 170, 228
Mukherjee, D., et al. 2021, MNRAS [https://doi.org/10.1093/mnras/stab1327]
Rossi, P., Bodo, G., Capetti, A., & Massaglia, S. 2017, A&A, 606, A57
Vaidya, B., Mignone, A., Bodo, G., Rossi, P., & Massaglia, S. 2018, ApJ, 865, 144

The Predictive Power of Computational Astrophysics as a Discovery Tool
Proceedings IAU Symposium No. 362, 2023
D. Bisikalo, D. Wiebe & C. Boily, eds.
doi:10.1017/S1743921322001557

Radiation Magnetohydrodynamic Simulations of Soft X-ray Emitting Regions in Active Galactic Nuclei

Taichi Igarashi[1] , **Yoshiaki Kato**[2], **Hiroyuki R. Takahashi**[3], **Ken Ohsuga**[4], **Yosuke Matsumoto**[1] and **Ryoji Matsumoto**[1]

[1]Chiba University, email: `igarashi.taichi@chiba-u.jp`

[2]RIKEN

[3]Komazawa University

[4]University of Tsukuba

Abstract. We present the results of global three-dimensional radiation magnetohydrodynamic simulations of the formation of soft X-ray emitting regions in active galactic nuclei by applying a radiation magnetohydrodynamic code based on the M1-closure scheme. The effect of Compton cooling is taken into account. When the surface density of the accretion flow exceeds the upper limit of the radiatively inefficient accretion flow (RIAF), the optically thin, hot accretion flow near the black hole co-exists with the soft X-ray emitting, warm ($T = 10^6 - 10^7$ K) Comptonized region around $r = 20 - 40 r_s$, where r_s is the Schwarzschild radius. Numerical results indicate that when the accretion rate approaches the Eddington accretion rate, the warm Comptonized region stays in optically thin for effective optical depth, Thomson thick, and radiation pressure dominant state. This region is found to oscillate between a geometrically thin, cool state and a geometrically thick state inflated by radiation pressure. The time variability of the accretion flow is consistent with that of the narrow-line Seyfert 1 galaxies.

Keywords. (magnetohydrodynamics:) MHD, radiative transfer, galaxies: active

1. Introduction

Soft X-ray excess is observed in luminous active galactic nuclei (AGN) such as Seyfert 1 galaxies. Since the temperature of the optically thick, standard disk around a supermassive black hole is in the order of 10^5 K, the soft X-rays should be emitted from a region different from standard disks. Recently, the appearance and disappearance of a soft X-ray excess component have been observed in changing look AGN (CLAGN), in which broad emission lines and soft X-ray excess are observed when the luminosity exceeds 0.5% of the Eddington luminosity (e.g., Noda & Done 2018). This transition can be explained by the state transition from RIAF to a radiatively cooled accretion flow, which occurs when the accretion rate exceeds the upper limit for RIAF.

Recently, 1-hour quasi-periodic oscillations (QPOs) have been observed in narrow-line Seyfert 1 (NLS1) galaxy RE J1034+396. The QPOs are superimposed on lower frequency time variabilities (e.g., Chaudhury et al. 2018). NLS1s show more rapid time variabilities than Seyfert 1 galaxies because their black hole mass is smaller. The accretion rate in NLS1 can exceed the Eddington accretion rate. The QPOs in RE J1034+396 are similar to the high-frequency QPOs observed in galactic black hole candidates in which QPOs are observed when their luminosity approaches 10% of the Eddington luminosity. The

1 hour period around a $10^6 M_\odot$ black hole corresponds to 30 Hz for $10 M_\odot$ black hole, which is the same order as the 67 Hz QPO observed in a galactic microquasar GRS 1915+105 (e.g., Morgan et al. 1997).

In addition to the high-frequency QPOs, lower frequency heart-beat oscillations are observed in GRS 1915+105. Honma et al. (1991) carried out 1-dimensional simulations of radiation pressure dominant, thermally unstable disk. They showed that limit-cycle oscillations between a radiation pressure dominant slim disk and a gas pressure domi- nant disk take place. Ohsuga (2006) reproduced heart-beat X-ray luminosity variations observed in GRS 1915+105 by 2-dimensional global radiation hydrodynamic simulations.

In the following, we present the results of global three-dimensional radiation magne- tohydrodynamic simulations of accretion flows onto a supermassive black hole when the accretion rate approaches the Eddington accretion rate.

2. Basic Equations and Numerical Setup

We solve the RMHD equations, consisting of resistive-magnetohydrodynamic equations coupled with the 0th and 1st moments of radiation transfer equations in cylindri- cal coordinates (r, φ, z). General relativistic effects are considered using the pseudo Newtonian potential $\phi_{PN} = -GM_{BH}/(R - r_s)$, where $M_{BH} = 10^7 M_\odot$ is the black hole mass, $R\,(=\sqrt{r^2 + z^2})$ is the distance from the black hole, $r_s\,(= 2GM_{BH}/c^2 = 3 \times 10^{12}$ cm) is the Schwarzschild radius, and G is the gravitational constant. We normalized phys- ical quantities by the Schwarzschild radius r_s, speed of light $c = 3 \times 10^{10}$ cm/s, and $t_0 = r_s/c = 100$ s. The mass accretion rate is normalized by the Eddington mass accretion rate $\dot{M}_{Edd} = L_{Edd}/c^2$, where L_{Edd} is the Eddington luminosity.

The basic equations are expressed as follows:

$$\frac{\partial \rho}{\partial t} + \nabla \cdot (\rho \mathbf{v}) = 0, \tag{2.1}$$

$$\frac{\partial \rho \mathbf{v}}{\partial t} + \nabla \cdot (\rho \mathbf{v}\mathbf{v} + p_t \mathbf{I} - \mathbf{B}\mathbf{B}) = -\rho \nabla \phi_{PN} - \mathbf{S}, \tag{2.2}$$

$$\frac{\partial E_t}{\partial t} + \nabla \cdot [(E_t + p_t)\mathbf{v} - \mathbf{B}(\mathbf{v} \cdot \mathbf{B})] = -\nabla \cdot (\eta \mathbf{j} \times \mathbf{B}) - \rho \mathbf{v} \cdot \nabla \phi_{PN} - cS_0, \tag{2.3}$$

$$\frac{\partial \mathbf{B}}{\partial t} + \nabla \cdot (\mathbf{v}\mathbf{B} - \mathbf{B}\mathbf{v} + \psi \mathbf{I}) = -\nabla \times (\eta \mathbf{j}), \tag{2.4}$$

$$\frac{\partial \psi}{\partial t} + c_h^2 \nabla \cdot \mathbf{B} = -\frac{c_h^2}{c_p^2} \psi, \tag{2.5}$$

where ρ, $\mathbf{v}$, $\mathbf{B}$, and $\mathbf{j} = \nabla \times \mathbf{B}$ are the mass density, velocity, magnetic field, and current density, respectively. In addition, $p_t = p_{gas} + B^2/2$ is the total pressure, and $E_t = \rho v^2/2 + p_{gas}/(\gamma - 1) + B^2/2$ is the total energy, where $\gamma = 5/3$ is the specific heat ratio. We apply the so-called anomalous resistivity. The resistivity η becomes large when the drift velocity v_d exceeds the critical velocity v_c (Yokoyama & Shibata 1994). In equations (2.4) and (2.5), ψ is introduced so that the divergence-free magnetic field is maintained within minimal errors during time integration where c_h and c_p are constants. These equations are solved by the MHD code CANS+ (Matsumoto et al. 2019).

In equations (2.2) and (2.3), $\mathbf{S}$ and S_0 are the radiation momentum and the radiation energy source terms, respectively, and are derived by solving the frequency-integrated

0th and 1st moments of the radiation transfer equation expressed in the following form.

$$\frac{\partial E_r}{\partial t} + \nabla \cdot \mathbf{F}_r$$

$$= \rho \kappa_{ff} c (a_r T^4 - E_r) + \rho (\kappa_{ff} - \kappa_{es}) \frac{\mathbf{v}}{c} \cdot [\mathbf{F}_r - (\mathbf{v} E_r + \mathbf{v} \cdot \mathbf{P}_r)] + \rho \kappa_{es} c E_{r0} \frac{k_B (T_e - T_r)}{m_e c^2} = c S_0,$$

$$\frac{1}{c^2} \frac{\partial \mathbf{F}_r}{\partial t} + \nabla \cdot \mathbf{P}_r = \rho \kappa_{ff} \frac{\mathbf{v}}{c} (a_r T^4 - E_r) - \rho (\kappa_{ff} + \kappa_{es}) \frac{1}{c} [\mathbf{F}_r - (\mathbf{v} E_r + \mathbf{v} \cdot \mathbf{P}_r)] = \mathbf{S}. \qquad (2.6)$$

The M1-closure is applied to relate the radiation stress tensor $\mathbf{P}_r$ and radiation energy density E_r (see e.g., Lowrie et al. 1999). In Equations (2.6), $\kappa_{ff} = 1.7 \times 10^{-25} m_p^{-2} \rho T^{-7/2}$ cm^2/g is the free-free absorption opacity where m_p is the proton mass and $\kappa_{es} = 0.4$ cm^2/g is the electron scattering opacity. The gas temperature, T, is related to the gas pressure and density by $p_{gas} = \rho k_B T/(\mu m_p)$, where k_B, and $\mu = 0.5$ are the Boltzmann constant and mean molecular weight, respectively. $T_e = \min(T, 10^9 \text{ K})$ and $T_r = (E_r/a_r)^{1/4}$ are the electron temperature and radiation temperature, respectively. For the initial conditions (Figure 1(a)), we set the rotational equilibrium torus embedding weak poloidal magnetic field (Kato et al. 2004). Initially, we calculate without radiation term until a quasi-steady state is attained. After that, we include radiation terms from $t = 1.1 \times 10^4 t_0$. We normalized density such that the accretion rate when the cooling term is included is close to the Eddington accretion rate. The over-dense region simulates the dense blob in the accretion flow (see Igarashi et al. 2020). The computational domain of our simulation is $0 \leq r < 2000 r_s$, $0 \leq \varphi < 2\pi$, and $|z| < 2000 r_s$; and the number of grid points is $(n_r, n_\varphi, n_z) = (464, 32, 464)$. Grid spacing is $0.1 r_s$ in the radial and vertical directions when $r < 20 r_s$ and $|z| < 5 r_s$ and increases outside the region. The absorbing boundary condition is imposed at $R = 2 r_s$, and the outer boundaries are free boundaries where waves can be transmitted.

3. Numerical Results of RMHD Simulations

The upper panel of Figure 1(b) shows the time-averaged distribution of azimuthally averaged density (top), temperature (middle), and the ratio of the radiation pressure to gas pressure + magnetic pressure in the poloidal plane after radiative cooling is switched on. Numerical results indicate that radiation pressure dominant torus is formed, oscillating in the vertical direction. At the early stage (left panels), the disk inflates due to radiation pressure because the radiation is trapped in the Thomson thick disk. The Thomson optical depth computed by

$$\tau_{es} = \int_{\pm 100 r_s}^{0} \rho \kappa_{es} dz. \qquad (3.1)$$

exceeds 10, so that radiation energy density quickly increases, and radiation pressure drives expansion of the disk both in vertical and radial direction. The gas temperature of the disk quickly decreases from 10^{10} K to 10^6 K in the radiation pressure supported torus. In the inner region $r < 15 r_s$, the gas temperature above or below the disk is around 10^8 K, and the Thomson optical depth exceeds 10 in the disk. However, the disk is optically thin for the effective optical depth ($\tau_{eff} = \sqrt{\tau_{es}(\tau_{es} + \tau_{ff})} < 0.01$). Thus we expect soft X-ray emission by thermal bremsstrahlung radiation and inverse Compton scattering from this region. The appearance of such a Thomson thick, warm Compton region is discussed by Kumar and Yuan (2021) in their search for steady transonic solutions for sub-Eddington accretion flows.

Rapid mass accretion takes place from the dense torus to the black hole. As the density of the disk decreases by expansion, Thomson optical depth decreases. This enables radiation to escape from the disk and enhances the radiative cooling rate. When the

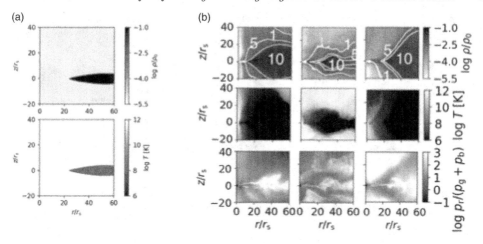

Figure 1. (a) Initial distribution of the gas density and gas temperature. (b) Azimuthally averaged gas density (top), gas temperature (middle), and the ratio of radiation pressure to gas pressure + magnetic pressure (bottom). The quantities are averaged over the time range $1.1 \times 10^4 t_0 < t < 1.175 \times 10^4 t_0$, $1.445 \times 10^4 t_0 < t < 1.475 \times 10^4 t_0$, and $1.675 \times 10^4 t_0 < t < 1.745 \times 10^4 t_0$, respectively from left to right. Contours in the upper panel show $\tau_{es} = 10, 5$, and 1. We include the radiation term from $1.1 \times 10^4 t_0$.

radiative cooling rate exceeds the heating rate of the disk, the disk shrinks in the vertical direction. Middle panels of figure 1(b) show that geometrically thin disk is formed in $r < 20 r_s$. The radiation pressure is comparable to the gas+magnetic pressure in this region. Subsequently, as the mass accumulates in the disk by accretion in $r > 20 r_s$, the radiation pressure of the disk becomes dominant.

Figure 2 shows the space-time plot of the radial velocity (left), surface density (middle), and radiation temperature (right), respectively. The surface density Σ is calculated by

$$\Sigma = \int_{z1}^{z2} \rho dz, \qquad (3.2)$$

where $z1$ and $z2$ are the negative and positive z where the Thomson optical depth is $\tau_{es} = 1$. The left panel of figure 2 shows that the sign of the radial velocity outside $40 r_s$ reverses on a time scale of $\sim 3000 t_0$. The middle panel of figure 2 shows that dense torus is formed in $20 r_s < r < 60 r_s$ around $t = 1.1 \times 10^4 t_0$ and $t = 1.6 \times 10^4 t_0$. During the period $1.2 \times 10^4 t_0 < t < 1.5 \times 10^4 t_0$, the surface density of the disk decreases due to rapid mass accretion of the disk. In the region $r < 20 r_s$, low-density region is formed. In this region, radial oscillation of the disk is excited. The typical period of this oscillation is around $200 t_0$, which corresponds to several hours when the black hole mass is $10^7 M_\odot$. The oscillation frequency is around 50 Hz when the black hole mass is $10 M_\odot$.

4. Thermal Limit Cycle Oscillation

When the accretion rate onto a supermassive black hole is close to the Eddington accretion rate, the outer region around $r = 40 r_s$ oscillates between a geometrically thick disk and a radiatively efficient cool disk. This oscillation is similar to the thermal limit cycle oscillations in luminous accretion disks around a stellar-mass black hole. However, the normalized oscillation period ($\sim 3000 t_0$) is much shorter than the oscillations in a microquasar GRS 1915+105, whose period is $10^5 - 10^6 t_0$.

Figure 3 (a) shows the trajectory of the state transition of the disk in our RMHD simulation at $r = 40 r_s$. The horizontal axis shows surface density, and the vertical axis shows radiation pressure vertically integrated with the region where $\tau_{es} > 1$. The filled

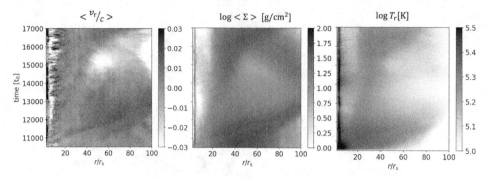

Figure 2. The space time diagram of the azimuthally averaged radial velocity, surface density, and radiation temperature, respectively.

circles plot the surface density and radiation pressure, and the greyscale shows time. As the radiation pressure increases, the accretion flow transits to a geometrically thick disk inflated by radiation pressure and stays in the state in $1.1 \times 10^4 t_0 < t < 1.3 \times 10^4 t_0$. During this stage, the surface density of the disk decreases by rapid accretion. Therefore, the radiation becomes easier to escape, and the radiative cooling rate increases. When the surface density becomes smaller than a threshold, the radiation pressure quickly decreases ($1.3 \times 10^4 t_0 < t < 1.4 \times 10^4 t_0$). Subsequently, surface density increases by mass accretion, and the next cycle starts around $t = 1.6 \times 10^4 t_0$.

Figure 3 (b) and (c) show the scatter plot of surface density and radiation flux F_z at $\tau_{es} = 1$, and the heating rate evaluated by $Q_{heat} = 3\dot{M}(1 - \sqrt{3r_s/r})/(8\pi r^3)$, respectively. Heating balances with radiative cooling during the cool stage when the surface density increases by accretion ($1.3 \times 10^4 t_0 < t < 1.5 \times 10^4 t_0$). As the radiation pressure increases, heating exceeds cooling around $t = 1.5 \times 10^4 t_0$, and the disk expands again by radiation pressure.

The grey dots in figure 3(d) show the trajectory of the limit-cycle oscillation shown in the left panel. The vertical axis is the total pressure. Solid curves show the thermal equilibrium curves. The limit cycle oscillation is different from the oscillation between a gas pressure dominant disk and slim disk studied by Honma et al. (1991). The dotted curve shows the state where the radiation pressure equals the gas pressure. The dashdotted curve shows $\tau_{eff} = 1$, and the dashed curve shows $p_{rad} = p_{mag} + p_{gas}$. Dark dots show the trajectory of the lower accretion rate model reported in Igarashi et al. (2020).

Numerical results shown in figure 2 and 3 indicate that during the transition from a geometrically thick disk inflated by radiation pressure to a radiatively efficient cool disk, the disk stays in a radiation pressure dominant state. Therefore, a slight increase in surface density by mass accretion enables photons to be trapped again and enhances the radiation pressure. Thus, the time scale for the transition from the radiatively efficient cool state to the geometrically thick state becomes shorter than the viscous time scale. Azimuthal magnetic field enhanced by the vertical contraction of the disk prevents the disk collapsing to a gas pressure dominant, cool disk (see middle panel of figure 1 and the dashed curve in the figure 3(d)).

5. Summary

Results of three-dimensional global RMHD simulations of accretion flows when the accretion rate is comparable to the Eddington accretion rate are presented. Our numerical results indicate that limit cycle oscillations can be excited in Thomson thick, optically thin for effective optical depth but radiation pressure dominant disk. This mechanism

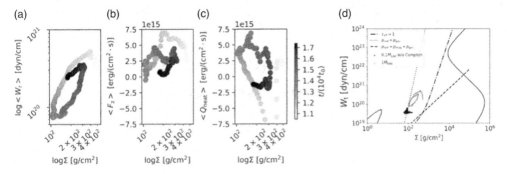

Figure 3. The scatter plot of the surface density and (a) vertically integrated radiation pressure, (b) radiative flux in z−direction at $\tau_{es} = 1$, and (c) heating rate $Q_{heat} = 3\dot{M}(1 - \sqrt{3r_s/r})/(8\pi r^3)$ at $r = 40r_s$. The radiation pressure is integrated with the region where $\tau_{es} > 1$. The greyscale of each point shows time. (d) The scatter plot of the surface density and vertically integrated total pressure at $r = 40r_s$. The total pressure is integrated in the region where $\tau_{es} > 1$. The dark dots show the results of Igarashi et al. (2020), and the grey dots show the results presented in this paper. The solid black curves show the thermal equilibrium curve. The dotted, dashed, and dash-dotted lines show the state where the gas pressure equals radiation pressure, magnetic pressure equals gas + radiation pressure, and effective optical depth $\tau_{eff} = 1$, respectively.

excites shorter time scale limit-cycle oscillation than those between gas pressure dominant disk and radiation pressure dominant slim disk. The appearance of this limit cycle oscillation is not clear in RMHD simulations by Jiang et al. (2019) probably because they started their simulation from a state nearly in thermal equilibrium.

The time scale of the oscillation in the outer region is $\sim 3000t_0$. Since the unit time $t_0 = 10$ sec when the black hole mass is $10^6 M_\odot$ the period of the thermal oscillation is 3×10^4 s ~ 8 hours. This time scale is close to the luminosity variations observed in NLS1 RE J1034+396 (Chaudhury et al. 2018). Numerical results indicate that higher frequency disk oscillation appears in the inner disk at $r < 20r_s$. The time scale of the oscillation is typically $\sim 200t_0 \sim 2000$ s ~ 0.5 hours. The oscillation frequency is close to the QPO observed in RE J1034+396.

Numerical simulations are carried out using XC50 at the Center for Computational Astrophysics, National Astronomical Observatory, Japan. This work is supported by JSPS KAKENHI 20H01941 (PI: R.M.)

References

Belloni, T. M., Bhattacharya, D., Caccese, P., et al. 2019, *MNRAS*, 489, 1037

Biang, W., & Zhao, Y. 2004, *MNRAS*, 352, 823

Chaudhury, K., Chitnis, V. R., Rao, A. R., et al. 2018, *MNRAS*, 478, 4830

Done, C., Davis, S. W., Jin, C., Blaes, O., & Ward, M. 2012, *MNRAS*, 420, 1848

Honma, F., Matsumoto, R., & Kato, S. 1991, *PASJ*, 43, 147

Igarashi, T., Kato, Y., Takahashi, H. R., et al. 2020, *ApJ*, 902, 103

Jiang, Y.-F., Stone, J. M., & Davis, S. W. 2019, *ApJ*, 880, 67

Kato, Y., Mineshige, S., & Shibata, K. 2004, *ApJ*, 605, 307

Kumar, R. & Yuan, Y. 2021, *ApJ*, 910, 9

Lowrie, R. B., Morel, J. E., & Hittinger, J. A. 1999, *ApJ*, 521, 423

Matsumoto, Y., Asahina, Y., Kudoh, Y., et al. 2019, *PASJ*, 71, 83

Morgan, E. H., Remillard, R. A., & Greiner, J. 1997, *ApJ*, 482, 993

Noda, H., & Done, C. 2018, *MNRAS*, 480, 3898

Ohsuga, K. 2006, *ApJ*, 640, 923

Takahashi, H. R., Ohsuga, K., Kawashima, T., & Sekiguchi, Y. 2016, *ApJ*, 826, 23

Yokoyama, T., & Shibata, K. 1994, *ApJ*, 436, L197

The Predictive Power of Computational Astrophysics as a Discovery Tool
Proceedings IAU Symposium No. 362, 2023
D. Bisikalo, D. Wiebe & C. Boily, eds.
doi:10.1017/S1743921322001181

Cold Gas in Outflow: Evidence for Delayed Positive AGN Feedback

Yu Qiu

Kavli Institute for Astronomy and Astrophysics, Peking University,
5 Yiheyuan Road, Haidian District, Beijing 100871, China

Abstract. Multiphase outflows driven by active galactic nuclei (AGN) have a profound impact on the evolution of their host galaxies. The effects of AGN feedback are especially prominent in the brightest cluster galaxies (BCGs) of cool-core clusters, where there is a concentration of gas in all phases, ranging from cold molecular gas to hot, $> 10^7$ K ionized plasma. In this proceeding I describe recent simulation efforts to understand the formation and evolution of the 10-kpc-scale Hα-emitting filaments driven by AGN activities. Combined with observed star formation regions co-spatial with the filaments, this feedback mechanism can directly contribute to the growth of the central galaxy, albeit delayed by the characteristic radiative cooling timescale, ~ 10 Myr, of the outflowing plasma.

Keywords. ISM: jets and outflows, galaxies: active, cooling flows, methods: numerical

1. Introduction

The rich concentration of gas in the brightest cluster galaxies (BCGs) of cool-core clusters provides some of the best environments for observing and understanding the impact of AGN feedback. Due to the radiative cooling of the hot intracluster medium (ICM), a cooling flow of $\sim 10 - 100\, M_\odot\, \mathrm{yr}^{-1}$ is expected to continuously feed the growth of the BCG (Fabian 1994). In reality, the star formation rates of these galaxies are around 10% or less of those inferred from cooling (McDonald et al. 2018). X-ray cavities filled with radio emission indicate that AGN jets can inflate bubbles and do mechanical work to heat the intracluster medium (Bîrzan et al. 2004; McNamara and Nulsen 2007). On the other hand, cold gas often resides in elongated filaments radially connected to the central supermassive black hole (SMBH; Conselice et al. 2001; McDonald et al. 2012; Gendron-Marsolais et al. 2018, see also Fig. 1), suggesting that the AGN is also responsible for the formation and distribution of the Hα-emitting filaments.

In particular, in the nearby Perseus cluster, two irregular filament structures, a.k.a the horseshoe filament and the blue loop, are located about 20 kpc on opposite sides of the central SMBH, indicating that they were driven by the same episode of AGN activity, such as jets or winds. Unlike other radial filaments, however, these two structures have azimuthal components that are not readily explained by outflows driven from the central AGN. Meanwhile, some cold gas in these filaments has recently collapsed and formed stars, with typical ages less than 10 Myr (Canning et al. 2014). If the filaments indeed were driven by AGN activities, their formation constitutes a mechanism for AGNs to delay and offset the central star formation fueled by the cooling flow.

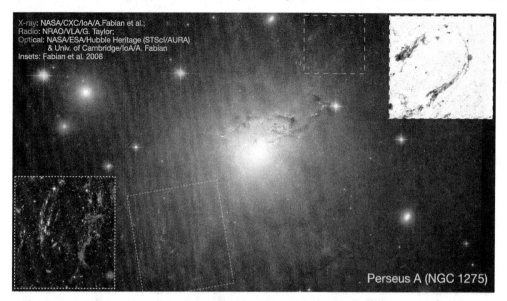

Figure 1. A composite image showing the various observational signatures of AGN feedback in the center of the cool-core Perseus cluster. Two loop-like cold filament structures are located on opposite sides of the central AGN (insets; Fabian et al. 2008), indicating that they were driven by the same bipolar AGN outburst. The star formation regions co-spatial with the Hα filaments (e.g., the "blue loop" in the bottom left inset) also indicate that AGN feedback may contribute positively to the galaxy growth, offset from the nucleus.

2. Simulations of AGN Feedback in Galaxy Clusters

In order to study the impact of AGN feedback on the evolution of galaxy clusters, as well as to understand the origin of the cold filaments, in both radial and azimuthal formations, we performed a series of simulations using Enzo (Bryan et al. 2014) + Moray (Wise and Abel 2011). In the large-scale ($\sim$ Mpc), long-duration (~ 10 Gyr) radiation-hydrodynamical simulations, we found that the AGN drives multiphase outflows that are responsible for both the X-ray cavities and the extended cold filaments in BCGs (Qiu et al. 2019b). Positive correlations can be drawn between AGN luminosity and properties of the filaments, such as filament mass, spatial extent, as well as Hα luminosity (Qiu et al. 2019a). The velocity dispersion of the filaments, however, depends primarily on the turbulent interactions between the ICM and the outflows, and are therefore not strongly correlated with the AGN luminosity (e.g., the standard deviation of cold gas velocity in Fig. 3, $\sim 100 - 200$ km s^{-1}, similar to the observed filament velocity dispersion values in the Perseus cluster; Gendron-Marsolais et al. 2018).

The similarities between the simulated filamentary nebula and those in observations of the Perseus cluster motivated us to examine more closely the formation mechanism of the cold gas in a follow-up study (Qiu et al. 2020). A strong constraint from observations of the filaments is that the line-of-sight velocities are mostly below a few hundred km s^{-1}. Apart from the projection effect, if the cold gas were directly ejected from the nucleus, it requires both > 1000 km s^{-1} speeds and unrealistically strong shielding from the shock heating and the hot ambient ICM to reach beyond 10 kpc. Our analysis however indicates that the filaments form out of ionized AGN-driven outflows launched from the central 1 kpc†, with temperature $\lesssim 10^7$ K and characteristic cooling time below 10 Myr. Although

---

† The origin of the filament gas is confirmed using a tracer fluid injected in the central 1 kpc of the simulated cluster. From this region, the outflowing gas may include a mixture of injected AGN wind plasma, shocked and entrained circumnuclear gas, as well as the ICM. Higher-resolution simulations focused on this central region are required to further dissect and understand these components.

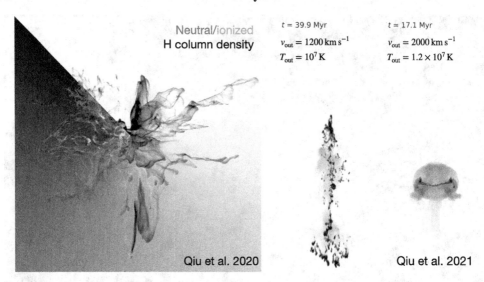

Figure 2. Left: A composite image showing the ionized and neutral hydrogen column densities, as well as the synthetic Hα emissivity, in the simulated galaxy cluster undergoing an AGN outburst (Qiu et al. 2020). Filaments of Hα-emitting gas extends ∼ 30 kpc from the central AGN. Right: Snapshots from a high-resolution follow up simulation study focusing on the individual filament morphology as a function of initial outflow properties (velocity v_{out} and temperature T_{out}; Qiu et al. 2021a). Both longitudinal and transverse filaments may form with respect to the outflow direction (up).

the initial speeds are high, the ram pressure deceleration from the ICM significantly reduces the outflow velocity before cold gas forms about 10 kpc away from the nucleus (Figs. 2, 3).

While the simulated filaments are visually similar to the observed filaments, the resolution required to study the filaments is much higher than those employed in cluster-scale simulations (∼ 0.1 kpc, which is similar to or larger than the width of individual filaments). In order to characterize the filament evolution, as well as to resolve the inter-actions between the filament and the ambient plasma, we use high-resolution simulations (with the smallest resolution size of ∼ 30 pc) to further study the dynamical and mor-phological evolution of the cold gas in radiatively cooling AGN-driven outflows embed in the ICM (Qiu et al. 2021a). Using the parameter space found in Qiu et al. (2020) and the line-of-sight velocity constraints from observations, we vary the initial velocity and temperature of the outflowing plasma to examine the structure of the emergent cold gas. The evolution of the center of mass can be reasonably well described by a 1D model comprising radiative cooling and ICM ram pressure (Fig. 3). In addition to longitudinal filaments parallel to the outflow direction, this study uncovered a mechanism for trans-verse filaments perpendicular to the direction of motion to form – when the outflowing plasma is marginally thermally stable between radiative cooling and heating through mixing with the ambient ICM (Fig. 2). The ring of cold gas in the right panel of Fig. 2 can potentially explain the origin of the loop-like filaments and star formation region in the Perseus cluster.

Besides the positive AGN feedback mechanism revealed by the irregularly shaped fil-aments, the location and mass of the filaments can be used to constrain the energetics of past AGN outbursts. Observationally, large amounts of molecular gas have been dis-covered co-spatial with the Hα filaments. Using cold gas mass measurements (∼ 10^8 $M_\odot$;

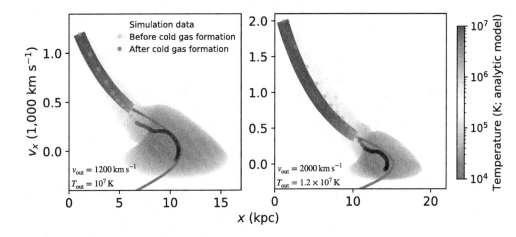

Figure 3. Evolution of the center of mass velocity as a function of distance traveled for the simulated outflows in Qiu et al. (2021a), compared with the 1D model in Qiu et al. (2020). The size of the ellipses centered on the data points represents the standard deviation ($\pm\sigma$) of the simulated cold gas properties. The initial velocity and temperature of the outflows ($v_{\rm out}$, $T_{\rm out}$) are shown in each panel, corresponding to the two cases in Fig. 2. Cold gas forms $\sim 10\,{\rm Myr}$ after the outflows are launched, consistent with the radiative cooling timescale of the $\sim 10^7\,{\rm K}$ plasma.

Salomé et al. 2008, 2011) and the initial outflow velocity inferred from the simulations, the AGN kinetic output can be estimated as

$$E_{\rm kin} \approx 10^{57} \frac{M_{\rm out}}{10^8\,M_\odot} \left(\frac{v_{\rm out}}{1000\,{\rm km\,s^{-1}}} \right)^2 {\rm erg}. \qquad (2.1)$$

Combined with the cooling timescale of the original outflow ($\sim 10\,{\rm Myr}$), and the dynamical timescale for different generations of filaments to evolve to their observed morphological states ($\sim 10 - 100\,{\rm Myr}$), these simulations can also help to reconstruct the history and constrain the duty cycle of the AGN activity in the past few hundred million years.

3. Summary

Using a series of simulations centered on the BCGs of cool-core clusters, we study the interplay between AGN feedback and the multiphase gas in the ICM. Besides the hot outflows which maintain the thermal balance of the plasma and prevent it from collapsing and fueling a central starburst (see Qiu et al. 2021b, for a study on the heating of the ICM delivered by hot outflows), in this proceeding I focus mainly on the formation and evolution of the cold gas driven by AGN activities. Our simulations indicate that the component of the outflow with characteristic cooling timescale below 10 Myr is responsible for the formation of the extended filamentary nebulae. Depending sensitively on the initial properties, both longitudinal and transverse filaments may form in the outflows. Evidently in the nearby Perseus cluster, some of the cold gas in these filaments has collapsed and formed stars in the recent few Myr. The feedback mechanisms uncovered in these works therefore demonstrate that AGN activities in gas rich environments, such as in the BCGs of galaxy clusters, can tightly control both the thermal balance of the hot gas reservoir and the scattered star formation in and around the host galaxies.

Acknowledgements

Y.Q. acknowledges support from the National Natural Science Foundation of China (12003003, 12073003) and the China Postdoctoral Science Foundation (2020T130019).

References

Bîrzan, L., Rafferty, D. A., McNamara, B. R., Wise, M. W., & Nulsen, P. E. J. 2004, A Systematic Study of Radio-Induced X-ray Cavities in Clusters, Groups, and Galaxies. *The Astrophysical Journal*, 607(2), 800–809.

Bryan, G. L., Norman, M. L., O'Shea, B. W., Abel, T., Wise, J. H., Turk, M. J., Reynolds, D. R., Collins, D. C., Wang, P., Skillman, S. W., Smith, B., Harkness, R. P., Bordner, J., Kim, J.-h., Kuhlen, M., Xu, H., Goldbaum, N., Hummels, C., Kritsuk, A. G., Tasker, E., Skory, S., Simpson, C. M., Hahn, O., Oishi, J. S., So, G. C., Zhao, F., Cen, R., & Li, Y. 2014, Enzo: An Adaptive Mesh Refinement Code for Astrophysics. *The Astrophysical Journal Supplement Series*, 211(2), 19.

Canning, R. E., Ryon, J. E., Gallagher, I. S., Kotulla, R., O'Connell, R. W., Fabian, A. C., Johnstone, R. M., Conselice, C. J., Hicks, A., Rosario, D., & Wyse, R. F. 2014, Filamentary Star Formation in NGC 1275. *Monthly Notices of the Royal Astronomical Society*, 444(1), 336–349.

Conselice, C. J., Gallagher, J. S., & Wyse, R. F. G. 2001, On the Nature of the NGC 1275 System. *The Astronomical Journal*, 122(5), 2281–2300.

Fabian, A. C. 1994, Cooling Flows in Clusters of Galaxies. *Annual Review of Astronomy and Astrophysics*, 32(1), 277–318.

Fabian, A. C., Johnstone, R. M., Sanders, J. S., Conselice, C. J., Crawford, C. S., Gallagher, J. S., & Zweibel, E. 2008, Magnetic Support of the Optical Emission Line Filaments in NGC 1275. *Nature*, 454(7207), 968–970.

Gendron-Marsolais, M., Hlavacek-Larrondo, J., Martin, T. B., Drissen, L., McDonald, M., Fabian, A. C., Edge, A. C., Hamer, S. L., McNamara, B., & Morrison, G. 2018, Revealing the Velocity Structure of the Filamentary Nebula in NGC 1275 in its Entirety. *Monthly Notices of the Royal Astronomical Society: Letters*, 479(1), L28–L33.

McDonald, M., Gaspari, M., McNamara, B. R., & Tremblay, G. R. 2018, Revisiting the Cooling Flow Problem in Galaxies, Groups, and Clusters of Galaxies. *The Astrophysical Journal*, 858(1), 45.

McDonald, M., Veilleux, S., & Rupke, D. S. N. 2012, Optical Spectroscopy of $H\alpha$ Filaments in Cool Core Clusters: Kinematics, Reddening, and Sources of Ionization. *The Astrophysical Journal*, 746(2), 153.

McNamara, B. & Nulsen, P. 2007, Heating Hot Atmospheres with Active Galactic Nuclei. *Annual Review of Astronomy and Astrophysics*, 45(1), 117–175.

Qiu, Y., Bogdanović, T., Li, Y., & McDonald, M. 2019,a Using $H\alpha$ Filaments to Probe Active Galactic Nuclei Feedback in Galaxy Clusters. *The Astrophysical Journal Letters*, 872a(1), L11.

Qiu, Y., Bogdanović, T., Li, Y., McDonald, M., & McNamara, B. R. 2020, The Formation of Dusty Cold Gas Filaments from Galaxy Cluster Simulations. *Nature Astronomy*,.

Qiu, Y., Bogdanović, T., Li, Y., Park, K., & Wise, J. H. 2019,b The Interplay of Kinetic and Radiative Feedback in Galaxy Clusters. *The Astrophysical Journal*, 877b(1), 47.

Qiu, Y., Hu, H., Inayoshi, K., Ho, L. C., Bogdanović, T., & McNamara, B. R. 2021,a Dynamics and Morphology of Cold Gas in Fast, Radiatively Cooling Outflows: Constraining AGN Energetics with Horseshoes. *The Astrophysical Journal Letters*, 917a(1), L7.

Qiu, Y., McNamara, B. R., Bogdanović, T., Inayoshi, K., & Ho, L. C. 2021,b On the Mass Loading of AGN-driven Outflows in Elliptical Galaxies and Clusters. *The Astrophysical Journal*, 923b(2), 256.

Salomé, P., Combes, F., Revaz, Y., Downes, D., Edge, A. C., & Fabian, A. C. 2011, A Very Extended Molecular Web around NGC 1275. *Astronomy & Astrophysics*, 531(5), A85.

Salomé, P., Combes, F., Revaz, Y., Edge, A. C., Hatch, N. A., Fabian, A. C., & Johnstone, R. M. 2008, Cold gas in the Perseus cluster core: excitation of molecular gas in filaments. *Astronomy & Astrophysics*, 484(2), 317–325.

Wise, J. H. & Abel, T. 2011, Enzo+Moray: Radiation Hydrodynamics Adaptive Mesh Refinement Simulations with Adaptive Ray Tracing. *Monthly Notices of the Royal Astronomical Society*, 414(4), 3458–3491.

The Predictive Power of Computational Astrophysics as a Discovery Tool
Proceedings IAU Symposium No. 362, 2023
D. Bisikalo, D. Wiebe & C. Boily, eds.
doi:10.1017/S1743921322001314

A New Code for Relativistic Hydrodynamics and its Application to FR II Radio Jets

Jeongbhin Seo[1] , Hyesung Kang[1] and Dongsu Ryu[2]

[1]Department of Earth Sciences, Pusan National University, Busan 46241, Korea

[2]Department of Physics, UNIST, Ulsan, 44919, Korea

Abstract. To study the dynamics of relativistic flows in astrophysical objects such as radio jets, we have developed a new special relativistic hydrodynamic (RHD) code based on the weighted essentially non-oscillatory (WENO) scheme, a high-order finite difference scheme. The code includes different WENO versions, and high-order time integration methods such as the 4th-order accurate Runge-Kutta (RK4) and strong stability preserving RK (SSPRK), as well as the equations of state (EOSs) that closely approximate the EOS of the single-component perfect gas in relativistic regime. Additionally, it is optimized for the reproduction of complex structures in multi-dimensional flows, and implements a modification of eigenvalues for the acoustic modes to effectively control carbuncle instability. As the first application of the code, we have simulated ultra-relativistic jets of FR II radio galaxies, and studied the nonlinear flow structures, such as shocks, velocity shear, and turbulence, through large-scale.

Keywords. hydrodynamics, galaxies: jets, methods: numerical, relativistic processes

---

1. Introduction

Relativistic jets are widely involved in high-energy astrophysical phenomena, such as, pulsar wind nebulae (PWNs), gamma-ray bursts (GRBs), and radio-loud active galactic nuclei (AGNs) (see Hardcastle & Croston 2020, for reviews). Many studies of relativistic jets have been done through relativistic hydrodynamics (RHD) simulations (English et al. 2016; Li et al. 2018; Matthews et al. 2019). Those studies adopted various codes that were developed from non-relativistic Newtonian hydrodynamics codes (see Martí & Müller 2003, 2015, for reviews). A popular numerical scheme for these codes is the weighted essentially non-oscillatory (WENO) scheme based on the upwind method, which is designed to achieve a high-order accuracy in smooth flows and avoid spurious oscillations near discontinuities, e.g., shocks and contact discontinuities. The first WENO scheme was introduced by Liu et al. (1994), which is a finite volume (FV) scheme. Later, a 5th-order accurate, finite difference (FD) WENO scheme was introduced by Jiang & Shu (1996), where the fluxes at the cell interfaces are reconstructed using the point values of the physical fluxes with weight functions. After that, different FD WENO versions have been proposed, achieving higher accuracies for smooth flows and/or smaller dissipation near discontinuities. We have developed a new code based on the WENO scheme, including three versions of WENO, WENO-JS (Jiang & Shu 1996), WENO-Z (Borges et al. 2008), and WENO-ZA (Liu et al. 2018).

The WENO scheme has been combined with high-order time integration methods, such as the classical Runge-Kutta (RK) method (e.g., Shu & Osher 1988, 1989; Jiang & Shu 1996). To enhance the nonlinear stability and reduce spurious oscillations near discontinuous structures, Spiteri & Ruuth (2002) introduced an improved RK, called the strong-stability-preserving Runge-Kutta (SSPRK) method. Our code includes the 4th-order accurate RK and SSPRK time integration methods.

In terms of thermodynamics, fluids that have thermal speeds approaching the speed of light manifest relativistic effects. In our code, we consider the equation of state (EOS) for "single-component" fluids. The EOS of the single-component, perfect gas, called RP (relativistic perfect) here, contains modified Bessel functions, which are difficult to be efficiently implemented in numerical codes (e.g., Synge et al. 1957). Hence, RHD simulations have been performed typically with simplified EOSs. The most popular EOS is the ID (ideal) EOS, which assumes a constant adiabatic index γ. However, the ID EOS cannot reproduce the transition from subrelativistic temperature of $\Theta \equiv p/\rho c^2 < 1$ to fully relativistic temperature of $\Theta > 1$. Here, Θ, p, and ρ are a temperature-like variable, the isotropic gas pressure, and the rest mass density, respectively. On the other hand, EOSs that closely approximate RP have been introduced, including TM (after Taub-Mathews, see Taub 1948; Mathews 1971; Mignone et al. 2006) and RC (after Ryu-Chattopadhyay, see Ryu et al. 2006). Our code contains these EOSs, as well as the ID EOS.

For multi-dimensional problems, the high-order accuracy in FD schemes can be preserved with the dimension-by-dimension method. Recently, Buchmüller & Helzel (2014) and Buchmüller et al. (2016) proposed a modified dimension-by-dimension method that implements the averaging of the state and flux vectors along the transverse direction, and demonstrated higher accurate reproduction of complex, nonlinear multi-dimensional structures. We adopt this method in our code.

It is know that the Carbuncle instability, which deforms grid-aligned, slow-moving shocks, is easily generated in high-accuracy, shock-capturing, upwind codes (e.g., Peery et al. 2018; Dumbser et al. 2008). This instability appears due to the insufficient numerical dissipation at such shocks. Fleischmann et al. (2020) introduced a method which efficiently mitigates the carbuncle instability by modifying eigenvalues for the acoustic modes. Our code includes this fix, as well.

Using our RHD code, we have simulated ultra-relativistic jets injected into the intracluster medium, with the parameters relevant to FR-II radio galaxies. In this paper, we briefly describe the code and also the flow structures in a FR II jet, such as, shocks, shear, and turbulence. The complete description of the code and jet simulations can be found in Seo et al. (2021a) and Seo et al. (2021b).

2. Description of the Code

2.1. RHD Equations

The code solves the conservation equations of special relativistic hydrodynamics, which are expressed in the laboratory frame as,

$$\frac{\partial D}{\partial t} + \vec{\nabla} \cdot (D\vec{v}) = 0, \tag{2.1}$$

$$\frac{\partial \vec{M}}{\partial t} + \vec{\nabla} \cdot (\vec{M}\vec{v} + p) = 0, \tag{2.2}$$

$$\frac{\partial E}{\partial t} + \vec{\nabla} \cdot [(E + p)\vec{v}] = 0. \tag{2.3}$$

Here, the conserved quantities, the mass, momentum, and total energy densities, are given as $D = \Gamma\rho$, $\vec{M} = \Gamma^2\rho(h/c^2)\vec{v}$, and $E = \Gamma^2\rho h - p$, respectively (e.g Landau & Lifshitz 1959), where c, ρ, $\vec{v}$, p, Γ, and h are the speed of light, the rest mass density, the fluid three-velocity, the isotropic gas pressure, the Lorentz factor, and the specific enthalpy.

2.2. Equation of State

The EOS of the single-component, perfect gas, RP, is given as

$$h(p, \rho) = \frac{K_3(1/\Theta)}{K_2(1/\Theta)}, \tag{2.4}$$

where K_α is the modified Bessel function of the second kind of order α (e.g., Synge et al. 1957). As h contains Bessel functions, the inversion of the conserved quantities to get the primitive variables, ρ, $\vec{v}$, and p, is computationally expensive. Here, we list three approximate EOSs, ID, TM, and RC, implemented in the code, respectively:

$$h = 1 + \frac{\gamma\Theta}{\gamma - 1}, \quad h = \frac{5}{2}\Theta + \frac{3}{2}\sqrt{\Theta^2 + \frac{4}{9}}, \quad h = 2\frac{6\Theta^2 + 4\Theta + 1}{3\Theta + 2}, \tag{2.5}$$

where γ is the adiabatic index. As mentioned in the introduction, ID cannot treat the transition from subrelativistic to fully relativistic. According to Ryu et al. (2006), RC reproduces RP slightly better than TM, so we have adopted RC in our jet simulations.

2.3. Spatial Integration

By using the dimension-by-dimension method in Cartesian geometry, Equations (2.1) - (2.3) are solved numerically, as follows,

$$
\begin{aligned}
q'_{i,j,k} = q_{i,j,k} &- \frac{\Delta t}{\Delta x}\left(F_{i+\frac{1}{2},j,k} - F_{i-\frac{1}{2},j,k}\right) \\
&- \frac{\Delta t}{\Delta y}\left(G_{i,j+\frac{1}{2},k} - G_{i,j-\frac{1}{2},k}\right) - \frac{\Delta t}{\Delta z}\left(H_{i,j,k+\frac{1}{2}} - H_{i,j,k-\frac{1}{2}}\right),
\end{aligned}
\tag{2.6}
$$

where q, F, G, and H are the state vector and the flux vectors along the x, y, and z-directions. Here, i, j, and k are the grid indices along the x, y, and z-directions, and the unprimed and primed quantities are defined at t and $t + \Delta t$, respectively. The 5th-order accurate FD WENO scheme is used to estimate the cell interface fluxes, $F_{i\pm\frac{1}{2},j,k}$, $G_{i,j\pm\frac{1}{2},k}$, and $H_{i,j,k\pm\frac{1}{2}}$, (see Jiang & Wu 1999, for detail). The code includes three versions of WENO, the original WENO-JS of Jiang & Shu (1996), and WENO-Z (Borges et al. 2008) and WENO-ZA (Liu et al. 2018), both of which are designed to improve the performance over the original WENO-JS. We have tested the three variants with various RHD test problems. While WENO-JS is the most stable in various conditions, it gives the most diffusive solutions. On the other hand, WENO-ZA gives less diffusive solutions, but they often break down when there are steep gradients. We have found that WENO-Z is best fitted because of its balance between stability and accuracy, and hence we employed it for our jet simulations.

2.4. Time Integration

The 4th-order RK method has been commonly used with the 5th-order WENO scheme (e.g., Shu & Osher 1988, 1989; Jiang & Shu 1996). Its time-stepping from q^n to q^{n+1} is given as

$$
\begin{aligned}
q^{(0)} &= q^n, \quad q^{(1)} = q^{(0)} + \frac{\Delta t}{2}\mathcal{L}^{(0)}, \quad q^{(2)} = q^{(0)} + \frac{\Delta t}{2}\mathcal{L}^{(1)}, \\
q^{(3)} &= q^{(0)} + \Delta t\mathcal{L}^{(2)}, \quad q^{n+1} = \frac{1}{3}\left(-q^{(0)} + q^{(1)} + 2q^{(2)} + q^{(3)}\right) + \frac{\Delta t}{6}\mathcal{L}^{(3)}.
\end{aligned}
\tag{2.7}
$$

Here, n is the timestep. In the 4th-order accurate, 5-stage SSPRK method (e.g., Spiteri & Ruuth 2002, 2003; Gottlieb 2005), the time-stepping is given as

$$q^{(0)} = q^n, \qquad q^{(l)} = \sum_{m=0}^{l-1} (\chi_{lm} q^{(m)} + \Delta t \beta_{lm} \mathcal{L}^{(m)}), \quad l = 1, 2, \cdots, 5, \qquad q^{n+1} = q^{(5)}, \quad (2.8)$$

where $\mathcal{L}_{i,j,k}^{(l)}$ is

$$\mathcal{L}_{i,j,k}^{(l)} = -\frac{F_{i+\frac{1}{2},j,k}^{(l)} - F_{i-\frac{1}{2},j,k}^{(l)}}{\Delta x} - \frac{G_{i,j+\frac{1}{2},k}^{(l)} - G_{i,j-\frac{1}{2},k}^{(l)}}{\Delta y} - \frac{H_{i,j,k+\frac{1}{2}}^{(l)} - H_{i,j,k-\frac{1}{2}}^{(l)}}{\Delta z}, \quad (2.9)$$

χ_{lm} and β_{lm} are the coefficients (Spiteri & Ruuth 2002).

For the stability in time integration, the time step, Δt, is restricted by the so-called Courant-Friedrichs-Levy (CFL) condition,

$$\Delta t = \text{CFL} / \left[\frac{\lambda_x^{\max}}{\Delta x} + \frac{\lambda_y^{\max}}{\Delta y} + \frac{\lambda_z^{\max}}{\Delta z} \right], \quad (2.10)$$

where $\lambda^{\max}$'s are the maxima of the cell-centered eigenvalues. We use CFL=0.8 in our code, which seems to be a good compromise for stability and speed. RK4 and SSPRK give similar results in tests of moderate RHD problems, but in the shock tests where the initial velocity perpendicular to the shock normal, $v_\perp$, is close to c, SSPRK gives more stable solutions than RK4. In relativistic jets, strong shears appear across the jet and backflows, and shocks with large $v_\perp$ could form, so we use SSPRK for the time integration.

2.5. Averaging of Fluxes along Transverse Directions

Our code includes transverse-flux averaging in the stage of the calculation of WENO fluxes, which improves the performance of multi-dimensional problems. The 4th-order accurate averaging scheme, given as (Buchmüller & Helzel 2014; Buchmüller et al. 2016),

$$\bar{q}_{i,j,k} = q_{i,j,k} - \frac{1}{24} \left(q_{i,j-1,k} - 2q_{i,j,k} + q_{i,j+1,k} \right) - \frac{1}{24} \left(q_{i,j,k-1} - 2q_{i,j,k} + q_{i,j,k+1} \right), \quad (2.11)$$

and

$$\begin{aligned}
\bar{F}_{i\pm\frac{1}{2},j,k} = F_{i\pm\frac{1}{2},j,k} &+ \frac{1}{24} \left(F_{i\pm\frac{1}{2},j-1,k} - 2F_{i\pm\frac{1}{2},j,k} + F_{i\pm\frac{1}{2},j+1,k} \right) \\
&+ \frac{1}{24} \left(F_{i\pm\frac{1}{2},j,k-1} - 2F_{i\pm\frac{1}{2},j,k} + F_{i\pm\frac{1}{2},j,k+1} \right),
\end{aligned} \quad (2.12)$$

is used. We have adopted this scheme that has the same order of accuracy as the time integration. We found that with this averaging, complex structures like vortices in the relativistic Kelvin-Helmholtz test are better reproduced (see Seo et al. 2021a).

2.6. Suppression of Carbuncle Instability

Low Mach number shocks that are aligned with the computational grid and move slowly are susceptible to carbuncle instability, owing to the lack of sufficient dissipation. Fleischmann et al. (2020) suggested that a modification of eigenvalues for two acoustic modes,

$$c_s' = \min(\phi|v_x|, c_s), \qquad \lambda_{1,5} = v_x \pm c_s', \quad (2.13)$$

could suppress this instability in Newtonian hydrodynamic codes. Here, ϕ is a positive number of order $\mathcal{O}(1)$. This method alleviates the instability problem by increasing the

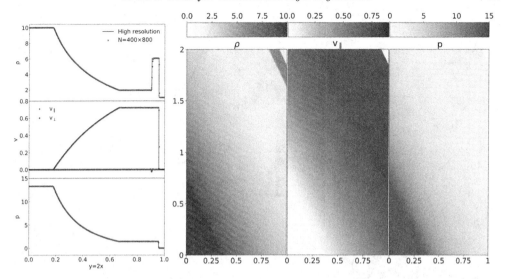

Figure 1. 2D relativistic shock tube test. The results of a simulation using 400×800 grid zones are shown $t = 0.45 \times \sqrt{5}$. *Left panels:* The flow quantities along $y = 2x$ (red dots) are compared to those from 1D high-resolution results with 20,000 grid zones for a converged benchmark (black solid lines). Here, $v_\parallel$ and $v_\perp$ are the velocity parallel and perpendicular to the normal of the shock and contact discontinuity, respectively. *Right panels:* The corresponding 2D images are shown.

acoustic Mach number to $M \geq 1/\phi$. We adopt this idea to our RHD code, modifying the eigenvalues of two acoustic modes as,

$$c'_s = \min(\phi|v_x|, c_s), \qquad \lambda_{1,5} = \frac{(1 - c'^2_s)v_x \pm c'_s/\Gamma\sqrt{\mathcal{Q}}}{1 - c'^2_s v^2}, \qquad (2.14)$$

where $\mathcal{Q} = 1 - v_x^2 - c'^2_s(v_y^2 + v_z^2)$. In simulations of RHD jets, As shown in the Figure 9 of Seo et al. (2021a), a part of the bowshock may be subject to carbuncle instability. We have found that this modification efficiently suppresses the instability.

2.7. Code test : 2D Relativistic Shock Tube

As the verification of the code, we present the results of a two-dimensional (2D) shock tube test. The initial condition is given as $u_L = (10, 0, 0, 13.3)$ for $y < 2(1-x)$ and $u_R = (1, 0, 0, 10^{-6})$ for $y > 2(1-x)$, where $u = (\rho, v_x, v_y, p)$. As shown in Figure 1, the shock and contact discontinuity are resolved with $2-3$ cells, and the rarefaction wave are well reproduced. We also see that the velocity perpendicular to the normal of the shock and contact discontinuity, $v_\perp$, which is an indication of numerical error in this test, is small with $|v_\perp/v_\parallel| \approx 1 - 2\%$.

The full description of the code along with a number of tests is given in Seo et al. (2021a).

3. Application to FR II Jets

As the first application of our RHD code, we have simulated three-dimensional (3D) FR II jets, which have well collimated structures and bright heads. In Figure 2, we present the 2D slice distributions of the density and vertical velocity in one of the simulated jets. The length scale of the shown jet is about 60kpc with the jet radius of $r_j = 1$kpc. The jet power, the energy injection rate through the jet radius, is matched to the value of typical FR II jets, $Q \approx 3.34 \times 10^{46}$erg/s. We have found that our code well reproduces

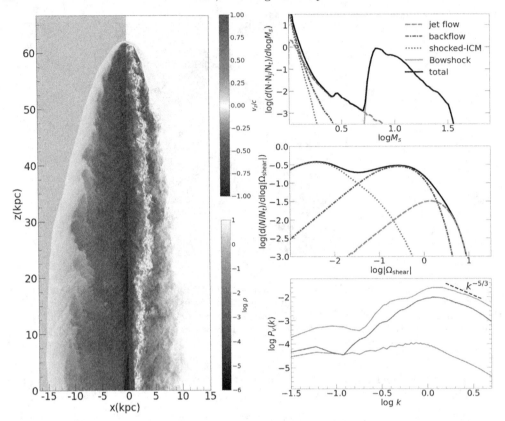

Figure 2. *left panels*: Density (left) and velocity(right) slice distributions at $t = 1.5$Myr in a FR II jet simulation with $(600)^2 \times 1200$ grid zones. *Right panels*: The PDF of the Mach number of the shocks generated in the jet-induced flow (top), the PDF of shear, $\Omega_{\text{shear}} = \partial v_z / \partial r$, where r is the radial distance from the jet spine (middle), and the velocity power spectrum of turbulence in the jet-induced flow (bottom).

the details of nonlinear structures, such as, shocks, shear, and turbulence, and hence we have been able to analyze the characteristics of flow structures. For instance, the Mach numbers of shocks have a power-law distribution except the bowshock (see the right-top panel of Figure 2), indicating that most of them are originated from turbulence. Strong shear is generated in the jet and backflow boundary (see the right-middle panel). And the turbulence in the jet-induced flows seems to follow the Kolmogorov spectrum (see the right-bottom panel).

The full description of the jet flow structures is given in Seo et al. (2021b).

This work was supported by the National Research Foundation (NRF) of Korea through grants 2016R1A5A1013277, 2020R1A2C2102800, and 2020R1F1A1048189. The work of J.S. was also supported by the NRF through grant 2020R1A6A3A13071702. Some of simulations were performed using the high performance computing resources of the UNIST Supercomputing Center.

References

Borges, R., Carmona, M., Costa, B., & Don, W. S. 2008, J. Comput. Phys., 227, 3191
Buchmüller, P., Dreher, J., & Helzel, C. 2016, ApMaC, 272, 460
Buchmüller, P., & Helzel, C. 2014, JSCom, 61, 343
Dumbser, M., Moschetta, J.-M., & Gressier, J. 2004, JCoPh, 197, 647

English, W., Hardcastle, M. J., & Krause, M. G. H. 2016, *MNRAS*, 461, 2025

Fleischmann, N., Adami, S., Hu, X. Y., & Adams, N. A. 2020, JCoPh, 401, 109004

Gottlieb, S. 2005, Journal of Scientific Computing, 25, 10

Hardcastle, M. J. & Croston, J. H. 2020, *New Astron. Revs*, 88, 101539

Jiang, G.-S., & Shu, C.-W. 1996, J. Comput. Phys., 126, 202

Jiang, G.-S., & Wu, C.-C. 1999, J. Comput. Phys., 150, 561

Landau, L. D., & Lifshitz, E. M. 1959, Fluid mechanics

Li, Y., Wiita, P. J., Schuh, T., et al. 2018, *ApJ*, 869, 32

Liu, X.-D., Osher, S., & Chan, T. 1994, J. Comput. Phys., 115, 200

Liu, S., Shen, Y., Zeng, F., & Yu, M. 2018, International Journal for Numerical Methods in Fluids, 87, 271

Mathews, W. G. 1971, *ApJ*, 165, 147

Matthews, J. H., Bell, A. R., Blundell, K. M., et al. 2019, *MNRAS*, 482, 4303

Martí, J. M. & Müller, E. 2003, Living Reviews in Relativity, 6, 7

Martí, J. M. & Müller, E. 2015, Living Reviews in Computational Astrophysics, 1, 3

Mignone, A., Plewa, T., & Bodo, G. 2005, *ApJS*, 160, 199

Peery, K., & Imlay, S. 1988, in 24th Joint Propulsion Conf. (Reston, VA:AIAA Journal), 2904

Ryu, D., Chattopadhyay, I., & Choi, E. 2006, *ApJS*, 166, 410

Seo, J, Kang, H., & Ryu, D. 2021b, *ApJ*, 920, 144

Seo, J, Kang, H., Ryu, D., Ha, S., & Chattopadhyay, I. 2021a, *ApJ*, 920, 143

Shu, C.-W., & Osher, S. 1988, J. Comput. Phys., 77, 439

Shu, C.-W., & Osher, S. 1989, J. Comput. Phys., 83, 32

Spiteri, R. J., & Ruuth, S. J. 2002, SIAM Journal on Numerical Analysis, 40, 469

Spiteri, R. J., & Ruuth, S. J. 2003, Mathematics and Computers in Simulation, 62, 125

Synge, J. L. 1957, The Relativistic Gas, Series in Physics (Amsterdam: North-Holland)

Taub, A. H. 1948, Phys. Rev., 74, 328

The Predictive Power of Computational Astrophysics as a Discovery Tool
Proceedings IAU Symposium No. 362, 2023
D. Bisikalo, D. Wiebe & C. Boily, eds.
doi:10.1017/S1743921322001685

Studying Magnetic Field Amplification in Interacting Galaxies Using Numerical Simulations

Simon Selg and Wolfram Schmidt

Hamburger Sternwarte, Universität Hamburg, Gojenbergsweg 112, D-21029 Hamburg,
Germany
email: `simon.selg@hs.uni-hamburg.de`

Abstract. There are indications that the magnetic field evolution in galaxies might be massively shaped by tidal interactions and mergers between galaxies. The details of the connection between the evolution of magnetic fields and that of their host galaxies is still a field of research.

We use a combined approach of magnetohydrodynamics for the baryons and an N-body scheme for the dark matter to investigate magnetic field amplification and evolution in interacting galaxies.

We find that, for two colliding equal-mass galaxies and for varying initial relative spatial orientations, magnetic fields are amplified during interactions, yet cannot be sustained. Furthermore, we find clues for an active mean-field dynamo.

Keywords. galaxies: interactions, turbulence, magnetic fields, MHD, methods: numerical

1. Introduction

It remains unclear how primordial magnetic fields of order $\ll 10^{-9}$ G are amplified inside galaxies to values of order more than 10^{-6} G around $z = 0$. Possible solutions are the Biermann battery, a dynamo or galaxy interactions (e.g. Beck *et al.* 1996; Beck 2015; Brandenburg & Subramanian 2005; Subramanian 2016, and references therein). The generation and action of dynamos has been studied in simulations of isolated galaxies (e.g. Ntormousi *et al.* 2020; Schober *et al.* 2013; Steinwandel *et al.* 2019).

Interactions between and mergers of galaxies are assumed to be an integral part of galaxy evolution. Within the framework of ΛCDM it is assumed that hierarchical structure formation leads to larger galaxies being formed through subsequent mergers of smaller galaxies (e.g. Helmi 2020, and references therein).

Drzazga *et al.* (2011) give evidence for temporally enhanced magnetic fields in observational data. They find that the magnetic field strength in interacting galaxies – both at close encounters and especially at coalescence – is enhanced by a factor of 2-3. Simulations performed by Kotarba *et al.* (2010) with the SPH code Gadget have confirmed this. Renaud *et al.* (2015, 2019) investigate starbursts in hydrodynamical simulations using adaptive mesh refinement (AMR) and Rodenbeck *et al.* (2016) simulate the evolution of magnetic fields in interacting galaxies using the AMR N-body code Enzo (Bryan *et al.* 2014).

However, Rodenbeck *et al.* (2016) lack a correct treatment of the dark matter. We therefore model magnetic field amplification in interacting galaxies subject to varying initial conditions, with each galaxy residing in a live dark matter halo.

In Section 2 we introduce our numerical model and the suite of simulations we performed while in Section 3 we highlight and discuss some of the results.

2. Numerical Methods

2.1. *Galaxies*

Galaxy mergers can be simulated either in large quantities within cosmological simulations (e.g. Patton *et al.* 2020) or much smaller quantities, i.e. mostly pair-wise (e.g. Rodenbeck *et al.* 2016, but see Kotarba *et al.* (2011) for the exception of three galaxies). For reasons of higher numerical resolution, we choose to use a 2^3 kpc^3 computational domain with one pair of galaxies.

Different from using galaxy formation recipes of cosmological galaxy formation (e.g. Pillepich et al. 2018) we choose an approach to initialise galaxies a priori. This gives us more control over the exact choice of initial conditions for both galaxies. In our model each galaxy is comprised of a gas disk which is embedded inside a dark matter halo. We use a modified version of the methods developed by Rodenbeck *et al.* (2016) and Wang *et al.* (2010) in order to initialise a three-dimensional gas disk in equilibrium that is magnetised and hosted by a dark matter halo. While Wang *et al.* (2010) have introduced an equilibrium disk model that supports the addition of external potentials to represent the gravitational influence of stars and dark matter it has been extended by Rodenbeck *et al.* (2016) to include magnetic fields. Since Rodenbeck *et al.* (2016) do not have dark matter halos in their simulations of galaxy interactions we extend their model and that of Wang *et al.* (2010) to include live dark matter halos which are sampled via an N-body simulation of 2×10^6 particles per halo.

Wang *et al.* (2010) describe how to compute the density profile and the rotation curve using an iterative approach. Rodenbeck *et al.* (2016) include magnetic fields:

$$B = B_{\text{tor}} = \sqrt{8\pi\epsilon_{\text{mag}}\rho\, c_{\text{s}}^2},$$
(2.1)

where ϵ_{mag} is the inverse of the plasma beta which describes the ratio of thermal and magnetic pressure.

Wang *et al.* (2010) show that the velocity profile of the entire three-dimensional disk can be specified in the midplane

$$v_{\text{rot}}^2(r, z) = v_{\text{thin}}^2(r) + (1 + \epsilon_{\text{mag}})c_{\text{s}}^2 \frac{\partial \ln \rho}{\partial \ln r}\bigg|_{z=0},$$
(2.2)

where the first term on the right hand side is described by a thin disk approximation (see for details e.g. Binney & Tremaine 2008; Freeman 1970) including the additional magnetic pressure introduced by Rodenbeck *et al.* (2016). The gas density distribution is computed with an iterative scheme proposed by Wang *et al.* (2010):

$$\rho_{\text{gas}}(r, z) = \rho_0(r) \exp\left(-\frac{\Phi_z(r, z)}{(1 + \epsilon_{\text{mag}})c_{\text{s}}^2}\right).$$
(2.3)

Here, Φ_z denotes the vertical difference of the total gravitational potential ($\Phi = \Phi_{\text{gas}} + \Phi_{\text{DM}}$) with respect to its midplane value.

The dark matter halo is modelled by the following density distribution (Hernquist 1990):

$$\rho_{\text{DM}} = \frac{M_{\text{DM}}}{2\pi} \frac{a}{r} \frac{1}{(r + a)^3},$$
(2.4)

with the scale length a. Furthermore, a connection between the Hernquist profile and an NFW profile (Navarro *et al.* 1996) is established by relating the scale lengths a and r_{s} as Springel *et al.* (2005) propose:

$$a = r_\mathrm{s}\sqrt{2(\ln(1+c) - c/(1+c))}, \tag{2.5}$$

where c is the concentration parameter. We use the method developed by Drakos et al. (2017) to find a stable N-body representation of a live dark matter halo.

2.2. Simulations

Inside the cosmic web galaxies move on orbits of different shape and orientation. We try to account for this fact in our simulations by performing a parametric study where we vary the relative orientations between the galaxies by an inclination angle i and an impact parameter which can also be described by an angle α_b. The relative inclination is computed from the individual galaxies' inclinations: $i = |i_2 - i_1|$. Each galaxy's inclination is computed as the angle between its angular momentum vector and the relative bulk velocity vector, $\mathbf{v} = \mathbf{v}_\mathrm{rel} = \mathbf{v}_1 - \mathbf{v}_2$:

$$i_\mathrm{j} = \angle(\mathbf{L}_\mathrm{j}, \mathbf{v}_\mathrm{rel}), \quad j \in \{1, 2\}. \tag{2.6}$$

Additionally, spatial orientation is determined through an impact parameter b which is the vertical distance between the centre of the primary galaxy and the line in the direction of the initial velocity of the secondary galaxy. At the beginning of each simulation, the two galaxies have a separation corresponding to the cut-off radius ($d_\mathrm{sep} = 209\,\mathrm{kpc}$) of a dark matter halo. Thus, $\sin\alpha_\mathrm{b} = b/d_\mathrm{sep}$ defines the angle associated with the impact parameter. In our simulations we realised a total of 26 different initial configurations covering inclinations between $0°$ and $90°$ and impact parameter angles between $0°$ and $45°$. Each galaxy consists of a gas disk of $10^{10}\,\mathrm{M}_\odot$ and a dark matter halo of $10^{12}\,\mathrm{M}_\odot$. The radial scale length is $3.5\,\mathrm{kpc}$ following the galaxy description in Wang et al. (2010). We solve the equations of ideal MHD and the evolution of the live halos with the AMR & N-body code Enzo (Bryan et al. 2014) and employ a subgrid-scale model for unresolved MHD turbulence (Grete et al. 2017, 2019). An adiabatic equation of state is used.

3. Results & Discussion

In the following, we limit our presentation of results to those of one simulation in particular: $i_1 = i_2 = 90°$ and $\alpha_\mathrm{b} = 20°$†. For our analysis, trajectories of both galaxies are computed and at the centre of mass of the primary galaxy a cylindrical region of $20\,\mathrm{kpc}$ radius and $8\,\mathrm{kpc}$ thickness is positioned. In a fashion similar to Drzazga et al. (2011) we further subdivide this region into a central cylinder of $5\,\mathrm{kpc}$ radius – the centre region – and a thick shell, $5 < r \leq 20\,\mathrm{kpc}$ – the off-centre region. Within these regions mass-weighted quantities are computed.

The magnetic field strength displayed in Fig 1 experiences only a moderate rise at the first encounter where only the outer parts of the galaxies interact. Considerable amplification by a factor of ~ 2 at second and third encounters is accompanied by a stronger penetration of both galaxies. In Fig. 3 we show gas density slices along the disks' midplanes: the first encounter of the galaxies at $762.5\,\mathrm{Myr}$ and a snapshot at $887.5\,\mathrm{Myr}$ while they are moving apart. During the first encounter a shock front forms where the outer disks interact. Using an implementation of the line integral convolution method in yt (Cabral & Leedom 1993; Turk et al. 2011) we illustrate the transition from ordered to random fields. At the first encounter the magnetic field is still toroidal outside of the collision zone while it appears random along the shock front. Furthermore, when the galaxies retract we can clearly recognise the turbulent pattern of the magnetic field in the bridge region.

† We plan the presentation of results of the other simulations in a future publication.

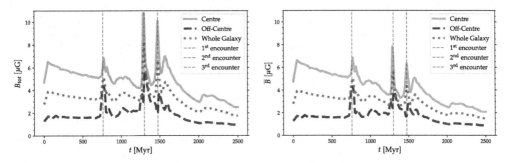

Figure 1. Timeseries of magnetic field evolution in three regions of the primary galaxy: the total magnetic field strength B_{tot} and the filter-averaged magnetic field strength $\overline{B}$ (see Eq. 3.1) are displayed on the left and right image, respectively. We highlight times of three encounters between the two galaxies at 762.5, 1300, and 1475 Myr.

In order to study possible dynamo activity we apply mean-field electrodynamics (e.g. Parker 1970; Steenbeck *et al.* 1966) to our data in order to investigate the possible presence of a mean field dynamo (Parker 1971) in interacting galaxies. In mean-field electrodynamics both the magnetic field and the velocity of the fluid can be decomposed into fluctuating and mean components, respectively:

$$\mathbf{B}_{\text{tot}} = \overline{\mathbf{B}} + \mathbf{B}_{\text{turb}} \qquad \& \qquad \mathbf{v}_{\text{tot}} = \overline{\mathbf{v}} + \mathbf{v}_{\text{turb}}. \tag{3.1}$$

To compute this we follow a similar approach as Ntormousi *et al.* (2020). To separate velocities and magnetic fields in the fashion of Eq. 3.1 a mass-averaged filter with a scale of $\Delta x = 300\,\text{pc}$ (5^3 cells) is applied. Substitution of Eq. 3.1 into the induction equation and subsequent averaging it yields the mean-field induction equation

$$\frac{\partial \overline{\mathbf{B}}}{\partial t} = \boldsymbol{\nabla} \times (\overline{\mathbf{v}} \times \overline{\mathbf{B}} + \mathcal{E}), \tag{3.2}$$

where we assume isotropic turbulence and ignore the magnetic diffusivity. We identify the electromotive force:

$$\mathcal{E} = \overline{\mathbf{v}_{\text{turb}} \times \mathbf{B}_{\text{turb}}}. \tag{3.3}$$

Following e.g. Ntormousi *et al.* (2020) $\mathcal{E}$ can be expanded in a series which yields after truncation

$$\overline{\mathbf{v}_{\text{turb}} \times \mathbf{B}_{\text{turb}}} = \alpha \overline{\mathbf{B}} - \eta_{\text{T}} \boldsymbol{\nabla} \times \overline{\mathbf{B}}, \tag{3.4}$$

where the transport coefficients α and η_{T} describe the α-effect and the turbulent magnetic diffusivity, respectively.

Eq. 3.4 describes the evolution of the averaged magnetic field. Comparing total and filtered fields in Fig. 1 we note that although peaks at second and third encounters are less pronounced, they still indicate significant amplification.

We show the evolution of the electromotive force in Fig. 2. We have normalised $\mathcal{E}$ by its value attained at $t = 500\,\text{Myr}$ which is still during the initial approach phase of the galaxies. Therefore, it is easier to observe changes in $\mathcal{E}$ with respect to the state of the galaxies before the first encounter when $\mathcal{E}/\mathcal{E}_{500} \approx 1$. Afterwards, the electromotive force keeps growing continuously until the third encounter, followed by a gradual decline until pre-interaction values are reached in the central region towards the end of the simulation. Analogous to the magnetic field amplification observed in Fig. 1 events of interaction are marked by peaks with $\mathcal{E}/\mathcal{E}_{500} > 10^2$ in the off-centre region.

In general, the transport coefficients α and η_{T} are tensors and complicated to evaluate. However, if we assume isotropic turbulence and neglect effects of Lorentz forces,

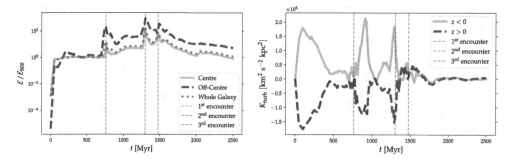

Figure 2. *Left*: Timeseries of the electromotive force $\mathcal{E}$ normalised by $\mathcal{E}_{500}$ in three regions of the primary galaxy. *Right*: Time evolution of the turbulent kinetic helicity above $(z > 0)$ and below $(z < 0)$ the primary galaxy's midplane. We highlight times of three encounters between the two galaxies (see Fig. 1).

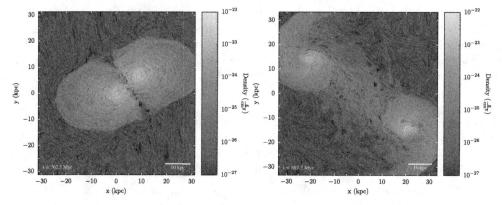

Figure 3. Two density slices along the midplane of the interacting galaxy pair overlaid with a representation of magnetic field lines provided by line integral convolution. The image on the left displays the first encounter and the image on the right the retracting galaxies between the first and second encounter.

α is proportional to the turbulent kinetic helicity (see e.g. Ntormousi *et al.* 2020, and references therein)

$$K_{\mathrm{turb}} = \int \mathbf{v}_{\mathrm{turb}} \cdot (\boldsymbol{\nabla} \times \mathbf{v}_{\mathrm{turb}}) \mathrm{d}V. \tag{3.5}$$

The time evolution of the turbulent kinetic helicity in Fig. 2 presents us with three remarkable events where K_{turb} reaches maximum values thus offering particularly favourable conditions for a mean-field dynamo: (1) the initial phase $t < 500\,\mathrm{Myr}$, (2) the timespan between the first and the second encounters, and (3) the third encounter. The strong initial growth and the subsequent decrease of K_{turb} during the first 500 Myr is the result of a relaxation process within the gas disk. This is because during initialisation the DM halo is not adjusted to the gravitational potential of the gas disk. The second case could be associated with strong helical turbulence inside the tidal arms and the tidal bridge connecting the retracting collision partners. It is only at the third encounter where the maximum turbulent kinetic helicity coincides with the maximum magnetic field strength.

In summary, we find that while turbulence is induced by galaxy encounters and there is indication of a mean-field dynamo that also amplifies the magnetic field during these encounters, at the same time this amplification is not persistent. Magnetic field strength, turbulent kinetic helicity and the electromotive force are dissipated after the third encounter, suggesting that the dynamo is not sustained.

We summarise the discussion following the presentation given at the symposium. A question addressed the transport coefficients in Eq. 3.4 and whether we have computed η_T. This is planned in future work.

The authors acknowledge funding from DFG grant SCHM 2135/6-1. The work was supported by the North-German Supercomputing Alliance (HLRN). We would like to thank the reviewer for their helpful comments.

References

Beck, R. 2015, *A&A Rev.*, 24, 4

Beck, R., Brandenburg, A., Moss, D., *et al.* 1996, *ARAA*, 34(1), 155–206

Binney, J, Tremaine, S 2008, Galactic Dynamics

Brandenburg, A., Subramanian, K. 2005, *Phys. Rep.*, 417, 1–209

Bryan, G. L, Norman, M. L, O'Shea, B. W., *et al.* 2014, *ApJS*, 211(2), 19

Cabral, B., Leedom, L. C. 1993, *Proceedings of the 20th Annual Conference on Computer Graphics and Interactive Techniques (New York)*

Drakos, N. E., Taylor, J. E., Benson, A. J. 2017, *MNRAS*, 468(2), 2345–2358

Drzazga, R. T., Chyży, K. T., Jurusik, W., *et al.* 2011, *A&A*, 533, A22

Freeman, K. C. 1970, *ApJ*, 160, 811

Grete, P., Latif, M. A., Schleicher, D. R. G., *et al.* 2019, *MNRAS*, 487(4), 4525–4535

Grete, P., Vlaykov, D. G., Schmidt, W., *et al.* 2017, *Phys. Rev. E.*, 95, 033206

Helmi, A. 2020, *ARAA*, 58, 205–256

Hernquist, L. 1990, *ApJ*, 356, 359

Kotarba, H., Karl, S. J., Naab, T., *et al.* 2010, *ApJ*, 716(2), 1438–1452

Kotarba, H., Lesch, H., Dolag, K., *et al.* 2011, *MNRAS*, 415(4), 3189–3218

Navarro, J. F., Frenk, C. S., White, S. D. M. 1996, *ApJ*, 462, 563

Ntormousi, E., Tassis, K., Del Sordo, F., *et al.* 2020, *A&A*, 641, A165

Parker, E. N. 1970, *ApJ*, 160, 383

Parker, E. N. 1971, *ApJ*, 163, 255

Patton, D. R., Wilson, K. D., Metrow, C. J., *et al.* 2020, *MNRAS*, 494(4), 4969–4985

Pillepich, A., Springel, V., Nelson, D., *et al.* 2018, *MNRAS*, 473(3), 4077–4106

Renaud, F., Bournard, F., Agertz, O., *et al.* 2019, *A&A*, 625, A65

Renaud, F., Bournard, F., Duc, P.-A. 2015, *MNRAS*, 446(2), 2038–2054

Rodenbeck, K., Schleicher, D. R. G. 2016, *A&A*, 593, A89

Schober, J., Schleicher, D. R. G., Klessen, R. S. 2013, *A&A*, 560, A87

Springel, V., Di Matteo, T., Hernquist, L. 2005, *MNRAS*, 361 (3), 776–794

Steenbeck, M., Krause, F., Rädler, K.-H. 1966, *Zeitschrift für Naturforschung A*, 21(4), 369–376

Steinwandel, U. P., Beck, M. C., Arth, A., *et al.* 2019, *MNRAS*, 438(1), 1008–1028

Subramanian, K. 2016, *Reports on Progress in Physics*, 79(7), 076901

Turk, M. J., Smith, B. D., Oishi, J. S., *et al.* 2011, *ApJS*, 192(1), 9

Wang, H.-H., Klessen, R. S., Dullemond, C. P., *et al.* 2010, *MNRAS*, 407(2), 705–720

The Predictive Power of Computational Astrophysics as a Discovery Tool
Proceedings IAU Symposium No. 362, 2023
D. Bisikalo, D. Wiebe & C. Boily, eds.
doi:10.1017/S1743921322001624

X-ray spectral and image spatial models of NGC 3081 with *Chandra* data

O.V. Kompaniiets[1,2]**, Iu.V. Babyk**[1]®**, A.A. Vasylenko**[1]**,**
I.O. Izviekova[1,3] **and I.B. Vavilova**[1]®

[1]Main Astronomical Observatory of the NAS of Ukraine, 27 Akademik Zabolotny Str., Kyiv, 03143, Ukraine

[2]Institute of Physics of the NAS of Ukraine, 46 av. Nauky, Kyiv, 03028, Ukraine

[3]ICAMER, NAS of Ukraine, 27 Akademik Zabolotny Str., Kyiv, 03143, Ukraine

Abstract. The physical properties of AGN such as accretion rate, column density, temperature of hot corona and other characteristics can be found from X-ray spectral data. We present the results of spatial and spectral analysis for Sy2 type galaxy NGC 3081 obtained with different mathematical tools of the Chandra Interactive Analysis of Observations software. We found evidence of extended emission in 0.5-3.0 keV as well as derived parameters for model A: photon index $\Gamma = 1.65^{+0.1}_{-0.9}$, column density $N_H = 57.5^{+5.7}_{-2.8} cm^{-2}$, warm component $kT_1 = 0.16^{+0.1}_{-0.02}$ and hot component $kT_2 = 1.0^{+3.0}_{-0.1}$ keV. We detected the presence of a component of the reflection spectrum, Fe K_α emission line with $E_{line} = 6.39^{+0.06}_{-0.02}$ keV and $EW = 50^{+0.01}_{-0.01}$ eV.

Keywords. methods: data analysis – galaxies: active galactic nuclei – NGC 3081

---

1. Introduction

Active galactic nuclei (AGNs) are one of the most powerful sources in the Universe. The main parts of the AGNs, such as an accretion disc, a torus, and a supermassive black hole (SMBH), testify themselves by different kinds of radiation (see, the Unification Scheme by Antonucci & Miller (1985)). The optical and UV emission from AGN arises due to the accretion process onto the SMBH, while the X-ray emission originates by Compton upscattering of optical/UV photons from the hot corona (Haardt & Maracchi 1991, 1993). This X-ray radiation is typically described by a power-law model and an exponential cut-off at high energies. In addition to these components of spectra there are a reflection spectrum appearing as a "reflection hump" at $\sim 20 - 30$ keV and a Fe K_α line near 6.4 keV (Mushotzky et al. (1993), Ramos & Ricci (2017); Kompaniiets & Vasylenko (2020)).

The AGN's drive is fueled and evolved by both intrinsic processes and environmental influence. To distinguish these factors, we propose considering isolated galaxies with AGNs, which had not undergone tidal effects during at least 3 Gyrs. Our sample of 61 isolated AGNs was formed by cross-matching the 2MIG (2MASS Isolated Galaxy) catalogue with the Veron-Cetty catalogue of quasars and AGNs, where the restriction was used for $K_s \leq 12.0$ mag and $V_r < 15\,000$ km/s (Pulatova et al. (2015); Karachentseva et al. (2010); Véron-Cetty & Véron (2010)). They belong to the Local Universe and are located not in dense regions, like Virgo and Fornax clusters or galaxy groups of the Local Supercluster (Karachentseva & Vavilova (1994); Karachentseva & Vavilova (1995); Vavilova et al. (2005)). For the first time, we revealed that the host isolated galaxies with AGNs of Sy1 type (without faint companions) appear to possess the bar morphological features, which provide transfer of gas and dust from galaxy's disc to AGN region (Pulatova et al. 2015) Therefore, the interaction with neighbouring galaxies is not a necessary condition for BLR formation in comparison with dual AGNs triggered by tidally induced gas inflows (Smirnova et al. (2010); Gross et al. (2019); Guainazzi et al. (2021)). Withal, the isolated AGN and AGNs in pairs (or mergers) show similar distributions in their global stellar mass, star-formation rate, and central [O III] surface brightness (Jin et al. (2021)).

We performed the X-ray spectral analysis for isolated AGNs with the XMM-Newton, Chandra, and Swift/XRT data in their computational software (Vavilova et al. (2015); Vasylenko (2020). It allowed revealing that isolated AGNs mostly have smaller luminosity $L_{2-10keV} \sim 10^{42}$ erg/s in comparison with the typical luminosity for the Seyfert galaxies (see, for example, Volvach et al. (2011). The spectra of NGC 5347 and MCG-02-09-040 show the neutral Fe K_α emission line. The X-ray spectrum of NGC 5347 is described by a pure reflection model with $E_{cut} \sim 117$ keV and with no signs of transmitted radiation, while MCG-02-09-040 shows the presence of heavy neutral obscuration of $N_H \sim 10^{24}$ cm^{-2}.

In this work, we present the results of the X-ray spectral analysis of yet another low-luminosity isolated S0/a galaxy NGC 3081 with Sy2 nucleus based on the Chandra data. It's worth noting that NGC 3081 has a complicated structure: a weak large scale bar and a nuclear bar, and four resonance rings - two outer rings, an inner ring, and a nuclear ring (Schnorr-Müller et al. (2016)).

2. X-ray spectral analysis of NGC 3081

The software for Chandra X-ray observatory called **CIAO** is available through (https://cxc.cfa.harvard.edu/ciao/). Observations of NGC 3081 were done in 2018 (OBSID 20622) with a total exposure 22.3 ks. We reprocessed these data by standard script called **chandra_repro**. Spectrum was extracted by **specextract** tool with radius 1.6783″. The background spectrum was created for annulus with the radii $R_1 = 4.873″$ and $R_2 = 9.743″$ and centered for the source region. To take into account telescope response we created the Auxiliary Response Files and the Redistribution Matrix Files. To apply χ^2 statistics the spectrum was grouped with **group_counts** with a 20 counts per been (Fig. 1). Finally, the spectrum was constructed by CIAO modeling and fitting application called **Sherpa** (Freeman et al. 2001; Doe et al. 2007).

A simple phenomenological model (model A) in **Sherpa** terminology is as follows:

```
Model A = xstbabs × (xsapec + xsapec + xsztbabs × xszpowerlw + xszgauss)
```

It includes power-law continuum (**xszpowerlaw**), neutral absorption (**xsztbabs**), emission line with a Gaussian profile (**xszgauss**), and two thermal components (**xsapec**). Galactic absorption was fixed with value $N_H = 0.04 \cdot 10^{22} cm^{-2}$.

We obtained photon index $\Gamma = 1.65^{+0.1}_{-0.9}$, and column density is $N_H = 57.5^{+5.7}_{-2.8} \ cm^{-2}$. Two different thermal components were also detected: a warm one with temperature $kT_1 = 0.16^{+0.10}_{-0.02}$ and a hot one with temperature $kT_2 = 1.0^{+3.0}_{-0.1}$ keV. The Fe K_α emission line is $E_{line} = 6.39^{+0.06}_{-0.02}$keV and equivalent width is $EW = 50^{+0.01}_{-0.01}$ eV. It's suggested that the Fe K_α emission line is originated in moderate density environment.

3. Image spatial analysis of NGC 3081

The Chandra telescope has the best angular resolution among all the X-ray space observatories (0.492 arcsec). This makes possible to use its observational data for spatial analysis too. The point spread function (PSF) describes the shape and size of the image produced by a delta function (a point) source. The shape and size of the High-Resolution

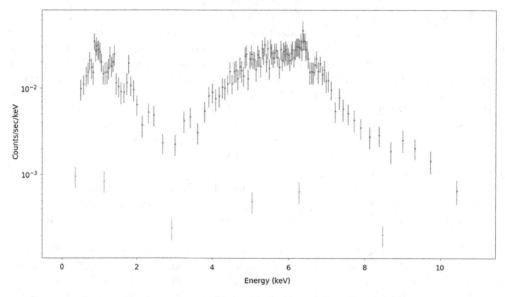

Figure 1. The source and background spectrum for NGC 3081 with Chandra software.

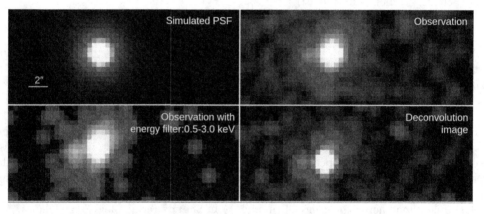

Figure 2. Image spatial analysis of NGC 3081. Top left panel: the MARX simulated PSF, top right panel: observable image, left down panel: observable image with energy filter 0.5-3.0 keV, right down panel: deconvolved image with bin size 4923″.

Mirror Assembly (HRMA) PSF vary significantly with source location in the telescope field of view and spectral energy distribution. The PSF should be simulated with **MARX**, where the input spectrum is created with a baseline model (for example, power-law and absorption). Parameters for this model were estimated from a simple fit for the source spectrum. The next step after the PSF simulation is to restore image resolution. For this procedure, the Lucy-Richardson deconvolution algorithm is applied, which is implemented in the `arestore` tool. In the case of NGC 3081, we used results from Model A. The obtained PSF, observational and deconvolutional images are given in Fig. 2. These results were put on the sub-pixel image with 1/2 and 1/8 bin size of the Chandra ACIS instrument (Fig. 3).

4. Conclusion

We presented the results of X-ray spectral and spatial analysis for NGC 3081 based on the Chandra observational data and Chandra Interactive Analysis of Observations

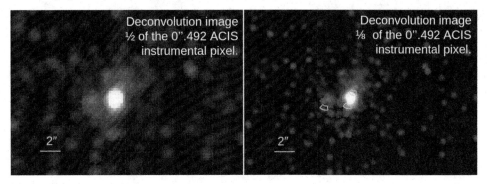

Figure 3. Image spatial analysis of NGC 3081. Left panel: sub-pixel deconvolved image with bin size $2463''$. Right panel: sub-pixel deconvolved image with bin size $61.53''$, where green regions characterise the HRMA artifact indicating that this part is not related to the source.

software. We found an evidence of extended emission in 0.5-3.0 keV and obtained parameters for model A: photon index $\Gamma = 1.65^{+0.1}_{-0.9}$, column density $N_H = 57.5^{+5.7}_{-2.8} cm^{-2}$, warm component with $kT_1 = 0.16^{+0.1}_{-0.02}$ and hot component with $kT_2 = 1.0^{+3.0}_{-0.1}$ keV. We detected the presence of one of the components of the reflection spectrum such as Fe K_α emission line with $E_{line} = 6.39^{+0.06}_{-0.02}$keV and $EW = 50^{+0.01}_{-0.01}$ eV.

Acknowledgements

This work has used data obtained from the Chandra Data Archive and the Chandra Source Catalog, and software provided by the Chandra X-ray Center (CXC) in the application packages CIAO and Sherpa. The authors note the support of the Grant for young research laboratories of the NAS of Ukraine (No07/01-2021(3)).

References

Antonucci R., Miller J. 1985, *ApJ*, 297, 621
Chesnok N.G., Sergeev S.G., Vavilova I.B. 2009, *Kinemat. Physics Celest. Bodies*, 25, 107.
Doe S., Nguyen D., Stawarz C. et al. 2007, *ADASS XVI*, 376, 543.
Freeman P., Doe S., Siemiginowska A. 2001, *Astronomical Data Analysis*, 4477, 76.
Gross A. et al. 2019, *AAS Meeting Abstracts*, #233.
Guainazzi M., De Rosa A., Bianchi S. et al. 2021, *MNRAS*, 504, 393.
Haardt F., Maraschi L. 1991, *ApJ*, 380, 51
Haardt F., Maraschi L. 1993, *ApJ*, 413, 507
Jin G., Dai Y.S., Pan H.A. et al. 2021, *Astrophys. J.*, 923, 6.
Karachentseva V.E. and Vavilova I.B. 1994, *Bull.Special Astrophys. Observatory* 37, 98.
Karachentseva V.E. and Vavilova I.B. 1995, *Kinemat. Phys. Celestial Bodies* 11, 38.
Karachentseva V.E., Mitronova S.N., Melnyk O.V. et al 2010, *Astrophys. Bull.*, 65, 1.
Kompaniiets O.V. & Vasylenko A.A. 2020, *Astrophysics*, 63, 307.
Mushotzky R.F., Done C., Pounds K.A. 1993 Annual Rev. Astron. Astrophys.31, 717
Pulatova N.G., Vavilova I.B., Sawangwit U., et al. 2015, *Mon. Not. R. Astron. Soc.*, 447, 3, 2209.
Ramos A.C. & Ricci C. 2017, *Nature Astronomy*, 1, 679.
Schnorr-Müller A. et al. 2016, *MNRAS*, 457, 972.
Smirnova A.A., Moiseev A.V., Afanasiev V.L. 2010, *Hunting for the Dark: the Hidden Side of Galaxy Formation*, 1240, 297.
Vasylenko A.A., Vavilova I.B., Pulatova N.G. 2020, *Astron. Nachr.*, 341, 8, 801.

Vavilova I.B., Karachentseva V.E., Makarov D.I., and Melnyk O.V. 2005, *Kinemat. Fiz. Nebesnykh Tel* 21, 3.

Vavilova I.B., Vasylenko A.A., Babyk Iu.V., Pulatova N.G. 2015, *Odessa Astron. Publ.*, 28(2), 150.

Véron-Cetty M.P., Véron P. 2010, *Astron. Astrophys.*, 518, A10.

Vol'Vach, A.E., Vol'Vach L.N., Kut'kin A.M. et al. 2011, *Astron. Reports* 55, 608.

The Predictive Power of Computational Astrophysics as a Discovery Tool
Proceedings IAU Symposium No. 362, 2023
D. Bisikalo, D. Wiebe & C. Boily, eds.
doi:10.1017/S1743921322001892

Simulating star formation in spiral galaxies

Steven Rieder[1,2], **Clare Dobbs**[2], **Thomas Bending**[2],
Kong You Liow[2] **and James Wurster**[2,3]

[1]Geneva Observatory, University of Geneva, Sauverny, Switzerland
email: `steven@stevenrieder.nl`

[2]School of Physics and Astronomy, University of Exeter, Exeter, United Kingdom

[3]Scottish Universities Physics Alliance (SUPA), School of Physics and Astronomy,
University of St. Andrews, United Kingdom

Abstract. We present Ekster, a new method for simulating the formation and dynamics of individual stars in a relatively low-resolution gas background. Here, we use Ekster to simulate star cluster formation in two different regions from each of two galaxy models with different spiral potentials. We simulate these regions for 3 Myr to study where and how star clusters form. We find that massive GMC regions form more massive clusters than sections of spiral arms. Additionally we find that clusters form both by accreting gas and by merging with other proto-clusters, the latter happening more frequently in the denser GMC regions.

Keywords. star cluster formation, star formation modeling

1. Introduction

Stars form in galaxies, from collapsing molecular clouds (Lada and Lada 2003). We see this happen mostly on the scales of individual Giant Molecular Clouds (GMCs) or molecular cloud complexes and the arms of spiral galaxies.

Simulations of star clusters generally focus on either starting from a single cloud (e.g. Bate et al. 2003), or take a distribution of stars as their starting point (e.g. Aarseth 1974). Such simulations either ignore the galactic environment completely, or include it only in rudimentary form, e.g. as a galactic tidal field.

Ideally, to simulate the formation of star clusters self-consistently one would run a full galaxy simulation with enough resolution to form individual stars. Computational limits make this unfeasible. Recently approaches have been made to more fully represent stellar populations with individual star particles (e.g. Wall et al. 2019) even if the gas resolution is not correspondingly high. These allow for studying the dynamics of the stars and the gas simultaneously.

We use AMUSE(Portegies Zwart and McMillan 2018) to combine SPH with multi-scale *N*-body dynamics and stellar evolution in a new simulation method, which we name Ekster. With this method we can simulate the formation of individual stars, while it also allows us to take the galactic environment into account.

Here, we follow cluster formation with full N-body dynamics. We start our simulations by extracting gas form a section centred on a GMC and a section of a spiral arm. We do this using two different spiral galaxy models, with different strength spiral arms, as a means of examining the role of spiral arms.

2. The Ekster method

We design our method Ekster†(Rieder and Liow 2021) to combine an SPH simulation at relatively low resolution with a method to form individual stars from star
forming regions ("sinks") and dynamically evolve these. Ekster employs AMUSE as
the environment that combines these elements. We use Phantom (Price et al. 2018)
for gas hydrodynamics, Petar (Wang et al. 2020) for stellar dynamics and SeBa
(Portegies Zwart and Verbunt 1996) for stellar evolution. We couple the gravitational
dynamics between stars and gas using Bridge (Portegies Zwart et al. 2020), using a
global timestep of 0.0025 Myr.

2.1. *Gas*

Gas particles in Phantom are integrated on individual time steps, while a synchronisa-
tion timestep of half the global timestep is used. All gas particles in Phantom have an
equal mass (here: $1M_\odot$ per particle). When a star formation region forms, gas particles
accreted by this region are removed from Phantom.

2.2. *Star formation*

When gas reaches a density of $(10^{-18}$g cm^{-3} and passes additional checks following the
method in Price et al. (2018, paragraph 2.8.4), a sink particle will form and accrete all
gas particles within a radius of 0.25 parsec (similar to the method in Wall et al. (2019)).
This will accrete approximately $200M_\odot$, enough to probe the IMF without a dearth of
high-mass stars. The position and velocity of the sink particle are taken as the centre-
of-mass and centre-of-mass-velocity of the gas particles, while the velocity dispersion of
the gas is also saved as a property of the sink. This sink will then start forming stars
by drawing a random mass from a Kroupa (2001) initial mass function, creating a star
only if its mass is still higher than the mass of the star. The new star will be placed at
a random position within the accretion radius with a velocity drawn from a Gaussian
distribution centred on the velocity dispersion of the sink, and its mass is subtracted
from the sink's mass. This process continues until the sink no longer has enough mass to
form the next star.

2.3. *Stars*

Once stars have formed, they are added to both the stellar evolution and stellar gravity modules. Stellar gravity is integrated using a combined tree/direct N-body method,
with additional support for algorithmic regularisation to integrate binary stars. Stars are
integrated without any softening, allowing for the dynamical formation of binary stars.

3. Simulations

Our initial conditions are based on snapshots from each of two galaxy scale simulations,
one of which is the simulation shown in Dobbs and Pringle (2013). The other is identical
apart from the stronger spiral potential, and ran to provide initial conditions for this
work.

From each simulation, we take two regions. One region, which we denote 'cloud', centres
on a massive GMC, whilst the other region, which we denote 'arm', centres on a section
of spiral arm with a number of lower mass clouds. Thus we have two 'arm' regions, one
in each simulation and two 'cloud' regions, again one in each simulation.

---

† publicly available at https://github.com/rieder/ekster

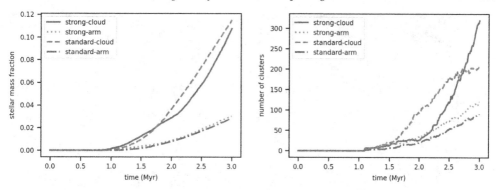

Figure 1. Left: Stellar mass fraction over time in the four simulations. Right: The number of clusters formed in the different models over time.

We use the two different models to investigate how cluster formation depends on the different galaxy models, and the different morphologies of the GMCs which form.

3.1. *Running the simulation*

After selecting the particles in our regions of interest, we re-sample the SPH particles following the method in Bending et al. (2020). Each original particle is split into 311 new particles of $1M_\odot$ each.

We run simulations of each of these regions with `Ekster`, using isothermal gas at 30 K. To preserve the large-scale environment of the original galactic simulation, we include the same tidal field used in the galaxy simulations.

Since our simulations do not include feedback, we limit our simulations to the embedded phase of star cluster formation, i.e. up to 3 Myr (Lada and Lada 2003). We save a snapshot every 0.01 Myr.

4. Results and discussion

We find that in all four simulations, star formation starts after $\sim$ 1 Myr (Figure 1, left panel). Both of the 'strong' models produce more clusters than their 'standard' equivalent by 3 Myr (Figure 1, right panel). Star formation is strongest in the 'cloud' regions of each simulation, but there is not a great difference between either the 'cloud', or the 'arm', regions from the 'strong' and 'standard' simulations. For the 'cloud' simulations, over 10% of the gas is converted to stars, whereas in the 'arm' regions it is around 3%, although we would expect that feedback may decrease these numbers.

In Figure 2 we show images of the column density of the four models in the final snapshot, with an inset showing the most massive cluster in each simulation. As expected, in the 'cloud' models stars have formed primarily towards the centre of the clouds, in what appears by eye to be more massive clusters. In the 'arm' models, clusters are more spread out along the total length of the arm. By eye, there is little obvious difference between the 'strong' and 'standard' models.

4.1. *Cluster evolution over time*

We compare the evolution of the four largest star clusters between similar regions in the two galaxies, as well as between different regions in the same galaxy. Figure 3 shows that cluster masses in the 'cloud' models grow to considerably larger values than in the 'arm' models.

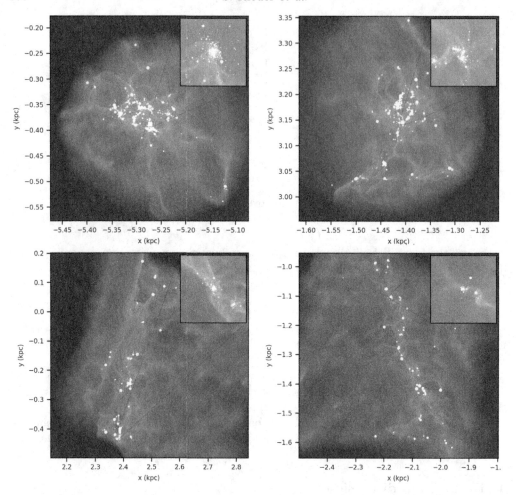

Figure 2. The four regions are shown at 3 Myr. The inset regions are 10 parsec wide and zoom in on the largest star cluster of each simulation. Top row: strong-cloud (left) and standard-cloud (right). Bottom row: strong-arm (left) and standard-arm (right).

Initially, the mass of the clusters increases linearly, indicating a steady conversion of gas into stars. Mergers of clusters happen more frequently in the 'cloud' simulations than the 'arm' simulations.

In the 'arm' simulations, growth slows down for most clusters after ∼ 1Myr, while in the 'cloud' simulations clusters continue to grow. This is caused by the stars having used up all the gas in their surroundings or because the stars have decoupled from the gas. In the 'cloud' simulations, this decoupling seems to not take place and as a result the clusters can grow larger.

4.2. *Cluster properties*

In Figure 4 we plot the cluster mass versus the half-mass radii, again at 3 Myr. Generally there is not a strong dependence of radius versus mass, as seen in observations (Portegies Zwart et al. 2010), though it is consistent with the relation in Marks and Kroupa (2012, eq. 7). We also show the young massive clusters Portegies Zwart et al. (2010, and references therein) plotted for comparison as open circles, which are similarly aged to our clusters ($<= 3.5$ Myr). We see again that the

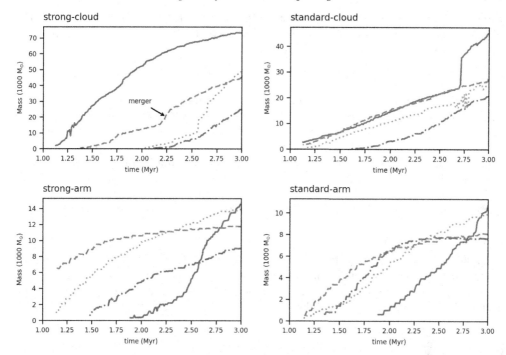

Figure 3. The mass of the four largest clusters from the four different simulations. Top left (green line) indicates an example of a merger. Wiggles in the orange line from the top right figure originate from a similar merger where the two progenitors have similar mass.

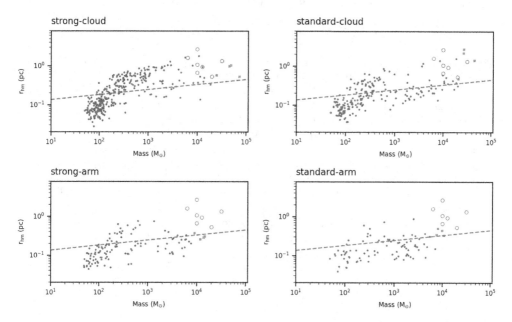

Figure 4. Mass versus half-mass radius for clusters at 3.0 Myr. The most massive clusters (orange squares) have properties comparable to the similarly aged observed young massive clusters (open circles). The dashed line indicates the relation from Marks and Kroupa (2012, Eq. 7).

'cloud' regions produce more massive star clusters than the 'arm' regions. When comparing our simulated clusters to the observed ones, we find that the more massive of our clusters have similar masses and half-mass radii.

5. Conclusions

We have simulated the formation of star clusters using the `Ekster` method that combines high-precision N-body dynamics with SPH and stellar evolution. By comparing two GMC regions and two spiral arm regions from two different galaxies, we find that the GMCs are able to form larger star clusters in a shorter time compared to typical molecular clouds. This is independent of the galaxy scale simulation.

We also find that clusters partially grow by merging with other (proto-)clusters. We find that our more massive clusters have similar properties to observed young massive clusters, including a fairly constant mass radius relation.

Acknowledgements

SR acknowledges funding from STFC Consolidated Grant ST/R000395/1 and the European Research Council Horizon 2020 research and innovation programme (Grant No. 833925, project STAREX). CLD acknowledges funding from the European Research Council for the Horizon 2020 ERC consolidator grant project ICYBOB, grant number 818940.

References

S. J. Aarseth. *A&A*, 35(2):237–250, Oct. 1974.

M. R. Bate, I. A. Bonnell, and V. Bromm. *MNRAS*, 339(3):577–599, Mar. 2003.

T. J. R., Bending, C. L., Dobbs, and M. R. Bate. *MNRAS*, 495(2):1672–1691, June 2020.

C. L. Dobbs and J. E. Pringle. *MNRAS*, 432(1):653–667, Jun 2013.

P. Kroupa. *MNRAS*, 322(2):231–246, Apr 2001.

C. J. Lada and E. A. Lada. *ARA&A*, 41:57–115, Jan. 2003.

M. Marks and P. Kroupa. *A&A*, 543:A8, July 2012.

S. Portegies Zwart and S. McMillan. 2514-3433. IOP Publishing, 2018. ISBN 978-0-7503-1320-9. URL http://dx.doi.org/10.1088/978-0-7503-1320-9.

S. Portegies Zwart, I. Pelupessy, C. Martínez-Barbosa, A. van Elteren, and S. McMillan. *Communications in Nonlinear Science and Numerical Simulations*, 85:105240, June 2020.

S. F. Portegies Zwart and F. Verbunt. *A&A*, 309:179–196, May 1996.

S. F. Portegies Zwart, S. L. W. McMillan, and M. Gieles. *ARA&A*, 48:431–493, Sep 2010.

D. J. Price, J. Wurster, T. S. Tricco, C. Nixon, S. Toupin, A. Pettitt, C. Chan, D. Mentiplay, G. Laibe, S. Glover, C. Dobbs, R. Nealon, D. Liptai, H. Worpel, C. Bonnerot, G. Dipierro, G. Ballabio, E. Ragusa, C. Federrath, R. Iaconi, T. Reichardt, D. Forgan, M. Hutchison, T. Constantino, B. Ayliffe, K. Hirsh, and G. Lodato. *Publ. Astron. Soc. Australia*, 35:e031, Sep 2018.

S. Rieder and K. Y. Liow, Sept. 2021. URL https://doi.org/10.5281/zenodo.5520944.

J. E. Wall, S. L. W. McMillan, M.-M. Mac Low, R. S. Klessen, and S. Portegies Zwart. *ApJ*, 887(1):62, Dec. 2019.

L. Wang, M. Iwasawa, K. Nitadori, and J. Makino. *MNRAS*, 497(1):536–555, Sept. 2020.

The Predictive Power of Computational Astrophysics as a Discovery Tool
Proceedings IAU Symposium No. 362, 2023
D. Bisikalo, D. Wiebe & C. Boily, eds.
doi:10.1017/S1743921322001259

The CNN classification of galaxies by their image morphological peculiarities

D. Dobrycheva[1], V. Khramtsov[2], M. Vasylenko[1,3] and I. Vavilova[1]

[1]Main Astronomical Observatory of the NAS of Ukraine, 27 Akademik Zabolotny Str., Kyiv, 03143, Ukraine

[2]Institute of Astronomy, V.N. Karazin Kharkiv National University, 35 Sumska St., Kharkiv, 61022, Ukraine

[3]Institute of Physics of the National Academy of Sciences of Ukraine, 46 avenue Nauka, Kyiv, 03028, Ukraine

Abstract. Multidimensional mathematical analysis, like Machine Learning techniques, determines the different features of objects, which is difficult for the human mind. We create a machine learning model to predict galaxies' detailed morphology ($\sim$300000 SDSS-galaxies with $z < 0.1$) and train it on a labeled dataset defined within the Galaxy Zoo 2 (GZ2). We use convolutional neural networks (CNNs) to classify the galaxies into five visual types (completely rounded, rounded in-between, smooth cigar-shaped, edge-on, and spiral) and 34 morphological classes attaining $> 94\%$ of accuracy for five-class morphology prediction except for the cigar-shaped ($\sim$87%) galaxies.

Keywords. methods: data analysis – galaxies: general– surveys – methods: convolutional neural networks, etc.

1. Introduction

The morphological classifications of galaxies play a vital role in reflecting the evolutionary history of various types of galaxies and the large-scale structure of the Universe as a whole (Barrow and Saich (1993); Peng et al. (2010); Reid et al. (2012); Dobrycheva et al. (2018); Vavilova et al. (2021a, 2020a); Elyiv et al. (2020)).

Basically, the morphological classification of galaxies is manual and requires extensive use of human resources or from highly qualified professionals, or in some cases, amateur astronomers and volunteers (for example, the Galaxy Zoo project, GZ, Willett et al. (2013)). Current and near-term galaxy observational surveys are approaching the Exabyte scale multiwavelength databases of hundreds of millions of galaxies, which is impossible to classify manually (Vavilova et al. (2020b)). That magnifies the interest to use the alternatives in the form of machine learning (ML) techniques, including deep learning (DL), for classification of galaxies by their features.

We found that Support Vector Machine gives the highest accuracy for binary morphological classification with photometry-based approach, namely 96.1 % early E and 96.9 % late L types of galaxies (Vavilova et al. (2021b)). We exploited different galaxy classification techniques (human labeling, multi-photometry diagrams, and five supervised ML methods). The photometry-based approach was applied to the SDSS DR9 dataset, which contained $\sim 310\,000$ of galaxies with redshifts of $0.02 < z < 0.1$ and absolute stellar magnitudes of $-24^m < M_r < -19.4^m$ (see, more details about this sample, data cleaning, and

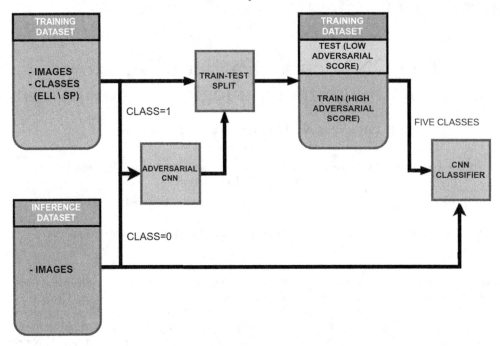

Figure 1. Scheme of our approach for morphological classification of galaxies with CNN.

correction procedures by Dobrycheva (2013); Dobrycheva et al. (2018); Vasylenko et al. (2019); Vasylenko (2020); Vavilova et al. (2021)). We tested such photometry parameters of galaxies as stellar magnitudes, color indices, inverse concentration indexes, which well correlate with the morphological type.

The aim of this work is to present the results of applying the machine learning model to the labeled images of galaxy dataset for prediction of the morphological type (morphological peculiarities) of galaxies.

2. Methods and Results

We continue to work with the above-mentioned sample, which included $\sim 310\,000$ galaxies (Dobrycheva (2013); Vavilova et al. (2021)). First of all, we have estimated how many galaxies from our sample belonged to the Galaxy Zoo 2 (GZ2) sample. We determined that more than a half of our sample do match with GZ2. So we divided our sample into two sub-samples (Fig. 1): **Training sample**: $\sim 170\,000$ galaxies (which do match GZ2 dataset); **Target sample (inference)**: $\sim 140\,000$ galaxies (which do not match GZ2 dataset).

Next, we used an adversarial neural networks to compare these two subsamples (Training and Inference). We trained the Convolutional Neural Network (CNN) on all the galaxy images of our sample, passing the class '0' for inference dataset and class '1' for the training one. The main idea of this procedure is to analyze potential differences between images from two datasets, because a classifier trained on one domain may behave incorrectly on another one. It turned out that the inference dataset can be easily separated from the GZ2 dataset with using ResNet-50 CNN. The accuracy of such separation is $\approx 90\%$. This occurred mostly because the galaxies from the sample of $\sim 310\,000$ were pre-selected via $m_r < 17.7$ limitations by stellar magnitude in r-band. Analyzing properties of the galaxies from both datasets, we observed that the galaxies from the target (inference) sample are, on average, fainter and smaller (90% Petrosian flux) than

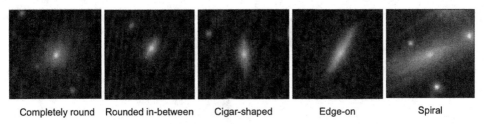

Completely round   Rounded in-between   Cigar-shaped   Edge-on   Spiral

Figure 2. Examples of galaxy images from the GZ2 randomly selected from each of five morphological classes.

the galaxies from the training GZ2 sample. So, we did not use the same architectures for the adversarial CNN and final CNN classifiers. The main aim of adversarial validation is to investigate differences between training and inference galaxy samples.

So, according to the adversarial result, we can conclude that our training sample contains galaxies, properties of which are not common with inference one. This means that any validation of the morphological classifier has to be done with the galaxies from the training set, which have a low adversarial score. But we have found a way out of this obstacle.

We took into consideration only those galaxies from the training sample for which GZ2's volunteers gave the most votes for a more accurate result. It turned out to be $\sim 72\,000$ galaxies. To do so, we randomly choose $\sim 9\,000$ galaxies with adversarial score less than 0.7 from the training sample of $\sim 72\,000$ galaxies (comprising five different morphological classes, Fig. 2). Within this train-test split, the test part of training galaxies ($\sim 9\,000$) was used to validate the morphology by CNN classifier, and the rest part of galaxies ($\sim 63\,000$) to train the CNN. We have added the augmentation procedures to increase the validation sample for prediction of the classes of fainter and smaller galaxies.

Also, we did work on the selection of the best neural network for our task (described in more detail in the works by Khramtsov et al. (2019); Vasylenko (2020)). We used `DenseNet-201` and images of galaxies. Images were requested from the SDSS cutout server. We have retrieved RGB images composed of *gri* bands colour scaling, each of $100\times100\times3$ pixels (39.6×39.6 arcsec in each channel of the RGB image, respectively).

Confusion matrix for the classification CNN model for the pre-selected test sample of $\sim 9\,000$ galaxies is presented in Fig. 3. Each row represents the fraction of galaxies from a certain class classified as galaxies from other classes. One can see that the cigar-shaped sample has the lower accuracy. We see at least two reasons. First of all, the votes of volunteers for cigar-shaped galaxies were quite different, see, for example, their forum ZooUniverse. The opinions on classifying such a galaxy as the "flat but with an angle toward the tips and no structures like a bulge visible" are varied from the "cigar-shape elliptical" to the "spiral edge-on bulge". Secondly, this misclassification becomes more evident for small-sized galaxies at higher redshifts.

We have found that the inference catalog comprises 27 378 completely round, 59 194 round in-between, 18 862 cigar-shaped, 7 831 edge-on, and 23 119 spiral galaxies (see, examples, in Fig. 2).

3. Conclusion

Our approach and developed method, which includes CNN model and adversarial validation, allows one to classify galaxies with the SDSS images into five classes automatically (completely round, round in-between, cigar-shaped, edge-on, and spiral). It has the state-of-art performance giving $> 94\%$ of accuracy for all classes, except cigar-shaped galaxies ($\sim 87\%$). Each of the five classes has an uneven number of samples that could led to

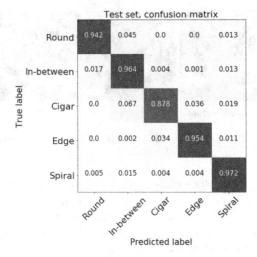

Figure 3. Confusion matrix for the classification CNN model for the pre-selected test sample of 9,000 galaxies. Each row represents the fraction of galaxies from a certain class (defined at the horizontal axis), classified as galaxies from other classes.

the categorical bias and measurable impact on overall accuracy. Due to the uneven class distributions, an overall accuracy metric may be not the best option. But we do not use it providing the accuracy scores per each class and noting that the accuracy is > 94% for all the classes except cigar-shaped galaxies. The preliminary visual inspection of classification shows the excellent agreement between the estimated classes and morphological parameters of the galaxies with their corresponding images.

4. Acknowledgements

This work was done in frame of the budgetary program "Support for the development of priority fields of scientific research" of the NAS of Ukraine ("e-Astronomy", CPCEL 6541230, 2020-2021). Vavilova I.B. thanks the Wolfgang Pauli Institute, Vienna, Austria, for the support in frame of "The Pauli Ukraine Project" (2022) under the "Models in plasma, Earth and space science" program.

References

Barrow, J.D. and Saich, P. 1993, *MNRAS*, 3, 717
Dobrycheva, D.V. 2013, *Odessa Astron. Publ.*, 26, 187
Dobrycheva, D.V., Vavilova, I.B., Melnyk, O.V. et al. 2018, *Kinemat. Phys. Cel. Bodies*, 34, 290
Elyiv, A.A., Melnyk, O.V., Vavilova, I.B. et al. 2020, *AA*, 635, A124
Khramtsov, V., Dobrycheva, D.V., Vasylenko, M.Y. et al. 2019, *Odessa Astron. Publ.*, 32, 21
Peng, Y., Lilly, S.J., Kovac, K. et al. 2010, *ApJ*, 721, 193
Reid, B.A., Samushia, L., White, M. et al. 2010, *MNRAS*, 426, 2719
Vasylenko, M. Y., Dobrycheva, D. V., Vavilova, I. B. et al. 2019, *Odessa Astron. Publ.*, 32, 46
Vasylenko, M., Dobrycheva, D., Khramtsov, V., et al. 2020, *Communications of BAO*, 67, 354
Vavilova, I., Dobrycheva, D., Vasylenko, M. et al. 2020, in: P. Skoda & F. Adam, *Knowledge Discovery in Big Data from Astronomy and Earth Observation* (Elsevier), p. 307–323
Vavilova, I., Pakuliak, L., Babyk, I. et al. 2020, in: P. Skoda & F. Adam, *Knowledge Discovery in Big Data from Astronomy and Earth Observation* (Elsevier), p. 57–102
Vavilova, I., Elyiv, A., Dobrycheva, D., Melnyk, O. 2021, in: I. Zelinka, M. Brescia, & D. Baron *Intelligent Astrophysics* (Springer), p. 57–79
Vavilova, I.B., Dobrycheva, D.V., Vasylenko, M.Y. et al. 2021, *AA*, 648, A122

Vavilova, I.B., Dobrycheva, D.V., Vasylenko, M.Yu. et al. 2018, *VizieR Online Data Catalog*, J/A+A/648/A122

Willett, K.W., Lintott, C.J., Bamford, S.P. et al. 2013, *MNRAS*, 435, 2835

Discussion

D. BISIKALO: I have question on accuracy of this method.

D. DOBRYCHEVA: Accuracy classification of galaxies of five classes is 94 % for completely round, 96 % for round in-between, 87 % for cigar-shaped, 95 % for edge-on, and 97 % for spiral galaxies.

D. BISIKALO: Does it mean that 94 % of all objects are classified correctly?

D. DOBRYCHEVA: Yes, correctly. For classification of features of galaxies is the same situation, likely for arms, bars, merging.

T. HANAWA: Could you show the confusion matrix again? The confusion comes from the data quiality or not? As you mentioned, some galaxies are less bright. And do you find any tendency that less bright galaxies are classified worse? I suppose that dim galaxies are hardly classified.

D. DOBRYCHEVA: We had a big sample of 170000 galaxies from Galaxy Zoo. We have done data cleaning. Then we have 72000 galaxies, which were very well classified by volunteers. So, we didn't work with galaxies with low score. We selected randomly 9000 galaxies and used them as a test dataset for CNN.

T. HANAWA: I am interested in whether the confused galaxies are less bright or as bright as on average.

C. BOILY: I have a similar question. I understand that eventually fainter galaxies should be more and more of those detected and, so, they will be more difficult to classify. I would imagine for all sorts of reasons, a lower signal noise and that kind of issue.

But my question was more about the SDSS dataset, which has something likely, I think, 4 or 5 wave bands. Are you able to remind what are the actual wave bands you used to define the reference datasets? And do you think that there are some biases (that are possible) if you choose one wave band, one color as opposed to another, for example, to look at the images of the galaxies because the morphology is depending on the color.

D. DOBRYCHEVA: As about deep galaxies, we worked with a redshift less than 0.1 and in this range ...

T. HANAWA: I suppose that Chris means that irregular galaxies are bluer than round ones.

C. BOILY: There would be a bias due to the color, for example, of the stellar population.

The Predictive Power of Computational Astrophysics as a Discovery Tool
Proceedings IAU Symposium No. 362, 2023
D. Bisikalo, D. Wiebe & C. Boily, eds.
doi:10.1017/S1743921322001521

Velocity-space substructures and bar resonances in an N-body Milky Way

Tetsuro Asano[1]†, Michiko S. Fujii[1], Junich Baba[2], Jeroen Bédorf[3,5], Elena Sellentin[3,4] and Simon Portegies Zwart[3]

[1]Department of Astronomy, Graduate School of Science, The University of Tokyo, 7-3-1 Hongo, Bunkyo-ku, Tokyo 113-0033, Japan

[2]National Astronomical Observatory of Japan, Mitaka-shi, Tokyo 181-8588, Japan

[3]Leiden Observatory, Leiden University, NL-2300RA Leiden, The Netherlands

[4]Mathematical Institute, Leiden University, NL-2300RA Leiden, The Netherlands

[5]Minds.ai, Inc., Santa Cruz, the United States

Abstract. The velocity-space distribution of the solar neighborhood stars shows complex substructures (moving groups) including the well-known Hercules stream. Recently, the Gaia observation revealed their detailed structures, but their origins are still in debate. We analyzed a high-resolution N-body simulation of a Milky Way (MW)-like galaxy. To find velocity-space distributions similar to that of the solar neighborhood stars, we used Kullback-Leibler divergence (KLD), which is a metric to measure similarities between probability distributions. The KLD analysis shows the time evolution and the spatial variation of the velocity-space distribution. Velocity-space distributions with small KLDs (i.e. high similarities) are frequently but not always detected around $(R, \phi) = (8.2\,\mathrm{kpc}, 30°)$ in the simulated MW. In the velocity-map with smallest KLD, the velocity-space substructures are made from bar resonances.

Keywords. Galaxy: disk, Galaxy: kinematics and dynamics, (Galaxy:) solar neighborhood, Galaxy: structure, methods: numerical

1. Introduction

The *Gaia* (Gaia Collaboration *et al.* 2016) observation has revealed the detailed phase-space distribution of the stars in the Milky Way (MW). Fig. 1 shows the velocity-space distribution of the solar neighbor stars observed in the *Gaia* Early Data Release 3 (EDR3; Gaia Collaboration *et al.* 2021). The names and the locations of the major substructures are indicated in the figure. The most prominent substructure is the Hercules stream. Dehnen (2000) demonstrated that the Hercules stream can be explained by trapping in the 2:1 outer Lindblad resonance (OLR) of the bar using test particle simulations. This model requires the fast rotating bar with the pattern speed of $\Omega_{\rm b} \gtrsim 50\,\mathrm{km\,s^{-1}\,kpc^{-1}}$, but the recent pattern speed measurements of the MW's bar suggest the slower bar of $\Omega_{\rm b} \lesssim 40\,\mathrm{km\,s^{-1}\,kpc^{-1}}$ (e.g. Bovy *et al.* 2019; Clarke *et al.* 2019; Sanders *et al.* 2019).

In the case of the slow bar, trapping in the corotation resonance (CR) well explains the structures of the Hercules stream (e.g. Pérez-Villegas *et al.* 2017; Binney 2020; Chiba *et al.* 2021). Higher order resonances are also the candidates for the origins of the Hercules stream and the other substructures (e.g. Hunt & Bovy 2018; Hattori *et al.* 2019; Monari *et al.* 2019; Kawata *et al.* 2021; Trick *et al.* 2021). Some studies suggested

† email: t.asano@astron.s.u-tokyo.ac.jp

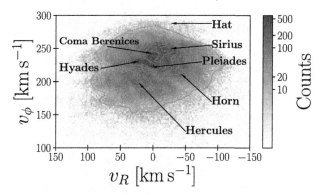

Figure 1. Solar neighbourhood star distribution in radial velocity versus azimuthal velocity $(v_R\text{-}v_\phi)$ space.

that the velocity-space substructures originate from the spiral arms (e.g. Hunt *et al.* 2019; Khoperskov *et al.* 2021) or external perturbations by the satellite galaxies (e.g. Laporte *et al.* 2019; Hunt *et al.* 2021).

Most of the previous studies use the test particle simulations. They assume static and symmetric gravitational potentials. Note that Chiba *et al.* (2021) analytically modeled the influences of the bar's slowing down.

Asano *et al.* (2020) used an *N*-body MW model simulated by Fujii *et al.* (2019) and found a Hercules-like stream made from 4:1 OLR and 5:1 OLR. This study confirmed that bar resonances form velocity-space substructures even in the non-static *N*-body models. We analyze the same simulation data but put the focus on the time evolution and the spatial variation of the velocity-space distribution. To evaluate the similarity of the velocity-space distribution in the simulation and that in the *Gaia* observation, we use the Kullback-Leibler divergence.

2. Methods

We analyzed the model MWaB, which is one of the *N*-body models simulated by Fujii *et al.* (2019). The numbers of the bulge, disk, and dark-matter halo particles are 30M, 208M, and 4.9B, respectively. The simulations were performed using the parallel GPU tree-code, `BONSAI` (Bédorf *et al.* 2012; Bédorf *et al.* 2014). Fig. 2 shows the face-on view of the model at $t = 10$ Gyr. Fig. 3 shows the bar's pattern speed and the bar's strength as functions of time. We determined them using the Fourier decomposition:

$$\frac{\Sigma(R,\phi)}{\Sigma_0(R)} = \sum_{m=0}^{\infty} A_m(R) \exp\{im[\phi - \phi_m(R)]\}, \qquad (2.1)$$

where $\Sigma(R,\phi)/\Sigma_0(R)$ is the disk surface density normalized at each radius. The bar pattern speed (Ω_b) and the strength are the rate of change of $\phi_2(R < 3\,\mathrm{kpc})$ and A_2, respectively. For details, see Fujii *et al.* (2019) and Asano *et al.* (2020).

We use Kullback-Leibler divergence (KLD) to quantitatively evaluate the similarities between the velocity-space distributions in the simulation and that in the *Gaia* EDR3 observation. To compute the KLD, we first determine a probability distribution in the velocity space for the stars within 200 pc from the Sun. We divide the $\hat{v}_R$ versus $\hat{v}_\phi$ space in a range of $(-0.5, 0.5) \times (-0.333, 0.333)$ into 48×32 cells, where the $\hat{v}_R$ and $\hat{v}_\phi$ are the relative velocities defined as $\hat{v}_R \equiv (v_R - \bar{v}_R)/\bar{v}_\phi$ and $\hat{v}_\phi \equiv (v_\phi - \bar{v}_\phi)/\bar{v}_\phi$, respectively. The mean velocities $(\bar{v}_R, \bar{v}_\phi)$ are determined for the stars within 200 pc from the Sun. We obtain the probability of which we find a star in the i-th cell, p_i, diving the number of the stars in the cell by the total number of the stars. We also determine the counterpart

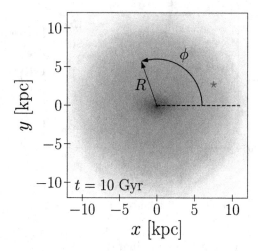

Figure 2. Faceon view of the N-body model. The star indicates the position of $(R, \phi) = (8\,\mathrm{kpc}, 20°)$.

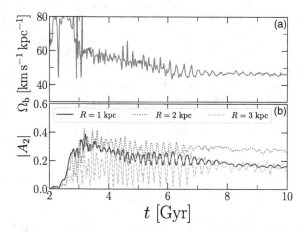

Figure 3. (a) Time evolution of the bar's pattern speed. (b) Time evolution of the bar's strength $|A_2|$ at $R = 1$, 2, and 3 kpc.

of p in the N-body model, q, which is evaluated at the 648 points of

$$
\begin{cases}
R = (6 + 0.5i)\,\mathrm{kpc} & (i = 0, \ldots, 8), \\
\phi = (-180 + 5j)° & (j = 0, \ldots, 71), \\
z = 0\,\mathrm{kpc},
\end{cases}
\tag{2.2}
$$

in each snapshot. The KLD between p and q is defined by

$$
\mathrm{KLD}(p\|q) = \sum_{i=1}^{N} p_i \log \frac{p_i}{q_i}.
\tag{2.3}
$$

It is a quantity like 'distance' between p and q since it satisfies (i) $\mathrm{KLD}(p\|q) \geq 0$ and (ii) $\mathrm{KLD}(p\|q) = 0$ if and only if p and q are identical. Therefore, The smaller the KLD is, the more q is similar to p.

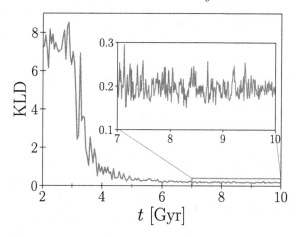

Figure 4. Time evolution of the KLD of the velocity-space distribution at $(R, \phi) = (8\,\text{kpc}, 20°)$.

3. Results

3.1. *Time evolution of the velocity-space distribution*

Fig. 4 shows the KLD of the velocity-space distribution at $(R, \phi) = (8\,\text{kpc}, 20°)$ as a functions of time. We select this position as a reference point because Asano *et al.* (2020) found a Hercules-like stream by eyes in the final snapshot ($t = 10\,\text{Gyr}$). The KLD is smaller in the later epochs of the simulation than in the earlier epochs. This is because Fujii *et al.* (2019) adjusted the initial parameters of the model so that the disk and bulge properties at the end of the simulation well reproduce the observed ones. The KLD drops rapidly at $t \simeq 3\,\text{Gyr}$, at which a clear bar structure appears. From $t \simeq 3\,\text{Gyr}$ to $t \simeq 7\,\text{Gyr}$, the KLD decreases as the bar's pattern speed slows down. The correlation of the KLD and the bar evolutions indicates that bar affects the velocity-space stellar distribution. The KLD oscillates after $t \simeq 7\,\text{Gyr}$, which means the velocity-space distribution at $(R, \phi) = (8\,\text{kpc}, 20°)$ is not always similar to that of the solar neighborhood.

3.2. *Spatial variation of the velocity-space distribution*

In this section, we investigate where in the disk we frequently detect velocity-space distributions similar to that of the solar neighborhood. We define that a velocity-space distribution is similar to that of the solar neighborhood if its KLD is less than 0.2, which is the KLD of the velocity-space distribution at $(R, \phi) = (8, \text{kpc}, 20°)$ at $t = 10\,\text{Gyr}$. Here, we analyzed the snapshots after $t = 7\,\text{Gyr}$. The panel (a) in Fig. 5 shows how many times the velocity-space distributions with KLD < 0.2 are detected at each angle with respect to the bar. The bar's major axis is aligned with the x-axis. We map the angles of $\phi < 0$ to $\phi + 180°$ in the figure since we focus only on the relative angles with respect to the major axis. The numbers of the detections of the small KLDs at $R = 7.5$ and $9\,\text{kpc}$ are more than four times smaller than those at $R = 8$, 8.2, and $8.5\,\text{kpc}$. The peaks of the histograms differ by R. The peak moves in the positive direction of ϕ as R increases. The histogram at $R = 8.2\,\text{kpc}$, which is the distance between the Sun and the Galactic center, has a peak at $\phi \simeq 30°$. This angle matches the observationally estimated bar angle (Bland-Hawthorn & Gerhard 2016). The R and ϕ dependence of the KLD indicates that bar resonances impact the velocity-space space distribution because the spacial distribution of the resonantly trapped stars is also dependent on R and ϕ (e.g. Ceverino & Klypin 2007; Khoperskov *et al.* 2020).

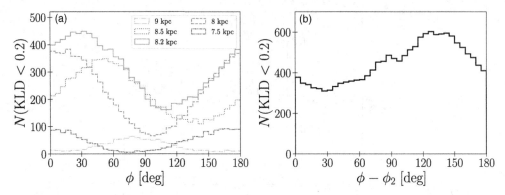

Figure 5. Histograms of the angles at which the KLD is less then 0.2. (a) The angle with respect to the bar. The yellow, red, green, blue, and purple lines shows the histograms at $R = 9$, 8.5, 8.2, 8, and 7.5 kpc, respectively. (b) The angle with respect to the spiral arms.

   The panel (b) in Fig. 5 shows the histogram of the angles as the panel (a) but with respect to the spirals ($\phi - \phi_2$). We define the spiral position as a phase angle of the Fourier $m = 2$ mode $\phi_2(R)$. Here, we limit the analysis at $R = 8$ and 8.5 kpc. The histogram shows the peak at $\phi - \phi_2 \simeq 130°$. This is the inter-arm region since the position of spiral arms are $\phi - \phi_2 = 0°$ and 180°. This is consistent with the observations of the spiral arms of the MW. The Sun locates in the inter-arm regions of the MW's spiral arms such as Perseus and Sagittarius-Carina (Reid *et al.* 2019). We note that the Sun is close to the 'Local arm', but that it is relatively weak.
   The fluctuation of the KLD seen in Fig. 4 may be due to the spiral arms. Dynamic transient spiral arms efficiently cause the radial migrations of stars (e.g. Baba *et al.* 2013). Spiral arms may make the stars escape from the bar resonances and weaken the bar's underlying impacts on the velocity-space distribution.

4. Summary

   We have evaluated the similarities between the velocity-space distributions in the simulation and that in the observation of the solar neighborhood stars using the KLD. We have obtained the following results.

(a) The time evolution of the KLD is linked with the bar's evolution. The KLD at $(R, \phi) = (8 \text{ kpc}, 20°)$ drops rapidly at the bar formation epoch. It decreases while the bar is decelerating.

(b) The KLD fluctuates even after the bar's slowdown, which indicates that the velocity-space distribution at $(R, \phi) = (8 \text{ kpc}, 20°)$ is not always similar to that of the solar neighborhood stars.

(c) The detection counts of the velocity-space distributions with small KLDs are strongly dependent on R and ϕ. One of the positions where the small KLDs are most frequently detected is $(R, \phi) = (8.2 \text{ kpc}, 30°)$, which matches the Sun's position in the MW.

(d) The detection frequency is higher in the inter-arm regions than the arm regions.

These results indicate that the bar resonances have significant impact on the velocity-space distribution even though the gravitational potential is not static. In Asano *et al.* (2021) we see that bar resonances form the substructures in the velocity-space distributions with small KLDs.

References

Asano, T., Fujii, M. S., Baba, J., Bédorf, J., Sellentin, E., Portegies Zwart, S., 2020, *MNRAS* 499, 2416

Asano, T., Fujii, M. S., Baba, J., Bédorf, J., Sellentin, E., Portegies Zwart, S., 2021, arXiv e-prints, arXiv:2112.00765

Baba, J., Saitoh, T. R., Wada, K., 2013, *ApJ* 763, 46

Bédorf, J., Gaburov, E., Portegies Zwart, S., 2012, *Journal of Computational Physics* 231, 2825

Bédorf, J., Gaburov, E., Fujii, M. S., Nitadori, K., Ishiyama, T., Portegies Zwart, S., 2014, in *International Conference for High Performance Computing, Networking, Storage and Analysis*, Proc. SC'14, p. 54

Binney, J., 2020, *MNRAS* 495, 895

Bland-Hawthorn, J., Gerhard, O., 2016, *ARAA* 54, 529

Bovy, J., Leung, H. W., Hunt, J. A. S., Mackereth, J. T., García-Hernández, D. A., Roman-Lopes, A., 2019, *MNRAS* 490, 4740

Ceverino, D., Klypin, A., 2007, *MNRAS* 379, 1155

Chiba, R., Friske, J. K. S., Schönrich, R., 2021, *MNRAS* 500, 4710

Clarke, J. P., Wegg, C., Gerhard, O., Smith, L. C., Lucas, P. W., Wylie S. M., 2019, *MNRAS* 489, 3519

Dehnen, W., 2000, *AJ* 119, 800

Fujii, M. S., Bédorf, J., Baba, J., Portegies Zwart, S., 2019, *MNRAS* 482, 1983

Gaia Collaboration *et al.* 2016, *A&A* 595, A1

Gaia Collaboration *et al.* 2018, *A&A* 616, A11

Gaia Collaboration *et al.* 2021, *A&A* 649, A1

Hattori, K., Gouda, N., Tagawa, H., Sakai, N., Yano, T., Baba, J., Kumamoto, J., 2019, *MNRAS* 484, 4540

Hunt, J. A. S., Bovy, J., 2018, *MNRAS* 477, 3945

Hunt, J. A. S., Bub, M. W., Bovy, J., Mackereth, J. T., Trick, W. H., Kawata, D., 2019, *MNRAS* 490, 1026

Hunt, J. A. S., Stelea, I. A., Johnston, K. V., Gandhi, S. S., Laporte, C. F. P., Bédorf, J, 2021, *MNRAS* 508, 1459

Kawata, D., Baba, J., Hunt, J. A. S., Schönrich, R., Ciucă, I., Friske, J., Seabroke, G., Cropper, M., 2021, *MNRAS* 508, 728

Khoperskov, S., Gerhard, O., Di Matteo, P., Haywood, M., Katz, D., Khrapov, S., Khoperskov, A., Arnaboldi, M., 2020, *A&A* 634, L8

Khoperskov, S., Gerhard, O., 2021, arXiv e-prints, arXiv:2111.15211

Laporte, C. F. P., Minchev, I., Johnston, K. V., Gómez, F. A., 2019, *MNRAS* 485, 3134

Monari, G., Famaey, B., Siebert, A., Wegg, C., Gerhard, O., 2019, *A&A* 626, A41

Pérez-Villegas, A., Portail, M., Wegg, C, Gerhard, O., 2017, *ApJ* 840, L2

Reid, M. J., *et al.*, 2019, *ApJ* 885, 131

Sanders, J. L., Smith, L., Evans, N. W., 2019, *MNRAS* 488, 4552

Trick, W. H., Fragkoudi, F., Hunt, J. A. S., Mackereth, J. T., White, S. D. M., 2021, *MNRAS* 500, 2645

The Predictive Power of Computational Astrophysics as a Discovery Tool
Proceedings IAU Symposium No. 362, 2023
D. Bisikalo, D. Wiebe & C. Boily, eds.
doi:10.1017/S1743921322001338

Fate of escaping orbits in barred galaxies

Debasish Mondal[1,a] and Tanuka Chattopadhyay[1,b]

[1]Department of Applied Mathematics, University of Calcutta,
92 A. P. C. Road, Kolkata 700009, India

E-mail: [a]dmappmath_rs@caluniv.ac.in (✉), [b]tchatappmath@caluniv.ac.in

Abstract. In the present work, we have developed a two-dimensional gravitational model of barred galaxies to analyse the fate of escaping stars from the central barred region. For that, the model has been analysed for two different bar profiles viz. strong and weak. Here the phenomena of stellar escape from the central barred region have been studied from the perspective of an open Hamiltonian dynamical system. We observed that the escape routes correspond to the escape basins of the two index-1 saddle points. Our results show that the formation of spiral arms is encouraged for the strong bars. Also, the formation of grand design spirals is more likely for strong bars if they host central super massive black holes (SMBHs). In the absence of central SMBHs, the formation of less-prominent spiral arms is more likely. Again, for weak bars, the formation of inner disc rings is more probable.

Keywords. Galaxy: evolution, Galaxy: kinematics and dynamics, Galaxies: bar, Chaos

1. Introduction

Stellar bars are quite common among the late-type galaxies. Observational studies reveal that bars are found in nearly 70% of the local disc galaxies (Eskridge et al. 2000). These robust stellar bodies are the result of rotational instabilities that arise in the galactic centre due to the density waves radiating from the core. These instabilities redistribute the stellar trajectories to generate a self-stabilising structure in the form of the bar (Bournaud & Combes 2002). Stellar bars have a paramount role behind the dynamical evolution of galaxies (Navarro & Henrard 2001; Ernst & Peters 2014).

Modelling of galaxies with a rotating bar component can be done via conservative or Hamiltonian dynamical systems. More specifically, stellar escapes from the bar ends can be studied from the viewpoint of escape phenomena observed in open Hamiltonian systems. An open Hamiltonian system is a dynamical system where for energies above an escape threshold, the energy shell is non-compact, and as a result a part of the stellar orbits explores (in our case from potential holes to saddles) an infinite part of the position space (Aguirre et al. 2001). In this dynamical set-up, the overall nature of the stellar orbits has categorized into following categories – (i) escaping and (ii) bounded. Bounded orbits are trapped inside the potential interior and exhibit both periodic (more generally quasi-periodic) and chaotic motions. On the other hand, escaping orbits are generally chaotic in nature. Now, the domains of bounded and chaotic motions in the $x - y$ plane have been observed via Poincaré surface section maps (Strogatz 1994) for different escape energies (i.e. energies higher than the escape threshold).

The orbital and escape dynamics of barred galaxies have been studied in the recent past and results are mainly concentrated towards the computation of the chaotic invariant manifolds near the bar ends and their role behind subsequent structure formations (Jung & Zotos 2016). In our work, we studied a two dimensional (2D) gravitational model of barred galaxies to study the orbital and escape dynamics of stars in the central region and also analysed the fate of escaping stellar orbits for two different bar profiles viz. strong (i.e. cuspy type) and weak (i.e. flat type).

2. Gravitational Model

To study the orbital and escape dynamics in barred galaxies, we developed a four component gravitational model in 2D (i.e. along the plane of the bar). This model consists of – (a) bulge, (ii) bar, (iii) disc and (iv) dark matter halo. In Cartesian coordinates (x, y), for a test particle of unit mass ($m = 1$), we construct the following quantities –

$$\text{total potential: } \Phi_t(x, y) = \Phi_B(x, y) + \Phi_b(x, y) + \Phi_d(x, y) + \Phi_h(x, y), \tag{2.1}$$

$$\text{effective potential: } \Phi_{\text{eff}}(x, y) = \Phi_t(x, y) - \frac{1}{2}\Omega_b^2(x^2 + y^2), \tag{2.2}$$

$$\text{Hamiltonian: } H(x, y, \dot{x}, \dot{y}) = \frac{1}{2}(\dot{x}^2 + \dot{y}^2) + \Phi_{\text{eff}}(x, y), \tag{2.3}$$

where Φ_B, Φ_b, Φ_d, Φ_h are potentials corresponding to the bulge, bar, disc and dark matter halo respectively, $\vec{\Omega}_b \equiv (0, 0, \Omega_b)$ (in clockwise sense) is the pattern speed of the bar and H is the Hamiltonian of this system. This gravitational model resembles a conservative dynamical system, in that case H is a constant of motion and equivalent to the total energy of the system (E). Hence, the equation of motion in the rotating reference frame of the bar is –

$$\ddot{\vec{r}} = -\vec{\nabla}\Phi_t - 2(\vec{\Omega}_b \times \dot{\vec{r}}) - \vec{\Omega}_b \times (\vec{\Omega}_b \times \vec{r}), \tag{2.4}$$

where $\vec{r} \equiv (x, y)$ and $\dot{\vec{r}} \equiv (\dot{x}, \dot{y})$ are position and linear momentum of the test particle at time t respectively, and $\vec{\nabla} \equiv (\frac{\partial}{\partial x}, \frac{\partial}{\partial y})$. Now, the Lagrangian (or equilibrium) points of the system are solutions of –

$$\frac{\partial \Phi_{\text{eff}}}{\partial x} = 0, \frac{\partial \Phi_{\text{eff}}}{\partial y} = 0. \tag{2.5}$$

Now, forms of the gravitational potentials for the bulge, bar, disc and dark matter halo are –

• Bulge: $\Phi_B(x, y) = -\frac{GM_B}{\sqrt{x^2+y^2+c_{B2}}}$ (Plummer 1911), where M_B is the bulge mass and c_B is the scale length.

• Strong Bar (Model 1): $\Phi_b(x, y) = -\frac{GM_b}{\sqrt{x^2+(\alpha y)^2+c_b^2}}$ (Caranicolas 2002), where M_b is the bar mass, α is the flattening parameter and c_b is the scale length.

• Weak Bar (Model 2): $\Phi_b(x, y) = \frac{GM_b}{2a} \ln(\frac{x-a+\sqrt{(x-a)^2+y^2+c_b^2}}{x+a+\sqrt{(x+a)^2+y^2+c_b^2}})$ (Jung & Zotos 2015), where M_b is the bar mass, a is the semi-major axis length and c_b is the scale length.

• Disc: $\Phi_d(x, y) = -\frac{GM_d}{\sqrt{x^2+y^2+(k+h)^2}}$ (Miyamoto & Nagai 1975), where M_d is the disc mass and k, h are the horizontal and vertical scale lengths, respectively.

• Dark matter halo: $\Phi_h(x, y) = \frac{v_0^2}{2} \ln(x^2 + \beta^2 y^2 + c_h^2)$ (Zotos 2012), where v_0 is the circular velocity, β is the flattening parameter and c_h is the scale length.

We analysed our gravitational model separately for two different bar profiles, namely – (i) strong bar (model 1) and (ii) weak bar (model 2). Now, without loss of any generality, we set G (gravitational constant) $= 1$ and other parameter values are given in Table 1 (Zotos 2012; Jung & Zotos 2016).
For these parameter values, we adopt the following scaling relations – unit of length: 1 kpc, unit of mass: $2.325 \times 10^7 M_\odot$, unit of time: 0.9778×10^8 yr, unit of velocity:

Table 1. Parameter values.

Parameter	Value	Parameter	Value
M_B	$9.3 \times 10^9 M_\odot$	M_d	$162.75 \times 10^9 M_\odot$
c_B	0.25 kpc	k	3 kpc
M_b	$81.375 \times 10^9 M_\odot$	h	0.175 kpc
c_b	1 kpc	v_0	150 km s^{-1}
α	2 kpc	β	1.3 kpc
a	10 kpc	c_h	20 kpc
Ω_b	12.5 km s^{-1} kpc^{-1}		

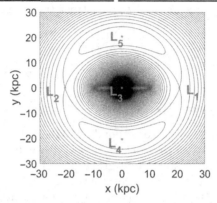

Figure 1. The isocontours of the effective potential—$\Phi_{\text{eff}}(x, y)$ in the $x - y$ plane, where locations of the Lagrangian points are marked in red.

10 km s^{-1}, Unit of angular momentum per unit mass: 10 km s^{-1} kpc^{-1} and unit of energy per unit mass: 100 km^2 s^{-2} (Jung & Zotos 2016). Now, from Eqs. 2.5, we calculate the locations of the Lagrangian points for both the strong and weak bar models. In both the models, system has five Lagrangian points, namely L_1, L_2, L_3, L_4 and L_5 (viz. Fig. 1). Among these Lagrangian points only L_1 and L_2 (classified as index-1 saddle points of the system) are corresponding to the bar ends i.e. responsible for stellar escapes. For the strong bar model the locations of L_1 and L_2 are $(\pm 20.23113677, 0)$ respectively, and for the weak bar model that locations are $(\pm 20.82978638, 0)$. Clearly, the bar area of model 2 is bigger than the model 1.

3. Computational Results

The effective potential term (Φ_{eff}) (viz. Eq. 2.1) is symmetric about the y axis and $E_{L_1} = E_{L_2}$, where E_{L_1} and E_{L_2} denotes the energy values of L_1 and L_2 respectively. Hence, studying the dynamics near either of L_1 and L_2 is sufficient to analyse the system. Now, escape of stars from the central barred region is only possible in the energy range: $E > E_{L_1}$, and for $E \leqslant E_{L_1}$ orbits are bounded inside the central barred region. To study the escaping motion, we integrate the Eq. 2.4 for a time-scale of 10^2 time units, which is equivalent to 10 Gyr i.e. the typical age of the bars (James & Percival 2018). For this orbit integration we use the `ode45` package of MATLAB. Further, in order to simplify the calculations, we adopt the dimensionless energy parameter $C = \frac{E_{L_1} - E}{E_{L_1}}$ (Jung & Zotos 2016). Hence, orbital escapes are possible for $C > 0$.

3.1. 2D Orbits

To study the nature of orbits near the Lagrangian point L_1 for both the bar models, we choose an initial condition $x_0 = 10$, $y_0 = 0$, $\dot{x}_0 = 15$, where $\dot{y}_0$ is evaluated from Eq. 2.3, and the corresponding trajectories in the $x - y$ plane are drawn for several values of $C > 0$

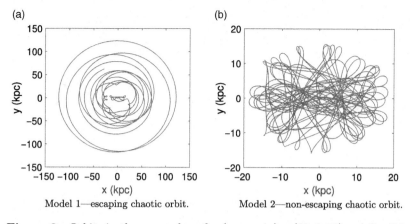

(a)

(b)

Model 1—escaping chaotic orbit.

Model 2—non-escaping chaotic orbit.

Figure 2. Orbits in the $x-y$ plane for $(x_0, y_0, \dot{x}_0) \equiv (10, 0, 15)$ and $C = 0.1$.

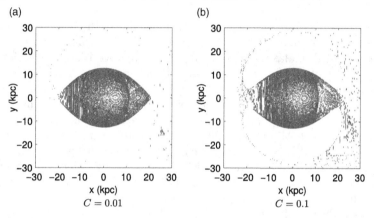

(a)

(b)

$C = 0.01$

$C = 0.1$

Figure 3. Model 1—Poincaré surface sections of $\dot{x} = 0$ and $\dot{y} \leqslant 0$.

(see Fig. 2). Any other initial condition in the suitable neighbourhoods of the aforesaid initial condition will follow the similar trend.

3.2. *Poincaré Maps*

Poincaré surface section maps in the $x-y$ plane for both the bar models are shown in Fig. 3 and Fig. 4 for several values of $C > 0$. In order to construct the Poincaré surface section maps, we choose a 43×43 gird of initial conditions i.e. (x_0, y_0) in the $x-y$ plane with restriction: $(x_0^2 + y_0^2) < r_{L_1}^2$, where r_{L_1} is the radial length of L_1. Also, $\dot{x}_0 = 0$ and $\dot{y}_0(>0)$ is evaluated from Eq. 2.3. In these maps our chosen surface cross sections are $\dot{x} = 0$ and $\dot{y} \leq 0$.

4. Conclusions

From all the above analyses our findings are –

• Model 1 (strong bar): For galaxies with strong bars, escape of stars from the central region has been encouraged (Fig. 2(a)). Again, the amount of escape from the bar ends has been increased with increment in the escape energy (Figs. 3(a) and 3(b)). Hence, for strong bars, the formation of spiral arms is more likely as a result of escape. Here, the increment in escape for higher escape energy values have been interpreted by the violence (viz. baryonic feedbacks from supernova, shocks etc.) occurring in the central region (Melia & Falcke 2001). We know, the central black hole has a strong influence on

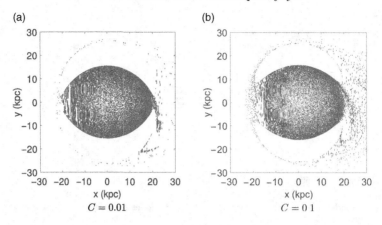

Figure 4. Model 2—Poincaré surface sections of $\dot{x} = 0$ and $\dot{y} \leqslant 0$.

the kinetic energy of random motions inside the bulge due to the central violence. Further, these kinetic energy of random motions is strongly related with spiral arm strength (Al-Baidhany et al. 2014). Galaxies with central Super Massive Black Holes (SMBHs) have more kinetic energy of random motions inside the bulge and spiral arms with small pitch angles (Seigar et al. 2008). Hence, for strong barred galaxies with central SMBHs, the formation of tightly wound or full-fledged spiral arms is more likely. This is the case of giant spiral galaxies, where grand design spiral arms are observed. Few examples of such giant spiral galaxies are Milky Way, NGC 1300 (Helou et al. 1991) etc. While, in the absence of central SMBHs, galaxies have competitively lesser kinetic energy of random motions inside the bulge and spiral arms with competitively higher pitch angles. Hence, for strong barred galaxies without central SMBHs, the formation of less prominent spiral arms is more likely.

• Model 2 (weak bar): For galaxies with weak bars, escape of stars from the central region has not been encouraged (Fig. 2(b)). Here also, the amount of escape from the bar ends has been increased with increment in the escape energy (Figs. 4(a) and 4(b)). Hence, for weak bars, the formation of inner disc rings is more likely as a result of escape. While, the increment in violence occurring in the central region may strengthen the ring patterns. Hence, for weak barred galaxies, the formation of inner disc rings is more likely. Few examples of such ring galaxies are NGC 1533, NGC 6028 (Helou et al. 1991) etc.

References

Aguirre, J., Vallejo, J. C., Sanjuán, M. A. F. 2001, *Phys. Rev. E*, 64, 066208
Al-Baidhany, I., Seigar, M., Treuthardt, P., *et al.* 2014, *J. Ark. Acad. Sci.*, 68, 25
Bournaud, F., Combes, F. 2002, *A&A*, 392, 83
Caranicolas, N. D. 2002, *J. Astrophys. Astron.*, 23, 173
Ernst, A., Peters, T. 2014, *MNRAS*, 443, 2579
Eskridge, P. B., Frogel, J. A., Pogge, R. W., *et al.* 2000, *AJ*, 119, 536
Helou, G., Madore, B. F., Schmitz, M., *et al.* 1991, *Databases & On-Line Data in Astronomy*, 171, 89
James, P. A., Percival, S. M. 2018, *MNRAS*, 474, 3101
Jung, C., Zotos, E. E. 2015, *PASA*, 32, e042
Jung, C., Zotos, E. E. 2016, *MNRAS*, 457, 2583
Melia, F., Falcke, H. 2001, *ARA&A*, 39, 309
Miyamoto, M., Nagai, R. 1975, *PASJ*, 27, 533
Navarro, J. F., Henrard, J. 2001, *A&A*, 369, 1112
Plummer, H. C. 1911, *MNRAS*, 71, 460

Seigar, M. S., Kennefick, D., Kennefick, J., Lacy, C. H. 2008, *ApJ*, 678, L93

Strogatz, S. H. 1994, *Nonlinear Dynamics and Chaos: With Applications to Physics, Biology, Chemistry, and Engineering*, Addison-Wesley

Zotos, E. E. 2012, *Res. Astron. Astrophys.*, 12, 500

Discussion

MELLEMA: Question: The rotation pattern of the bar does not affect the position of resonances, bar mass etc. Could you comment on this?

Answer: Here we only consider a fixed bar pattern speed and focus completely on the overall nature of stellar escapes for different bar types. In our upcoming works, we are planning to study the effect of the bar pattern speed on the position of bar resonances, bar mass etc. to classify the stellar orbits under different bar types.

Question: What will be the outcome if the system has more than one pattern speed?

Answer: If the system has more than one bar pattern speed, then in general the potential becomes time-dependent. So, the system becomes non-conservative and our Hamiltonian approach fails. In that case, we can still use our Hamiltonian approach, if both the bars are aligned perpendicularly to each other and have the same pattern speed.

The Predictive Power of Computational Astrophysics as a Discovery Tool
Proceedings IAU Symposium No. 362, 2023
D. Bisikalo, D. Wiebe & C. Boily, eds.
doi:10.1017/S1743921322001661

Dynamical evolution modeling of the Collinder 135 & UBC 7 binary star cluster

Marina Ishchenko[1] , Peter Berczik[2,3] and Nina Kharchenko[1]

[1]Main Astronomical Observatory, National Academy of Sciences of Ukraine,
27 Akademika Zabolotnoho St, 03143 Kyiv, Ukraine

[2]Astronomisches Rechen-Institut, Zentrum für Astronomie, University of Heidelberg,
Mönchhofstrasse 12-14, 69120, Heidelberg, Germany

[3]Konkoly Observatory, Research Centre for Astronomy and Earth Sciences, Eötvös Loránd
Research Network (ELKH), MTA Centre of Excellence, Konkoly Thege Miklós út 15-17, 1121
Budapest, Hungary

Abstract. The purpose of the present work is a detailed investigation of the dynamical evolution of Collinder 135 and UBC 7 star clusters. We present a set of dynamical numerical simulations using realistic star cluster N-body modeling technique with the forward integration of the star-by-star cluster models to the present day, based on best-available 3D coordinates and velocities obtained from the latest Gaia EDR3 data release. We have established that Collinder 135 and UBC 7 are probably a binary star cluster and have common origin. We carried out a full star-by-star N-body simulation of the stellar population of both clusters using the new algorithm of Single Stellar Evolution and performed a comparison of the results obtained in the observational data (like cumulative number counts), which showed a fairly good agreement.

Keywords. galaxies: star clusters, methods: n-body simulations

1. Introduction

Recently, based on analysis of positions and kinematic data of Gaia space mission (Gaia Collaboration et al. 2016), several pairs of star clusters showed a probable common origin (for example: Bisht et al. 2021; Pang et al. 2020; Zhong et al. 2019; Kovaleva et al. 2020). Collinder 135 (hereafter Cr 135) and newly discovered (Castro-Ginard et al. 2018) UBC 7 are located in the Vela-Puppis region which attracts active attention since Gaia data allowed to appreciate its complicated history of evolution, space and kinematic structure (Beccari et al. 2018; Cantat-Gaudin et al. 2019b,a; Beccari et al. 2020). We have demonstrated (Kovaleva et al. 2020) that these two clusters locations and velocities suggest that they might have been closer to each other at their initial history, ≈ 50 Myr ago. Backward orbital integration indicates also that the clusters might have been even gravitationally bound in the past, assuming that they were significantly more massive before their violent relaxation (Kovaleva et al. 2020).

We used Gaia DR2 and EDR3 data to restore the most probable members of Cr 135 and UBC 7, using the method presented in Kharchenko et al. 2012. Based on the obtained dataset, we constrain present day parameters of the two clusters, such as space positions, space velocities, masses and density profiles. These data are used further as boundary conditions for star-by-star numerical simulation of dynamical evolution of this pair of star clusters.

Table 1. Initial position and velocity values for Cr 135 and UBC 7 center of mass in Cartesian Galactic coordinates. Taken from `#(53,61)` model, Kovaleva et al. 2020.

Cluster	X, pc	Y, pc	Z, pc	V_x, km/s	V_y, km/s	V_z, km/s
Cr 135	-1061.421	-8382.545	-22.70332	-230.559	33.1769	5.52892
UBC 7	-1065.137	-8386.942	-14.5731	-230.792	34.0251	5.86780

2. N-body modeling of the star clusters evolution

2.1. Numerical method

We used the φ–GPU code for the numerical solution of the equations of motion. The φ–GPU package uses a high order Hermite integration scheme and individual block time steps (the code supports time integration of particle orbits with schemes of 4th, 6th and even 8th order). Such a direct N-body code evaluates in principle all pairwise forces between the gravitating particles, and its computational complexity scales asymptotically with N^2. We refer the more interested readers to a general discussion about different N-body codes and their implementation in Spurzem et al. 2011a, Spurzem et al. 2011b.

The φ–GPU code is fully parallelized using the MPI library. This code is written from scratch in `C++` and is based on earlier CPU serial N-body code, Nitadori and Makino 2008. The MPI parallelization was done in the same j particle parallelization mode as in the earlier φ–GRAPE code, Harfst et al. 2007. The current version of the code uses a native GPU support and direct code access to the GPU's using the NVIDIA native CUDA library. The multi GPU support is achieved through global MPI parallelization. Simultaneously, our code effectively exploits also the current CPU's OpenMP parallelization. More details about the GPU code public version and its performance are presented in Spurzem et al. 2012 and Berczik et al. 2013. The present code is well tested and has already been used to obtain important results in our earlier large scale (up to few million body) simulations, for more details see Khan et al. 2018, Panamarev et al. 2019, Shukirgaliyev et al. 2017 and Ernst et al. 2011.

2.2. Initial parameter space

We were looking for the best-fitting King models (King 1966) for the current observations from Gaia DR3. We assume that the clusters age is exactly 50 Myr. Our main goal was to reproduce the final cumulative mass profiles $M(\text{r})$ for both objects. For the Cr 135 we used the range within $0 < r < 20$ pc and for UBC 7 – $0 < r < 15$ pc. These limits correspond to the clusters' current Jacobi radii. Because the initial masses of the clusters are quite uncertain, we used for the modelling the initial masses as one of the initial fitting parameters. Since clusters are formed in molecular clouds with a low star formation efficiency, they are most probably supervirial after the initial gas expulsion phase (Shukirgaliyev et al. 2021). For the initial mass function we used the Kroupa 2001 approximation, with the lower mass $m_l = 0.1~M_\odot$ and the upper mass $m_h = 10~M_\odot$ limits. The other two main parameters for the cluster initial models was a R_{core} and the King concentration parameter W_0, individually for ach clusters (see Table 2). For the stellar metallicity we used the value $Z = 2\%$ (assumed as a Solar value) for both clusters.

For the initial positions and orbital velocities of the star clusters centre we used the selected `#(53,61)` model from our Kovaleva et al. 2020 paper. The initial conditions for the clusters center of mass coordinates and velocities are taken from this model (see Table 1).

More than 50 individual models with stellar evolution have been computed. The total running time for one typical model on a AMD 3600X 4.1 GHz CPU with a GeForce RTX 2600 Super GPU card was about 2 min. Minimizing the difference between the

Table 2. Initial values of physical parameters for Cr 135 and UBC 7.

Cluster	$M_\odot$	N	R, pc	W_0
Cr 135	230	442	10	3
UBC 7	200	384	7	11

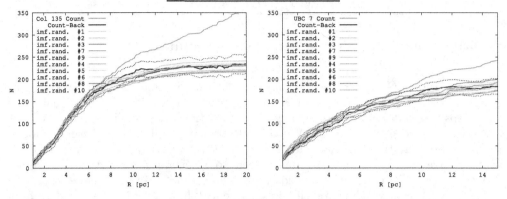

Figure 1. Cumulative number distribution of stars for clusters Cr 153 (left panel) and UBC 7 (right panel). The different color lines represents the set of randomization for initial mass function when the coordinates and velocities of the stars are fixed.

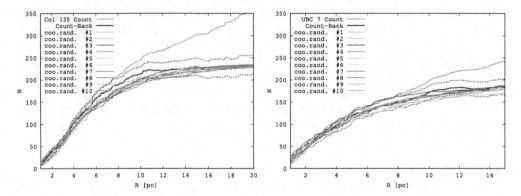

Figure 2. Same as Fig. 1 but randomization for initial coordinates and velocities was done with the fixed initial stellar mass function.

cumulative number distributions of the observed clusters and the numerical models we find simultaneously the best-fit parameters for both clusters, see Table 2.

After the first set of fitting procedure we run extra 20 more numerical models (with the same cluster physical parameters but using different randomization parameters). First we generated 10 random sets with different initial mass function (different color lines), keeping fixed the initial positions and velocities of the stars, see Fig. 1. For the second 10 random sets we used one selected initial mass function and randomize the stars positions and velocities (different color lines), see Fig. 2. On these figures we present as a black thick line the observed cumulative number distribution of stars for both clusters. The total cumulative number distributions of stars including the stellar background are presented on the figures as constantly growing gray lines. Because the observations have own limitations due to the Gaia satellite specifications, we select from the numerical models only the stars which are inside the specific stellar mass range - from 0.28 to 4.0 $M_\odot$. On our comparison figures we also exclude the neutron stars and black holes. The dotted gray lines on the figures represents the one $\pm\sigma$ difference levels from observed cumulative number distribution of stars.

Table 3. Comparison of position and velocity values for Cr 135 and UBC 7 center mass in Cartesian Galactic coordinates at 50 Myr with numerical simulation and observations.

Cluster	Type	X, pc	Y, pc	Z, pc	V_x, km/s	V_y, km/s	V_z, km/s
Cr 135	Sim	−8282.94	−284.32	−34.25	−7.65	237.32	−5.01
	Obs	−8282.14	−271.00	−36.16	−7.38	237.94	−4.85
UBC 7	Sim	−8284.14	−251.01	−45.80	−7.43	237.61	−4.65
	Obs	−8276.21	−250.87	−43.63	−7.00	238.13	−4.27

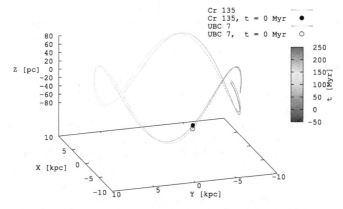

Figure 3. 3D orbits evolution of the Cr 135 (solid line) and UBS 7 (dotted line) up to 250 Myr. Black filled and black open circles are position of clusters at 0 Myr.

As we can see from figures Fig. 1 and Fig. 2 both sets of randomization of our best fitted physical model for clusters Cr 135 and UBC 7 are well inside the one $\pm\sigma$ gap.

3. Results

We present a numerical simulations using realistic star cluster N-body modelling by integrating a star-by-star cluster models in the analytic Milky Way potential to the present day. The code takes in to account up to date stellar evolution models (Banerjee et al. 2020). The average model relative errors between the observations and numerical simulations are better than 1%. This is a remarkable small error taking in account that the observational average line of site velocity error is around 10%.

The orbits integration with a simple integrator yielded the initial position of the Cr 135 ans UBS 7 at the time of their formation (see Table 1). The present-day position and velocity of the Cr 135 and UBC 7 obtained from numerical simulation are given in Table 3. The full 3D orbits of the evolution are shown in Fig. 3. It should be noted that the clusters rotate around each other during their orbital motion.

4. Conclusions

We present a numerical simulations using realistic star cluster N-body modelling by integrating a star-by-star cluster models in the analytic Milky Way potential to the present day. We were looking for the best-fitting King models for the observations from Gaia DR3 after 50 Myr of evolution. The comparative result of observational data and simulations for 50 Myr showed a fairly good agreement. The probability of a random coincidence chance is only about 2%.

5. Acknowledgements

This work has made use of data from the European Space Agency (ESA) mission *Gaia* (https://www.cosmos.esa.int/gaia), processed by the *Gaia* Data Processing and Analysis

Consortium (DPAC, https://www.cosmos.esa.int/web/gaia/dpac/consortium). Funding for the DPAC has been provided by national institutions, in particular the institutions participating in the *Gaia* Multilateral Agreement. The work of PB and MI was supported by the Volkswagen Foundation under No. 97778. The work of PB was also supported by the Volkswagen Foundation under the special stipend No. 9B870.

References

Banerjee, S., Belczynski, K., Fryer, C. L., Berczik, P., Hurley, J. R., Spurzem, R., & Wang, L. 2020, BSE versus StarTrack: Implementations of new wind, remnant-formation, and natal-kick schemes in NBODY7 and their astrophysical consequences. *A&A*, 639, A41.

Beccari, G., Boffin, H. M. J., & Jerabkova, T. 2020, Uncovering a 260 pc wide, 35-Myr-old filamentary relic of star formation. *MNRAS*, 491(2), 2205–2216.

Beccari, G., Boffin, H. M. J., Jerabkova, T., Wright, N. J., Kalari, V. M., Carraro, G., De Marchi, G., & de Wit, W.-J. 2018, A sextet of clusters in the Vela OB2 region revealed by Gaia. *MNRAS*, 481(1), L11–L15.

Berczik, P., Spurzem, R., Wang, L., Zhong, S., & Huang, S. Up to 700k GPU cores, Kepler, and the Exascale future for simulations of star clusters around black holes. In *Third International Conference "High Performance Computing", HPC-UA 2013, p. 52–59* 2013, pp. 52–59.

Bisht, D., Zhu, Q., Yadav, R. K. S., Ganesh, S., Rangwal, G., Durgapal, A., Sariya, D. P., & Jiang, I.-G. 2021, Multicolour photometry and Gaia EDR3 astrometry of two couples of binary clusters (NGC 5617 and Trumpler 22) and (NGC 3293 and NGC 3324). *MNRAS*, 503(4), 5929–5947.

Cantat-Gaudin, T., Jordi, C., Wright, N. J., Armstrong, J. J., Vallenari, A., Balaguer-Núñez, L., Ramos, P., Bossini, D., Padoan, P., Pelkonen, V. M., Mapelli, M., & Jeffries, R. D. 2019,a Expanding associations in the Vela-Puppis region. 3D structure and kinematics of the young population. *A&A*, 626a, A17.

Cantat-Gaudin, T., Mapelli, M., Balaguer-Núñez, L., Jordi, C., Sacco, G., & Vallenari, A. 2019,b A ring in a shell: the large-scale 6D structure of the Vela OB2 complex. *A&A*, 621b, A115.

Castro-Ginard, A., Jordi, C., Luri, X., Julbe, F., Morvan, M., Balaguer-Núñez, L., & Cantat-Gaudin, T. 2018, A new method for unveiling open clusters in Gaia. New nearby open clusters confirmed by DR2. *A&A*, 618, A59.

Ernst, A., Just, A., Berczik, P., & Olczak, C. 2011, Simulations of the Hyades. *A&A*, 536, A64.

Gaia Collaboration, Prusti, T., de Bruijne, J. H. J., Brown, A. G. A., & et al. 2016, The Gaia mission. *A&A*, 595, A1.

Harfst, S., Gualandris, A., Merritt, D., Spurzem, R., Portegies Zwart, S., & Berczik, P. 2007, Performance analysis of direct N-body algorithms on special-purpose supercomputers. *NewAstr*, 12, 357–377.

Khan, F. M., Capelo, P. R., Mayer, L., & Berczik, P. 2018, Dynamical Evolution and Merger Timescales of LISA Massive Black Hole Binaries in Disk Galaxy Mergers. *ApJ*, 868(2), 97.

Kharchenko, N. V., Piskunov, A. E., Schilbach, E., Röser, S., & Scholz, R.-D. 2012, Global survey of star clusters in the Milky Way. I. The pipeline and fundamental parameters in the second quadrant. *A&A*, 543, A156.

King, I. R. 1966, The structure of star clusters. III. Some simple dynamical models. *AJ*, 71, 64.

Kovaleva, D. A., Ishchenko, M., Postnikova, E., Berczik, P., Piskunov, A. E., Kharchenko, N. V., Polyachenko, E., Reffert, S., Sysoliatina, K., & Just, A. 2020, Collinder 135 and UBC 7: A physical pair of open clusters. *A&A*, 642, L4.

Kroupa, P. 2001, On the variation of the initial mass function. *MNRAS*, 322(2), 231–246.

Nitadori, K. & Makino, J. 2008, Sixth- and eighth-order Hermite integrator for N-body simulations. *NewAstr*, 13, 498–507.

Panamarev, T., Just, A., Spurzem, R., Berczik, P., Wang, L., & Arca Sedda, M. 2019, Direct N-body simulation of the Galactic centre. *MNRAS*, 484(3), 3279–3290.

Pang, X., Li, Y., Tang, S.-Y., Pasquato, M., & Kouwenhoven, M. B. N. 2020, Different Fates of Young Star Clusters after Gas Expulsion. *ApJL*, 900(1), L4.

Shukirgaliyev, B., Otebay, A., Sobolenko, M., Ishchenko, M., Borodina, O., Panamarev, T., Myrzakul, S., Kalambay, M., Naurzbayeva, A., Abdikamalov, E., Polyachenko, E., Banerjee, S., Berczik, P., Spurzem, R., & Just, A. 2021, Bound mass of Dehnen models with a centrally peaked star formation efficiency. *A&A*, 654, A53.

Shukirgaliyev, B., Parmentier, G., Berczik, P., & Just, A. 2017, Impact of a star formation efficiency profile on the evolution of open clusters. *A&A*, 605, A119.

Spurzem, R., Berczik, P., Berentzen, I., Ge, W., Wang, X., Schive, H.-Y., Nitadori, K., & Hamada, T. Supermassive Black Hole Binaries in High Performance Massively Parallel Direct N-body Simulations on Large GPU Clusters. In Dubitzky, W., Kurowski, K., & Schott, B., editors, *Large Scale Computing Techniques for Complex Systems and Simulations* 2011,a, Wiley Publishers, pp. 35–58.

Spurzem, R., Berczik, P., Hamada, T., Nitadori, K., Marcus, G., Kugel, A., Männer, R., Berentzen, I., Fiestas, J., Banerjee, R., & Klessen, R. 2011,b Astrophysical Particle Simulations with Large Custom GPU Clusters on Three Continents. *Computer Science - Research and Development (CSRD)*, 26b, 145–151.

Spurzem, R., Berczik, P., Zhong, S., Nitadori, K., Hamada, T., Berentzen, I., & Veles, A. Supermassive Black Hole Binaries in High Performance Massively Parallel Direct N-body Simulations on Large GPU Clusters. In Capuzzo-Dolcetta, R., Limongi, M., & Tornambè, A., editors, *Advances in Computational Astrophysics: Methods, Tools, and Outcome* 2012, volume 453 of *Astronomical Society of the Pacific Conference Series*, 223.

Zhong, J., Chen, L., Kouwenhoven, M. B. N., Li, L., Shao, Z., & Hou, J. 2019, Substructure and halo population of Double Cluster h and χ Persei. *A&A*, 624, A34.

6. Discussion

Q: Thank you for presenting the simulation results. I have a question about how the comparison of the results of the simulation with the observational data was carried out? Such as figure 1 or 2 (Christian Boily) A: To compare our results with observational data, we performed a transformation of the galactocentric coordinates into equatorial coordinate system. (Marina Ishchenko)

The Predictive Power of Computational Astrophysics as a Discovery Tool
Proceedings IAU Symposium No. 362, 2023
D. Bisikalo, D. Wiebe & C. Boily, eds.
doi:10.1017/S174392132200117X

Multiparticle collision simulations of dense stellar systems and plasmas

P. Di Cintio[1,2,3], **M. Pasquato**[4,5], **L. Barbieri**[2,3,6], **H. Bufferand**[7], **L. Casetti**[2,3,6], **G. Ciraolo**[7], **U. N. di Carlo**[8], **P. Ghendrih**[7], **J. P. Gunn**[7], **S. Gupta**[9], **H. Kim**[10], **S. Lepri**[1,3], **R. Livi**[2,3,1], **A. Simon-Petit**[2,3], **A. A. Trani**[11,12] and **S.-J. Yoon**[10]

[1]Consiglio Nazionale delle Ricerche, Istituto dei Sistemi Complessi via Madonna del piano 10, I-50019 Sesto Fiorentino, Italy
email: pierfrancesco.dicintio@cnr.it

[2]Dept. of Physics & Astronomy, University of Florence, via G. Sansone 1, I-50019 Sesto Fiorentino (FI), Italy

[3]INFN, Sezione di Firenze, via G. Sansone 1, I-50019 Sesto Fiorentino (FI), Italy

[4]Dept. of Physics & Astronomy Galileo Galilei, University of Padova, Vicolo dell'Osservatorio 3, I-35122, Padova, Italy

[5]Département de Physique, Université de Montréal, Montreal, Quebec H3T 1J4, Canada

[6]INAF, Osservatorio Astrofisico di Arcetri, Largo Enrico Fermi, 5, I-50125 Firenze, Italy

[7]IRFM, CEA, F-13108 St Paul Lez Durance, France

[8]McWilliams Center for Cosmology and Department of Physics, Carnegie Mellon University, Pittsburgh, PA 15213, USA

[9]Tata Institute of Fundamental Research, Homi Bhabha Road, Colaba Mumbai 400005, India

[10]Department of Astronomy & Center for Galaxy Evolution Research, Yonsei University, Seoul 120-749, Republic of Korea

[11]Department of Earth Science and Astronomy, College of Arts and Sciences, The University of Tokyo, 3-8-1 Komaba, Meguro-ku, Tokyo 153-8902, Japan

[12]Okinawa Institute of Science and Technology, 1919-1 Tancha, Onna-son, Okinawa 904-0495, Japan

Abstract. We summarize a series of numerical experiments of collisional dynamics in dense stellar systems such as globular clusters (GCs) and in weakly collisional plasmas using a novel simulation technique, the so-called Multi-particle collision (MPC) method, alternative to Fokker-Planck and Monte Carlo approaches. MPC is related to particle mesh approaches for the computation of self consistent long-range fields, ensuring that simulation time scales with $N \log N$ in the number of particles, as opposed to N^2 for direct N-body. The collisional relaxation effects are modelled by computing particle interactions based on a collision operator approach that ensures rigorous conservation of energy and momenta and depends only on particles velocities and cell-based integrated quantities.

Keywords. methods: n-body simulations, methods: numerical, gravitation, plasmas

---

1. Introduction

The evolution of self-gravitating systems and plasmas is mainly dominated by collective mean-field processes due to their large number of particles N and the long-range nature of the $1/r^2$ gravitational or Coulombian force. The typical dynamical scale t_{dyn} of such systems is of the order of $1/\sqrt{G\bar{\rho}}$, where G is the gravitational constant and $\bar{\rho}$ is some mean mass density, for the self-gravitating systems, and proportional to the inverse of the plasma frequency $\omega_P = \sqrt{4\pi n_e e^2/m_e}$, where n_e, e and m_e are the number density, charge and mass of the electrons, for plasmas.

Collisional processes typically contribute to the dynamics on a time scale that is a function† of N as $t_c \approx t_{\mathrm{dyn}} \ln N/N$, that in the case of gravitating systems may exceed several times the age of the Universe (see Binney & Tremaine (2008)), while in plasma physics is always regulated by temperature and mean free path. For these reasons, the numerical modelling of systems governed by $1/r^2$ forces is usually carried out by means of *collisionless* approaches such as the widely uses particle-mesh or particle-in-cell (PIC) schemes (see e.g. Hockney & Eastwood (1988); Dehnen & Read (2011)). However, there are several examples of of systems that are (at least partially) dynamically regulated by collisions. In particular, dense stellar systems such as globular clusters or galactic cores can be largely in collisional regimes, while their modelling in terms of "honest" direct $N-$body simulations with a one-to-one correspondence between stars and simulation particles remain still challenging due to the (relatively) large size N of the order of 10^6. In plasma physics, collisional systems can be found in trapped one component ion or electron plasmas (see Dubin & O'neil (1999)) or ultracold neutral plasmas (see Killian *et al.* (2005)), while a transition between collisional and collisionless plasma regimes is expected in the scrape-off layer of tokamaks (see Fundamenski (2005)).

Numerical simulations of collisional stellar systems usually employ the Monte Carlo technique or hybrid particle-mesh direct $N-$body schemes. Vice versa, PIC plasma codes based on the solution of the Vlasov-Maxwell equations on some reduced geometry rely on analytic or semi-analytic methods to reconstruct the collision integral in the kinetic picture.

In a series of papers, we have implemented a novel approach to the simulation of both gravitational and Coulomb collisional systems based on the so-called multiparticle collision operator (hereafter MPC), originally developed by Malevanets & Kapral (1999) in the context of mesoscopic fluid dynamics (see also Gompper *et al.* (2009) and references therein for an extensive review). Here we give a brief introduction of the method and highlight the major results obtained so far.

2. Overview of the method

The MPC scheme in a three-dimensional simulation domain containing N particles partitioned in N_c cells, amounts to a cell-dependent rotation of an angle α_i of the particle's relative velocity vectors $\delta \mathbf{v}_j = \mathbf{v}_j - \mathbf{u}_{\mathrm{com,i}}$ in the centre of mass frame of each cell i, so that

$$\mathbf{v}'_j = \mathbf{u}_{\mathrm{com,i}} + \delta \mathbf{v}_{j,\perp}\cos(\alpha_i) + (\delta \mathbf{v}_{j,\perp} \times \mathbf{R}_i)\sin(\alpha_i) + \delta \mathbf{v}_{j,\|}. \qquad (2.1)$$

In the formulae above $\mathbf{R}_i$ is a random axis chosen for the given cell, $\delta \mathbf{v}_{j,\perp}$ and $\delta \mathbf{v}_{j,\|}$ are the relative velocity components perpendicular and parallel to $\mathbf{R}_i$, respectively and

$$\mathbf{u}_{\mathrm{com,i}} = \frac{1}{m_{\mathrm{tot,i}}} \sum_{j=1}^{n_i} m_j \mathbf{v}_j; \quad m_{\mathrm{tot,i}} = \sum_{j=1}^{n_i} m_j. \qquad (2.2)$$

For each cell, if the rotation angle α_i is chosen randomly by sampling a uniform distribution, after the vectors $\delta \mathbf{v}_j$ are rotated back around $\mathbf{R}_i$ by $-\alpha_i$, the linear momentum and the kinetic energy are conserved *exactly* in each cell, from which follows the conservation

---

† Note that N corresponds to the total number of particles of the systems in the gravitational case, while it is the average number of particles, ions or electrons, withing a Debye length for plasmas.

of the *total* parent quantities (see e.g. Di Cintio *et al.* (2017)), at variance with other previously implemented collision schemes, such as for example the Nambu (1983) method that imposes the conservation of the kinetic energy only, while breaking the conservation of linear momentum (that is preserved only in time average).

By introducing an additional constraint on the rotation angles α_i, one can add the conservation of a component of the total angular momentum of the cell $\mathbf{L}_i$ (see e.g. Noguchi & Gompper (2008)). In this case α_i is given by

$$\sin(\alpha_i) = -\frac{2a_i b_i}{a_i^2 + b_i^2}; \quad \cos(\alpha_i) = \frac{a_i^2 - b_i^2}{a_i^2 + b_i^2}, \tag{2.3}$$

where

$$a_i = \sum_{j=1}^{N_i} [\mathbf{r}_j \times (\mathbf{v}_j - \mathbf{u}_i)]\,|_z; \quad b_i = \sum_{j=1}^{N_i} \mathbf{r}_j \cdot (\mathbf{v}_j - \mathbf{u}_i). \tag{2.4}$$

In the expression above $\mathbf{r}_j$ are the particles position vectors, and with $[\mathbf{x}]|_z$ we assume that we are considering (without loss of generality) the component of the vector $\mathbf{x}$ parallel to the z axis of the simulation's coordinate system, which implies that in our case the z component of the cell angular momentum is conserved.

The exact conservation of the total angular momentum inside the cell can be enforced by choosing $\mathbf{R}_i$ parallel to the direction of the cell's angular momentum vector $\mathbf{L}_i$ and using the definition of a_i accordingly.

In our implementation of the MPC method, the collision step is always conditioned to a cell-dependent probability accounting for "how much the system is collisional" locally. The latter is defined as

$$p_i = \text{Erf}\left(\beta \Delta t \nu_c\right), \tag{2.5}$$

where Δt is the simulation timestep, ν_c is the collision frequency, β is a dimensionless constant of the order of twice the number of the simulation cells, and $\text{Erf}(x)$ is the standard error function. In Equation (2.5) the collision frequency is given by

$$\nu_c = \frac{8\pi G^2 \bar{m}_i^2 \bar{n} \log \Lambda}{\sigma_i^3} \tag{2.6}$$

for the gravitational case, where $\bar{n}$ is the mean stellar number density, $\bar{m}_i$ and σ_i are the average particle mass and the velocity dispersion in the cell, respectively, and by

$$\nu_c = \frac{4\pi e^4 n_e \log \Lambda}{m_e^{1/2} T^{3/2}} \tag{2.7}$$

for the plasma case, where T is the electron temperature. In both cases the Coulomb logarithm $\log \Lambda$ of the maximum to minimum impact parameter is usually taken of order 10 for the systems of interest. Otherwise, it can also be evaluated cell-dependently.

Once Equation (2.5) is evaluated in each cell, a random number p_{*i} is sampled from a uniform distribution in the interval $[0, 1]$ and the multi-particle collision is applied for all cells for which $p_{*i} \le p_i$.

Between two applications of the MPC operator, the particles are propagated under the effect of the self-consistent gravitational or electromagnetic fields, obtained by solving on a grid the Poisson or Maxwell equations with standard finite elements schemes. From the point of view of the computational cost, for a given initial condition and fixing the order of the particle integrator and time step, the MPC simulations are on average a factor 5 faster than direct N-body calculation in the range of N between 10^3 to 10^6. Adding the conservation of angular momentum and including the binaries (or processes of electron

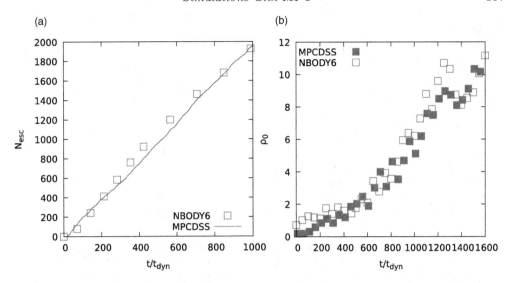

Figure 1. a) Number of escapers as function of time for a cluster with $N = 32000$ particles propagated with MPCDSS (solid line) and NBODY6 (squares). b) Evolution of the central density for a core collapsing cluster evolved with MPCDSS (filled symbols) and NBODY6 (empty symbols).

capture or collisional ionization in the plasma codes) typically reduces the computationl gain of a factor two with respect to other codes.

3. Results and discussion

3.1. *Evolution of Globular clusters*

Di Cintio *et al.* (2021) and Di Cintio *et al.* (2022a) performed a wide range of hybrid MPC-Particle-mesh simulations of globular clusters with the MPCDSS code, investigating the effect of a mass spectrum on the dynamics of core collapse, with particle numbers up to $N = 10^6$, different radial anisotropy profiles while adopting the widely used Plummer density distribution. Low resolution simulations with $N = 32000$ have been compared to their counterparts performed with the direct $N-$body code NBODY6 (see Nitadori & Aarseth (2012)) using the exact same initial conditions. Overall we observed a good agreement between the two approaches in the low N limit in the evolution of indicators such as the fraction of escapers (i.e. particles with *positive* total energy being outside a given cut-off radius) shown in panel a) of Fig. 1 or the mean density ρ_0 evaluated within a fixed radius of the order one tenth of the initial scale radius, see panel b) of the same figure. The slight discrepancy between the evolution of said central density observed at late stages of the collapse can be ascribed to the fact that the MPC and the $N-$body simulations have radically different force-evaluation schemes, and, due to the intrinsically chaotic nature of the $N-$body problem, particle trajectories can differ sensibly even starting with analogous initial conditions. This reflects in the fact that denser regions where particle encounters happen more frequently could show rather complex fluctuations.

In models with a mass spectrum we confirm the theoretical self-similar contraction picture but with a dependence on the slope of the mass function. Moreover, the time of core collapse shows a non-monotonic dependence on the slope that holds also for the depth of core collapse and for the dynamical friction timescale of heavy particles. Cluster density profiles at core collapse show a broken power law structure, suggesting that central cusps are a genuine feature of collapsed cores.

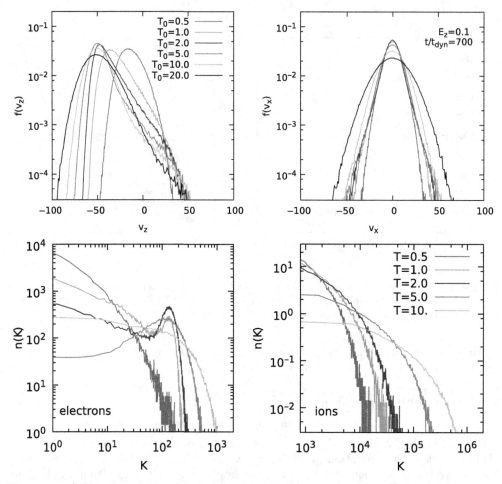

Figure 2. a) Electron velocity distribution along the direction of the electric field. b) Electron velocity distribution in the transverse direction. c) Electron kinetic energy distribution. d) Ion kinetic energy distribution. All distributions are evaluated at $t \sim 700/\omega_P$ for different values of T_0 (indicated in figure) and $E_z = 0.1$ in computer units.

In addition we also investigated the dynamics of a central intermediate mass black hole (IMBH) finding that the latter, independently on the structure of the mass spectrum and the anisotropy profiles accelerates the core collapse while making it shallower. In general we also observe that the presence of a mass spectrum results in a way large wander radius of the IMBH for fixed total stellar mass and different masses of the IMBH itself.

The results on the structure and evolution of velocity dispersion and anisotropy profiles of cluster with central IMBHs are to be published elsewhere (Di Cintio *et al.* (2022b)).

3.2. *Weakly collisional plasmas*

The most part of the MPC simulations performed with the TROPIC³O code (see Di Cintio *et al.* (2015), Di Cintio *et al.* (2017)) where devoted to the investigation of the transition between different regimes of anomalous energy transport in low dimensional (1D and 2D, see Lepri *et al.* (2019)) toy set-ups aiming at shedding some light on the heat flux profile structure along magnetic field lines crossing strong temperature gradients in weakly collisional plasmas, relevant for magnetic fusion devices. Ciraolo *et al.* (2018) found a surprisingly good agreement between the electron temperature profiles between a hot source and a cold interface obtained by MPC simulations with the corresponding profiles evaluated by 1D fluid codes with a semi-analytical collisional closure of the fluid equations and a non-local definition of the heat flux accounting for suprathermal particles.

Lepri *et al.* (2021) further investigated the nonequilibrium steady states of a 1D models of finite length L in contact with two thermal-wall heat reservoirs, finding a clear crossover from a kinetic transport regime to an anomalous (hydrodynamic) one over a characteristic scale proportional to the cube of the collision time among particles. In addition, test simulations of models with thermal walls injecting particles with given non thermal velocity. showed that for fast and relatively cold particles, smaller systems never establish local equilibrium keeping non-Maxwellian velocity distributions.

Di Cintio *et al.* (2018) used MPC simulations to study the non thermal profiles of the weakly ionized media in filamentary structures in molecular clouds finding that strong collisions in dense regions enforce the production of high velocity tails in the particle's distribution, able to climb the gravitational potential well of the filament, thus forming hot and diffuse external envelopes without additional heat sources from the intergalactic media or star formation feedback. Density and temperature profiles obtained in these numerical experiments qualitatively mach the observed transverse profiles of both gas density and temperature.

The possibility to incorporate an energy and momentum preserving particle-based collisional operator in plasma codes opens the possibility to study problems involving the presence of different species or complex non-thermal processes without making strong assumptions on a (possibly unknown) phase-space distribution function. In particular, we aim at studying the mechanism of electron run-away in presence of net electrostatic fields and/or violation of the plasmas quasineutrality. We performed some preliminary test simulations in a 3D periodic geometry aiming at observing the formation of a suprathermal tail in the velocity distribution when applying a constant electric field $\mathbf{E}$ along one of the three spatial coordinates, for different values of the initial electron temperature T_0 at fixed density n_e and assuming that electrons and ions are initially at equilibrium. In Fig. 2 we show the electron velocity distributions along the direction of $\mathbf{E}$ (z, without loss of generality) and one of the transverse directions $f(v_z)$ and $f(v_x)$, panels a) and b). It appears clearly that for fixed density and therefore fixed mean free path, systems with lower initial electron temperature are less and less prone to develop non thermal tails along the direction of the fixed external field as particles accelerated by such field suffer a stronger dynamical friction effect due to the lower velocity dispersion, as theorized by Dreicer (1959). Further increasing T_0 produces more and more deformed $f(v_z)$ while the velocity distribution in the transverse directions $f(v_x)$ regains a thermal structure when T_0 is larger than some critical value for which the hotter systems are basically collisionless and the different degrees of freedom are decoupled. This is also evident from the full differential kinetic energy distribution $n(K)$ (see panel c, same figure), where the "monochromatic" peak at $K \approx 1.5 \times 10^2$ in computer units appears only for systems starting with somewhat intermediate values of T_0. Being heavier (in this case with a mass ratio of 2×10^3), and thus having lower mobility, ions do not bare any peculiar structure in their kinetic energy distributions even at late times of the order of thousand of plasma oscillations, such that their $n(K)$ corresponding to different T_0 can be collapsed onto one another (panel d).

References

Binney, J.; Tremaine, S. 2008 Galactic dynamics (2nd edition, Princeton University Press)

Ciraolo, G., Bufferand, H., Di Cintio, P., Ghendrih, P., Lepri, S., Livi, R., Marandet, Y., Serre, E., Tamain, P. & Valentinuzzi, M. 2018, *Contributions to Plasma Physics*, 58, 457

Dehnen, W. & Read, J.I. 2011, *EPJ Plus* 126, 55

Di Cintio, P., Livi, R., Bufferand, H., Ciraolo, G. Lepri, S. & Straka M.J. 2015, *Phys.Rev.E*, 92, 62108

Di Cintio, P., Livi, R., Lepri, S. & Ciraolo, G. 2017, *Phys.Rev.E*, 95, 43203

Di Cintio, P., Gupta, S., & Casetti, L. 2018, *MNRAS*, 475, 1137

Di Cintio, P., Pasquato, M., Kim, H., & Yoon, S.-J. 2021, *A&A*, 649, A24

Di Cintio, P., Pasquato, M., Simon-Petit, A., & Yoon, S.-J. 2022 (in press), *A&A*

Di Cintio, P., Pasquato, M., Barbieri, L., Trani, A.A., & di Carlo, U.N. 2022 (in preparation)

Dreicer, H. 1959, *Physical Review*, 115, 238

Dubin, D. H. & O'neil, T. M. 1999 *Rev. Mod. Phys.*, 71, 87

Fundamenski, W. 2005 *Plasma Phys. and Contr. Fus.*, 47, 163

G. Gompper, T. Ihle, D. M. Kroll, and R. G. Winkler 2009, *Multi-Particle Collision Dynamics: A ParticleBased Mesoscale Simulation Approach to the Hydrodynamics of Complex Fluids*, in Adv Polym Sci 221, 1

Hockney, R. W., Eastwood, J. W. 1988 Computer simulations using particles (Taylor & Francis eds.)

Killian, T. C., Pattard, T., Pohl, T. & Rost, J.-M. 2005 *Physics Reports*, 449,77

Lepri, S., Bufferand, H., Ciraolo, G., Di Cintio, P., Ghendrih, P. & Livi, R. 2019, in Stochastic Dynamics Out of Equilibrium, ed. G. Giacomin, S. Olla, E. Saada, H. Spohn, & G. Stoltz (Cham: Springer International Publishing), 364

Lepri, S., Ciraolo, G., Di Cintio, P., Gunn, J.P. & Livi, R. 2021, *Phys.Rev.Research*, 3, 13207

Malevanets, A. & Kapral, R. 1999, *J. Chem. Phys*, 110, 8605

Nambu, K. 1983, *J. Phys. Soc. Jap.*, 52, 3382

Nitadori, K., & Aarseth, S. J. 2012 *MNRAS* 424, 545

Noguchi, H. & Gompper, G. 2008 *Phys. Rev. E* 78, 016706

The Predictive Power of Computational Astrophysics as a Discovery Tool
Proceedings IAU Symposium No. 362, 2023
D. Bisikalo, D. Wiebe & C. Boily, eds.
doi:10.1017/S1743921322001703

A novel generative method for star clusters from hydro-dynamical simulations

Stefano Torniamenti[1,2,3]

[1]Physics and Astronomy Department Galileo Galilei, University of Padova, Vicolo dell'Osservatorio 3, I–35122, Padova, Italy

[2]INFN- Sezione di Padova, Via Marzolo 8, I–35131 Padova, Italy

[3]INAF, Osservatorio Astronomico di Padova, vicolo dell'Osservatorio 5, 35122 Padova, Italy
email: `stefano.torniamenti@studenti.unipd.it`

Abstract. Most stars form in clumpy and sub-structured clusters. These properties also emerge in hydro-dynamical simulations of star-forming clouds, which provide a way to generate realistic initial conditions for $N-$body runs of young stellar clusters. However, producing large sets of initial conditions by hydro-dynamical simulations is prohibitively expensive in terms of computational time. We introduce a novel technique for generating new initial conditions from a given sample of hydro-dynamical simulations, at a tiny computational cost. In particular, we apply a hierarchical clustering algorithm to learn a tree representation of the spatial and kinematic relations between stars, where the leaves represent the single stars and the nodes describe the structure of the cluster at larger and larger scales. This procedure can be used as a basis for the random generation of new sets of stars, by simply modifying the global structure of the stellar cluster, while leaving the small-scale properties unaltered.

Keywords. galaxies: star clusters, stellar dynamics, methods: numerical, methods: statistical

---

1. Introduction

A large fraction of star formation happens in clusters or associations Lada & Lada (2003). These star-forming systems are characterized by complex phase-space distributions, where sub-structures Larson (1995), fractality (Cartwright 2009; Kuhn *et al.* 2019), relative sub-clump motions Cantat-Gaudin *et al.* (2019), expansion (due to gas expulsion, Hills 1980), and possibly rotation Hénault-Brunet *et al.* (2012) are observed. The early evolution of these stellar systems is of fundamental importance for the comprehension of the present-day structure of older open and globular clusters, and can be modeled in a realistic way by means of direct $N-$body simulations. However, adequate initial conditions, able to take into account the observed phase-space complexity, are necessary for a correct comprehension of this evolutionary phase. In this sense, spherical equilibrium models like Plummer spheres Plummer (1911) or King models King (1966) cannot represent the initial spatial and kinematic distributions of young star clusters. A natural way to reproduce the observed properties of these stellar systems is by means of hydro-dynamical simulations of collapsing molecular clouds, where they naturally emerge (Ballone *et al.* 2020, 2021; Bate *et al.* 2009; Klessen & Burkert 2000; Wall et al. 2019). However, running hydro-dynamical simulations including all the relevant physics is computationally very expensive, and producing large sets of initial conditions would turn to be prohibitive in terms of computational time.

Table 1. Properties of the stellar clusters and their parent molecular cloud.

	m1e4	m2e4	m3e4	m4e4	m5e4	m6e4	m7e4	m8e4	m9e4	m1e5
$N_{\rm s}$	2523	2571	2825	2868	2231	3054	4214	2945	3161	3944
$M_{\rm s}\,[10^3\,{\rm M}_\odot]$	4.2	6.7	10.3	14.4	14.1	20.4	31.5	28.3	30.5	38.0
$M_{\rm mc}\,[10^3\,{\rm M}_\odot]$	10.0	20.0	30.0	40.0	50.0	60.0	70.0	80.0	90.0	100.0
$\epsilon_{\rm sf}$	0.42	0.33	0.34	0.36	0.28	0.34	0.45	0.35	0.34	0.38

Columns: [1] number and [2] total mass of the sink particles, [3] mass of the parent molecular cloud, and [4] the resulting star formation efficiency

Here, we introduce a novel method for producing new initial conditions for $N-$body runs without running additional independent hydro-dynamical simulations. Our approach relies on applying a clustering algorithm to an existing set of initial conditions. In particular, we adopt a hierarchical clustering algorithm, which learns a tree-like representation of the stellar cluster, describing its structure at different scales. In our generative model, new stellar clusters are then obtained by modifying selected nodes of the hierarchical tree.

2. Sink particle distributions

We consider the sink particle distributions (also named as stars in the following) from 10 smoothed-particle hydro-dynamical simulations of molecular clouds, performed by Ballone *et al.* (2020). These simulations are initialized as spherical molecular clouds with total gaseous mass ranging between $10^4\,{\rm M}_\odot$ and $10^5\,{\rm M}_\odot$, uniform temperature $T = 10\,{\rm K}$ and uniform density $\rho = 2.5 \times 10^2\,{\rm cm}^{-3}$. Star formation is implemented during the simulation by means of a sink particle algorithm Bate *et al.* (1995). The star clusters are the result of the instantaneous gas removal at 3 Myr, mimicking the impact of the first supernova explosions (no stellar feedback was included in the simulations). We refer to Ballone et al. (2020,2021) (Ballone *et al.* 2020, 2021) for more details about the hydro-dynamical simulations. Table 1 resumes the main properties of the sink particle distributions under consideration.

3. Hierarchical clustering

Clustering algorithms are a class of unsupervised machine learning methods. In general, clustering identifies similar instances in a given sample and assigns them to groups (or clusters). In the specific case of hierarchical clustering, the algorithm proceeds in a hierarchical way, by connecting the most similar pair of clusters, starting from the individual instances (in this case the single stars), until a certain number of groups is reached† Kaufman *et al.* (1990). To identify similar elements in the sample, hierarchical clustering is provided with a similarity prescription, called linkage. For this case, we make use of Ward's linkage, which merges two clusters such that the variance within all clusters increases the least. This often leads to clusters that are relatively equally sized. For more details on such choice, we refer the reader to Torniamenti *et al.* (2022). Also, we use the implementation offered by the SCIKIT-LEARN library Pedregosa (2011).

3.1. *Application to stellar clusters*

We applied hierarchical clustering to the stellar clusters introduced in Section 2. Before applying the algorithm, we scaled the positions and the velocities by their standard deviations. Figure 1 shows different levels of the cluster hierarchy for the m1e4 cluster. The clumps are organized into a hierarchical tree-like structure $\mathcal{T}$, where the trunk

† When the hierarchy is built from the bottom up, the algorithm is also referred to as agglomerative clustering algorithm.

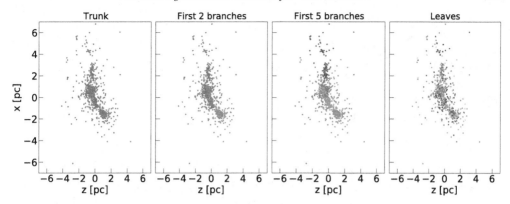

Figure 1. Different levels in the hierarchical tree of the m1e4 simulation. The panels show different levels of the the tree: [1] the trunk, [2] the first two branches, [3] the first five branches, and [4] the leaves.

contains the whole set of stars and each subsequent node is a two-way split (with each branch being a sub-clump), down to the leaves, representing individual stars.

The hierarchical construction allows to identify groups of similar instances in the distribution of stars as well as drawing information about the structure of the star system at different scales. To describe the relevant physical properties of the cluster by means of the tree formalism, at each node we evaluate the distance vector between the centres of mass of the clumps l_i, their relative velocity vector u_i, and the mass ratio between the two clumps q_i, defined as the ratio between the lightest of the two resulting groups and the total mass of the node.

4. Generative method

Each node of the tree, $\mathcal{T}_i$, describes the relations between two sub-clumps departing from a parent branch. As a consequence, the relevant quantities l_i, u_i, and q_i can be used as instructions to progressively split clumps of stars in the phase space, starting from one reference mass. During this procedure, we can modify some selected nodes of the tree to obtain a new and different realization of the stellar cluster. The way in which the nodes are modified depends on what scales of the systems we want to preserve or alter. In our case, we aim to obtain new macroscopic configurations by preserving the small scale properties (such as their complex fractal structure), which make our clusters so realistic. For this reason, to obtain a new realization, we replace the first nodes of a reference tree, which describe the large scale distribution of sub-clumps, with the same quantities drawn from other trees. The method consists in the following steps:

• We consider a reference tree $\mathcal{T}$, and we replace the quantities l_i, u_i and q_i, associated to the first $i < k$ nodes, with the same-level quantities from the tree $\mathcal{T}'$, learned from another set of sink particles. Here, we consider $k = 3$. For more details on such choice, we refer the reader to Torniamenti *et al.* (2022) Torniamenti *et al.* (2022).

• We consider one particle containing the total mass of the cluster we want to generate, placed at rest in the origin of the coordinate system. This particle is split into two particles, such that the resulting mass ratio is q_1 (the mass ratio relative to the first node of the new tree). The positions and velocities of the new particles are assigned such that their centre of mass is at rest in the origin of the system, and their distance and relative velocity vectors are l_1 and u_1, respectively.

• At each step i, we split a chosen particle into two new particles with mass ratio q_i and place them at a distance l_i from each other, moving with relative velocity u_i. The particle-to-split is chosen by considering the same order of splitting as the original

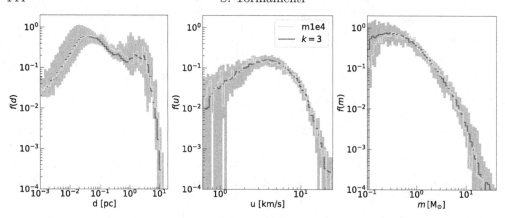

Figure 2. Distributions of inter-particle distances $f(d)$ (left), velocities $f(v)$ (center), and masses $f(m)$ (right) for the sink particles taken from the m1e4 simulation (thick yellow line) and the distributions of new generations obtained by replacing the first 2 nodes (corresponding to $k = 3$, purple). The shaded area encloses the distribution of the new generations, and the solid line is the median of the distribution.

reference tree. This splitting procedure is then repeated until a cluster with the same number of particles as the reference one is obtained.

• Finally, we remove the very low-mass stars (which may result in planet-sized objects) by setting a cutoff mass to the minimum mass of the original stars on which $\mathcal{T}$ was learned.

In Fig. 2 we compare the distance, velocity and mass distributions of m1e4 to those of a set of new generations. All the distributions of the new realizations are consistent with those of the original simulation. In particular, the new distance distributions are altered at large scales but, moving toward small distances, they recover the same trend as the original simulation, as meant for this method. Figure 3 and 4 show the spatial distributions of the original cluster and of three new generations per each, for all the sink particle distributions of our sample. The new generations are qualitatively indistinguishable from the original clusters.

5. Summary

We introduced a new method for generating a number of new realizations from a given set of initial conditions from hydro-dynamical simulations. This method is based on a hierarchical clustering algorithm, which learns a tree-like representation of the stellar system. This tree encodes the structural properties of the sink particle distributions and can be turned into new macroscopic realizations by modifying its first branches. This procedure results in different large scale realizations (e.g., the number of main clumps and their distances), while approximately preserving the characteristics of the small scale structure responsible for most of dynamical evolution. The new realizations are qualitatively similar to the original simulations when visualized in the three-dimensional space, and present consistent velocity, mass and pairwise distance distributions.

This generative method leads to a speedup in computation of several orders of magnitude: generating initial conditions from hydro-dynamical simulations, in fact, requires hundreds of thousands core hours per simulation, while our procedure takes about some core seconds to generate a new realization. Also, our scheme is very flexible, allowing to set how deep we alter the tree structure by choosing the number of initial branches we modify.

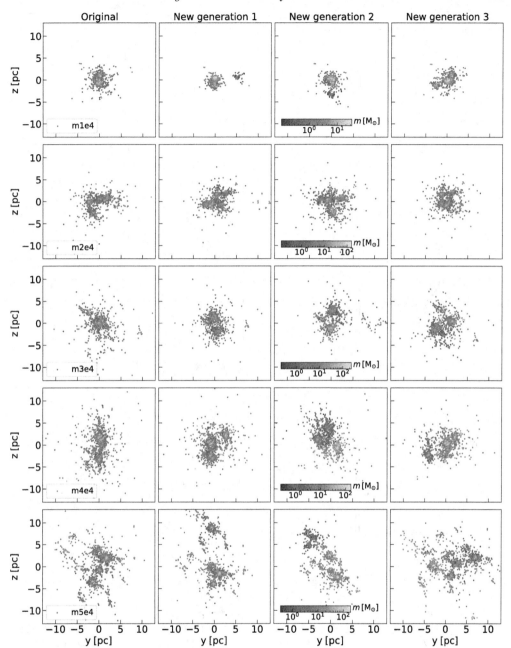

Figure 3. Projections in $y - z$ of the 5 least massive star clusters (left), and of three different generated clusters per each. The colour map marks the mass of the individual stars.

6. Acknowledgements

This project is partially supported by European Research Council for the ERC Consolidator grant DEMOBLACK, under contract no. 770017, and by the European Unions Horizon 2020 research and innovation program under the Marie Skłodowska-Curie grant agreement No.896248.

S. Torniamenti

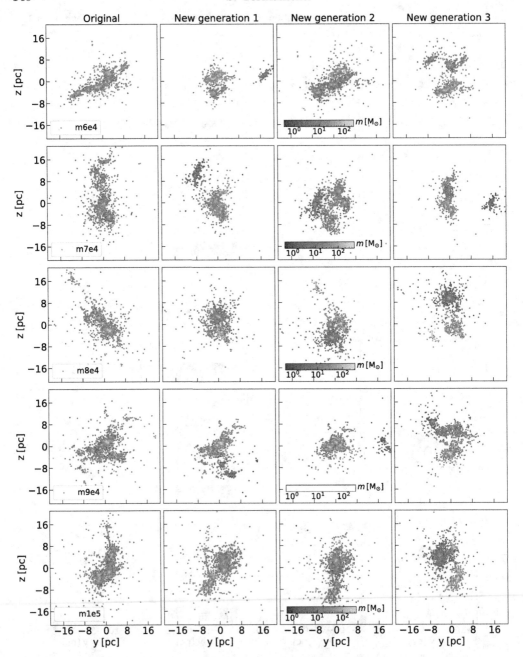

Figure 4. Projections in $y - z$ of the 5 most massive star clusters (left), and of three different generated clusters per each. The colour map marks the mass of the individual stars. The colour map marks the mass of the individual stars.

References

Ballone A., Mapelli M., Di Carlo U. N., Torniamenti S., Spera M., Rastello S., 2020, *MNRAS*, 496, 49

Ballone A., Torniamenti S., Mapelli M., Di Carlo U. N., Spera M., Rastello S., Gaspari N., Iorio G., 2021, *MNRAS*, 501, 2920

Bate M. R., Bonnell I. A., Price N. M., 1995, MNRAS, 277, 362B

Bate M. R., 2009, *MNRAS*, 392, 1363

Cantat-Gaudin T., et al., 2019, *A&A*, 626, A17

Cartwright A., 2009, *MNRAS*, 400, 1427

Hénault-Brunet V., et al., 2012, *A&A*, 545, L1

Hills J. G., 1980, *ApJ*, 235, 986

Kaufman L., Rousseeuw P. J., 1990, Finding groups in data. an introductionto cluster analysis, Wiley Series in Probability and Statistics.

King I. R., 1966, *AJ*, 71, 64

Klessen R. S., Burkert A., 2000, *ApJS*, 128, 287

Kuhn M. A., Hillenbrand L. A., Sills A., Feigelson E. D., Getman K. V., 2019, *ApJ*, 870, 32

Lada, C. J., Lada, E. A., 2003, *ARAA* 41, 57

Larson R. B., 1995, 272, 213

Pedregosa F., et al., 2011, *Journal of Machine Learning Research*, 12, 2825

Plummer H. C., 1911, *MNRAS*, 71, 460

Torniamenti S., Pasquato M., Di Cintio P., Ballone A., Iorio G., Mapelli M., 2022, *MNRAS*, 510, 2097

Wall J. E., McMillan S. L. W., Mac Low M.-M., Klessen R. S., PortegiesZwart S., 2019, ApJ, 887, 62

The Predictive Power of Computational Astrophysics as a Discovery Tool
Proceedings IAU Symposium No. 362, 2023
D. Bisikalo, D. Wiebe & C. Boily, eds.
doi:10.1017/S1743921322001430

Stellar Population Photometric Synthesis with AI of S-PLUS galaxies

Vitor Cernic⬛, for the S-PLUS collaboration.

Universidade de São Paulo, Instituto de Astronomia, Geofísica e Ciências Atmosféricas,
Rua do Matão 1226, CEP 05508-090, São Paulo, SP, Brazil
email: vitorcernic@usp.br

Abstract. We trained a Neural Network that can obtain selected STARLIGHT parameters directly from S-PLUS photometry. The training set consisted of over 55 thousand galaxies with their stellar population parameters obtained from a STARLIGHT application by Cid Fernandes *et al.* (2005). These galaxies were crossmatched with the S-PLUS iDR 3 database, thus, recovering the photometry for the 12 band filters for 55803 objects. We also considered the spectroscopic redshift for each object which was obtained from the SDSS. Finally, we trained a fully connected Neural Network with the 12-band photometry + redshift as features, and targeted some of the STARLIGHT parameters, such as stellar mass and mean stellar age. The model performed very well for some parameters, for example, the stellar mass, with an error of 0.23 dex. In the future, we aim to apply the model to all S-PLUS galaxies, obtaining never-before-seen photometric synthesis for most objects in the catalogue.

Keywords. galaxies, stellar populations, machine learning

1. Introduction

As astronomical surveys get more complex, the amount of data produced increases year after year. This has created a need for robust statistical tools that can operate on massive amounts of data and reliably find relationships between data parameters without dealing with each object separately.

The knowledge of intrinsic parameters of galaxies, such as stellar mass and mean stellar age, is a necessary part of understanding how a galaxy operates. Such parameters can be obtained through spectral energy distribution (SED) fitting. Although this process yields significant results, obtaining the spectrum in the first place can be a difficult task. Photometric surveys tackle this observational problem by providing the photometry of each object in several bands. While the spectrum has more information, the photometry of galaxies can be measured much easily in photometric surveys like the Southern Photometric Local Universe Survey (S-PLUS, Mendes de Oliveira *et al.* 2019). In this work, we aim to create a tool that can predict stellar population parameters directly from S-PLUS photometry using machine learning techniques.

2. Methodology and Results

To create this machine learning tool, we have used supervised learning. We needed large amounts of robust data, both for the inputs (photometry) and the outputs (stellar population parameters). This model is intended to be applied to S-PLUS data, thus the photometry was gathered directly from the S-PLUS iDR3. The redshift plays a significant role in determining such parameters, so we also collected the Sloan Digital Sky Survey (SDSS)-IV (Blanton *et al.* 2017) spectroscopic redshift for each object from the

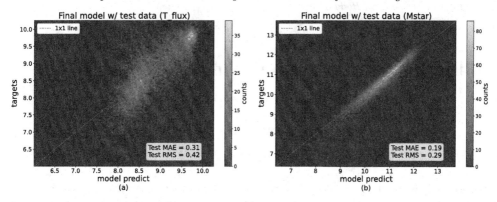

Figure 1. Prediction plots for: (a) mean star age weighted by flux; (b) stellar mass. The MAE and RMS of the test set is depicted, as well as the 1 x 1 line.

extended Baryon Oscillation Spectroscopic Survey (eBOSS, Dawson *et al.* 2016). We used the stellar population parameters from a STARLIGHT (Cid Fernandes *et al.* 2005) application to the SDSS. We decided to use 11 different parameters that could be interesting for astronomers, including age, metallicity, stellar mass, and others. After crossmatching all databases, we ended up with data for 55803 objects. Each object is detected in the 12 S-PLUS photometric bands, is matched with an SDSS spectroscopic redshift, and is associated with STARLIGHT fits to the 11 target parameters.

Our objective was to solve a regression problem. While we can use many different models, we eventually found from testing that a Neural Network had the best performance. We chose a fully-connected feed-forward architecture for this network with just two hidden layers. The input layer had 13 nodes (12 magnitudes + 1 redshift), and the output had a single node (one parameter). We used the Keras framework (Keras *et al.* 2015) and evaluated the performance of each network using the Mean Average Error (MAE) and Root Mean Square Error (RMS).

Out of the 11 parameters we predicted, 8 of them presented satisfying results with an adequate MAE and RMS by comparing it to a baseline zero rule prediction algorithm. The 8 parameters are: visible absorption, velocity dispersion, mean star age weighted by flux and by mass, metallicity weighted by flux and by mass, stellar mass and stellar star formation rate. The results for age and stellar mass are depicted in Figure 1.

3. Conclusions

In this work, we created a machine learning tool that can reliably predict eight stellar population parameters directly from S-PLUS photometry. Better network performance can still be achieved, but we are now interested in applying each network to the whole S-PLUS database using photometric redshifts.

In the following months, we will achieve new, never-before-seen parameters for thousands of galaxies in the S-PLUS catalogue and study the properties of the galaxy population as described by these parameters. The method used here can be generalized to other photometric surveys, such as J-PAS, J-PLUS, and others.

I would also like to thank CAPES for the financial support throughout the project.

References

Blanton M. R., et al., 2017, *AJ*, 154, 28. doi:10.3847/1538-3881/aa7567
Chollet, F., & others., 2015. Keras. GitHub. Retrieved from https://github.com/fchollet/keras
Cid Fernandes et al., 2005., *MNRAS*, 358(2), 363378. doi:10.1111/j.1365-2966.2005.08752.x
Kyle S. Dawson, *et al.*, 2016, *AJ*, 151, 44. doi:10.3847/0004-6256/151/2/44
Mendes de Oliveira C., et al., 2019, *MNRAS*, 489(1), 241267. doi:10.1093/mnras/stz1985

The Predictive Power of Computational Astrophysics as a Discovery Tool
Proceedings IAU Symposium No. 362, 2023
D. Bisikalo, D. Wiebe & C. Boily, eds.
doi:10.1017/S1743921322001697

Clustering stellar pairs to detect extended stellar structures

Sergey Sapozhnikov🆔 and Dana Kovaleva🆔

Institute of Astronomy RAS, Pyatnitskaya 48., Moscow, Russia

Abstract. Gaia data allows for search for extended stellar structures in phase (coordinates plus velocities) space. We describe a method of using DBSCAN clustering algorithm, which is used to group closely-packed-together data points, to a list of preliminary selected pairs of stars, with parameters expected to be found within stellar streams and comoving groups: loose structures in which stars are not gravitationally bound, but do share motion and evolutionary properties. To test our approach, we construct a model population of background stars, and use pair-constructing and clustering algorithms on it. Results show that transitioning to a list of pairs sharply reveals structures not presented in background model, which then become more apparent targets in coordinate-velocity phase space for DBSCAN algorithm thanks to now increased relative density of the extended stellar structure.

Keywords. stars: kinematics, galaxies: structure, galaxies: star clusters

---

Gaia data have many applications, in particular, they can be used to search for star clusters and extended stellar structures: loose groups of stars, which are not bound gravitationally, but do exhibit similarities in motion and age suggesting that they share a similar genesis, like shown in Harshil Kamdar *et al.* (1999). Approaches to find extended stellar structures include, for example, search in the vicinity of stellar clusters for kinematically similar stars (see Jerabkova, Tereza *et al.* (2021); Siegfried Roser *et al.* (2019)), or using clustering algorithms to detect new objects in the field Hunt, Emily L. *et al.* (2021). The latter approach is somewhat complicated by the fact that such structures are very stretched on the sky, with many unrelated stars in their vicinity. We aim to test the approach of using our algorithm initially developed to find ultra-wide binary stars to search for stellar pairs with common motion, and feed this catalogue of pairs to the DBSCAN clustering algorithm. By adjusting the restrictions on pairs properties, we aim to highlight the comoving groups of stars and make them stand out more clearly from the background and make it easier for clustering algorithm to recognize.

Creation of pairs method is applied to a large sample of stars from Gaia DR2 data, 30x30 deg on the sky and 100-1000 pc distance from Sun. Criteria for pairs are selected according to Harshil Kamdar *et al.* (1999) and are the following:

- projected separation $< 1\ pc$
- projected relative motion $< 3\ km/s$
- proper motion difference $< 6\ mas/yr$
- parallax consistency within $-0.1 < \pi_{cons} < 1$

Here, "parallax consistency" is

$$\pi_{cons} = 3(\sigma_{\pi_1} + \sigma_{\pi_2}) - |(\pi_1 - \pi_2)|. \qquad (0.1)$$

Clustering is performed using DBSCAN clustering algorithm in four-parameter space (2 coordinates + 2 velocities). We use small sample of region in Coma Berenices to test

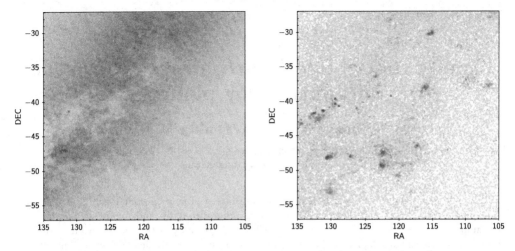

Figure 1. Scatter plots showing the sky positions of stars (left) and pairs (right) from the Gaia data in the selected region. More densely populated regions are closer to orange color. These scatter plots show how the translation from stars to pairs increases the contrast of the stellar structures.

for false negative result. Different parameters of DBSCAN clustering algorithm yield different amount of clusters when applied to the same set of pairs from Coma Ber. We compare our results with both clustering of pairs created from model uniform star distribution, and direct clustering of stars from that model, and find that clustering of stars from model devoid of clusters produce false positives much more easily compared to clustering of pairs created from the same sample. Also, clustering of pairs (compared to clustering of stars) for real data allow for easier recognition of clusters (in larger area of DBSCAN hyperparameters).

We conclude that:

• Transition from stars to pairs makes stellar structures stand out sharply against the stellar background. These structures are labelled by DBSCAN algorithm.

• We check for false positives using uniform stellar background model. No clusters are found in this sample, as expected.

• We applied our method to Coma Ber comoving group and verified that it correctly finds existing stellar structure there.

• Comparison with direct clustering of stars reveals that our method of pair clustering is less prone to false positives creation.

References

Harshil Kamdar, Charlie Conroy et al. 2019, *Astrophys. J*, 884(2):L42
Jerabkova, Tereza and Boffin, Henri M. J. et al. 2021, *Astronomy and Astrophysics*, 646:A104
Röser, Siegfried and Schilbach, Elena 2019, *Astronomy and Astrophysics*, 627:A4
Hunt, Emily L. and Reffert, Sabine 2021, *Astronomy and Astrophysics*, 646:A104

The Predictive Power of Computational Astrophysics as a Discovery Tool
Proceedings IAU Symposium No. 362, 2023
D. Bisikalo, D. Wiebe & C. Boily, eds.
doi:10.1017/S1743921322001843

Kinetic modeling of auroral events at solar and extrasolar planets

Valery I. Shematovich and Dmitry V. Bisikalo

Institute of Astronomy of the RAS, 48 Pyatnitskaya str., 119017, Moscow, Russian Federation
email: shematov@inasan.ru

Abstract. Auroral events are the prominent manifestation of solar/stellar forcing on planetary atmospheres and are closely related to the energy deposition by and evolution of planetary atmospheres. Observations of auroras are widely used to analyze the composition, structure, and chemistry of the atmosphere under study, as well as charged particle and energy fluxes that affect the atmosphere. Numerical kinetic Monte Carlo models had been developed allowing us to study the processes of precipitation of high-energy electrons, protons and hydrogen atoms into the planetary atmospheres on molecular level of description, taking into account the stochastic nature of collisional scattering at high kinetic energies. Such models are used to study auroras at both magnetized and non-magnetized planets in the Solar and extrasolar planetary systems. The current status of the kinetic model is illustrated in application to the auroral events at Mars.

Keywords. solar system: general, stars:planetary systems, radiation mechanisms: nonthermal

1. Introduction

The interaction of the solar wind with atmospheres of the terrestrial planets played an important role in their atmospheric loss over billions of years of its existence (Ramstad & Barabash 2021). The values of the current losses by Mars, Venus, and Earth are comparable, but for the total mass and composition of the main components for Venus and Earth, the losses caused by the solar wind are insignificant. This is not a case for Mars because of significant difference between Mars and Venus in the processes of interaction of the solar wind with the atmosphere. Mars has an extended exosphere, the outer part of the atmosphere that extends for thousands of kilometers. This is due to the fact that the gravity of Mars is small. In addition, the processes of recombination dissociation of O_2^+ ions (see, e.g., Groeller et al. 2014) lead to formation of an extended oxygen exosphere. The relatively low gravity and the extended exosphere lead to the fact that kinetic processes on Mars play a major role in the interaction of the solar wind with the planet and in atmospheric processes.

The Sun influences the upper layers of the terrestrial planet atmosphere through both the radiation absorbed in the soft X-ray and extreme UV wavelength ranges and the solar wind plasma forcing, which results in the formation of the extended neutral corona populated by suprathermal (hot) H, C, N, and O atoms (see, e.g., Shematovich & Bisikalo 2021a). For example, one of the important results of the current Mars Atmosphere and Volatile Evolution (MAVEN) mission was that the Imaging UV Spectrograph (IUVS) observations confirmed the presence of the extended corona composed of hydrogen, carbon, and oxygen atoms (Deighan et al. 2015). Such extended corona contains both thermal and hot fractions (see, e.g., Shematovich & Bisikalo 2021a) and, in turn, is changed due to the solar wind plasma inflow and the local fluxes of ions picked up from the ionosphere to the planetary exosphere. This inflow leads to the formation of superthermal atoms (energetic neutral atoms – ENAs) escaping from the neutral atmosphere of Mars because of the charge exchange with precipitating high-energy ions.

Proton auroral events are indicators of enhanced solar activity and are accompanied by rather intense losses of neutrals from the upper atmospheres of the terrestrial planets. Proton auroras at the Earth were first observed in the polar atmosphere by the Doppler shift of spectral hydrogen lines H-β and H-α. These auroras are strongly controlled by our planet's magnetosphere: protons penetrate the atmosphere along the terrestrial magnetic field lines and lose their initial energy near the magnetic poles, where the proton auroras are observed (Frey et al. 2003). Proton auroras are among a few observable Martian phenomena which result from direct interaction of solar wind protons with an extended hydrogen corona. Studying Martian proton auroras can also supply an additional understanding of the evolution and loss of the atmosphere, since the processes responsible for the formation of Martian auroral events (for example, interactions between the solar wind and the extended hydrogen corona) are also responsible for the loss of the neutral atmosphere (Shematovich 2021). There were attempts to estimate the contribution of atmospheric sputtering during proton auroral events at Mars to the atomic oxygen loss rate (see, e.g., Shematovich 2021, and references therein). These attempts have a particular importance, because this process was ignored in the recent analysis of the atmospheric loss of Mars based on the MAVEN data (Jakosky et al. 2018).

Proton auroras were recently discovered in observations of the enhanced Ly-α emission of atomic hydrogen in the dayside atmosphere of Mars (Deighan et al. 2018; Hughes et al. 2019) and they are induced by the fluxes of high-energy hydrogen atoms (ENA-Hs) penetrating into the atmosphere (Deighan et al. 2018). We use a set of the kinetic Monte Carlo (kMC) models (Shematovich et al. 2019, 2021; Shematovich 2021) to calculate the the energy deposition rates by neutral atmosphere and source functions of suprathermal hydrogen and oxygen atoms formed in the precipitation processes. For example, kMC1 model (Shematovich et al. 2021) makes it possible to analyze the charge-exchange process for solar-wind protons in the extended hydrogen corona of Mars and to obtain the spectra of hydrogen atoms penetrating into the atmosphere through the boundary of the induced magnetosphere of Mars. The calculated spectra and fluxes of ENA-Hs are strongly non-equilibrium and were used as an upper boundary condition for the kMC2 model (Shematovich et al. 2019) of precipitation of energetic hydrogen atoms into the upper atmosphere allowing us to calculate the characteristics (Shematovich & Bisikalo 2021b) of proton auroral events at Mars. Specifically, we obtain the formation rate and energy spectra of suprathermal hydrogen and oxygen atoms produced in elastic and inelastic collisions between atmospheric hydrogen and oxygen atoms and ENA-Hs penetrated into the atmosphere. And, finally, these formation rates of suprathermal hydrogen and oxygen atoms are used as a source function in the kMC3 model (Shematovich 2021) aimed to study the kinetics and transport of suprathermal hydrogen and oxygen atoms in the upper atmosphere, which results in the formation of extended hot atomic coronas of the terrestrial planet.

In this paper, we present a set of the MC kinetic models that describe the influence of the proton flux of the undisturbed solar wind on the upper atmosphere of the terrestrial planet in the Solar and extrasolar planetary systems. With this set, the the change of the solar-wind proton spectrum due to the cascade charge-exchange process in the extended hydrogen corona of Mars has been modeled to obtain the energy fluxes and energy spectra of hydrogen atoms penetrating into the daytime upper atmosphere through the boundary of the induced magnetosphere. These characteristics make it possible to estimate the parameters of proton auroral events observed in the upper atmosphere of the teresstrial planet, and for Mars to validate the models by comparison of kinetic calculations with observations by the IUVS instrument onboard the MAVEN spacecraft.

2. Results of the kinetic modeling of auroral events

The process of proton aurora formation at Mars is markedly different from the scenario for Earth. The exosphere (or corona) of Mars is mostly populated by atomic and molecular hydrogen and extends over several Martian radii. The absence of a global internal magnetic field on Mars allows the solar wind to interact with the hydrogen corona on the illuminated hemisphere of Mars, starting from the bow shock region and flowing around the planet along the boundary of the induced magnetosphere (IMB). The source of proton auroras at Mars is a flux of energetic neutral hydrogen atoms that enter the atmosphere through the IMB and participate in collisions with neutral particles in the lower atmosphere. It is known that the instruments onboard the MAVEN spacecraft do not have the ability to measure the energy spectra of neutral particles, so a kinetic model is required to calculate both the ENA-H flux and its energy spectrum. We present the kinetic calculations of the ENA-H flux penetration, which is formed due to the charge exchange between solar-wind protons and thermal hydrogen atoms in the extended hydrogen corona of Mars, into the planetary upper atmosphere. The estimates of the energy flux and the energy spectra of hydrogen atoms and protons, which are formed due to the charge exchange in the Martian upper atmosphere, have been obtained. The calculation results of the auroral event parameters are presented for the basic case, within which the boundary of the induced magnetosphere is at an altitude of 820 km. The calculations were performed for a basic case (M1), with the profiles of the temperature and the number density of the main components (CO_2 and O) of the upper atmosphere corresponding to a low level of solar activity. The distribution of hydrogen atoms in the extended corona of Mars was specified by the Chamberlain model for a planetary exosphere, the parameters of which were chosen as follows: the exobase height is $h_{exo} = 200$ km, at which the temperature is $T(h_{exo}) = 179$ K and the number density of hydrogen atoms is $n_H(h_{exo}) = 1.48 \times 10^6$ cm^{-3} according to the results of Chaffin et al. (2018). Then, as a boundary condition for the kMC1 model at an altitude of 3000 km, we used the energy flux and the energy spectrum of protons of the undisturbed solar wind measured with the SWIA/MAVEN instrument for the orbit on February 27, 2015 (Halekas et al. 2015). The calculated spectra of hydrogen atoms were assumed to be an upper boundary condition in the kMC2 model (Shematovich et al. 2019) for the precipitation of high-energy hydrogen atoms into the daytime upper atmosphere through the induced magnetosphere boundary (IMB). Observations onboard the MEX and MAVEN spacecraft resulted in detecting significant variations of the atomic hydrogen content in the corona of Mars (Chaffin et al. 2018; Girazian & Halekas 2021). These variations induce changes in the efficiency of the charge exchange between protons of the undisturbed solar wind and hydrogen atoms in the corona (Shematovich et al. 2021). Consequently, analogous calculations were performed for the case, within which the variation in the column density of hydrogen in the corona of Mars was taken into account (case M2). Namely, under the same exobase parameters, the number density of hydrogen atoms $n_H(h_{exo})$ increased twofold at the exobase level, which corresponds to the position of the IMB at an altitude of 1260 km (for details, see Shematovich et al. 2021). In case M2, the charge-exchange efficiency approaches a value of 6%, i.e., the energy flux of ENA-Hs penetrating into the upper atmosphere is 1.5 times higher than the corresponding value in case M1. The calculations were made for the solar zenith angle equal to zero.

The fluxes and energy spectra of high-energy hydrogen atoms and protons, which were obtained with the kinetic Monte Carlo model, allow us to calculate all of the necessary parameters of proton auroral phenomena in the upper atmosphere of Mars. For example,

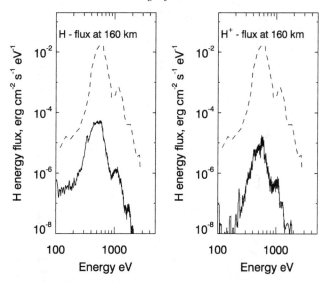

Figure 1. The energy spectrum of downward fluxes of ENA-H (left panel) and protons (right panel) at height of 160 km, at which the peak of energy deposition of ENA-H flux is usually found. Dashed line – the energy spectrum of protons of the undisturbed solar wind measured with the SWIA/MAVEN instrument for the orbit on February 27, 2015 (Halekas et al. 2015), which was taken as the upper boundary condition of kMC1 model.

in Fig. 1, we show the energy spectrum of downward fluxes of ENA-H (left panel) and protons (right panel) at height of 160 km, at which the peak of energy deposition of ENA-H flux is usually found. It is seen that both spectra still keep the form and structure of the energy spectrum of protons of the undisturbed solar wind shown by dashed line. These spectra were calculated for the case M1 with the IMB at 820 km of the precipitation of high-energy hydrogen atoms ENA-Hs penetrating into the Martian atmosphere through the IMB calculated for cases M1 (red line) and M2 (brown line). Such energy distributions of the ENA-H flux allow us to calculate the H Ly-α excitation rates and in conjunction with the radiative transfer model to estimate the excess in the H Ly-α emission during a proton auroral event in the Martian atmosphere (Gerard et al. 2019).

An additional source of hot oxygen atoms - collisions with the momentum and energy transfer from the flux of precipitating ENA-H particles with high kinetic energies to atomic oxygen in the upper atmosphere of Mars - is included in the Boltzmann kinetic equation, the solution of which was obtained using the kMC3 model (Shematovich 2021). This allows us to determine self-consistently the sources of suprathermal oxygen atoms, as well as their kinetics and transport, and to estimate the population of the hot oxygen corona of Mars and the atmospheric losses of atomic oxygen during proton auroral events. We derived the kinetic-energy distribution functions for suprathermal oxygen atoms in the thermosphere-to-exosphere transition region of the penetration of undisturbed solar-wind protons into the illuminated atmosphere of Mars due to the charge exchange in the extended hydrogen corona (Shematovich et al. 2021). It has been found that the exosphere is inhabited by a significant number of suprathermal oxygen atoms with kinetic energies up to the escape energy of 2 eV, i.e., the hot oxygen corona (Fig. 2) of Mars forms due to the interaction with solar wind. It is seen that height scale for hot oxygen distribution for cases M1 (red line) and M2 (brown line) is larger than the one (black line) for thermal O atoms, therefore hot O corona extends into the upper regions of the Martian corona and should be taken into account in the consideration of the solar wind interaction with upper atmosphere of Mars. The values of $(3.5 - 5.8) \times 10^7$ cm^{-2}s^{-1} of the loss rate for

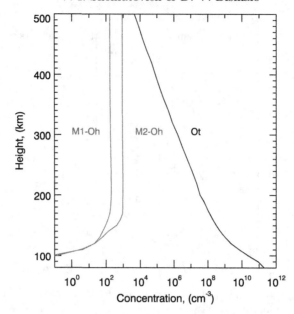

Figure 2. Height profiles of suprathermal oxygen concentration in the upper atmosphere during proton auroral events at Mars. O_h indicates the calculated height distribution of hot O atoms, and O_t – of thermal O atoms in the upper atmosphere of Mars.

oxygen atoms from the Martian atmosphere during proton auroral events on Mars were obtained in self-consistent calculations and are comparable to the oxygen loss rate of $(1.5 - 3.5) \times 10^7$ cm^{-2}s^{-1} due to the exothermic photochemistry (Groeller et al. 2014; Jakosky et al. 2018). Though proton auroras are sporadic events, the kinetic calculations showed that the precipitation-induced flux of escaping hot oxygen atoms may become dominant under conditions of extreme solar events: solar flares and coronal mass ejections (Deighan et al. 2015; Jakosky et al. 2018; Shematovich et al. 2021). The analysis of the atomic oxygen loss due to atmospheric sputtering during proton aurorae on Mars should be taken into account especially when studying the climate evolution of the planet on geological time scales.

3. Conclusions

We have developed a set of kinetic Monte Carlo models, which could be considered as a discovery tool to investigate the auroral events caused by the precipitation processes of high-energy charged and neutral particles from the solar/stellar wind into the upper atmospheres of terrestrial planets in the Solar and extrasolar systems. This set of kinetic MC models was used to study and interpret the recently discovered discrete, diffuse and proton auroras at Mars observed and measured by the Mars Express, MAVEN spacecraft. Such application allowed us to evaluate both the characteristics of the precipitating high-energy particles from the solar wind plasma, and the space distribution and brightness of ultraviolet and optical emissions in the auroras induced by the precipitation processes. Estimates of the neutral atmosphere loss due to the forcing of solar wind plasma onto the upper atmosphere of Mars were also obtained. By studying auroral events we can better understand the relations between energy fluxes coming from the Sun/host star, the extended hydrogen corona, and the circumplanetary plasma environment. In addition, the appearance and the intensity of proton polar auroras can indirectly indicate the changes occurring in the dynamics of underlying atmospheric layers such as the neutral atmosphere chemical composition and the input of the dust activity. These kinetic models would be used in the studies of the stellar wind forcing on the upper atmospheres of terrestrial type planets orbiting other stars (Savanov & Shematovich 2021).

The authors are grateful to the Government of the Russian Federation and the Ministry of Higher Education and Science of the Russian Federation for grant support no. 075-15-2020-780 (no. 13.1902.21.0039).

References

Chaffin, M. S., Chaufray, J. Y., Deighan, J., Schneider, N. M., ..., & Clarke, J. T. 2018, *J. Geophys. Res.: Planets*, 123, 2192

Deighan, J., Chaffin, M. S., Chaufray, J.-Y., Stewart, A. I. F., ..., & Jakosky, B. M. 2015, *Geophysical Research Letters*, 42, 8902

Deighan, J., Jain, S. K., Chaffin, M. S., Fang, X., ... & Jakosky, B. M. 2018, *Nature Astronomy*, 2, 802

Halekas, J. S., Lillis, R. J., Mitchell, D. L., Cravens, T. E., ..., & Ruhunusiri, S. 2015, *Geophys. Res. Lett.*, 42, 9805

Hughes, A., Chaffin, M., Mierkiewicz, E., Deighan, J., ..., & Jakosky, B. M. 2019, *J. Geophys. Res.: Space Physics*, 124, 10,533

Frey, H. U., Mende, S. B., Immel T. J., Gérard, J.-C., ..., & Shematovich, V. I. 2003, *Space Science Reviews*, 109, 255

Gérard, J.-C., Hubert, B., Ritter, B., Shematovich, V. I., & Bisikalo, D. V. 2019, *Icarus*, 321, 266

Groeller, H., Lichtenegger, H., Lammer, H., & Shematovich, V. I. 2014, *Planet. Space Sci.*, 98, 93

Girazian, Z., & Halekas, J. 2021, *J. Geophys. Res.: Planets*, 126, e06666

Jakosky, B. M., Brain, D., Chaffin, M., Curry, S., Deighan, J., ..., & Zurek, R. 2018, *Icarus*, 315, 146

Ramstad, R., & Barabash, S. 2021, *Space Science Reviews*, 217, id. 36

Savanov, I. S., & Shematovich, V. I. 2021, *Astrophysical Bulletin*, 76, 450

Shematovich, V. I. 2021, *Solar System Research*, 55, 322

Shematovich, V. I., & Bisikalo, D. V. 2021a, *Oxford Research Encyclopedia of Planetary Science*, Ed. by P. Read, et al., Oxford Univ. Press, 104

Shematovich, V. I., & Bisikalo, D. V. 2021b, *Astronomy Reports*, 65, 869

Shematovich, V. I., Bisikalo, D. V., Gérard, J.-C., & Hubert, B. 2019, *Astronomy Reports*, 63, 835

Shematovich, V. I., Bisikalo, D. V., & Zhilkin, A. G. 2021, *Astronomy Reports*, 65, 203.

The Predictive Power of Computational Astrophysics as a Discovery Tool
Proceedings IAU Symposium No. 362, 2023
D. Bisikalo, D. Wiebe & C. Boily, eds.
doi:10.1017/S1743921322001260

Stellar activity effects on the atmospheric escape of hot Jupiters

Hiroto Mitani[1] , **Riouhei Nakatani[2], and Naoki Yoshida[1,3,4]**

[1]Department of Physics, School of Science, The University of Tokyo, 7-3-1 Hongo, Bunkyo, Tokyo 113-0033
email: `hiroto.mitani@phys.s.u-tokyo.ac.jp`

[2]RIKEN Cluster for Pioneering Research, 2-1 Hirosawa, Wako, Saitama 351-0198, Japan

[3]Kavli Institute for the Physics and Mathematics of the Universe (WPI), UT Institutes for Advanced Study, The University of Tokyo, Kashiwa, Chiba 277-8583, Japan

[4]Research Center for the Early Universe, School of Science, The University of Tokyo, 7-3-1 Hongo, Bunkyo, Tokyo 113-0033

Abstract. Transit observations have revealed the existence of atmospheric escape in several hot Jupiters. High energy photons from the host star heat the upper atmosphere and drive the hydrodynamic escape. The escaping atmosphere can interact with the stellar wind from the host star. We run radiation hydrodynamics simulations with non-equilibrium chemistry to investigate the wind effects on the escape and the transit signature. Our simulations follow the planetary outflow driven by the photoionization heating and the wind interaction in a dynamically coupled, self-consistent manner. We show that the planetary mass-loss rate is almost independent of the wind strength, which however affects the Ly-α transit depth considerably. But the Hα transit depth is almost independent of the wind strength because it is largely caused by the lower hot layer. We argue that observations of both lines can solve the degeneracy between the EUV flux from the host and the wind strength.

Keywords. hydrodynamics – methods: numerical – planets and satellites: atmospheres – planets and satellites: physical evolution

---

1. Introduction

For close-in exoplanets, the extreme irradiation from the host star can drive the atmospheric escape. Such escape process can be important in planetary evolution and can even shape the statistical properties of observed close-in exoplanets (sub-Jovian desert; Szabó and Kiss (2011), sub-Neptune desert; Fulton et al. (2017)). Transit observations have revealed the extended atmosphere for close-in exoplanets (Vidal-Madjar et al. 2003; Ehrenreich et al. 2015).

Extreme-Ultraviolet (EUV $>$ 13.6 eV) photons ionize the hydrogen atoms and heat the atmospheric gas through thermalization of the photoelectrons. The photoionization heating contributes to atmospheric escape. Radiation hydrodynamics simulations identified the important physical processes and allowed detailed studies of the atmospheric structure (Murray-Clay et al. 2009). Interaction with the stellar wind is also investigated in several studies (Bisikalo et al. 2013, 2018; Cherenkov et al. 2018; Vidotto and Cleary 2020; Carolan et al. 2021). Often the Ly-α transit depth is considered because of its large absorption. Ly-α photons from the host star can be easily absorbed by the interstellar medium between the star-plane system and the earth, which makes it difficult to observe directly. Recent observations by ground-based telescopes use other lines (e.g. Helium triplet line, Hα line) which are more useful to detect the extended atmosphere.

Recent observations have also detected hot Jupiters around young active stars. Vigorous activities of such young stars can cause a strong influence on the planetary atmosphere, especially on the close-in planets. Strong winds can confine the upper atmosphere and reduce both the mass loss and the transit depth. The wind effect is expected to be important particularly in young systems with strong activities.

Numerical simulations so far that have used to investigate the wind confinement and Ly-α and Hα transits do not treat photoionization heating and the launched outflow in a self-consistent manner. To study the wind effect on the planetary atmosphere, self-consistent radiation hydrodynamics simulations are necessary to follow the launching of the outflow because Hα absorption may be significant in the lower atmospheric layers. We run simulations with varying the strength of the stellar wind and calculate the Ly-α and Hα transit depths. We discuss the possibility that the strong stellar activity changes the absorption signatures.

2. Methods

We first introduce our simulations. We use the hydrodynamics simulation code PLUTO (Mignone et al. 2007) with the EUV radiation transfer module (Nakatani et al. 2018). Detailed implementation is described in (Nakatani et al. 2018; Mitani et al. 2021). Our simulations solve the following 2D axisymmetric hydrodynamic equations and non-equilibrium chemistry:

$$\frac{\partial \rho}{\partial t} + \nabla \cdot \rho \vec{v} = 0 \tag{2.1}$$

$$\frac{\partial \rho v_R}{\partial t} + \nabla \cdot (\rho v_R \vec{v}) = -\frac{\partial P}{\partial R} - \rho \frac{\partial \Psi}{\partial R} \tag{2.2}$$

$$\frac{\partial \rho v_z}{\partial t} + \nabla \cdot (\rho v_z \vec{v}) = -\frac{\partial P}{\partial z} - \rho \frac{\partial \Psi}{\partial z} \tag{2.3}$$

$$\frac{\partial \rho E}{\partial t} + \nabla \cdot (\rho H \vec{v}) = -\rho \vec{v} \cdot \nabla \Psi + \rho (\Gamma - \Lambda) \tag{2.4}$$

$$\frac{\partial n_H y_i}{\partial t} + \nabla \cdot (n_H y_i \vec{v}) = n_H R_i \tag{2.5}$$

where $\rho, \vec{v}, P$ are density, velocity, pressure of the gas. The potential Ψ includes contributions of the star and the planet and also incorporates the centrifugal force due to the orbital motion. We also follow the non-equilibrium chemistry including photoionization of the hydrogen atoms which can be important in the hydrogen absorption signatures. $y_i = n_i/n_H$ and R_i represent the abundance and the reaction rate, respectively. The incorporated chemical species are H, H$^+$, H$_2$, e$-$.

The heating and cooling rates are denoted as Γ, Λ, respectively. We calculate the EUV photoionization heating rate by ray-tracing:

$$F_\nu = \frac{\Phi_\nu}{4\pi a^2} \exp[-\sigma_\nu N_{\mathrm{HI}}] \tag{2.6}$$

$$\Gamma_{\mathrm{ph}} = \frac{1}{\rho} n_{\mathrm{HI}} \int_{\nu_0}^{\infty} d\nu \, \sigma_\nu h(\nu - \nu_0) F_\nu \tag{2.7}$$

where σ_ν is the absorption cross section as a function of ν (Osterbrock and Ferland 2006) and N_{HI} is the column density of hydrogen atoms and $h\nu_0 = 13.6\,\mathrm{eV}$. We implement Ly-$\alpha$ cooling and hydrogen recombination cooling (Spitzer 1978; Anninos et al. 1997).

Table 1. Model parameters in the fiducial run.

Stellar parameters	
Stellar Mass M_*	$1\,M_\odot$
Stellar Radius R_*	$1\,R_\odot$
Stellar EUV photon emission rate Φ_ν	$1.4 \times 10^{38}\,\mathrm{s}^{-1}$
Stellar wind velocity	$540\,\mathrm{km/s}$
Stellar wind temperature	$2 \times 10^6\,\mathrm{K}$
Stellar wind density	$2.5 \times 10^3\,\mathrm{g/cm}^3$

Planetary parameters	
Planet Mass M_p	$0.3\,M_\mathrm{J}$
Planet Radius R_p	$1\,R_\mathrm{J}$
Semi-major axis a	$0.045\,\mathrm{AU}$

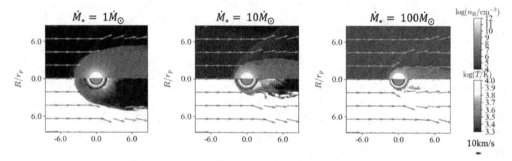

Figure 1. The atmospheric structure of our simulations. The EUV photons and the wind from the host star are injected from the left side of the figure. The stellar mass loss rates are $\dot{M}_* = 1\dot{M}_\odot\,(left)$, $10\dot{M}_\odot\,(middle)$, $100\dot{M}_\odot\,(right)$. In each figure, the upper panel shows the density and lower panel shows the temperature.

In our simulations, Ly-α cooling is a major radiative cooling process, and the adiabatic cooling dominates the overall cooling processes.

Table 1 shows the stellar and planetary parameters of our fiducial simulation.

We also run simulations with various stellar wind strength by varying the stellar wind density.

3. Results and Implications

Figure 1 shows the atmospheric structure of our simulations. The strong wind confines the outflow. The wind-outflow structure is shaped by the balance between the ram pressure of the wind and the thermal pressure of the outflow. The balanced point can be estimated as

$$k_\mathrm{B}\rho_\mathrm{p}(r)\,T_\mathrm{p}(r)/\mu m_H = \rho_*(r)\,v_*^2(r) \tag{3.1}$$

where $\rho_\mathrm{p}(r), T_\mathrm{p}(r)$ are the density and temperature of the planetary atmosphere at the contact point, μ is the mean molecular weight, m_H is the hydrogen atomic mass, and $\rho_*(r), v_*(r)$ are the density and velocity of the wind. The balanced point depends on both the velocity and density of the wind because the ram pressure determines the structure, implying that the mass-loss rate of the star does not uniquely determine the structure of the atmosphere and the observational signatures.

Table 2 shows that the mass-loss rates of the planet are almost independent of the wind strength and that the value is approximately $10^{10}\,\mathrm{g/s}$ because the wind can suppress the outflow only when it strongly affects the atmosphere around the launching point. This is achieved for $\dot{M}_* > 1000\,M_\odot$.

Table 2. Planetary Mass-loss rates with different stellar winds.

Stellar Wind strength	Mass-loss rate (g/s)
$1\,\dot{M}_\odot$	2.9×10^{10}
$10\,\dot{M}_\odot$	2.2×10^{10}
$100\,\dot{M}_\odot$	2.3×10^{10}

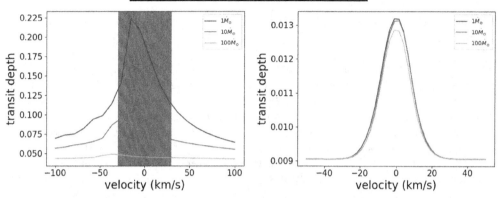

Figure 2. *left panel*: Ly-α transit depth at mid-transit. The shaded region is the line center region where the local interstellar medium can absorb. *right panel*: Hα transit depth.

Figure 2 shows the Ly-α and Hα transit of our outputs. The peak of the Ly-α transit is blue-shifted due to the wind. This is consistent with the previous studies (McCann et al. 2019). To calculate Hα absorption, we assume the $n = 2$ level population using the $2p, 2s$ population of Christie et al. (2013):

$$\frac{n_2}{n_1} = \frac{n_{2p} + n_{2s}}{n_1 s} \simeq 10^{-9} \left(\frac{5R_*}{a}\right)^2 e^{16.9 - (10.2\,\mathrm{eV}/k_B T_{\mathrm{Ly\alpha},*})}$$
$$+ 1.627 \times 10^{-8} \left(\frac{T}{10^4\,\mathrm{K}}\right)^{0.045} e^{11.84 - 118400\,\mathrm{K}/T}$$
$$\times \frac{8.633}{\log(T/T_0) - \gamma} \tag{3.2}$$

where $T_{Ly\alpha,*} \sim 7000\,\mathrm{K}$ is the excitation temperature for the solar Lyman-α, $T_0 = 1.02\,\mathrm{K}$ and $\gamma = 0.57721\ldots$ is the Euler-Mascheroni constant. In our simulations, the strong wind can reduce the Ly-α transit depth while the Hα signature is almost independent of the strength. The Hα absorption by the atmosphere is significant in lower hot region because the $n = 2$ level population is larger there ($n_2/n_1 \sim 10^{-9}$).

The transit signatures due to the extended atmosphere are also dependent on the EUV flux from the host star. The Ly-α transit depth depends on the EUV flux and the wind strength, whereas the Hα transit signature does not sensitively depend on the wind strength unless the wind is extremely strong with $\dot{M}_* > 1000\dot{M}_\odot$. Our simulations are 2D axisymmetric and neglect the tail contribution to the transit signatures. The tail contribution should be significant in Ly-α blue-wing but is likely unimportant in Hα absorption because the lower hot atmospheric layer matters. The difference between the wind effects on Ly-α and Hα would be essentially the same even if we consider the tail contribution to the signature. We argue that observations of both signatures can solve the degeneracy of the stellar EUV luminosity and the wind properties in close-in gas giants.

The stellar activities have a significant impact on the planetary atmosphere and the observational signatures (Zhilkin et al. 2020). The rate of the flare activities is well

162 H. Mitani, R. Nakatani & N. Yoshida

known for the Sun (Maehara et al. 2017). Strong flares are accompanied by coronal mass ejections (CMEs) in many cases, and thus we can estimate the rate of the strong mass-loss $\dot{M}_* > 10\,M_\odot$ due to the activities.

$$\int_{10^{32}\,\mathrm{erg}}^{\infty} f(E_{\mathrm{flare}})\,\mathrm{d}E_{\mathrm{flare}} \sim 1 - 100\,\mathrm{year}^{-1} \tag{3.3}$$

For solar type stars, the possibility that the strong CME changes the observational Ly-α is expected to be small.

For young stars, stellar activities should be stronger than that of the sun. Also the rate becomes higher. In the case of a very young host star ($< 50\,\mathrm{Myr}$), the probability becomes an order of magnitude larger (Feinstein et al. 2020). We note that, in many cases, the age of the host star in the observed exoplanets is older than 50 Myr and the effect can be small. Interestingly, the activities are also stronger (Maehara et al. 2014) for cooler stars. The spectral type dependence of the stellar activity is also investigated. The flare frequency in M dwarfs is a few orders of magnitude larger than in G-type stars. The strong CME frequency becomes larger in late-type stars, and the CME happens almost always in every transit around M dwarfs.

References

Anninos, P., Zhang, Y., Abel, T., & Norman, M. L. 1997, Cosmological hydrodynamics with multi-species chemistry and nonequilibrium ionization and cooling. *New Astron.*, 2(3), 209–224.

Bisikalo, D., Kaygorodov, P., Ionov, D., Shematovich, V., Lammer, H., & Fossati, L. 2013, Three-dimensional Gas Dynamic Simulation of the Interaction between the Exoplanet WASP-12b and its Host Star. *ApJ*, 764(1), 19.

Bisikalo, D. V., Shematovich, V. I., Cherenkov, A. A., Fossati, L., & Möstl, C. 2018, Atmospheric Mass Loss from Hot Jupiters Irradiated by Stellar Superflares. *ApJ*, 869(2), 108.

Carolan, S., Vidotto, A. A., Villarreal D'Angelo, C., & Hazra, G. 2021, Effects of the stellar wind on the Ly α transit of close-in planets. *MNRAS*, 500(3), 3382–3393.

Cherenkov, A. A., Bisikalo, D. V., & Kosovichev, A. G. 2018, Influence of stellar radiation pressure on flow structure in the envelope of hot-Jupiter HD 209458b. *MNRAS*, 475(1), 605–613.

Christie, D., Arras, P., & Li, Z.-Y. 2013, Hα Absorption in Transiting Exoplanet Atmospheres. *ApJ*, 772(2), 144.

Ehrenreich, D., Bourrier, V., Wheatley, P. J., Lecavelier des Etangs, A., Hébrard, G., Udry, S., Bonfils, X., Delfosse, X., Désert, J.-M., Sing, D. K., & Vidal-Madjar, A. 2015, A giant comet-like cloud of hydrogen escaping the warm Neptune-mass exoplanet GJ 436b. *Nature*, 522(7557), 459–461.

Feinstein, A. D., Montet, B. T., Ansdell, M., Nord, B., Bean, J. L., Günther, M. N., Gully-Santiago, M. A., & Schlieder, J. E. 2020, Flare Statistics for Young Stars from a Convolutional Neural Network Analysis of TESS Data. *AJ*, 160(5), 219.

Fulton, B. J., Petigura, E. A., Howard, A. W., Isaacson, H., Marcy, G. W., Cargile, P. A., Hebb, L., Weiss, L. M., Johnson, J. A., Morton, T. D., Sinukoff, E., Crossfield, I. J. M., & Hirsch, L. A. 2017, The California-Kepler Survey. III. A Gap in the Radius Distribution of Small Planets. *AJ*, 154(3), 109.

Maehara, H., Notsu, Y., Notsu, S., Namekata, K., Honda, S., Ishii, T. T., Nogami, D., & Shibata, K. 2017, Starspot activity and superflares on solar-type stars. *PASJ*, 69(3), 41.

Maehara, H., Shibayama, T., Notsu, Y., Notsu, S., Nagao, T., Honda, S., Nogami, D., & Shibata, K. Superflares on Late-Type Stars. In Haghighipour, N., editor, *Formation, Detection, and Characterization of Extrasolar Habitable Planets* 2014, volume 293, pp. 393–395.

McCann, J., Murray-Clay, R. A., Kratter, K., & Krumholz, M. R. 2019, Morphology of Hydrodynamic Winds: A Study of Planetary Winds in Stellar Environments. *ApJ*, 873(1), 89.

Mignone, A., Bodo, G., Massaglia, S., Matsakos, T., Tesileanu, O., Zanni, C., & Ferrari, A. 2007, PLUTO: A Numerical Code for Computational Astrophysics. *ApJS*, 170(1), 228–242.

Mitani, H., Nakatani, R., & Yoshida, N. 2021, Stellar Wind Effect on the Atmospheric Escape of Hot Jupiters. *arXiv e-prints*, arXiv:2111.00471.

Murray-Clay, R. A., Chiang, E. I., & Murray, N. 2009, Atmospheric Escape From Hot Jupiters. *ApJ*, 693(1), 23–42.

Nakatani, R., Hosokawa, T., Yoshida, N., Nomura, H., & Kuiper, R. 2018, Radiation Hydrodynamics Simulations of Photoevaporation of Protoplanetary Disks by Ultraviolet Radiation: Metallicity Dependence. *ApJ*, 857(1), 57.

Osterbrock, D. E. & Ferland, G. J. 2006,. *Astrophysics of gaseous nebulae and active galactic nuclei*.

Spitzer, L. 1978,. *Physical processes in the interstellar medium*.

Szabó, G. M. & Kiss, L. L. 2011, A Short-period Censor of Sub-Jupiter Mass Exoplanets with Low Density. *ApJ*, 727(2), L44.

Vidal-Madjar, A., Lecavelier des Etangs, A., Désert, J. M., Ballester, G. E., Ferlet, R., Hébrard, G., & Mayor, M. 2003, An extended upper atmosphere around the extrasolar planet HD209458b. *Nature*, 422(6928), 143–146.

Vidotto, A. A. & Cleary, A. 2020, Stellar wind effects on the atmospheres of close-in giants: a possible reduction in escape instead of increased erosion. *MNRAS*, 494(2), 2417–2428.

Zhilkin, A. G., Bisikalo, D. V., & Kaygorodov, P. V. 2020, Coronal Mass Ejection Effect on Envelopes of Hot Jupiters. *Astronomy Reports*, 64(2), 159–167.

Discussion

HANAWA: I am wondering whether your simulation resolved the bow shock and contact discontinuity.

HIROTO: We talked about the force balance which describes the shock front. Our simulations resolve the shock front and contact discontinuity.

BISIKALO: Do you take into account orbital motion of planets? Because I cannot see the influence of the Coriolis force.

HIROTO: Our simulations are 2D axisymmetric and we consider the centrifugal force but neglect the Coriolis force. The Ly-α transit signature can be affected by the 3D effect because of the tail contribution. That can be important. Hα transit absorption can be independent of the existence of the tail because the absorption is significant around a relatively lower region in which the gas temperature is high. The difference of the stellar wind strength dependence between Ly-α and Hα may be qualitatively similar to the 3D.

RONY: In the case of Jupiter, the strong magnetic field plays a role of having magnetic (ram) pressure, setting the size of the planetary interaction with the solar wind, will you eventually do MHD simulations as well? And what is your boundary condition at the planet: you just give a radial subsonic (or supersonic) outflow?

HIROTO: Magnetic pressure can indeed be important. I would run MHD simulations eventually but do not have an immediate plan currently. In our simulations, the EUV-driven outflow is excited from the hydrostatic, stratified gas, and the outflow base is well above the planet boundary. We use the boundary condition where the gas is hydrostatic across the boundary.

The Predictive Power of Computational Astrophysics as a Discovery Tool
Proceedings IAU Symposium No. 362, 2023
D. Bisikalo, D. Wiebe & C. Boily, eds.
doi:10.1017/S1743921322001855

Study of the non-thermal atmospheric loss for exoplanet π Men c

Anastasia A. Avtaeva[1,2] and Valery I. Shematovich[2]

[1]Sternberg Astronomical Institute, Moscow State University
Universitetsky pr., 13, Moscow 119234, Russia
email: `nastyaavt@inasan.ru`

[2]Institute of Astronomy, Russian Academy of Sciences
Pyatnitskaya str., 48, Moscow 119017,
email: `shematov@inasan.ru`

Abstract. We have studied the input of the exothermic photochemistry into the formation of the non-thermal escape flux in the transition $H_2 - H$ region of the extended upper atmosphere of the hot exoplanet - the sub-neptune π Men c. The formation rate and the energy spectrum of hydrogen atoms formed with an excess of kinetic energy due to the exothermic photochemistry forced by the stellar XUV radiation were calculated using a numerical kinetic Monte Carlo model of a hot planetary corona. The escape flux was estimated to be equal to 2.5×10^{12} $cm^{-2}s^{-1}$ for the mean level of stellar activity in the XUV radiation flux. This results in the mean estimate of the atmospheric loss rate due to the exothermic photochemistry equal to 6.7×10^8 $g\ s^{-1}$. The calculated estimate is close to the observational estimates of the possible atmospheric loss rate for the exoplanet π Men c in the range less than 1.0×10^9 gs^{-1}.

Keywords. exoplanet, numerical model, exoplanets atmosphere, non-thermal escape

---

We have studied the input of the exothermic photochemistry into the non-thermal atmospheric loss in the transition $H_2 - H$ region of the extended upper atmosphere of the hot exoplanet - the sub-neptune π Men c. This exoplanet is a typical example of the hot sub-Neptune (super-Earth), and is the first transit planet discovered by the TESS space telescope (Gandolfi *et al.* 2018). The atmosphere of this exoplanet was studied in HST observations - no absorption in $H\ Ly - \alpha$ line was detected (Garcia Munoz *et al.* 2020).

Stellar XUV flux was modeled basing on the solar XUV spectrum scaled to the orbit of π Men c with the semi-major axis ar=0.067 a.u. using the scaling coefficients from (Garcia Munoz *et al.* 2020) for stellar XUV flux. It currently corresponds to the mean level F10.7=100 of solar activity. Impact processes by the suprathermal photo-electrons were taken into account. Precipitation from stellar wind/magnetosphere was not considered. The formation rate and the energy spectrum of hydrogen atoms formed with an excess of kinetic energy due to the exothermic photochemistry forced by the stellar XUV radiation were calculated using a numerical kinetic Monte Carlo model (Avtaeva A. A. & Shematovich V. I. 2021) and are shown in Figures 1 and 2. It is seen, that there are slow and fast fractions of the fresh H atoms. As a result of model realization, we derive the kinetic-energy distribution functions for suprathermal hydrogen atoms in the H−2→H transition region. As an example, the energy spectrum of upward moving H atoms at height of 3.53 R₀ are shown in Figure 3 (R₀ is a photometric planet radius). It is seen that the suprathermal tails are formed in the energy distribution functions atomic

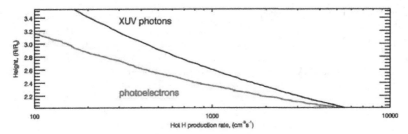

Figure 1. Height profiles of hot H production rates due to the photodissociation and dissociative ionization of molecular hydrogen by the stellar XUV radiation (black line) and by the impact of suprathermal photoelectrons (red line).

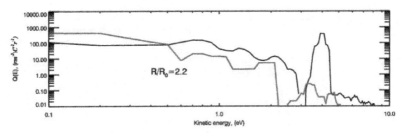

Figure 2. Energy spectra of the source functions for hot H atoms formed due to dissociation by stellar XUV flux (black line) and due to the impact by the suprathermal photoelectrons (red line).

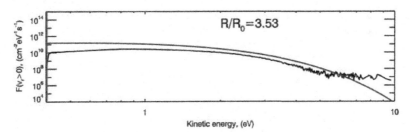

Figure 3. Energy specra of the upward moving H atoms at height of of 3.53 R$_0$. Blue line shows the energy spectrum of the upward moving thermal flux of H atoms, calculated with Maxwellian distribution for gas temperature taken from aeronomic model Shaikhislamov *et al.* (2020).

hydrogen in the transition region populating the hot fraction of the extended hydrogen corona of the sub-neptune π Men c.

Non-thermal escape flux in the planet-star direction is equal to $2.5 \times 10^{12}\ cm^{-2}s^{-1}$ for mean level of stellar activity. It is of the same value as the Jeans escape rate of $1.0 \times 10^{12}\ cm^{-2}s^{-1}$ for thermal H atoms. Non-thermal mass loss rate is about $6.7 \times 10^8\ gs^{-1}$. The calculations have shown that the non-thermal flux of hydrogen escape from the atmosphere is comparable to the thermal flux and should be taken into account in the current aeronomic models of hot exoplanets.

We acknowledge the support by Russian Science Foundation (Project 22-22-00909).

References

Shaikhislamov I. F., Fossati L., Khodachenko M. L., Lammer H., García Muñoz A., Youngblood A., Dwivedi N. K., & Rumenskikh M. S. 2020, *Astronomy & Astrophysics*, 639, id.A109(7 pp.)

Avtaeva, A.A. & Shematovich V. I. 2021, *Solar System Research*, 55, 150.

Gandolfi D., Barragán O., Livingston J. H., & 29 more 2018, *Astronomy & Astrophysics*, 619, id.L10(10 pp.).

Garcia Munoz A., Youngblood A., Fossati, L., Gandolfi D., Cabrera J., & Rauer H. 2020, *The Astrophysical Journal Letters*, 888, id. L21(12 pp.).

Bisikalo D. V., Shematovich V. I., Kaygorodov P. V., & Zhilkin A. G. 2021, *Physics - Uspekhy*, 64, 747.

The Predictive Power of Computational Astrophysics as a Discovery Tool
Proceedings IAU Symposium No. 362, 2023
D. Bisikalo, D. Wiebe & C. Boily, eds.
doi:10.1017/S1743921322001752

Multi-component MHD model for hydrogen-helium extended envelope of hot Jupiter

Y.G. Gladysheva[ID], A.G. Zhilkin[ID] and D.V. Bisikalo[ID]

Institute of Astronomy of the Russian Academy
of Sciences, 48 Pyatnitskaya st. 119017, Moscow, Russia
email: ygladysheva@inasan.ru

Abstract. We describe a numerical model of hot Jupiter extended envelope that interacts with stellar wind. Our model is based on approximation of multi-component magnetic hydrodynamic. The processes of ionization, recombination, dissosiation and chemical reactions in hydrogen-helium envelope are taken into account. In particular, the ionization of neutral hydrogen atoms takes place due to processes of photo-ionization, charge-exchange and thermal collisions. Further, this model is supposed to be used for research on biomarkers' dynamics in extended envelopes of hot Jupiters.

Keywords. magnetic hydrodynamics (MHD), hot Jupiters, chemical reactions

---

1. Introduction

Hot Jupiters (HJ) are giant exoplanets with mass of the order of Jupiter mass, located in the immediate vicinity of the host star. Due to the close location to the host and relatively large size, gas envelopes can overfill their Roche lobes. This expanding upper atmosphere is called an extended envelope of HJs. The structure of the extended envelope and its physical properties are determined by the influence of several forces: the planet gravity, gravitational force of the star, the orbital centrifugal force, the orbital Coriolis force and the forces determined by the interaction with stellar wind, the radiation of the star and the magnetic field.

The magnetic field in the vicinity of HJs can be determined by different sources: generated in the interior of the planet, induced by electrical currents in the upper layers of its atmosphere, by the stellar wind magnetic field and host star magnetic field. Magnetic field can play an important role in the process of the stellar wind flow around the atmosphere of the HJ. Therefore, we use MHD solution instead of hydrodynamic one for simulation (Zhilkin & Bisikalo 2021). In order to describe specific processes that lead to local changes in the concentration of components we take into account the chemical composition. The model that includes chemical components should be multi-fluid and based on an MHD model.

2. Model

We consider an approximation of 3D multi-component (multi-fluid) magnetic hydrodynamics, which uses equations for mass quantities (density ρ, average mass velocity v, average mass internal energy ε), the induction equation for magnetic field B, as well as the continuity equations for the components. The individual components (electrons, ions and neutrals of various kinds) of plasma are marked with the index s (Zhilkin & Bisikalo

2021). The continuity equations for each component of the kind s can be written as:

$$\frac{\partial}{\partial t}(\rho\xi_s) + \nabla \cdot (\rho\xi_s \boldsymbol{v}) = S_s, \quad s = 1, \dots, N, \tag{2.1}$$

where ξ_s is mass fraction of component s, S_s are source functions, which describe changes in the number of particles of the kind s due to chemical reactions, N is the number of components.

The 3D numerical simulation of the model takes into account the continuity equations with the help of a chemical module. Chemical constants from the UMIST database (McElroy et al. 2005) are used to calculate chemical reactions and processes of ionization, recombination and dissociation. We apply these reactions selectively: only the reactions concerned hydrogen-helium envelope of HJ (Shaikhislamov et al. 2020). We use 38 reactions for 9 components in this envelope. We add the module to the existing code of MHD model and solve the equations of chemical kinetics for the selected network of chemical reactions. As a result we have a number density of each component in the range of a quarter of period for HJ orbital motion. These calculations are made in every cell and at every integration step Δt.

3. Conclusion

We considered the multi-fluid MHD model for hydrogen-helium extended envelope of HJ. In this model plasma is described as a combination of components (electrons, ions, and neutrals of various kinds). The model takes into account chemical reactions, processes of ionization, recombination and dissociation of hydrogen molecules. We assume that the hydrogen-helium atmosphere has a homogeneous chemical composition. For numerical simulation we developed and added a chemical module into the MHD code with a specific selection of chemical reactions. Further we plan to extend the chemical model to include additional components and corresponding reactions. In particular, this will allow us to follow in more detail the dynamics of specific components (for example, biomarkers) in the envelopes of hot exoplanets.

The authors are grateful to the Government of the Russian Federation and the Ministry of Higher Education and Science of the Russian Federation for the support (grant 075-15-2020-780 (no. 13.1902.21.0039)).

References
Zhilkin, A.G., & Bisikalo, D.V. 2021, *Universe*, 7, 422
Shaikhislamov, I.F., Khodachenko, M.L., Lammer, H., Kislyakova, K.G., Fossati, L., Johnstone, C.P., Prokopov, P.A., Berezutsky, A.G., Zakharov, Yu.P., & Posukh, V. G. 2016, *Nature*, 173, 20
McElroy D., Walsh C., Markwick, A.J., Cordiner, M.A., Smith, K., & Millar, T.J. 2019, *A&A*, 550, 13

Discussion

PAUL ARGHYADEEP: I was wondering what is the velocity and the magnetic profile of the stellar wind? Also, how is the magnetic field of HJ modelled? Is it a dipole?

YULIYA GLADYSHEVA: The magnetic field is calculated from the wind model. The model corresponds to the paper (Weber & Davis, 1967). The magnetic field of HJ consists of two components: the intrinsic magnetic field of the planet (a dipole one) and the magnetic field induced by electrical currents in the upper layers of atmosphere (not dipole).

The Predictive Power of Computational Astrophysics as a Discovery Tool
Proceedings IAU Symposium No. 362, 2023
D. Bisikalo, D. Wiebe & C. Boily, eds.
doi:10.1017/S1743921322002861

Modeling Stellar Jitter for the Detection of Earth-Mass Exoplanets via Precision Radial Velocity Measurements

Samuel Granovsky[1,2,3] (ORCID), **Irina N. Kitiashvili**[1] (ORCID) and **Alan A. Wray**[1]

[1]NASA Ames Research Center, Moffett Field, MS 258-6, Mountain View, CA, USA

[2]New Jersey Institute of Technology, 323 Dr Martin Luther King Jr Blvd, Newark, NJ, USA

[3]Universities Space Research Association, 7178 Columbia Gateway Drive, Columbia, MD, USA

Abstract. The detection of Earth-size exoplanets is a technological and data analysis challenge. Future progress in Earth-mass exoplanet detection is expected from the development of extreme precision radial velocity measurements. Increasing radial velocity precision requires developing a new physics-based data analysis methodology to discriminate planetary signals from host-star-related effects, taking stellar variability and instrumental uncertainties into account. In this work, we investigate and quantify stellar disturbances of the planet-hosting solar-type star HD121504 (G2V spectral type) from 3D radiative modeling obtained with the StellarBox code. The model has been used for determining statistical properties of the turbulent plasma and obtaining synthetic spectroscopic observations for several Fe I lines at different locations on the stellar disk to mimic high-resolution spectroscopic observations.

Keywords. stars: individual (HD121504); line: profiles; techniques: radial velocities, spectroscopic; methods: numerical

---

1. Introduction

Due to the relatively large mass of Jupiter-size planets, the radial velocity (RV) method for detecting exoplanets is viable since such planets are able to apply a sufficient acceleration to their host star. This is especially true for Jupiter-mass planets which orbit close to their host stars. However, this is typically not the case for smaller Earth-mass exoplanets, especially those which orbit farther from their host stars. Variations in RV of the host star as it orbits the barycenter is comparable to the fluctuations in RV caused by stellar jitter, the noise caused by the movement of material on the star's surface. While noise caused by stellar jitter may be on the order of 100 m/s (Saar & Donahue 1997) in relation to the star's center of mass, the RV of the star due to the effects of an Earth-mass planet may be only 0.1 to 1 m/s (Plavchan et al. 2015). In particular, it was demonstrated that local variations of surface gravity due to granular motions could significantly contribute to RV signal (Bastien et al. 2016). Therefore, development of an accurate model of stellar jitter is essential for making RV measurements precise enough to detect Earth-mass exoplanets. Available capability to model stellar surface dynamics from first physical principles enables the synthesis of stellar spectra with high-degree realism. This approach has been used by (e.g., Cegla et al. 2013; Cegla et al. 2018), where

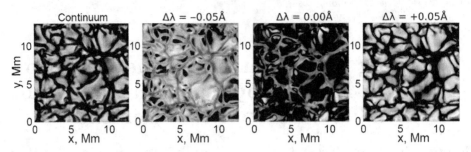

Figure 1. Intensity variations at different locations from the Fe I line core ($\lambda_{ref} = 6173.7$Å) show how the structure of the atmosphere changes with height. The presented synthetic images correspond to an area at the disk center.

modeling of the iron line (6302Å) has been performed at different locations on the stellar disk to estimate the contribution of different 'elements' of a magnetized stellar surface such as granulation, magnetic and non-magnetic intergranular lanes, and magnetic bright points, as well as center-to-limb effect. Due to the limited signal-to-noise ratio it is also important to consider the combination of many spectral lines. For instance, Dravins et al. (2017); Dravins et al. (2021a,b) analyzed a series of synthetic FeI and FeII disk-integrated spectral lines to characterize its properties for different types of stars. This work combines both approaches, including multi-wavelength analysis of the disk-integrated observables and analysis of individual lines to characterize stellar surface disturbances.

To investigate stellar jitter, we selected the G2V spectral type planet-host star HD121504 with the following parameters: 1.18 $M_\odot$, 1.7 $L_\odot$, $T_{eff} = 6067$K, $\log(g) = 4.36$. The Jupiter-mass planet orbiting around has been detected with the radial velocity method. We will use HD121504 as a testbed to study stellar jitter using a realistic-type modeling approach combined with the spectral line synthesis. This paper presents initial modeling results of the HD121504 star using 3D radiative simulations and a radiative transfer code to generate time series for synthetic observables.

2. Modeling of Synthetic Observables

To investigate convection-driven disturbances, we obtained a 3D radiative model of star HD121504 via the StellarBox code (Wray et al. 2015, 2018). The stellar surface area 12.8 Mm × 12.8 Mm wide with a spatial resolution of 50 km was computed centered on the following Fe I lines: 5247Å, 5250.6Å, 5251Å, 6173.7Å, 6301.9Å, and 6302.9Å. The spectral line synthesis has been performed using the SPINOR code (Frutiger et al. 2000) for every 10° in angular distance between −80° and +80° from the disk center and every 30° in polar angle between −60° and +90°. An example of a snapshot at disk center is shown in Figure 1 at various wavelengths relative to the reference line at 6173.7Å. Figure 2 illustrates how properties of a spectral line change between disk center and limb. In particular, there is a decrease of the continuum intensity and an increase of the full width at half maximum (FWHM) as flows tangent to the stellar surface gain a radial component relative to the observer.

Using the described process, a time-series comprising 492 minutes of stellar dynamics was computed for the six Fe I lines for each of the previously mentioned locations. For each line, a weighted disk-averaged profile was computed. From the center of mass of each profile, Doppler-shift time-series were obtained. Figure 3 shows an example of the disk-averaged Doppler shift derived from synthetic data of 6173.7Å. Preliminary results show Doppler shift fluctuations for HD121504 mostly in the range of ±500 m/s. The current computations of the disk-integrated Doppler shift fluctuations are computed from the time-series of tiles distributed across the stellar disk. Because for each tile, the

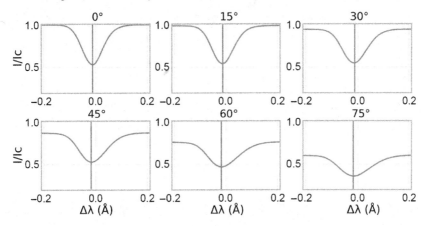

Figure 2. Spectral line profiles for the Fe I line ($\lambda_{\mathrm{ref}} = 6173.7$Å, red) at several angular distances from disk center. Vertical lines indicate the location of the line core.

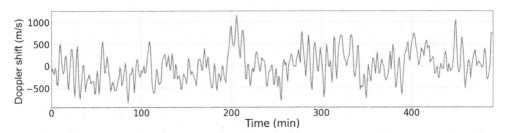

Figure 3. Disk-averaged Doppler-shift time variations obtained for star HD12504 from the synthesized time-series at various stellar disk locations. The presented disk-integrated fluctuations are amplified primarily by synchronous dynamics of simulated patches near the disk center.

photospheric variations are the same everywhere on the stellar disk and differ only in deviations related to the disk location, the amplitude of the disk-integrated Doppler shift fluctuations is larger than the observed amplitude. This effect is strongest for the tiles located near the disk center. In future work, we plan to apply a random time offset to the time series to resolve this issue.

The disk-averaged profiles were compared to observational data of HD121504 taken by the ESO HARPS spectrograph for each of the six lines, as shown in Figure 4. The observed profiles (blue curves) differ somewhat from the synthetic profiles (red), which is potentially due to low spacial resolution of the numerical model, as well as magnetic effects not yet being accounted for. There is also some uncertainty in the absolute iron abundance that should be used to perform line synthesis.

3. Discussion and Conclusions

Detection of Earth-mass exoplanets orbiting solar-type stars is challenging due to significant contamination of the RV signal with disturbances originating from turbulent photospheric dynamics. To model these disturbances, we use a 3D radiative model of the planet-hosting star HD121504 to generate synthetic observables and to characterize stellar convective motions. To characterize the stellar convective motions near the photosphere, we computed synthetic high-resolution spectra of six Fe I lines for different locations on the stellar disk to obtain disk-integrated observables. Further analysis using hydrodynamic and MHD simulations and synthetic data sets with higher spatial

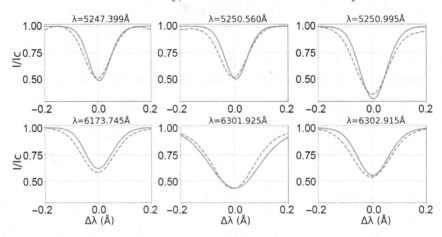

Figure 4. Comparison of observed (blue, dotted) and synthetic (red, solid) spectral lines of Fe I.

and temporal resolutions is required for a more realistic representation of observational data. For future disk-averaged profiles, we plan to apply a randomized temporal offset to the time-series for each location on the disk to prevent resonance affecting the Doppler data, as the current model uses a disk average which assumes every location on the disk behaves identically at any given time.

Acknowledgments

Observations used in the paper were made with HARPS spectrograph on the 3.6m ESO telescope at La Silla Observatory, Chile. This work is supported by the NASA Extreme Precision Radial Velocity Foundation Science Program.

References

Bastien, F. A., Stassun, K. G., Basri, G., & Pepper, J. 2016, ApJ, 818, 43
Cegla, H. M., Shelyag, S., Watson, C. A., & Mathioudakis, M. 2013, ApJ, 763, 95
Cegla, H. M., Watson, C. A., Shelyag, S., et al. 2018, ApJ, 866, 55
Dravins, D., Ludwig, H.-G., Dahlén, E., & Pazira, H. 2017, A&A, 605, A90
Dravins, D., Ludwig, H.-G., & Freytag, B. 2021a, A&A, 649, A16
Dravins, D., Ludwig, H.-G., & Freytag, B. 2021b, A&A, 649, A17
Frutiger, C., Solanki, S. K., Fligge, M., & Bruls, J. H. M. J. 2000, A&A, 358, 1109
Plavchan, P., Latham, D., Gaudi, S., et al. 2015, arXiv e-prints, arXiv:1503.01770
Saar, S. H. & Donahue, R. A. 1997, ApJ, 485, 319
Wray, A. A., Bensassi, K., Kitiashvili, I. N., Mansour, N. N., & Kosovichev, A. G. 2015, arXiv e-prints, arXiv:1507.07999
Wray, A. A., Bensassiy, K., Kitiashvili, I. N., Mansour, N. N., & Kosovichev, A. G. 2018, Realistic Simulations of Stellar Radiative MHD, ed. J. P. Rozelot & E. S. Babayev, 39

The Predictive Power of Computational Astrophysics as a Discovery Tool
Proceedings IAU Symposium No. 362, 2023
D. Bisikalo, D. Wiebe & C. Boily, eds.
doi:10.1017/S1743921322001533

Thermal atmospheric escape of close-in exoplanets

Eugenia S. Kalinicheva[iD] and Valery I. Shematovich[iD]

Institute of Astronomy, Russian Academy of Sciences
Pyatnitskaya str., 48, Moscow 119017, Russia
email: `kalinicheva@inasan.ru`

Abstract. In this work we applied the previously developed self-consistent 1D model of hydrogen-helium atmosphere with suprathermal electrons to close-in hot neptune GJ 436 b. The obtained height profile of density shows the two-scale structure of the planetary atmosphere. The mass-loss rate is found to be about $1.6 \times 10^9 g s^{-1}$.

Keywords. exoplanetary atmosphere, thermal atmospheric loss, numerical model

Many of recently discovered exoplanets with extended hydrogen-helium atmospheres are orbiting their host stars on very close orbits. Extremely high soft X-ray and ultraviolet (XUV) radiation causes the hydrodynamic escape of their gaseous envelopes. Such effects have been already observed for some planetary systems with either giant planets and smaller ones like super-Earths. Hot neptunes are extrasolar planets with masses and radii about the mass and radius of our Neptune in the Solar system, but are orbiting closer to their host stars (closer than 0.1 AU). The thermospheric temperatures of such planets are extremely high, up to thousands Kelvin. Such high temperature is caused by absorption of host-star XUV radiation.

In this work we present the atmosphere model of the well-known hot neptune GJ 436 b. The envelope was obtained using a previously developed self-consistent one-dimensional aeronomic model of the hydrogen-helium atmosphere, which includes the presence of suprathermal electrons (Ionov *et al.* 2017). This object was observed by the Hubble Space Telescope (HST) which showed the clear presence of extended gaseous envelope formation with the size comparable to the host star disk radius. The COS/HST independent transit observations have shown a 50% absorption in the Ly-α line in the Doppler velocity shift range (-120, -40) km/s (Ehrenreich *et al.* 2015, Lavie *et al.* 2017). The data also showed the presence of a dense cloud in front of the planet and its long gaseous tail.

The main advantage of the model used is the including of suprathermal particles contribution which leads to the more accurate calculation of atmospheric heating and, accordingly, clarifying the rate of its outflow Ionov *et al.* (2017), Ionov *et al.* (2018). Close-in exoplanets are exposed to the very high fluxes of XUV star radiation which makes the accurate heating calculations very important. High-energy radiation heats the upper atmosphere, ionizing atomic hydrogen and helium. Part of this radiation energy passes into the kinetic energy of reaction products. Usually, if the fresh photoelectron energy exeeds the thermal energy by several orders of magnitude (suprathermal particle), it can enter into a secondary reaction of ionization or excitation of other atmospheric particles. At the same time, the kinetic energy that the suprathermal electron had initially is expended. Taking into account these processes makes a significant contribution to the

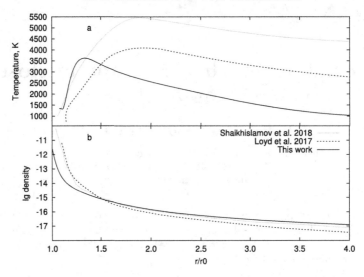

Figure 1. Outflow temperature and density of exoplanet GJ 436b as a function of height. Solid grey line represents Shaikhislamov *et al.* (2018), dotted black line represents Loyd *et al.* (2017), solid black line represents this work.

dynamics and energy of the exoplanetary atmosphere Ionov *et al.* (2017), Ionov *et al.* (2018).

The obtained height profiles of the atmospheric temperature (Figure 1a) and density (Figure 1b) of the simulated exoplanet were calculated, they differ significantly from the results of other authors, as the accounting of photoelectrons leads to a decrease in the rate of atmospheric heating and, accordingly, the rate of mass loss, which affects the evolution of the gaseous envelope of a hot exoplanet at astronomical times. The calculations revealed a two-level structure of the atmosphere under study. The lower part of the atmosphere is more massive and has an exponential density decrease (see distances less then 1.2 r/r_0). The density of the upper atmosphere, the corona, changes much slower (see, distances above 1.2 r/r_0) according to the height scale, which responds to a higher temperature at the peak of the gas heating. The preliminary calculated mass loss rate is found be $\dot{M} = 1.6 \times 10^9 gs^{-1}$. It is lower than Shaikhislamov *et al.* (2018) and Loyd *et al.* (2017) calculations ($\dot{M} = 3.1 \times 10^9 gs^{-1}$) and is close to the upper estimate in the Kulow *et al.* (2014) calculations ($\dot{M} = 3.7 \times 10^6 - 1.1 \times 10^9 gs^{-1}$).

The reported study was funded by RFBR according to the research project № 20-32-90149

References

I. F. Shaikhislamov, M. L. Khodachenko, H. Lammer, A. G. Berezutsky, I. B. Miroshnichenko, M. S. Rumenskikh 2018, *MNRAS* 481, 4, 5315

R. O. P. Loyd, T. T. Koskinen, K. France, C. Schneider, S. Redfield 2017, *ApJ* (Letters) 834, 2, L17

J. R. Kulow, K. France, J. Linsky, R. O. P. Loyd 2014, *ApJ* 786, 2, 132

D. E. Ionov, V. I. Shematovich, Ya. N. Pavlyuchenkov 2017, *Astron. Rep.* 61, 387

D. Ehrenreich, V. Bourrier, P. J. Wheatley, A. Lecavelier des Etangs, G. Hebrard, S. Udry, X. Bonfils, Xavier Delfosse, J.-M. Desert, D. K. Sing, A. Vidal-Madja 2015, *Nature* 522, 7557, 459

B. Lavie, D. Ehrenreich, V. Bourrier, A. Lecavelier des Etangs, A. Vidal-Madjar, X. Delfosse, A. Gracia Berna, K. Heng, N. Thomas, S. Udry, P. J. Wheatley 2017, *A&A* 605, L7

D. E. Ionov, Ya. N. Pavlyuchenkov, V. I. Shematovich 2018, *MNRAS* 476, 4, 5639

The Predictive Power of Computational Astrophysics as a Discovery Tool
Proceedings IAU Symposium No. 362, 2023
D. Bisikalo, D. Wiebe & C. Boily, eds.
doi:10.1017/S174392132200148X

Inference of Magnetic Fields and Space Weather Hazards of Rocky Extrasolar Planets From a Dynamical Geophysical Model

Varnana M. Kumar[1] ⓘ**, Thara N. Sathyan**[2]**, T.E. Girish**[3]**, P.E. Eapen**[4]**, Biju Longhinos**[5] **and J. Binoy**[2]

[1]Department of Physics Mar Ivanios College, Trivandrum 695015, India
email: mvarnana@gmail.com

[2]Department of Physics, Government College for Women, Trivandrum-695014, India
emails: tharasathyann@gmail.com; binoyjohndas@gmail.com

[3]Department of Physics, University College, Trivandrum-695034, India
email: tegirish5@yahoo.co.in

[4]Department of Physics, S.G College, Kottarakkara, Kerala-691531, India
email: peeapen@yahoo.co.in

[5]Department of Geology, University College, Trivandrum-695034, India
email: longhinos@universitycollege.ac.in

Abstract. In this paper we have inferred the magnetic shielding characteristics and space weather hazards of selected potentially habitable extrasolar planets using a dynamical geophysical model from calculations of internal heat, phases of volcanism and planetary magnetic moments. The space weather hazards on the extrasolar planet Kepler-452b orbiting around a Sun-like star are found to be a minimum which enhances the habitability probability of this planet.

Keywords. Extrasolar planets, Volcanism, Planetary magnetic fields, Space weather hazards

---

1. Introduction

The time variation of internal heat of planetary bodies causes changes in magnetic field of these planets. We suggest that the operations of the planetary dynamos are intimately connected to major geophysical transitions, mantle dynamics and volcanism. From our previous studies (Varnana *et al.* 2021a,b) we could develop a model in which we could find close associations between geological time evolution of internal heat, volcanism and magnetic fields in rocky planets in our solar system. These results are applied to infer magnetic shielding and space weather hazards of selected extrasolar planets with possible rocky composition.

2. Calculations and Results

Geological time evolution of internal heat flux S in a rocky planet is given by the relation

$$S(t) = So \exp(-\lambda t) \tag{2.1}$$

So is the internal heat flux during the planetary formation and $\lambda(1.5 \times 10^{-17} \text{ s}^{-1})$ is radioactive decay constant applied to chondrites (Turcotte and Schubert 1982). We have

Table 1. Inference of the phase of volcanism, magnetic shielding characteristics and space weather hazards in selected extrasolar planets.

Extrasolar planet	Spectral type and age of a host star	Phase of volcanism	Magnetic shielding possibility and moment $(A\text{-}m)^2$	Space weather hazard in the planet
Trappist 1-g	M, 7.6 Gyrs	Post-cessation	**Low**, 1.52×10^{20}	High
GJ 667C-c	M, 2 Gyrs	Early ascending	**Low**, 9.9×10^{20}	High
Trappist 1-e	M, 7.6 Gyrs	Post-cessation	**Low**, 8.2×10^{20}	High
Kepler 452-b	G, 6 Gyrs	Late ascending	**High**, 2.7×10^{23}	Low
GJ 667C-f	M, 2 Gyrs	Early ascending	**Low**, 4.77×10^{20}	High

inferred the time evolution of $S(t)$ in extrasolar planets using (2.1). Peak volcanism and cessation of major volcanism in solar system rocky planets happens when S reaches certain critical values in our model. Applying this to rocky extrasolar planets we can find different phases of volcanism in these planets. The magnitude of magnetic fields is significant only during active phases of volcanism in these planets. The average magnetic moment of an extrasolar planet depends on its angular momentum. From the best available data of mass and radii of selected potentially habitable extrasolar planets we have studied its thermal evolution, cessation and peak ages of volcanism and planetary magnetic moments (Durand-Manterola 2009). The space weather hazard on the extrasolar planet depends on both host star activity conditions (this is related to age for G type stars and rotation for M type stars) and planetary magnetic field shielding characteristics which is dynamical in nature. The results of our calculations are given in Table 1.

The main results of our study are:

(i) Using a dynamical geophysical model applicable for rocky planets we have inferred phases of volcanism and magnetic field characteristics of selected potentially habitable extrasolar planets.

(ii) The space weather hazards on these extrasolar planets are also inferred based on the knowledge of stellar activity conditions in their host stars and planetary magnetism.

(iii) The space weather hazard on the extrasolar planet Kepler-452 b orbiting around a Sun-like star is found to be the lowest which enhances the habitability probability of this planet.

References

Durand-Manterola, 2009, Planetary and Space Science, 57(12), 1405–1411.
Turcotte, D. L., and G. Schubert 1982, *Geodynamics*, Wiley, New York, pp 450.
Varnana, et al. 2021(a), arXiv preprint arXiv:2105.05676 (2021).
Varnana, et al. 2021(b), Bulletin of the American Astronomical Society 53(3), 1117.

The Predictive Power of Computational Astrophysics as a Discovery Tool
Proceedings IAU Symposium No. 362, 2023
D. Bisikalo, D. Wiebe & C. Boily, eds.
doi:10.1017/S1743921322001387

Hydrodynamical Simulations of Misaligned Accretion Discs in Binary Systems: Companions tear discs

S. Doğan[1], **C. J. Nixon[2]**, **A. R. King[2,3,4]**, **J. E. Pringle[2,5]** **and D. Price[6]**

[1]Department of Astronomy & Space Sciences, University of Ege, Bornova, İzmir, Turkey

[2]Department of Physics and Astronomy, University of Leicester, Leicester, LE1 7RH, UK

[3]Anton Pannekoek Institute, University of Amsterdam, Science Park 904, 1098 XH Amsterdam, Netherlands

[4]Leiden Observatory, Leiden University, Niels Bohrweg 2, NL-2333 CA Leiden, Netherlands

[5]Institute of Astronomy, Madingley Road, Cambridge, CB3 0HA, UK

[6]Monash Centre for Astrophysics (MoCA), School of Mathematical Sciences, Monash University, Vic. 3800, Australia
email: suzan.dogan@ege.edu.tr

Abstract. Accretion discs appear in many astrophysical systems. In most cases, these discs are probably not completely axisymmetric. Discs in binary systems are often found to be misaligned with respect to the binary orbit. In this case, the gravitational torque from a companion induces nodal precession in misaligned rings of gas. We first calculate whether this precession is strong enough to overcome the internal disc torques communicating angular momentum. For typical parameters, precession torque wins. To check this result, we perform numerical simulations using the Smoothed Particle Hydrodynamics code, PHANTOM, and confirm that sufficiently thin and sufficiently inclined discs can break into distinct planes that precess effectively independently. Disc tearing is widespread and severely changes the disc structure. It enhances dissipation and promotes stronger accretion onto the central object. We also perform a stability analysis on isolated warped discs to understand the physics of disc breaking and tearing observed in numerical simulations. The instability appears in the form of viscous anti-diffusion of the warp amplitude and the surface density. The discovery of disc breaking and tearing has revealed new physical processes that dramatically change the evolution of accretion discs, with obvious implications for observed systems.

Keywords. accretion, accretion discs — hydrodynamics — instabilities — black hole physics

---

1. Introduction

Accretion discs (e.g. Pringle 1981; Frank, King, & Raine 2002) are the essential ingredient for a vast range of astrophysical phenomena, including star and planet formation, X-ray binaries and active galactic nuclei (AGN). In most cases, discs are unlikely to be axisymmetric. The lack of symmetry produces a torque on misaligned rings of gas which makes their orbits precess differentially. Given a sufficiently strong viscosity communicating the precession between the rings, the disc warps. If the viscosity is too weak, or the external torque on the disc is sufficiently strong, the disc may instead break into distinct planes with only tenuous gas flows between them (Nixon & King 2012). If in addition these planes are sufficiently inclined to the axis of precession, they can precess until they are partially counterrotating, promoting angular momentum cancellation and rapid infall – disc tearing (Nixon et al. 2012; Nixon, King, & Price 2013).

The aim of this investigation is to find out if tearing can happen in circumprimary discs where the disc around one component is disrupted by the perturbation from a companion. This would have significant implications for all binary systems: e.g. fuelling SMBH during the SMBH binary phase and accretion outbursts in X-ray binaries. We first compare the disc precession torque with the disc viscous torque to determine whether the disc should warp or break. We check our findings by comparing our analytical reasoning with hydrodynamical simulations. Finally, we present an instability analysis which we perform to understand the underlying physics of 'disc breaking' (the case with no external torque) and 'disc tearing' (the case when the disc is subject to an external torque) observed in numerical simulations.

2. Disc Tearing

We consider binary systems with an initially planar disc around one component, misaligned with respect to the (circular) binary orbit. We expect the disc to break when the precession induced in the disc is stronger than any internal communication in the disc. The disc precession caused by the presence of a binary companion is retrograde, and has frequency (Bate et al. 2000)

$$\Omega_{\rm p} = \frac{3}{4} \frac{M_0}{M_1} \left(\frac{R}{a} \right)^3 \Omega \cos \theta. \qquad (2.1)$$

Here θ is the inclination angle between the disc plane and the binary orbital plane, M_0 & M_1 are the masses of each component of the binary with the disc around M_1, a is the binary separation, R (assumed $\ll a$) is the disc radius, and $\Omega = (GM_1/R^3)^{1/2}$ is the disc orbital frequency. Thus, the magnitude of the precession torque per unit area is

$$|\boldsymbol{G}_{\rm p}| = |\Omega_{\rm p} \times \boldsymbol{L}| = \frac{3}{4} \frac{M_0}{M_1} \left(\frac{R}{a} \right)^3 \Sigma R^2 \Omega^2 \cos \theta \sin \theta. \qquad (2.2)$$

On the other hand, by using the α-viscosity prescription by Shakura & Sunyaev (1973) $\nu_i = \alpha_i H^2 \Omega$, the total magnitude of the azimuthal and vertical viscous torques per unit area in a warped disc can be written as (Papaloizou & Pringle 1983)

$$|G_{\rm total}| = |G_{\nu_1}| + |G_{\nu_2}| = \frac{\Sigma R^2 \Omega^2}{2} \frac{H}{R} \left[3\alpha_1 + \alpha_2 |\psi| \right] \qquad (2.3)$$

where $|\psi|$ is the warp amplitude and defined as $|\psi| = R |\partial l / \partial R|$ (Ogilvie 1999). To break the disc the precession must be stronger than its viscous communication, i.e. $|G_{\rm p}| \gtrsim |G_{\rm total}|$. This comparison gives an idea of where in the disc we expect breaking to occur:

$$R_{\rm break} \gtrsim \left[\frac{4 \left(\alpha_1 + \frac{\alpha_2}{3} |\psi| \right)}{\sin 2\theta} \frac{H}{R} \frac{M_1}{M_0} \right]^{1/3} a. \qquad (2.4)$$

It is not straightforward to evaluate (2.4) as both α_1 and α_2 are strong functions of the warp amplitude $|\psi|$ (Ogilvie 1999, 2000) and the warp amplitude itself is unknown before performing a full 3D calculation of the disc evolution. For large $\alpha \gtrsim 0.1$ it is reasonable to exclude the α_2 term. Proceeding with the method of the earlier papers on disc tearing we get

$$R_{\text{break}} \gtrsim \left(\frac{4\alpha}{\sin 2\theta} \frac{H}{R} \frac{M_1}{M_0} \right)^{1/3} a. \tag{2.5}$$

We note that this equation is not relevant for $\alpha \ll 0.1$ as the vertical viscosity becomes important and small inclination angles where the strong vertical viscosity can result in rapid disc alignment. This tearing criterion is equivalent to requiring a minimum inclination of the disc to the binary orbit, θ_{min}, defined by

$$\sin 2\theta_{\text{min}} \gtrsim 4\alpha \frac{H}{R} \frac{M_1}{M_0} \left(\frac{a}{R_{\text{break}}} \right)^3. \tag{2.6}$$

We can simplify this formula in two limits. If the disc is around the less massive component we have $M_1 < M_0$ and the tidal limit on the disc size requires

$$\frac{a}{R_{\text{break}}} > 2.5 \left(\frac{M}{M_1} \right)^{1/3}, \tag{2.7}$$

where $M = M_1 + M_0$ is the total binary mass, so (2.6) becomes

$$\sin 2\theta \gtrsim 0.06 \left(\frac{\alpha}{0.1} \right) \left(\frac{H/R}{0.01} \right) \qquad (M_1 < M_0) \tag{2.8}$$

since $M \simeq M_0$ in this case. If instead the disc is around the more massive binary component we have $M_1 > M_0$ and the disc size is approximately $0.6a$ (Artymowicz & Lubow 1994). In this case, breaking occurs if

$$\sin 2\theta \gtrsim 0.18 \left(\frac{\alpha}{0.1} \right) \left(\frac{H/R}{0.01} \right) \left(\frac{M_1/M_0}{10} \right) \qquad (M_1 > M_0) \tag{2.9}$$

For typical black hole disc parameters $\alpha = 0.1$, $H/R \lesssim 10^{-2}$ almost all discs should break unless they are aligned to the binary plane within a few degrees. However, a very large mass ratio $M_1/M_0 \gg 1$ makes the perturbation by the smaller companion so weak that breaking would occur only after a very long interval.

To check this analytical reasoning, we perform 3D hydrodynamical numerical simulations using the PHANTOM smoothed particle hydrodynamics code (Price et al. 2018). The disc is initially planar and extends from $R_{\text{in}} = 0.1a$ to $R_{\text{out}} = 0.35a$ with a surface density profile $\Sigma = \Sigma_0(R/R_{\text{in}})^{-p}$ and locally isothermal sound speed profile $c_s = c_{s,0}(R/R_{\text{in}})^{-q}$, where we have chosen $p = 3/2$ and $q = 3/4$. This achieves a uniformly resolved disc with the shell-averaged smoothing length per disc scale-height $\langle h \rangle /H \approx$ constant. Σ_0 and $c_{s,0}$ are set by the disc mass, $M_d = 10^{-3}M$ and the disc angular semi-thickness, $H/R = 0.01$ (at $R = R_{\text{in}}$) respectively. Initially the disc is composed of 1 million particles, which for this setup gives $\langle h \rangle /H \approx 0.8$. The simulations use a disc viscosity with Shakura & Sunyaev $\alpha \simeq 0.1$ (which requires artificial viscosity $\alpha_{\text{AV}} = 1.2$; cf. Lodato & Price 2010) and $\beta_{\text{AV}} = 2$. We assume that the binary components, represented by two Newtonian point masses with $M_1 = M_2 = 0.5M$, accrete any gas coming within a distance $0.05a$ of them, and so remove this gas from the computation.

We perform our simulations for $\theta = 10°, 30°, 45°$ and $60°$. The complete results of our simulations are given in Doğan et al. (2015). When the initial inclination is small, the precession torque caused by the companion is weak. In this case, the disc evolves with a mild warp. The strength of the precession torque is higher when $\theta = 30°$, and it has its maximum value when $\theta = 45°$. As expected, we confirm disc breaking in our simulations

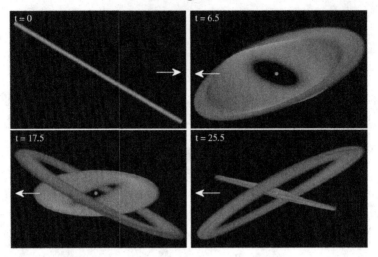

Figure 1. 3D surface rendering of the disc which was initially inclined at 30° to the binary plane with no warp. These snapshots are taken after 0, 6, 17.5 and 25.5 binary orbits. The disc is viewed along the binary orbital plane and the arrow points at the direction of the companion (see (Doğan et al. 2015) for details).

with initial inclinations of 30° and 45°. Fig. 1 shows a simulation with an initial inclination of 30°. Here the disc becomes significantly warped after a few orbits. Then the outer disc breaks off to form a distinct outer ring. The accretion rate through tearing discs is generally significantly enhanced. In the $\theta = 60°$ simulation, the disc quickly becomes a narrow and quite eccentric ring (through a strong interaction induced by tearing). The remaining eccentric disc then goes through strong Kozai-Lidov cycles producing a highly variable accretion (Doğan et al. 2015; Martin et al. 2014).

3. Instability Analysis of Warped Discs

The dynamical behavior of disk tearing has been explored by a series of numerical investigations in different contexts. Disc tearing has been shown to occur in (i) discs inclined to the spin of a central black hole (Nixon et al. 2012), (ii) circumbinary discs around misaligned central binary systems (Nixon, King, & Price 2013; Facchini, Lodato, & Price 2013; Aly et al. 2015), (iii) circumprimary discs misaligned with respect to the binary orbital plane (Doğan et al. 2015). In these studies, the criterion for disc tearing has been derived simplistically by comparing the viscous torque with the precession torque induced in the disc. Initially, the precession torque was compared to the torque arising from azimuthal shear (Nixon et al. 2012). This criterion is clearly insufficient as it does not account for the (more important) torque attempting to smooth the disc warp which arises from vertical shear. For small α and small disc inclination angles, the simulations of Doğan et al. 2015 showed that the inclusion of the vertical viscous torque, which depends on the disc structure through the warp amplitude, was required.

Further to these simulations of disc tearing, it has also been shown that warped discs may break without any external forcing (Lodato & Price 2010). When there is no external forcing, the disc evolution must be driven by the dependence of the effective viscosities on the disc structure. This instability occurs in a simpler environment than disc tearing, and likely underpins that process. Motivated by the breaking and tearing behaviour observed in numerical simulations of warped discs, we perform a stability analysis of the warped disc equations to connect these behaviours with disc instabilities and determine a general criterion for discs to break.

Here we provide a brief description of the relevant equations and refer the reader to Ogilvie (2000) and Doğan et al. (2018) for more details and Doğan & Nixon (2020) for the non-Keplerian case. The governing evolutionary equations are the conservation of mass equation,

$$\frac{\partial \Sigma}{\partial t} + \frac{1}{r}\frac{\partial}{\partial r}(r\bar{v}_r\Sigma) = 0, \tag{3.1}$$

and the conservation of angular momentum equation,

$$\frac{\partial}{\partial t}\left(\Sigma r^2\Omega l\right) = $$
$$\frac{1}{r}\frac{\partial}{\partial r}\left[Q_1\Sigma c_s^2 r^2 l + Q_2\Sigma c_s^2 r^3\frac{\partial l}{\partial r} + Q_3\Sigma c_s^2 r^3 l \times \frac{\partial l}{\partial r} - \left(\frac{\partial}{\partial r}\left[Q_1\Sigma c_s^2 r^2\right] - Q_2\Sigma c_s^2 r\left|\psi\right|^2\right)\frac{h}{h'}l\right]. \tag{3.2}$$

Here $\Sigma(R,t)$ is the disc surface density, $\bar{v}_r$ is the mean radial velocity, $\Omega(R)$ is the orbital angular velocity of each annulus of the disc, $l(R,t)$ is the unit angular momentum vector pointing perpendicular to the local orbital plane, c_s is the sound speed, $h = r^2\Omega$ is the specific angular momentum, $h' = dh/dr$ and Q_i are the dimensionless torque coefficients.

The stability is considered with respect to linear perturbations in $\delta\Sigma$ and δl. This yields a third-order dispersion relation:

$$s^3 - s^2\left[aQ_1 - 2Q_2 + \left|\psi\right|\left(aQ_1' - Q_2'\right)\right]$$
$$- s\left[2aQ_1Q_2 - Q_2^2 - Q_3^2 + \left|\psi\right|\left(aQ_1Q_2' - Q_2Q_2' - Q_3Q_3'\right)\right]$$
$$- a\left[Q_1(Q_2^2 + Q_3^2) + \left|\psi\right|\left(Q_1Q_2Q_2' - Q_1'Q_2^2 + Q_1Q_3Q_3' - Q_1'Q_3^2\right)\right] = 0. \tag{3.3}$$

Here, the prime on Q_i represents differentiation with respect to $\left|\psi\right|$, $a = h/rh' = \mathrm{d}\ln r/\mathrm{d}\ln h = 1/(2-q)$. We note that $a = 2$ for a Keplerian disc with $q = 3/2$. The dimensionless growth rate, s, is defined by

$$s = -\frac{i\omega}{\Omega}\left(\frac{\Omega}{c_s k}\right)^2. \tag{3.4}$$

We note that validity of the equations requires $k \lesssim 1/H$. Full solutions of (3.3) are provided in Doğan et al. (2018) with a more detailed investigation. The disc becomes unstable if any of the roots of (3.3) has a positive real part, i.e. $\Re(s) > 0$, as the perturbations then grow exponentially with time. The simplified criteria for instability can be expressed as follows: If

$$\left[a\frac{\partial}{\partial\psi}(Q_1\left|\psi\right|) - \frac{\partial}{\partial\psi}(Q_2\left|\psi\right|)\right] > 0, \tag{3.5}$$

the disc is unstable, or if

$$\left[a\frac{\partial}{\partial\psi}(Q_1\left|\psi\right|) - \frac{\partial}{\partial\psi}(Q_2\left|\psi\right|)\right] < 0, \quad\text{and}\quad 4a\left[(Q_1Q_2 + (Q_1Q_2' - Q_1'Q_2)\left|\psi\right|)\right] > 0 \tag{3.6}$$

the disc is also unstable. The instability criterion simply implies that if the maximum diffusion rate is not located at maxima in warp amplitude, then local maxima in warp amplitude will grow, and the disc will break. As this will result in rapid transfer of mass out of this region due to the large warp amplitude implying large torques, this will also be realized by a significant drop in local surface density. This resembles the Lightman-Eardley viscous instability but for a warped disc with the warp amplitude playing the

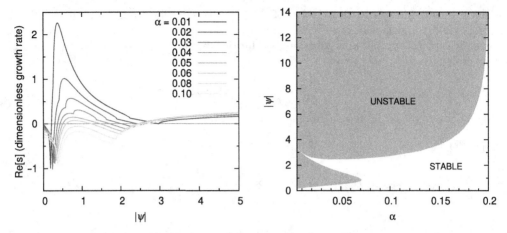

Figure 2. Left-hand panel shows the dimensionless growth rates $\Re(s)$ as functions of $|\psi|$ for different values of α. The grey line represents zero growth rate. The disc becomes unstable for sufficiently large warp amplitudes. The critical warp amplitudes where the disc becomes unstable are smaller and the growth rates of the instability are higher for low values of α. Right-hand panel shows the stable (white) and unstable (blue) regions in the $(\alpha, |\psi|)$ parameter space. This plot represents the critical warp amplitudes for instability to occur in discs with various α values. For a given value of α, there is always a minimum value of the warp amplitude, which gives rise to instability (see Doğan et al. 2018 for details).

role of the surface density. In Doğan et al. (2018), we showed that there is always a critical warp amplitude, $|\psi|_c$, which gives rise to instability for the parameters we have explored, with the exception of nearly flat discs with $|\psi| \lesssim 0.1$. The dimensionless growth rates and the critical warp amplitudes for the instability are shown in Fig. 2. The growth rates of the instability can be comparable with the dynamical rate ($\Re[s] \sim 1$).

4. Conclusion

We have shown that tilted discs inside a binary are susceptible to tearing from the outside in, because of the gravitational torque from the companion star. We have also connected the disk breaking and tearing behavior observed in numerical simulations with the instability of warped disks that was derived by Ogilvie (2000) through a local stability analysis of the warped disk equations. The necessary conditions for disk "breaking" have been derived. The instability occurs physically due to viscous anti-diffusion of the warp amplitude. This underlies the process of disc tearing which has the capacity to dramatically alter the instantaneous accretion rate and the observable properties of the disc on short time scales.

Acknowledgements

SD is supported by the Turkish Scientific and Technical Research Council (TÜBİTAK – 117F280).

References

Aly H., Dehnen W., Nixon C., King A., 2015, MNRAS, 449, 65. doi:10.1093/mnras/stv128
Artymowicz P., Lubow S. H., 1994, ApJ, 421, 651.
Bate M. R., Bonnell I. A., Clarke C. J., Lubow S. H., Ogilvie G. I., Pringle J. E., Tout C. A., 2000, MNRAS, 317, 773.
Doğan S., Nixon C., King A., Price D. J., 2015, MNRAS, 449, 1251.
Doğan S., Nixon C. J., King A. R., Pringle J. E., 2018, MNRAS, 476, 1519.
Doğan S., Nixon C. J., 2020, MNRAS, 495, 1148.

Facchini S., Lodato G., Price D. J., 2013, MNRAS, 433, 2142.

Frank J., King A., Raine D. J., 2002, Accretion Power in Astrophysics: 3rd Edn. Cambridge Univ. Press, Cambridge

Lodato G., Price D. J., 2010, MNRAS, 405, 1212.

Martin R. G., Nixon C., Lubow S. H., Armitage P. J., Price D. J., Doğan S., King A., 2014, ApJL, 792, L33.

Nixon C. J., King A. R., 2012, MNRAS, 421, 1201.

Nixon C., King A., Price D., Frank J., 2012, ApJL, 757, L24.

Nixon C., King A., Price D., 2013, MNRAS, 434, 1946.

Ogilvie G. I., 1999, MNRAS, 304, 557.

Ogilvie G. I., 2000, MNRAS, 317, 607.

Papaloizou J. C. B., Pringle J. E., 1983, MNRAS, 202, 1181.

Price D. J., Wurster J., Tricco T. S., Nixon C., Toupin S., Pettitt A., Chan C., et al., 2018, PASA, 35, e031.

Pringle J. E., 1981, ARA&A, 19, 137.

Shakura N. I., Sunyaev R. A., 1973, A&A, 24, 337.

The Predictive Power of Computational Astrophysics as a Discovery Tool
Proceedings IAU Symposium No. 362, 2023
D. Bisikalo, D. Wiebe & C. Boily, eds.
doi:10.1017/S1743921322001715

Hybrid magnetic structures around spinning black holes connected to a surrounding accretion disk

I. El Mellah⬡, B. Cerutti, B. Crinquand and K. Parfrey

Univ. Grenoble Alpes, CNRS, IPAG
School of Mathematics, Trinity College Dublin

Abstract. The hot accretion flow around Kerr black holes is strongly magnetized. Magnetic field loops sustained by a surrounding accretion disk can close within the event horizon. We performed particle-in-cell simulations in Kerr metric to capture the dynamics of the electromagnetic field and of the ambient collisionless plasma in this coupled configuration. We find that a hybrid magnetic topology develops with a closed magnetosphere co-existing with open field lines threading the horizon reminiscent of the Blandford-Znajek solution. Further in the disk, highly inclined open magnetic field lines can launch a magnetically-driven wind. While the plasma is essentially force-free, a current sheet forms above the disk where magnetic reconnection produces macroscopic plasmoids and accelerates particles up to relativistic Lorentz factors. A highly dynamic Y-point forms on the furthest closed magnetic field line, with episodic reconnection events responsible for transient synchrotron emission and coronal heating.

Keywords. acceleration of particles - magnetic reconnection - black hole physics - radiation mechanisms: non-thermal - methods: numerical

---

1. Introduction

Accretion and ejection have been found to be tightly linked around supermassive and stellar-mass black holes (BHs). Non-thermal flares from these systems are thought to be due to particles accelerated at relativistic speeds by shocks and/or magnetic reconnection. These particles then radiate through different leptonic (and, to a lesser extent, hadronic) emission/scattering processes like synchrotron and inverse Compton emission. The origin, properties and location of the particle acceleration sites around accreting BHs remain elusive. Multiple evidence indicate that a hot and collisionless corona surrounds accreting BHs (Cangemi et al. 2021). This environment is highly magnetized and threaded by large scale poloidal magnetic field lines (The Event Horizon Telescope collaboration 2021a). This highly ordered magnetic field is sustained by an accretion disk whose dynamics is well represented by the idealized magneto-hydrodynamics framework (The Event Horizon Telescope collaboration 2021b). In contrast, the corona is essentially force free. Last physical ingredient but not least, these BHs have jets which are thought to be rotation-driven. In the high-mass X-ray binary Cygnus X-1, some diagnostiscs based on thermal continuum fitting or X-ray reflection spectroscopy suggest that the stellar-mass BH might be maximally spinning (Miller-Jones et al. 2021). Things are less clear for the supermassive BHs Sagittarius A* (SgrA*) and M87* but the joint efforts of the Gravity and EHT collaborations could bear fruits in the years to come.

I first describe the model we rely on. Then, I present the results we obtained in terms of topology of the magnetic field in the corona and of particle acceleration driven by magnetic reconnection in the corona.

2. Model

2.1. *Kerr black hole magnetosphere*

We work on a stationary and axisymmetric Kerr spacetime produced by a rotating BH with mass M and dimensionless spin a. The problem is scale-invariant with the gravitational radius $r_g = GM/c^2$ as length scale (with c the speed of light and G the gravitational constant. We use the 3+1 formalism and Kerr-Schild spherical coordinates to alleviate the coordinate singularity which arises at the event horizon in other coordinate systems. We focus on the corona. The disk itself is not modeled in our particle-in-cell (PIC) simulations. Instead, it serves as a background steady environment where magnetic field lines are frozen and co-rotate with their footpoint on the disk. The angular speed profile in the disk is assumed to be Keplerian. We inject electron/positron pairs in the corona in a ad hoc manner such as (i) the charge density is high enough for the force-free regime to be possible, (ii) the magnetization remains $\gg 1$ and (iii) the plasma skin depth is resolved.

Since the seminal work by Blandford & Znajek in the 70's, most models of rotating BH magnetospheres focused on a specific configuration where magnetic field lines passing through the BH are open, a particularly convenient framework to launch jets and outflows. In contrast, fewer studies considered the alternative case: a Kerr BH surrounded by a geometrically thin accretion disk threaded by a large scale magnetic field connected to the BH. This coupling configuration has been studied by Uzdensky (2005) and more recently by Yuan et al. (2019) but they worked in the purely force-free approximation which cannot capture dissipative effects or changes of the topology of the magnetic field.

2.2. *Numerical setup*

We work on a 2D grid with axisymmetry around the BH spin axis. The disk is located in the equatorial plane of the BH and we only consider the part of the corona above the disk. The inner boundary is located within the event horizon to prevent any numerical artifact which might appear due to the boundary conditions to propagate upstream. The outer boundary is located at 30 r_g, with a region between 27 and 30 r_g where particles are deleted and the electromagnetic fields are damped. The resolution of the grid is 2,048 radial cells and 1,120 azimuthal cells. The resolution of the grid is high enough to capture the kinetic scales associated to the electron/positron pair plasma. The plasma injection method we use typically results in the simulation space being filled with a few 10^8 particles.

We use the fully relativistic `GRZeltron` code first introduced by Parfrey et al. (2019) and based on the `Zeltron` code (Cerutti et al. 2013). These PIC simulations rely on a hybrid Eulerian/Lagrangian approach (see Figure 1): electromagnetic fields on a grid are advanced in time by numerically solving Maxwell-Faraday and Maxwell-Ampère equations while charged particles move on this grid under the influence of the Lorentz force produced by the electromagnetic fields and of other user-defined forces. The plasma is collisionless in the sense that particles do not directly interact with each other but collective plasma effects can arise owing to the interplay with the electromagnetic fields. In `GRZeltron`, after particles were pushed, charges and currents are deposited on the grid following a first-order scheme. The lack of conservativity of this method is compensated by the resort to a divergence cleaning step based on an iterative Gauss-Seidel relaxation scheme such that Maxwell-Gauss equation is verified (Cerutti et al. 2015). `GRZeltron`

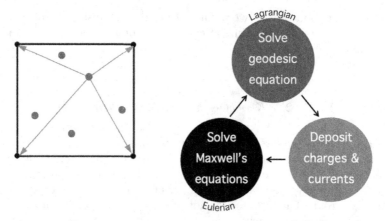

Figure 1. Principle of a PIC code, with the three main steps: (i) advancing the magnetic and electric fields on a grid by solving Maxwell-Faraday and Maxwell-Ampère equations respectively (in black), (ii) solving the equation of motion for particles (in blue) and (iii) depositing the charges and currents on the nodes of the grid to be used as source terms for Maxwell equations (in red).

uses a staggered grid which guarantees that an initially divergence-free magnetic field remains so to machine precision as the field evolves in time (Yee 1966).

In the initial state, the plasma is permeated with a poloidal magnetic field which threads the horizon and is anchored at its other end into a steady, geometrically thin, aligned and perfectly conducting disk in Keplerian rotation. The disk extends all the way down the innermost stable circular orbit (ISCO). In what follows, we explore the influence of the BH spin for $a \geq 0.6$ but we also ran simulations with $a = -0.8$ (counter-rotating disk) and for a slim accretion disk that we do not discuss in detail here. For complementary information, see El Mellah et al. (2021).

3. Results

3.1. *Hybrid magnetosphere*

First of all, let us inspect the structure of the magnetic field. As the initial condition relaxes, we observe a phenomena first highlighted by Uzdensky (2005). If the BH spins, magnetic field lines are sheared because in the ergosphere, they are dragged by the Lense-Thirring effect and rotate at a different speed from the one of their footpoint on the disk. Accordingly, we observe the following relaxation of the magnetic field from the initial state: a toroidal component grows, propagates outward and leads to the opening of magnetic field lines beyond a certain critical distance. It is the general relativistic analogue of the magnetic field lines around a spinning neutron star which open beyond the light cylinder. The main difference here is that a disk is needed since the BH cannot sustain its own magnetic field. In the left panel Figure 2, we represent the 3D topology of the magnetic field lines by assuming axisymmetry around the BH spin axis. The open magnetic field lines are either anchored in the disk or threading the event horizon. The latter are strongly twisted and will form the backbone of an electromagnetic jet à-la Blandford-Znajek (Blandford and Znajek 1977). The former are inclined enough to launch a magneto-centrifugal wind according to the Blandford-Payne criteria (Blandford and Payne 1982). In-between the two, the shearing is low enough that magnetic field lines coupling the disk to the BH remain and form a closed magnetosphere (see the two innermost magnetic field lines in the left panel in Figure 2). The smaller

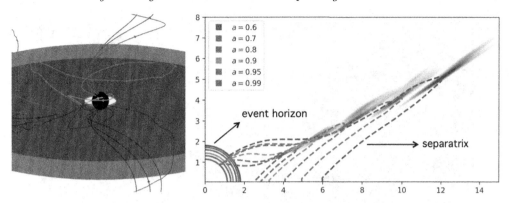

Figure 2. (Left) 3D representation of the magnetic field lines above and below the disk (black semi-transparent wedges). The central black sphere stands for the BH event horizon. (Right) Separatrix for different BH spins (dashed lines). Color-shaded regions stand for the probability map of the Y-point location. The x and y-axis are in units of the gravitational radius r_g.

the BH spin, the more extended the outermost closed magnetic field line (hereafter, the separatrix), as visible in the right panel in Figure 2.

We computed the jet power as the Poynting flux in the region of the open magnetic field lines threading the horizon. Provided we account for the correcting factor derived by Tchekhovskoy et al. (2011) at high spin, we obtain a dependence of the jet power on the spin which is fully in agreement with the force-free formula. It is consistent with the uniform angular speed of the open magnetic field lines threading the horizon which matches half of the BH angular speed. Through the coupling magnetic field lines, energy and angular momentum are transported from the BH to the disk (for $a \geq 0.6$). Energy is deposited at a rate of the order of 10% of the jet power and for high spin values, we compute angular momentum transfer rates high enough to modify the structure of the disk within an accretion time scale. Here again, the values we measure correspond to the force-free predictions.

3.2. *Particle acceleration*

At the intersection of the 3 aforementioned regions, a Y-point forms at the furthest point on the separatrix. From this point, a current sheet develops where magnetic reconnection happens and plasmoids form (see Figure 3). In this highly non force-free region, electromagnetic energy is converted into particle kinetic energy. Particle acceleration is more efficient for higher spin, with maximum Lorentz factors of the order of a few 100 for $a = 0.99$. Most of the dissipation occurs near the Y-point. The separatrix progressively stretches until a plasmoid detaches, the Y-point suddenly recedes and the cylce repeats, with a period ranging from 17 to $6r_g/c$ for a spin going from 0.6 to 0.99. The color-shaded regions in the right panel in Figure 2 correspond to the occurrence of the location of the Y-point at different times and for different spin values. This mechanism provides a natural heating source for the corona where relativistic electron-positron pairs are susceptible to upscatter disk photons via inverse Compton up to high energies. Its time scale is too short to account for the hour-long flares from SgrA* but for a stellar-mass BH, it corresponds to millisecond flares similar to the ones observed by Gierlinski et al. (2003) in Cygnus X-1. These dense, hot and macroscopic plasmoids could be responsible for enhanced localized emission as they propagate on a helicoidal trajectory. When seen face-on, they would mimic an orbiting hot spot similar to the one reported by The Gravity collaboration (2018).

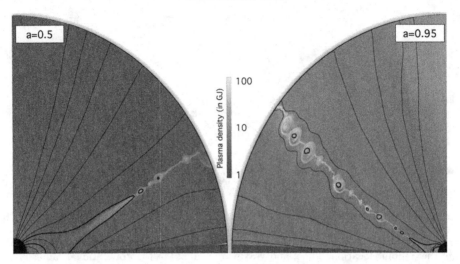

Figure 3. Charge density maps in units of Goldreich-Julian density for an intermediate (left panel) and high (right) BH spin. The disk is in red in the mid-plane. The poloidal magnetic field lines are in black and the red dot locates the Y-point on the separatrix i.e. the outermost closed magnetic field line. The dashed line is the ergosphere while the black disk is delimited by the event horizon.

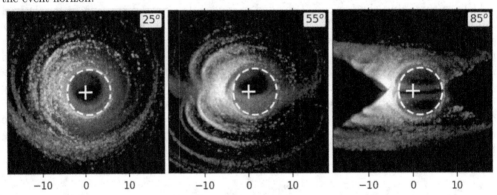

Figure 4. Intensity maps (logarithmic scale) for synchrotron emission from the current sheet, as seen from different viewing angles. The BH has a spin of 0.8 and is located at the white cross and its shadow is the white dashed line. The bottom axis is in units of r_g.

3.3. *Light emission*

Based on the kinetic energy distribution of particles in the current sheet, we derived synchrotron spectra for different BH spins. We identified a power-law component which could serve as the irradiating spectrum in order to lower the number of degrees-of-freedom of the X-ray reflection spectroscopy fits. Eventually, we computed synthetic synchrotron emission maps of the current sheet for different viewing angles using a forward ray-tracing version of the `geokerr` code (Dexter 2009), first used in Crinquand et al. (2021). In Figure 4, we can see arcs which are artefacts due to the axisymmetric assumption we rely on. However, for exposure times longer than a few hours, the current sheet would manifest itself as a hourglass-shaped bright region.

4. Conclusion

The corona around accreting BHs is a collisionless environment where non-ideal kinetic effects can be captured with PIC simulations. Closed and open magnetic field lines can co-exists, with some coupling the disk to the BH. While open magnetic field lines funnel a jet or a wind, the closed ones ensure energy and angular momentum exchanges between the BH and the disk. A current sheet forms above disks in prograde rotation around the BH where particles are accelerated via magnetic reconnection. Through synchrotron emission, they generate a high energy power-law component. Through inverse Compton, this population of non-thermal electrons and positrons provides a physically motivated irradiation source for hard X-rays above the disk, useful for reflection models. Episodic plasmoid ejection might explain millisecond flares observed in Cygnus X-1 in the high soft state. As they cascade back to the disk from the Y-point, relativistic particles could produce a hot spot at the footpoint of the separatrix.

5. Questions

Dmitry Bisikalo: What happens to the plasmoids as they move away from the BH?

Ileyk El Mellah: They progressively expand and dilute so their radiative efficiency drops.

Dmitry Bisikalo: Do you see collisions and shock waves between plasmoids?

Ileyk El Mellah: Good point. We do resolve secondary current sheets transverse to the first one which appear when two plasmoids collide in the main current sheet.

Christian Boily: What are your boundary conditions?

Ileyk El Mellah: Since the inner edge is within the event horizon, trivial boundary conditions (cancellation of electromagnetic fields and absorption of particles) are enough since no signal can propagate back above the horizon. At the outer edge, between 27 and $30r_g$, we have a buffer zone where particles are deleted and where the electromagnetic fields are damped.

Acknowledgments

This project has received funding from the European Research Council (ERC) under the European Union's Horizon 2020 research and innovation programme (grant agreement No 863412). Computing resources were provided by TGCC and CINES under the allocation A0090407669 made by GENCI. The authors thank Maïca Clavel, Antoine Strugarek, Guillaume Dubus and Geoffroy Lesur for fruitful discussions and constructive feedback.

References

Blandford, R. et al. 1977, MNRAS, 179
Blandford, R. et al. 1982, MNRAS, 199
Cangemi, F. et al. 2021, A&A, 650
Cerutti, B. et al. 2013, ApJ, 770
Cerutti, B. et al. 2015, MNRAS, 448
Crinquand, B. et al. 2021, A&A, 650
Dexter, J. and E. Agol 2009, ApJ, 696
El Mellah, I. et al. 2021, A&A (in press), arXiv:2112.03933
Gierlinski, M. et al. 2003, MNRAS, 343
The Event Horizon Telescope collaboration 2021, ApJ Letters, 910, 1, L12
The Event Horizon Telescope collaboration 2021, ApJ Letters, 910, 1, L13
The Gravity collaboration 2018, A&A, 618
Miller-Jones, J. et al. 2021, Science, 371
Parfrey, K. et al. 2019, PRL, 122
Tchekhovskoy, A. et al. 2011, MNRAS, 418
Uzdensky, D. 2005, ApJ, 620
Yee, K. 1966, IEEE Transactions on Antennas and Propagation, 14
Yuan, Y. et al. 2019, MNRAS, 484

The Predictive Power of Computational Astrophysics as a Discovery Tool
Proceedings IAU Symposium No. 362, 2023
D. Bisikalo, D. Wiebe & C. Boily, eds.
doi:10.1017/S1743921322001351

Mass ejection from neutron-star mergers

Masaru Shibata[1,2] ⓘ, Sho Fujibayashi[1], Kota Hayashi[2], Kenta Kiuchi[1,2] and Shinya Wanajo[1]

[1]Max Planck Institute for Gravitational Physics (Albert Einstein Institute), am Mühlenberg 1, Potsdam-Golm D-14476, Germany
email: mshibata@aei.mpg.de

[2]Center for Gravitational Physics, Yukawa Institute for Theoretical Physics, Kyoto University, Kyoto 606-8502, Japan

Abstract. Merger of binary neutron stars and black hole-neutron star binaries is the promising source of short-hard gamma-ray bursts, the most promising site for the r-process nucleosynthesis, and the source of kilonovae. To theoretically predict the merger and mass ejection processes and resulting electromagnetic emission, numerical simulation in full general relativity (numerical relativity) is the unique approach. We summarize our current understanding for the processes of neutron-star mergers and subsequent mass ejection based on the results of long-term numerical-relativity simulations. We pay particular attention to the electron fraction of the ejecta.

Keywords. Neutron-star merger, Kilonovae, Numerical Relativity, Nucleosynthesis

---

1. Introduction

Merger of neutron-star binaries (binary neutron stars (NS-NS) and black hole-neutron star (BH-NS) binaries) is one of the most promising sources of gravitational waves for ground-based advanced detectors, such as advanced LIGO and advanced Virgo (Abbott et al. 2021). These detectors have already observed gravitational waves from two NS-NS and three BH-NS binaries (Abbott et al. 2021), and thus, it is natural to expect that they will detect a number of gravitational-wave signals from NS binaries in the next decade.

In Shibata & Hotokezaka (2019), we summarized our understanding for the merger and mass ejection processes of NS binaries in 2018. Since then, several numerical simulations with all the relevant physical contents such as general relativity, neutrino cooling and heating, and magneto/viscous hydrodynamical angular momentum transport for a long timescale of 1 10 s have been performed for exploring the evolution of the post-merger systems, and as a result, our understanding has been deepened in the last three years. In particular, for the case that the merger remnant is a black hole, our understanding for the property of the post-merger ejecta has been updated. Thus, in this article we summarize our latest understanding for the ejecta paying particular attention to the update of our knowledge on the post-merger ejecta.

2. Merger remnants and their evolution

The fate of NS binaries depends on the mass (m_1, m_2) and spin of binary components, and the NS equation of state (EOS). For NS-NS binaries, for which the effect of their spin is minor, the total mass $(M_{\rm tot} = m_1 + m_2)$, the mass ratio $(q = m_2/m_1 \ (\leqslant 1))$ of the system, and the EOS are the key quantities for determining the merger remnant. For BH-NS binaries, the BH spin as well as the mass ratio and the EOS are the keys.

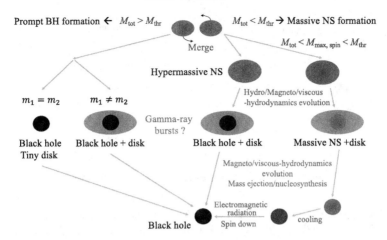

Figure 1. A summary for the merger and post-merger evolution of NS-NS binaries. M_{thr} and $M_{max,spin}$ denote the threshold mass for the prompt formation of a BH and the maximum mass of rigidly rotating cold NSs, respectively. Their values are likely to be $M_{thr} \gtrsim 2.8 M_\odot$ and $M_{max,spin} \gtrsim 2.4 M_\odot$. For the total mass $M_{tot} > M_{thr}$, a BH is formed in the dynamical timescale after the onset of merger, and for the nearly equal-mass case, $m_1 \approx m_2$, the mass of disks surrounding the BH is tiny $\ll 10^{-2} M_\odot$, while it could be $\gtrsim 10^{-2} M_\odot$ for a highly asymmetric system with $m_2/m_1 \lesssim 0.8$. For $M_{max,spin} < M_{tot} < M_{thr}$, a *hypermassive* neutron star (HMNS) is formed, and it subsequently evolves through several angular-momentum transport/dissipation processes, leading to eventual collapse to a BH surrounded by a disk (or torus). See, e.g., Shibata 2016 for the definition of the HMNS (and SMNS referred to below). For the case that M_{tot} is close to M_{thr}, the lifetime of the MNS is relatively short, while for smaller values of M_{tot} toward $M_{max,spin}$, the lifetime is longer. For the longer lifetime, the angular-momentum transport process works for a longer timescale, and the disk mass could be $\gtrsim 0.1 M_\odot$, whereas for a short lifetime, it could be $\sim 10^{-2} M_\odot$ or less. For $M_{tot} < M_{max,spin}$, a *supramassive* neutron star (SMNS) is formed and it will be alive for a dissipation timescale of its angular momentum which will be much longer than the neutrino cooling timescale of 1–10 s. Note that MNS denotes either an SMNS or an HMNS.

Figure 1 summarizes the possible remnants and their evolution processes for NS-NS mergers. Broadly speaking, there are two possible remnants formed immediately after the onset of merger; a BH and a massive neutron star (MNS). A BH is formed if the total mass (M_{tot}) is so high that the self gravity of the merger remnant cannot be sustained by the pressure associated primarily with the repulsive force among nucleons and centrifugal force due to rapid rotation resulting from the orbital angular momentum of the pre-merger binary. An important finding for many numerical-relativity simulations performed in the last two decades is that for $M_{tot} \lesssim 2.8 M_\odot$, the remnant is, at least temporarily, an MNS not a BH irrespective of the EOS that can reproduce $\gtrsim 2 M_\odot$ NSs. The total mass of NS-NS for GW170817 is $2.74^{+0.04}_{-0.01} M_\odot$ for a reasonable low spin prior (Abbott et al. 2017), and thus, for this event, an MNS is likely to be formed at least temporarily. By contrast, for GW190425 (Abbott et al. 2020) for which the total mass is $\sim 3.4 M_\odot$, a BH is likely to be formed immediately after the onset of the merger.

For the case of the prompt formation of a BH in NS-NS mergers, the dimensionless BH spin, χ, is approximately 0.8. The remnant BH in this formation channel is not surrounded by a massive disk if the mass ratio q is close to unity. The mass of the remnant disk around the BH increases with the decrease of q, and in the presence of a significant mass asymmetry, $q \lesssim 0.8$, the disk mass could be $\gtrsim 10^{-2} M_\odot$ (cf. Fig. 1). The disk around the BH is likely to be evolved by magnetohydrodynamics (MHD) processes, in particular by the effect of MHD turbulence induced by magnetorotational

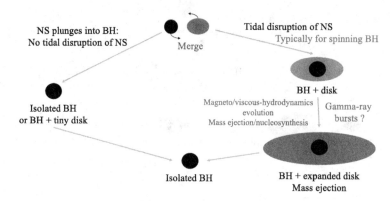

Figure 2. Two possible fates for the merger and post-merger evolution of BH-NS binaries; the NS is tidally disrupted (right path) or not (left path) by the companion BH. The tidal disruption is more subject for a spinning BH with the spin aligned with the orbital angular momentum. For the presence of tidal disruption, thus, the remnant is typically a rapidly spinning BH surrounded by a disk. The evolution process of the BH-disk system is essentially the same as that for NS-NS mergers.

instability (MRI) (Balbus & Hawley 1998) and winding, and resultant turbulent viscous process. During the MHD and viscous evolution of the disk, a short gamma-ray burst (sGRB) could be launched by the Blandford-Znajek mechanism (Blandford & Znajek 1977) associated with the strong magnetic field generated by the MHD turbulence which subsequently forms a BH magnetosphere. The viscous effect in the disk plays a central role for the post-merger mass ejection (see § 3.2).

For BH-NS binaries for which the NS is tidally disrupted at a close orbit, the remnant is also a BH surrounded by a disk. However, the disk mass can have a variety for this system. If the tidal disruption takes place far from the innermost stable circular orbit of the BH, the disk mass can be far beyond $10^{-2}M_\odot$. This is the case if the original BH is spinning rapidly and the spin is aligned with the orbital angular momentum (Kyutoku et al. 2021). Also in the presence of the tidal disruption of the NS, the remnant BH should be rapidly rotating as in the original BH, and thus, an sGRB associated with the Blandford-Znajek effect is also promising. For BH-NS binaries of low BH spin or of the high BH mass (say much larger than $10M_\odot$), the tidal disruption of the NS is unlikely to take place. In such cases, no or tiny mass of the disk and ejecta is expected (cf. Fig. 2).

For the MNS formation from NS-NS mergers, its evolution is determined by several processes. Soon after its formation, the gravitational torque associated with non-axisymmetric structure of the merger remnant plays an important role for transporting angular momentum from the MNS to matter surrounding it. This process reduces the angular momentum of the MNS while developing the disk formation. In addition, the MNS emits gravitational waves, and dissipates energy and angular momentum. Thus, if it is marginally stable against gravitational collapse, the MNS can collapse to a BH in ~ 100 ms by the angular momentum transport/dissipation. Numerical-relativity simulations have shown that the resulting system is a spinning BH of $\chi \sim 0.6$–0.7 surrounded by a disk of mass 10^{-2}–$10^{-1}M_\odot$ (see the references in Shibata & Hotokezaka (2019)). The disk mass is larger for the longer lifetime of the MNS and for more asymmetric systems of $q < 1$.

If the lifetime of the MNS is longer than ~ 100 ms, angular-momentum transport effects resulting from MHD turbulence are likely to play a role for the evolution of the MNS. At its formation, the MNS is differentially rotating. Furthermore, it should be strongly magnetized and in an MHD turbulence state with an excited turbulent viscosity, because at merger, a shear layer is formed at the contact surfaces of two NSs and the Kelvin-Helmholtz instability intensively occurs (Price & Rosswog 2006; Kiuchi et al. 2014). By this instability, a number of small-size vortexes are generated in the shear layer and magnetic fields are wound up by the vortex motion, which enhances the magnetic-field strength in a timescale much shorter than the dynamical timescale of the system ~ 0.1 ms. Note that the growth timescale of the Kelvin-Helmholtz instability is $\tau_{\mathrm{KH}} \sim 10^{-7}(\lambda/1\,\mathrm{cm})$ ms for the wavelength λ because the typically velocity at the onset of merger is $\sim 10^{10}$ cm/s. Because of the presence of the differential rotation and turbulent viscosity, the angular momentum in the MNS should be transported outward, and as a result, the inner part of the MNS is likely to settle to a nearly rigidly rotating state (e.g., Shibata et al. 2021). Simultaneously, a massive disk surrounding the MNS is developed because of the outward angular momentum transport in the outer region of the MNS. During the turbulent evolution in the MNS, the magnetic-field strength may be further amplified by the dynamo action. If global magnetic fields are generated during the dynamo action, the magnetic braking can also play a role for transporting the angular momentum of the MNS to the matter surrounding it.

If the lifetime of the MNS is even longer (i.e. MNS mass is not very large), it could be evolved by the viscous angular momentum transport to the disk from the contact region with the MNS and cooling by neutrino emission. The viscous timescale is approximately written as

$$\tau_{\mathrm{vis,disk}} \approx 0.2\,\mathrm{s} \left(\frac{\alpha_{\mathrm{vis}}}{10^{-2}}\right)^{-1} \left(\frac{c_s}{c/10}\right)^{-1} \left(\frac{R_{\mathrm{inn}}}{20\,\mathrm{km}}\right) \left(\frac{H/R_{\mathrm{inn}}}{1/3}\right)^{-1}, \qquad (2.1)$$

where α_{vis}, c_s, R_{inn}, and H are the dimensionless viscous parameter, the sound speed, the inner disk radius, and scale height there. The neutrino cooling timescale for MNSs is

$$\tau_\nu \approx \frac{U}{L_\nu} = 1\,\mathrm{s} \left(\frac{U}{10^{53}\,\mathrm{erg}}\right) \left(\frac{L_\nu}{10^{53}\,\mathrm{erg/s}}\right)^{-1}, \qquad (2.2)$$

where U is total thermal energy of the MNS and L_ν is total neutrino luminosity. Note that at the formation of the MNS (including the disk and envelope region), $L_\nu \gtrsim 10^{53}$ erg/s, because shock heating at merger significantly increases its temperature, but in ~ 100 ms after its formation, L_ν deceases to $O(10^{52})$ erg/s (e.g., Fujibayashi et al. 2018). Thus, if the viscous angular momentum transport or the neutrino cooling has a significant effect and it is marginally stable to gravitational collapse, the MNS would collapse to a BH in either of these timescales.

If the MNS mass is low enough, it will not collapse to a BH in ~ 10 s, but the MNS is likely to settle to a rigidly rotating cold NS (the so-called SMNS). The maximum mass of the SMNS is by up to ~ 0.3–$0.4 M_\odot$ larger than the maximum mass of the non-spinning cold NS, M_{max} (Cook et al. 1994). Thus, if the value of M_{max} is, e.g., $2.1 M_\odot$, the maximum mass for SMNSs would be $M_{\mathrm{max,spin}} \sim 2.4$–$2.5 M_\odot$, and hence, the strong self gravity of the SMNS could be sustained. However, the SMNS is magnetized. Thus, its rotational kinetic energy is inevitably dissipated by the magnetic dipole radiation or magnetic braking effect. Assuming the presence of a dipole magnetic radiation with the luminosity L_{B}, the spin-down timescale of the SMNS by the magnetic-dipole radiation is

$$\tau_{\mathrm{B}} \approx \frac{T_{\mathrm{rot}}}{L_{\mathrm{B}}} \approx 9 \times 10^2\,\mathrm{s} \left(\frac{B_p}{10^{15}\,\mathrm{G}}\right)^{-2} \left(\frac{M_{\mathrm{MNS}}}{2.5 M_\odot}\right) \left(\frac{R}{15\,\mathrm{km}}\right)^{-4} \left(\frac{\Omega}{6000\,\mathrm{rad/s}}\right)^{-2}, \qquad (2.3)$$

where $T_{\mathrm{rot}} (\sim 0.3 M_{\mathrm{MNS}} R^2 \Omega^2)$ is rotational kinetic energy, B_p the magnetic-field strength of the SMNS pole, and Ω the angular velocity of the SMNS. We assumed that the magnetic-field strength would be significantly enhanced at merger or during the dynamo action in the post-merger stage. This estimate shows that the rotational kinetic energy

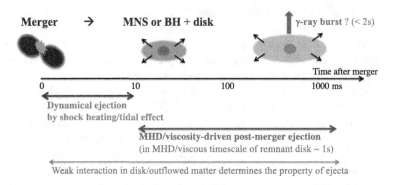

Figure 3. Mass ejection mechanisms during and after merger of NS binaries. Soon after the onset of merger, dynamical mass ejection takes place in the timescale of $\lesssim 10\,\mathrm{ms}$. Subsequently, MHD- or viscosity-driven mass ejection occurs. In the post-merger evolution, the weak interaction process is the key for determining the electron fraction of the ejecta.

could be dissipated in $\sim 10^3\,\mathrm{s}$. After the dissipation of its rotational kinetic energy, the SMNS should collapse to a BH.

3. Mass ejection

During the merger and post-merger evolution of merger remnants, neutron-rich matter is in general ejected. At the merger, the matter is dynamically ejected in the timescale of $\lesssim 10\,\mathrm{ms}$. Such mass ejection is referred to as the dynamical mass ejection. Subsequently, the mass ejection can proceed from the merger remnant, in particular from the remnant disk, through MHD and viscous processes. Such mass ejection is referred to as the post-merger mass ejection (see Fig. 3 for these mass ejection processes). In the following, we summarize the mechanisms for these two mass ejections separately. We focus in particular on the mass, velocity, and electron fraction of the ejecta because these quantities determine the property of the nucleosynthesis and electromagnetic counterparts associated with the ejecta.

3.1. *Dynamical mass ejection*

At merger of NS-NS binaries, strong shock waves are generated because two NSs with high relative velocity $\gtrsim 0.3c$ collide. In the shock waves, kinetic energy associated with the plunging motion is converted to thermal energy, which enhances thermal pressure and induces the ejection of the shocked matter. Also, if an MNS is the merger remnant at least temporally, the resulting non-axisymmetric MNS gravitationally exerts torque to the matter in the outer region and induces quick angular-momentum transport. By this process, a fraction of the matter gains energy sufficient for being ejected from the system. These two mechanisms drive dynamical mass ejection for NS-NS mergers. For the BH-NS case, the dynamical mass ejection takes place when the NS is tidally disrupted. During the tidal disruption process, the NS is deformed in a non-axisymmetric shape, which induces the angular momentum transport and resulting mass ejection. Also, during the dynamical infall of the NS matter into the BH, the spacetime structure is modified; e.g., the quadrupole moment of the entire system is reduced. With this dynamical effect, the matter outside the BH receives a gravitational impact and the mass ejection is induced. The timescale of the dynamical mass ejection is $\lesssim 10\,\mathrm{ms}$ both for the NS-NS and BH-NS cases. By the tidal torque, matter is ejected primarily in the equatorial direction, while by shock heating, the matter is ejected in a quasi-isotropic manner.

<u>Mass</u>: The mass of dynamical ejecta depends on the total mass, M_{tot}, mass ratio, q, and NS compactness (i.e., EOS) for NS-NS binaries (e.g., Hotokezaka et al. 2013; Shibata et al. 2017; Radice et al. 2018; Nedora et al. 2021). For the BH-NS case, the mass ratio and the BH spin are key parameters (Kyutoku et al. 2021). As summarized in Fig. 1, for $M_{\text{tot}} > M_{\text{thr}}$, a BH is promptly formed after the onset of NS-NS mergers. For $q \approx 1$ with $M_{\text{tot}} > M_{\text{thr}}$, $\geq 99.9\%$ of the NS matter is swallowed into the formed BH, and appreciable mass ejection cannot take place. For asymmetric NS-NS mergers, a fraction of matter may be dynamically ejected even for $M_{\text{tot}} > M_{\text{thr}}$, and for this case, short-term shock heating and tidal torque exerted by a deformed merger object collapsing to a BH are the engines for the dynamical mass ejection. However, numerical-relativity simulations indicate that for the dynamical mass ejection with mass $\geq 10^{-3} M_\odot$ only for $q \lesssim 0.8$ (e.g., Hotokezaka et al. 2013).

For the MNS formation case from NS-NS mergers, the dynamical ejecta mass depends strongly on the compactness (i.e., EOS) of NSs as well as M_{tot}. If the EOS is stiff (i.e., for a large NS radius), the relative velocity of two NSs at merger is relatively low, because the minimum orbital separation is relatively large, and thus, shock heating efficiency is relatively low. This results in small dynamical ejecta mass. Due to essentially the same reason, the dynamical ejecta mass depends on the total mass of the system, because for high total mass, shock heating efficiency can be high, resulting in large dynamical ejecta mass. Numerical-relativity simulations show that for EOSs with $R_{\text{NS}} \gtrsim 13$ km or for $M_{\text{tot}} \lesssim 2.6 M_\odot$, the dynamical ejecta mass is of $O(10^{-3}) M_\odot$ for $q \sim 1$. Even for $q \sim 0.8$, the dynamical ejecta mass could be at most $0.003 M_\odot$. On the other hand, for $R_{\text{NS}} \lesssim 12$ km with $M_{\text{tot}} \gtrsim 2.7 M_\odot$, the dynamical ejecta mass could be ~ 0.003–$0.01 M_\odot$ depending on q.

For BH-NS mergers, the mass of dynamical ejecta is determined by how tidal disruption of NSs proceeds. If an NS is tidally disrupted far from the innermost stable circular orbit around its companion BH, a large fraction of the NS matter (0.1–$0.3 M_\odot$) stays outside the BH horizon. This is in particular the case for rapidly spinning BHs. Numerical-relativity simulations show that for such a case, $\sim 10\%$–30% of the matter located outside the horizon can escape from the system as ejecta (for plausible BH mass of $\gtrsim 4 M_\odot$). Thus, larger disk mass results in larger dynamical ejecta mass, up to $\sim 0.1 M_\odot$ for the maximum case. By contrast, if NSs are not tidally disrupted, the dynamical ejecta mass (as well as the disk mass) is negligible. Thus, the dynamical ejecta mass is in a wide range of 0–$0.1 M_\odot$ for BH-NS mergers.

<u>Velocity</u>: Because the dynamical mass ejection proceeds near the last stable binary orbit of radius r, the velocity of the ejecta should be of the order of the escape velocity there, i.e., $\sim \sqrt{GM_{\text{tot}}/r} \approx 0.33c(M_{\text{tot}}/3M_\odot)^{1/2}(r/40 \text{ km})^{-1/2}$. Numerical-relativity simulations have shown that the typical average velocity is indeed 0.15–$0.25c$ for the MNS formation case (e.g., Shibata & Hotokezaka 2019). For this case, the dynamical mass ejection is induced both by the shock heating and tidal torque. For the component ejected by the shock heating, a fraction of matter could have a relativistic speed up to $\sim 0.8c$ and, in addition, the ejecta morphology is quasi-spherical. For the prompt formation of a BH from highly asymmetric NS-NS binaries, the average velocity of ejecta can exceed $\sim 0.3c$, because the dynamical mass ejection proceeds only for the matter in the vicinity of the object collapsing to a BH by a tidal torque exerted.

For BH-NS mergers, the average velocity of the dynamical ejecta is 0.2–$0.3c$; it is higher for the rapidly spinning BHs with the spin axis aligned with the orbital angular momentum. By contrast to the MNS formation case of NS-NS mergers, the maximum velocity is always $\lesssim 0.4c$. The maximum velocity is higher for the rapidly spinning BHs.

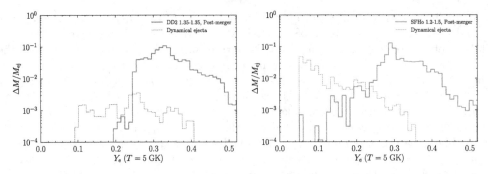

Figure 4. Mass histogram as a function of Y_e for the formation of a long-lived MNS (left) and a HMNS that collapses into a BH at $\sim 20\,\mathrm{ms}$ after the onset of merger (right). The dotted and solid curves denote the contribution from dynamical and post-merger ejecta, respectively.

<u>Electron fraction</u>: The electron fraction (Y_e) of ejecta is one of the key quantities for determining elements synthesized by the r-process nucleosynthesis. Because the abundance pattern of the r-process elements is the key input for determining the opacity of the kilonova emission from the merger ejecta, the electron fraction is one of the most important quantities for determining the output after NS mergers (Barnes & Kasen (2013); Tanaka & Hotokezaka (2013)).

Because the typical value of Y_e for cold NSs is quite low, 0.01–0.1, Y_e of the dynamical ejecta would be also low if the NS matter is ejected without undergoing weak interaction processes. However, the property of the ejecta could be influenced strongly by the weak processes. First, shock heating at merger and during subsequent evolution of the merger remnant increases the matter temperature beyond $10\,\mathrm{MeV}$ in particular for NS-NS mergers. In such a high-temperature environment, the electron-positron pair creation is enhanced. Then, neutrons efficiently capture positrons via $n + e^+ \rightarrow p + \bar{\nu}_e$. Because the inverse reaction, $p + e^- \rightarrow n + \nu_e$, also proceeds, these two reactions tend to bring material in a chemical equilibrium. Thus, in the presence of many positrons produced by the pair creation, the fraction of protons and Y_e are increased (the neutron richness is reduced).

In the presence of an MNS that is a strong neutrino emitter, the neutrino irradiation to the matter surrounding the MNS can change its composition. Since neutrons and protons absorb neutrinos via $n + \nu_e \rightarrow p + e^-$ and $p + \bar{\nu}_e \rightarrow n + e^+$, respectively, the fractions of neutrons and protons tend to approach the values in a chemical equilibrium. Because the luminosity and average energy of ν_e and $\bar{\nu}_e$ from the MNS are not significantly different, p/n ratio changes from ~ 0.1 toward 1, and thus, Y_e also increases in the presence of the strong neutrino irradiation.

As already mentioned, there are two engines for the dynamical mass ejection: shock heating and tidal torque. For the MNS formation case, both effects play an important role. By the shock heating and neutrino irradiation from the MNS, Y_e for a large fraction of ejecta is significantly increased. On the other hand, matter ejected by the tidal torque does not efficiently undergo the weak interaction and the low-Y_e state is preserved. Therefore, the dynamical ejecta for the MNS formation case in general has components with a wide variety of Y_e between ~ 0.03 (i.e., the nearly original minimum value of NSs) and ~ 0.4 (cf. Fig. 4). If the effect by the weak interaction is subdominant, the fraction of low-Y_e ejecta is richer.

For the prompt BH formation case (high-mass NS-NS and BH-NS mergers), the dynamical mass ejection is induced primarily by the tidal effect. Thus, the ejecta tend to have low values of Y_e. In particular, for BH-NS mergers, the dynamical ejecta is dominated by the low-Y_e components with $Y_e < 0.1$ (cf. Fig. 6). This result is highly different from that in MNS-forming NS-NS mergers. For high-mass asymmetric NS-NS mergers leading to the prompt formation of a BH, the low-Y_e components are also dominant for the dynamical ejecta. However, because of the presence of the shock heating effect at the onset of the merger, a small fraction of the ejecta has high values of Y_e up to ~ 0.4.

3.2. *Post-merger mass ejection*

After many of NS mergers, an MNS or a BH surrounded by a disk is likely to be formed. Both the MNS and disk at their formation are differentially rotating and likely to be strongly magnetized, and hence, MHD turbulence could be induced. Then, strong turbulent viscosity could be enhanced as already mentioned in § 2. This viscous effect induces the so-called viscosity-driven post-merger mass ejection (Fernández & Metzger 2013; Metzger & Fernández 2014; Just et al. 2015; Fujibayashi et al. 2018).

<u>Mechanism</u>: The post-merger mass ejection is driven primarily from the disk orbiting the central compact object. We first summarize how this mass ejection proceeds. In the presence of the viscosity (which is supposed to be from the MHD turbulence), the disk is subject to long-term viscous heating and angular momentum transport. As indicated in Eq. (2.1) (replacing R_{inn} to a typical disk radius of 20–100 km), the timescale for this is several hundreds ms to a few seconds for the central objects of mass ~ 2–$10 M_\odot$. In an early stage of the disk evolution, for which the initial maximum density and temperature are $\gtrsim 10^{11}$ g/cm^3 and several MeV, thermal energy generated by the viscous heating is consumed primarily by the neutrino emission. In this stage, the disk expands by the angular momentum transport but the thermal energy generated cannot contribute a lot to the disk evolution. With the viscous expansion of the disk, its density and temperature decrease. By the decrease of the temperature, T, the neutrino emissivity is steeply reduced because it is approximately proportional to T^6 (e.g., Ruffert & Janka 1996). The latest numerical simulations have shown (e.g., Fujibayashi et al. (2020b)) that after the maximum temperature of the disk decreases below ~ 2–3 MeV, neutrino cooling timescale is longer than the viscous heating one. In this late stage, the viscous heating is fully used for the evolution of the (torus-shape) disk. Because the viscous heating is most efficient in the innermost region of the disk, convective motion is enhanced in the disk and activates efficiently carrying the thermal energy outward, in particular toward the torus surface. Primarily by this effect, the outer part of the disk (torus), which was already expanded by the viscous angular momentum transport, gains the thermal energy, and as a result, the matter in the outer region eventually escapes from the system as ejecta.

In the presence of strong magnetic fields, not only the viscous effects but also purely MHD effects play a role for the post-merger mass ejection (Siegel & Metzger 2017; Fernández et al. 2019; Christie et al. 2019; Shibata et al. 2021). However, the latest MHD simulations (Christie et al. 2019; Shibata et al. 2021; Hayashi et al. 2022) have shown that this is a subdominant effect for a reasonable magnetic-field profile. The neutrino-irradiation effect could also play a role for enhancing mass ejection in the presence of an MNS which is a strong neutrino emitter, but the latest simulation shows that this gives a subdominant effect for increasing the ejecta mass (Fujibayashi et al. 2020a), although the neutrino irradiation is one of the important effects for determining the electron fraction of the ejecta (see below).

<u>Mass</u>: Long-term (~ 10 s-long) numerical simulations have shown that the ejecta mass could be more than 30% of the initial disk mass, resulting in $M_{\mathrm{eje,pm}} = 0.05$–$0.1 M_\odot$, in the presence of a long-lived MNS (Metzger & Fernández 2014; Fujibayashi et al. 2018, 2020a). For the case that the MNS is the central object, the mass accretion onto it proceeds only after the thermal energy of the matter is dissipated by neutrino cooling. Because it takes neutrino-cooling timescale (~ 1 s) for this, which is comparable to the

viscous timescale, the matter infall is suppressed, and consequently, the mass ejection is enhanced.

For the case that a BH is the central object, mass falling into it is appreciably larger than that of the outflow. However, numerical simulations for disks around spinning BHs show that 10–20% of the disk mass can be the ejecta component (e.g., Fernández & Metzger 2013; Just et al. 2015; Fujibayashi et al. 2020b,c; Just et al. 2021). MHD simulations show that in the presence of a strong poloidal magnetic field, the post-merger mass ejection is enhanced by the MHD process (Siegel & Metzger 2017; Fernández et al. 2019; Christie et al. 2019), but with initially toroidal-dominant magnetic field, which is likely to be more plausible, the enhancement of the mass ejection is not very significant (Christie et al. 2019; Shibata et al. 2021; Hayashi et al. 2022).

Velocity: The post-merger ejecta is launched primarily from the outer part of disks (tori). If the mass ejection takes place at a radius of $r \gtrsim 100 G c^{-2} M$ ($M = M_{\mathrm{MNS}}$ or M_{BH}), the characteristic velocity would be $\lesssim 0.1c$. Hence, the typical velocity for the viscosity-driven ejecta component is smaller than that for dynamical ejecta. However, the velocity may be significantly enhanced in the presence of an MNS and dynamo action, because MHD turbulence could be developed enhancing the magnetic-field strength and forming global field lines around the MNS. After the development of the global magnetic field, the magneto-centrifugal effect (Blandford & Payne 1982) associated with the MNS rotation can play a significant role for the angular momentum transport and ejecta acceleration (Shibata et al. 2021). Also, as mentioned in § 2, the remnant MNS is differentially rotating, and thus, in the presence of turbulent viscosity, the angular momentum may be transported in the MNS with a short timescale (substitute $R_{\mathrm{inn}} \sim 10\,\mathrm{km} \sim H$ into Eq. (2.1)). Then, the angular-velocity profile of the MNS can be rearranged to be a nearly rigidly rotating one, and associated with the resulting rearrangement of the density and pressure profiles, strong pressure waves can propagate outward providing the energy to the matter in the outer region (Fujibayashi et al. 2018).

Irrespective of the MHD/viscous mechanisms described in the previous paragraph, the engine of the mass ejection is rotational kinetic energy in the presence of an MNS, which is estimated by $T_{\mathrm{kin}} \sim 0.3 M_{\mathrm{MNS}} R^2 \Omega^2$ (Cook et al. 1994), where M_{MNS}, R, and Ω are the mass, equatorial radius, and angular velocity of the MNS, and thus,

$$T_{\mathrm{kin}} \sim 1.7 \times 10^{53} \left(\frac{M_{\mathrm{MNS}}}{2.6 M_\odot} \right) \left(\frac{R}{15\,\mathrm{km}} \right)^2 \left(\frac{\Omega}{7000\,\mathrm{rad/s}} \right)^2 \mathrm{erg}. \qquad (3.1)$$

If $\sim 1\%$ of this energy is transferred to the ejecta, the average velocity of the ejecta evaluated by $\sqrt{2 \times 0.01 T_{\mathrm{kin}} / M_{\mathrm{eje,pm}}}$ is $\sim 0.15c (T_{\mathrm{kin}}/2 \times 10^{53}\,\mathrm{erg})^{1/2} (M_{\mathrm{eje,pm}}/0.1 M_\odot)^{-1/2}$ for the post-merger ejecta mass of $M_{\mathrm{eje,pm}}$. If the efficiency of the energy transfer is $\sim 10\%$, the average velocity of the ejecta could be $\sim 0.5c$. Hence, the typical velocity for the post-merger ejecta may be enhanced by a factor of ~ 2–3 in the presence of a long-lived MNS. However, the dynamo effect and subsequent magnetic-field growth are not well understood to date. One of the important topics in numerical relativity is to clarify these effects by a first-principle MHD simulations.

Electron fraction: The understanding of the electron fraction for the post-merger ejecta has been modified in the last three years, because long-term numerical simulations incorporating all the relevant physics such as weak interaction physics including neutrino effects and general relativity with more realistic initial conditions have been done in detail (e.g., Fujibayashi et al. 2018; Fujibayashi et al. 2020a; Fujibayashi et al. 2020b; Fujibayashi et al. 2020c; Just et al. 2021). What is important is that in the viscosity-driven post-merger mass ejection, the electron fraction is determined primarily by the long-term viscous evolution of the disk, and irrespective of the central object (MNS or BH), the electron fraction of the disk increases significantly during the viscous evolution.

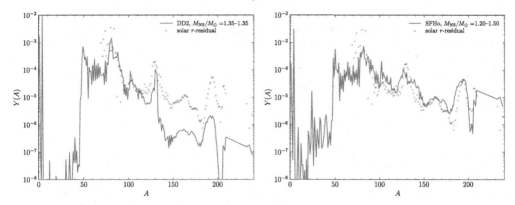

Figure 5. Elemental fraction for the r-process nucleosynthesis as a function of the mass number A for the models shown in Fig. 4. The small solid circles denote the solar-abundance. The solar abundance and numerical results are matched at $A = 90$. For the formation of a long-lived MNS (left), light r-process elements are overproduced (heavy elements are under-produced), and hence, the elemental pattern does not agree with the solar abundance pattern. By contrast, for the HMNS formation collapsing in ~ 20 ms after the onset of merger (right), the abundance pattern is in a good agreement with the solar abundance pattern.

As already mentioned, in its early stage, the density and temperature in the disk are high: The maximum density and temperature are $\gtrsim 10^{11}$ g/cm^3 and several MeV. In such conditions, the capture timescale of electrons and positrons by protons and neutrons and the timescale of the neutrino absorption by nucleons are shorter than the viscous timescale. That is, the following reactions are in equilibrium:

$$n + \nu_e \leftrightarrow p + e^-, \quad p + \bar{\nu}_e \leftrightarrow n + e^+, \tag{3.2}$$

and thus the chemical equilibrium $\mu_n + \mu_{\nu_e} = \mu_p + \mu_e$ is established (Beloborodov 2003). Here, μ_A ($A = n, p, \nu_e, e$) denotes the chemical potential of each particle. By the viscous angular momentum transport, subsequently, the density and temperature of the disk decrease, and thus, neutrino absorption timescale becomes longer. However, the electron/positron capture timescales are still shorter than the viscous timescale for a while, and thus, the following reactions are still in equilibrium:

$$p + e^- \rightarrow n + \nu_e, \quad n + e^+ \rightarrow p + \bar{\nu}_e. \tag{3.3}$$

In the early stage of the disk evolution, the density is high, and thus, μ_e is dominant associated with the strong electron degeneracy, resulting in the neutron-rich (low-Y_e) state with $\mu_n \gg \mu_p$ (μ_{ν_e} is negligible in the disk). However, with the decrease of the disk density due to the viscous expansion, μ_e (which is approximately proportional to $\rho^{1/3}$) decreases, and as a result, the neutron richness becomes lower. Therefore, with the viscous evolution of the disk, the value of Y_e increases.

The post-merger mass ejection sets in when the neutrino cooling timescale becomes longer than the viscous timescale, at which the maximum temperature of the disk is ~ 2–3 MeV (Fujibayashi et al. 2020b; Just et al. 2021). Broadly speaking, below this temperature, all the timescales of the weak interaction becomes longer than the viscous timescale. Thus, approximately at the same time, the reactions of Eq. (3.3) freeze out and the electron fraction is fixed. Thus, the typical electron fraction of the post-merger ejecta is approximately determined by that at the moment of the freeze-out of the weak interaction (Fernández & Metzger 2013). The latest numerical simulations show that the typical value of Y_e at the freeze-out is ~ 0.2–0.4 irrespective of the central object (either MNS or BH) (Fujibayashi et al. 2020a,b,c; Just et al. 2021). In the presence of the MNS,

Table 1. $M_{\rm ej,dyn}$ and $M_{\rm ej,vis}$: dynamical and post-merger ejecta mass in units of $M_\odot$, $Y_{e,\rm dyn}$: Y_e of dynamical ejecta, $Y_{e,\rm pm}$: Y_e of post-merger ejecta, $\langle v_{\rm ej,dyn} \rangle$: average velocity of dynamical ejecta in units of c. Low-m, Mid-m, and High-m imply that the remnants formed in ~ 1 ms after the merger are SMNS, HMNS, and BH.

Type of binary	Remnant	$M_{\rm ej,dyn}$	$M_{\rm ej,pm}$	$Y_{e,\rm dyn}$	$Y_{e,\rm pm}$	$\langle v_{\rm ej,dyn} \rangle / c$
Low-m NS-NS	SMNS	$O(10^{-3})$	0.05–0.1	0.03–0.4	0.2–0.5	0.15–0.20
Mid-m NS-NS (stiff EOS)	HMNS	$O(10^{-3})$	$O(10^{-2})$	0.03–0.4	0.2–0.5	0.15–0.25
Mid-m NS-NS (soft EOS)	HMNS	$\lesssim 10^{-2}$	$O(10^{-2})$	0.03–0.4	0.2–0.5	0.20–0.25
High-m NS-NS ($q \sim 1$)	BH	$< 10^{-3}$	$< 10^{-3}$	—	—	—
High-m NS-NS ($q < 1$)	BH	$O(10^{-3})$	$\lesssim 10^{-2}$	0.03–0.4	0.2–0.5	0.2–0.3
BH-NS	BH	0–0.1	0–0.1	0.03–0.1	0.1–0.4	0.2–0.3

which is the strong neutrino emitter, the electron fraction of the ejecta could be even higher due to the neutrino irradiation (Metzger & Fernández 2014; Fujibayashi et al. 2018), but this effect is subdominant. The actual electron fraction is widely distributed as in dynamical ejecta. However, the low end is at smallest ~ 0.2 for the viscous hydrodynamics with plausible viscous parameters. In the MHD simulations, the lower end can be smaller, up to ~ 0.15, because the purely MHD effects can eject matter within the viscous timescale. On the other hand, the high end is always $\gtrsim 0.5$.

The typical property of ejecta is summarized in Table 1, which shows that the ejecta property depends strongly on the binary parameters. In Fig. 4, we also show numerical results for the mass histogram as a function of Y_e for the formation of a long-lived MNS (left) and a HMNS that collapses into a BH at ~ 20 ms after the onset of merger (right). The dotted and solid curves denote the contributions from dynamical and post-merger ejecta, respectively, each of which results from the merger and subsequent post-merger simulations (Fujibayashi et al. 2022). It is found that the ratio of the masses of the dynamical ejecta to the post-merger ejecta is larger for the HMNS formation case while it is small (of $O(10^{-2})$) for the long-lived MNS formation case, as summarized in Table 1. The main reason is that for the formation of the long-lived MNS, the mass of the post-merger mass ejection becomes very large.

Figure 5 shows the results of a nucleosynthesis calculation for the models presented in Fig. 4 (Fujibayashi et al. 2022). For the long-lived MNS formation, a large amount of the light r-process elements with $A \lesssim 100$ are synthesized primarily from the post-merger ejecta while the amount of the heavy r-process elements ($A \gtrsim 130$) synthesized from the dynamical ejecta is quite small. For this case, the abundance pattern is significantly different from the solar abundance pattern. By contrast, for the formation of the HMNS with a short lifetime of $O(10)$ ms, the abundance pattern is close to the solar abundance pattern. Suppose that the NS mergers would be the major site for the r-process nucleosynthesis and the abundance pattern of the r-process elements would be universally similar to the solar pattern in the universe. Then, the frequent formation of the long-lived MNS in the NS mergers is disfavored, while the frequent formation of the short-lived HMNS is favored. Otherwise, the solar abundance pattern of the r-process elements cannot be reproduced.

A word of caution is appropriate here. To derive the quantitative results for the properties of the post-merger ejecta, self-consistent long-term simulations have been performed for the MNS formation (Fujibayashi et al. 2018, 2020a, 2022). However, for the case of prompt BH formation, the post-merger simulations have been performed starting from ad hoc initial conditions for the matter profile of disks and magnetic-field configurations or for the hypothetical viscous coefficient (e.g., Fernández & Metzger 2013; Just et al. 2015; Siegel & Metzger 2017; Christie et al. 2019; Fujibayashi et al. 2020b,c; Just et al. 2021). Thus, the results can depend significantly on the initial condition employed, and in the viscous-hydrodynamics simulations, they can depend on the viscous parameter.

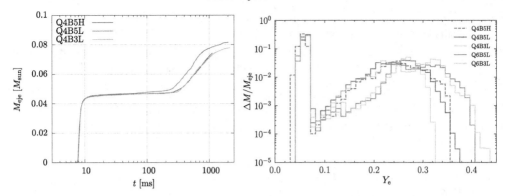

Figure 6. Left: Mass ejection history for BH-NS merger models with the BH mass of $5.4M_\odot$, BH dimensionless spin of 0.75, and NS mass of $1.35M_\odot$. Right: Mass histogram as a function of Y_e for the ejecta of BH-NS mergers. Q4 and Q6 denote the models with the BH mass of $5.4M_\odot$ and $8.1M_\odot$. See Hayashi et al. (2022) for more details.

To overcome these problems, we need to perform a long-term simulation from the merger to the post-merger stages self-consistently. The topic of the next section is on our latest effort to this direction.

4. First self-consistent simulation

As we described above, we need a long-term simulation which can fully explore the process from the merger up through the post-merger mass ejection stage, to establish a self-consistent picture for the entire mass ejection and resultant r-process nucleosynthesis. We have recently performed seconds-long simulations for BH-NS binaries for the first time (Hayashi et al. 2022). This section is devoted to briefly introducing the results of this work.

We performed a neutrino-radiation MHD simulation in full general relativity for the merger between BHs and an NS. The BH mass is $5.4M_\odot$ or $8.1M_\odot$ with the dimensionless spin of 0.75. The NS mass is $1.35M_\odot$ with a relatively stiff equation of state that gives the NS radius of 13.2 km. In this setting, the NS is tidally disrupted before it is swallowed into the BH, and at the merger, the matter with mass of 0.04–$0.05M_\odot$ is dynamically ejected in the first ~ 10 ms (see the left panel of Fig. 6) and a disk with the initial mass of 0.2–$0.3M_\odot$ is formed (for lower initial BH mass, the value is larger). The disk is subsequently evolved by the MHD effects. As described in § 2 and § 3, the MHD turbulence is developed primarily by the MRI and winding in the disk, and subsequently, effective viscosity is induced and drives the post-merger mass ejection. The post-merger mass ejection sets in at ≈ 0.3–0.5 s after the tidal disruption at which the neutrino luminosity becomes sufficiently low (see the left panel of Fig. 6 for the mass ejection history). The resulting post-merger ejecta mass is 0.02–$0.03M_\odot$ (i.e., about 10% of the initial disk mass). All these results for the post-merger mass ejection are in fair agreement with those for the latest long-term simulations to BH-disk systems in general relativity (e.g., Fujibayashi et al. 2020b,c).

The right panel of Fig. 6 shows the mass histogram of the ejecta as a function of Y_e for several models of BH-NS mergers with different BH mass and different initial magnetic-field strength. It is found that irrespective of the setting, two components emerge; one is the dynamical ejecta with $Y_e < 0.1$ and the other is the post-merger ejecta with $Y_e \approx 0.1$–0.4. The two-components distribution of Y_e agrees semi-quantitatively with that described in § 3. The values of Y_e are slightly smaller than those predicted by viscous hydrodynamics simulations (Fujibayashi et al. 2020b; Just et al. 2021). Our interpretation for this is that by the purely MHD effect, the matter in the inner region of disk, which has relatively low values of Y_e, can be ejected. However, besides this minor quantitative difference, the mass histogram of the post-merger ejecta is similar to that found in the post-merger simulations.

The seconds-long MHD simulation exploring both for the merger and post-merger stages is obviously suitable for fully understanding the mass ejection mechanisms. In the next decade, we need to focus on our effort along this line.

References

Abbott, R. et al. 2021, *arXiv: 2111.03606*

Abbott B.P. et al. 2017, *Phys. Rev. Lett.*, 119, 161101

Abbott B. et al. 2020, *Astrophys. J. Lett.*, 892, L3

Balbus, S.A. & Hawley, J.F. 1998, *Rev. Mod. Phys.*, 70, 1

Barnes, J. & Kasen, D. 2013 *Astrophys. J.* 775, 18

Beloborodov, A.M. 2003 *Astrophys. J.* 588, 931

Blandford, R.D. & Znajek, R.L. 1977, *Mon. Not. R. Astron. Soc.* 179, 433

Blandford, R. & Payne, D.G. 1982, *Mon. Not. R. Astron. Soc.* 199, 883

Christie, I.M., et al. 2019, *Mon. Not. R. Astron. Soc.* 490, 4811

Cook, G.B., Shapiro, S.L., & Teukolsky, S.A. 1994, *Astrophys. J.* 423, 823

Fernández, R. & Metzger, B.D. 2013, *Mon. Not. Royal Astron. Soc.* 435, 502

Fernández, R., Tchekhovskoy, A., Quataert, E., Foucart, F., & Kasen, D. 2019,*Mon. Not. R. Astron. Soc.* 482, 3373

Fujibayashi, S., et al. 2018, *Astrophys. J.* 860, 64

Fujibayashi, S., et al. 2020, *Astrophys. J.* 901, 122

Fujibayashi, S., et al. 2020, *Phys. Rev. D* 101, 083029

Fujibayashi, S., et al. 2020, *Phys. Rev. D* 102, 123014

Fujibayashi, S., et al. 2022, in preparation

Hayashi, K. et al. 2022 *Phys. Rev. D* in submission (arXiv: 2111.04621)

Hotokezaka, K, Kiuchi, K, Kyutoku, K., Okawa, H., Sekiguchi, Y., Shibata, M. & Taniguchi, K. 2013, *Phys. Rev. D* 87, 024001

Just, O., Bauswein, A., Pulpillo, R.A., Goriely, S., & Janka, H.-Th. *Mon. Not. Royal Astron. Soc.* 448, 541

Just, O. et al. 2021, *ArXiv: 2102.08387*

Kiuchi, K. et al. 2014, *Phys. Rev. D* 90, 041502

Kyutoku, K. et al. 2021, *Living Review Relativity* 24, 5

Metzger, B.D. & Fernández, R. 2014, *Mon. Not. Royal. Astron. Soc.* 441, 3444

Nedora, V. et al. 2021, *Astrophys. J.* 906, 98

Price, D.J., & Rosswog, S. 2006, *Science* 312, 719

Radice, D. et al. 2018, *Astrophys. J.* 869, 130

Ruffert, M. & Janka, H.-Th 1996, *Astron. Astrophys.* 344, 573

Shibata, M 2016 *Numerical Relativity (World Scientific, Singapore)*

Shibata, M., Fujibayashi, S., Hotokezaka, K., Kiuchi, K., Kyutoku, K., Sekiguchi, Y. & Tanaka, M. 2017, *Phys. Rev. D* 96, 123012

Shibata, M., & Hotokezaka, K. 2019, *Ann. Rev. Nucl. Part. Sci.*, 69, 41

Shibata, M. et al. 2021, *Phys. Rev. D* 104, 063026

Siegel, D.M. & Metzger, B.D. 2017, *Phys. Rev. Lett.* 119, 231102

Tanaka, M. & Hotokezaka, K. 2013, *Astrophys. J.* 775, 113

The Predictive Power of Computational Astrophysics as a Discovery Tool
Proceedings IAU Symposium No. 362, 2023
D. Bisikalo, D. Wiebe & C. Boily, eds.
doi:10.1017/S174392132200182X

Merging of spinning binary black holes in globular clusters

Margarita Sobolenko[1]⬤, Peter Berczik[1,2,3]⬤, Manuel Arca Sedda[2]⬤, Konrad Maliszewski[4], Mirek Giersz[5]⬤ and Rainer Spurzem[2,6]

[1]Main Astronomical Observatory, National Academy of Sciences of Ukraine,
27 Akademika Zabolotnoho St., 03143 Kyiv, Ukraine
email (MS): sobolenko@mao.kiev.ua,

[2]Astronomisches Rechen Institut - Zentrum für Astronomie der Universität Heidelberg,
Mönchhofstrasse 12-14, D-69120 Heidelberg, Germany

[3]Konkoly Observatory, Research Centre for Astronomy and Earth Sciences, Eötvös Loránd
Research Network (ELKH), MTA Centre of Excellence, Konkoly Thege Miklósút 15-17, 1121
Budapest, Hungary

[4]Astronomical Observatory, Warsaw University, al. Ujazdowskie 4, 00-478 Warsaw, Poland

[5]Nicolaus Copernicus Astronomical Center, Polish Academy of Sciences,
ul. Bartycka 18, 00-716 Warsaw, Poland

[6]Kavli Institute for Astronomy and Astrophysics, Peking University,
Yiheyuan Lu 5, Haidian Qu, 100871, Beijing, China

Abstract. Based on our current high resolution direct N-body modelling of the Milky Way typical Star Cluster systems dynamical evolution we try to numerically estimate the influence of individual spin values and orientations on gravitational wave (GW) waveforms and observed time-frequency maps during multiple cycles for binary black hole (BBH) mergers. In our up to date N-body dynamical simulations we use the high order relativistic post-Newtonian corrections for the BH binary particles (3.5 post-Newtonian (PN) terms including spin-spin and spin-orbit terms). In the current work, we present the GW waveforms catalogue which covers the large parameter space in mass ratios 0.05 - 0.82 and extreme possible individual spin cases.

Keywords. black hole physics, gravitational waves, methods: n-body simulations

---

1. Introduction

Currently, over 50 gravitational events from compact binaries were reported by the LIGO-Virgo-Kagra consortium (Abbott B. P. 2019; Abbott R. 2021; The LIGO Scientific Collaboration et al. 2021a)†. We see Gravitational Waves (GW) as a new and very powerful informational channel. The current GW observations contain the Binary Black Hole (BBH) systems key orbital parameters, such as mass, semi-major axis, eccentricity and even the possible spins of the BH's. The next 3G generation of ground-based observatories (Punturo et al. 2010; Abbott et al. 2017) will have the opportunity to work with GWs during multiple cycles. It can significantly improve the estimations of individual component parameters of BH's.

---

† Just before the conference started the new catalogue GWTC-3 was presented (The LIGO Scientific Collaboration et al. 2021b).

2. Binary black hole initial models

To be more physically motivated, we took initial physical parameters (such as individual masses, initial separation and eccentricity) for 16 BBHs from N-body (Arca-Sedda et al. 2021) (with Id 1-4 in Table 1) and MOCCA (Maliszewski et al. 2021) simulations (with Id 5-16 in Table 1). Gravitational wave transient catalogues GWTC-1 (Abbott B. P. 2019), GWTC-2 (Abbott R. 2021) and GWTC-2.1 (The LIGO Scientific Collaboration et al. 2021a) contain systems with more than 50 % credibility of non-zero individual BH spin which allows us to investigate such systems. Initial binary orbit lies on XY plane, and each BH has individual spin:

$$|\boldsymbol{S_{0,1}}| = \chi_{0,1} \frac{G m_{0,1}^2}{c}, \tag{2.1}$$

where G is gravitational constant, $m_{0,1}$ is individual BH mass, and $\chi_{0,1} \in [0,1]$ is dimensionless spin magnitude (shortly "spin" throughout this paper). The dimensionless spin magnitude was limited by values 0, 1 or -1 independently for two BHs in three directions x, y, z. The total number of relative spin combinations for two BHs is 49. Effective inspiral spin parameter (Damour 2001) measures the mass-averaged spin along the orbital momentum axis, that LIGO can infer from gravitational waveform:

$$\chi_{\text{eff}} = \frac{m_0 \chi_0 \cos \theta_0 + m_1 \chi_1 \cos \theta_1}{M}, \tag{2.2}$$

where $M = m_0 + m_1$ is total BBH mass, $\theta_{0,1}$ is angle between individual spin vector $\boldsymbol{S_{0,1}}$ and orbital angular momentum vector $\boldsymbol{L}$. Another value obtained from the observations is chirp mass (Cutler & Flanagan 1994):

$$\mathcal{M} = (m_0 m_1)^{3/5} (m_0 + m_1)^{-1/5}. \tag{2.3}$$

We also determine mass ratio as $q = m_1/m_0$, where $m_1 < m_0$.

For numerical dynamical evolution we used N-body ϕ-GPU code†, reducing number of particles to two BHs (Berczik et al. 2011). The acceleration for BH particle with post-Newtonian terms up to 3.5PN, spin-orbit (denoted SO) and spin-spin (denoted SS) terms (Blanchet 2006; Faye et al. 2006; Tagoshi et al. 2001; Buonanno et al. 2003) can be written in form:

$$\frac{d\boldsymbol{v}}{dt} = \boldsymbol{a}_{\text{N}} + \frac{\boldsymbol{a}_{\text{1PN}}}{c^2} + \frac{\boldsymbol{a}_{\text{1.5PN,SO}}}{c^2} + \frac{\boldsymbol{a}_{\text{2PN}}}{c^4} + \frac{\boldsymbol{a}_{\text{2PN,SS}}}{c^4} + \frac{\boldsymbol{a}_{\text{2.5PN}}}{c^5} + \frac{\boldsymbol{a}_{\text{2.5PN,SO}}}{c^5}$$
$$+ \frac{\boldsymbol{a}_{\text{3PN}}}{c^6} + \frac{\boldsymbol{a}_{\text{3.5PN}}}{c^7} + \mathcal{O}\left(\frac{1}{c^8}\right), \tag{2.4}$$

where $\boldsymbol{a}_{\text{N}}$ is Newtonian acceleration, $\boldsymbol{a}_{\text{1PN,2PN,3PN}}$ are conservative terms, $\boldsymbol{a}_{\text{2.5PN,3.5PN}}$ are dissipative terms, that respond to emission of GW. For the simple waveform calculation, we used the GW quadrupole term expression and obtained h_+ and $h_\times$ polarisation strains from h^{ij} tensor (Kidder 1995; Cutler 1998):

$$h^{ij} \approx \frac{4 G \mu}{D c^4} \left[v^i v^j - \frac{GM}{r} n^i n^j \right], \tag{2.5}$$

where $\mu = m_0 m_1 / (m_0 + m_1)$ is reduced binary mass, D is luminosity distance, c is light velocity, v^i and n^i are the relative velocity and normalized position vectors in this reference frame respectively.

3. Evolution of Spinning Black Holes

We obtained merging time T_{merge} for 16 BBHs (column (9) at Table 1) from simulations with low resolution in time. Spin post-Newtonian terms were neglected. BBHs were

† ftp://ftp.mao.kiev.ua/pub/berczik/phi-GPU/

Table 1. Parameters for simulated BBHs.

Id	M $\mathrm{M}_\odot$	$\mathcal{M}$ $\mathrm{M}_\odot$	m_0 $\mathrm{M}_\odot$	m_1 $\mathrm{M}_\odot$	q	a $\mathrm{R}_\odot$	e	T_{merge} yr	Reference
(1)	(2)	(3)	(4)	(5)	(6)	(7)	(8)	(9)	(10)
1	349	62.3	328	21	0.064	1.21	0.410	6.9×10^1	[1]
2	329	62.3	307	22	0.072	0.70	0.085	1.6×10^1	[1]
3	355	70.7	329	26	0.079	697.00	0.997	4.10948×10^5	[1]
4	307	60.4	285	22	0.077	36.70	0.955	4.1326×10^4	[1]
5	443.181	68.2	422.640	20.541	0.049	25.0410	0.95592	3.7787×10^3	[2]
6	595.418	168.9	510.350	85.068	0.167	633.5100	0.99914	3.340×10^2	[2]
7	143.036	41.4	121.800	21.236	0.174	24.7450	0.97453	6.2124×10^3	[2]
8	256.010	79.9	211.460	44.550	0.211	284.3800	0.99800	2.7372×10^3	[2]
9	349.159	113.7	282.670	66.489	0.235	562.5900	0.99886	2.1929×10^3	[2]
10	211.460	70.4	169.230	42.230	0.250	152.0200	0.99520	7.3900×10^3	[2]
11	216.321	81.6	157.980	58.341	0.369	28.3510	0.97000	3.3422×10^3	[2]
12	282.672	107.2	205.310	77.362	0.377	133.5300	0.99864	1.85×10^1	[2]
13	157.778	60.7	112.940	44.838	0.397	95.7140	0.99851	3.78×10^1	[2]
14	134.505	54.4	89.955	44.550	0.495	303.4000	0.99949	1.375×10^2	[2]
15	106.500	45.4	63.270	43.230	0.683	0.4666	0.16982	2.21×10^1	[2]
16	144.994	62.7	79.783	65.211	0.817	132.9000	0.99891	5.05×10^1	[2]

NOTE: Columns from left to right contain the following information: (1) - identifier for BBH; (2) - BBH total mass; (3) - BBH chirp mass; (4) and (5) - primary and secondary BH individual masses; (6) - mass ratio; (7) - initial binary separation; (8) - initial binary eccentricity; (9) - estimated merging time; (10) - source of the initial BBH data, where [1] is Arca-Sedda et al. (2021) and [2] is Maliszewski et al. (2021).

merged if separation fall below $5R_{\mathrm{Sch}}$, where $R_{\mathrm{Sch}} = 2GM/c^2$ is Schwarzschild radius. Simulations were rerun several times with higher time resolution each time for reaching the moment when separation fall below $\sim 1000R_{\mathrm{Sch}}$, this gives us $16 \times 3 = 48$ runs. From this point we turned on spin-spin and spin-orbit terms and simulated 16 binaries with 49 relative spin combinations with time resolution enough to have 100 points in each orbit, which gives us $16 \times 49 = 784$ runs. In summary, the number of simulations with low and high time resolutions is 832.

3.1. GW waveform and time frequency picture

For the BBH set the minimum merging time T_{merge} was observed for systems with spin direction opposite to orbital angular momentum: $\boldsymbol{S}_{0,1} \uparrow\downarrow \boldsymbol{L}$; the maximum merging time T_{merge} was observed for systems with same spin direction and orbital angular momentum: $\boldsymbol{S}_{0,1} \uparrow\uparrow \boldsymbol{L}$; which demonstrates known "hang up" effect† (Campanelli et al. 2006). A system with zero $\boldsymbol{S}_{0,1} = 0$ and antialigned spins $\boldsymbol{S}_{0,1} \swarrow\nearrow \boldsymbol{L}$ has an average merging time. The difference between maximum and minimum merging time is less than $\sim 10\%$ accounting just timescale from high time resolution runs.

The study aim was to understand how we can catch systems with non-zero individual spins. For each binary with different spin combinations, we obtained waveforms and time-frequency pictures for h_+ and $h_\times$ polarisations. The initial orbital angular moment $\boldsymbol{L}$ has a z-direction.

For purposes of illustration, we describe results for BBH Id 1 from Table 1. Resulting waveforms can be distinguish into two parts for evolution during multiple GW cycles. We see the continuously increasing waveforms till merging if a spin vector, at least of the more massive BH, has precisely the same or opposite direction than orbital angular momentum. But at the same time these systems have maximum value for effective spin parameter $\chi_{\mathrm{eff}} = 1$ in case of $\boldsymbol{S}_{0,1} \uparrow\uparrow \boldsymbol{L}$ and $\chi_{\mathrm{eff}} = -1$ in case of $\boldsymbol{S}_{0,1} \uparrow\downarrow \boldsymbol{L}$. This means that such systems can be easily specified from observations.

† Video illustration of this effect at the last 100 sec of merging for BBH Id 1 from Table 1 can be found by the link: `https://youtu.be/D0luUTebBzk`

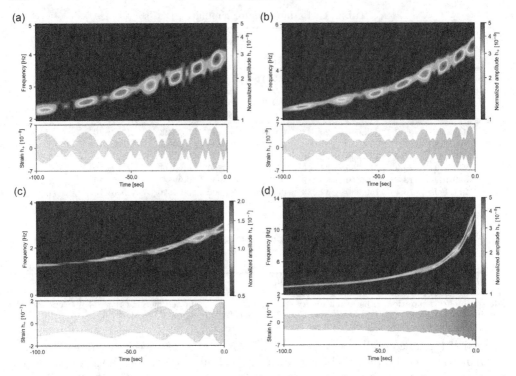

Figure 1. Time-frequency representation (top) of the strain data (bottom) for gravitational waveforms of h_+ polarisation from BBHs (Table 1): (a) - Id 5 ($q = 0.049$), (b) - Id 1 ($q = 0.064$), (c) - Id 6 ($q = 0.167$), (d) - Id 7 ($q = 0.174$). Data are depicted for the last 100 sec of merging. Individual dimensionless spin parameters are $\chi_0 = \chi_1 = [-1, 0, 0]$.

Other systems with antialigned spin vectors show continuously increase waveforms with bumps, we named this effect "waveforms beating". It's caused by more complex black holes orbits caused by the precession of the spin vectors. Also, this beating is observed in the time-frequency picture, where we see spotted-like-signal. It means that this type of system can also be specified from observations. It should be mentioned that such type of signal possibly can be caused by very eccentric binaries (in Keplerian sense) Romero-Shaw et al. (2021).

The remaining type is the systems with zero spin for more massive BH that have continuously increasing waveforms and $\chi_{\text{eff}} \approx 0$, which make them most hardly specifying from observations.

The described results are for a system with an extreme mass ratio $q = 0.064$. Systems with $q \approx 1$ do not show beating in the waveforms which make spin detections harder. We compared waveforms and time-frequency pictures for all our 16 binaries with mass ratios from 0.049 to 0.817 and found that from mass ratio 0.2 it is hard to observe waveforms beating (Fig. 1). Signal start to be more continuous at the time-frequency map with increasing mass ratio.

3.2. *Comparison with GWTC catalogues*

We plotted individual masses distribution for our BBH from Table 1 and observed merging BBH from GWTC-1, 2, 2.1 catalogues to find comparable systems. Three simulated systems Id 7, Id 15, Id 16 coincided in parameter space with observed events, and Table 2 contains combined literature data for them. Event GW190403_051519 has the

Table 2. Comparison of simulated and observed BBHs.

Id (1)	M, $M_\odot$ (2)	q (3)	χ_{eff} (4)	$\chi_{0,1}$ (5)
7	143.036	0.1743510		
GW190929_012149	$104.3^{+34.9}_{-25.2}$	$0.298^{+0.180}_{-0.613}$	$0.01^{+0.34}_{-0.33}$	
GW190403_051519	$110.5^{+30.6}_{-24.2}$	$0.251^{+0.138}_{-0.582}$	$0.70^{+0.15}_{-0.27}$	$\chi_1 = 0.92^{+0:07}_{-0.22}$
15	106.500	0.6832620		
GW190519_153544	$106.6^{+13.5}_{-14.8}$	$0.614^{+0.230}_{-0.340}$	$0.31^{+0.20}_{-0.22}$	
GW190701_203306	$94.3^{+12.1}_{-9.5}$	$0.757^{+0.319}_{-0.321}$	$-0.07^{+0.23}_{-0.29}$	
GW190706_222641	$104.1^{+20.2}_{-13.9}$	$0.570^{+0.265}_{-0.469}$	$0.28^{+0.26}_{-0.29}$	
16	144.994	0.8173550		
GW190521	$163.9^{+39.2}_{-23.5}$	$0.724^{+0.354}_{-0.476}$	$0.03^{+0.32}_{-0.39}$	$\chi_{i=\{1,2\}} > 0.8$ with 58% credibility.
GW190426_190642	$184.4^{+41.7}_{-36.6}$	$0.717^{+0.427}_{-0.542}$	$0.19^{+0.43}_{-0.40}$	

NOTE: Columns from left to right contain the following information: (1) - the identifier for BH pairs, where bold text denoted events from GWTC-2.1 (The LIGO Scientific Collaboration et al. 2021a), other events from GWTC-2 (Abbott R. 2021); (2) - BBH total mass; (3) - mass ratio; (4) - observed effective spin; (5) - observed individual spin.

highest credibility of non-zero effective spin, which can mean that at least more massive BH should have non-zero individual spin. Estimated individual spin for the more massive BH should be near the maximum value ~ 1. Another event GW190521 shows a non-zero individual spin. But high mass ratio $q \sim 0.7$ did not allow analysing the waveforms and time-frequency picture.

Also, we tried to analyse presented strain data by eyes and to find any type of beating at the waveforms. Events GW151226, GW170608, GW190412, GW190707_093326, GW190728_064510 were chosen as more promising. Only event GW190412 have a suitable parameters: possibility of non-zero effective spin $\chi_{\text{eff}} = 0.25^{+0.08}_{-0.11}$ and asymmetric mass ratio $q = 0.28^{+0.12}_{-0.06}$. Signal was observed during just several gravitational waves cycles (last 0.5 sec) which is very short to specify beating in waveforms. The next 3G generation of gravitational wave telescopes (Einstein Telescope (Punturo et al. 2010), Cosmic Explorer (Abbott et al. 2017)) will capture several thousands of cycles and we will have the opportunity to see the waveforms beating.

4. Conclusions

The main conclusions of this study can be summarized as follows.

(i) We obtained merging time T_{merge}, h_+ and $h_\times$ polarization waveforms and time-frequency maps for a set of BBHs with mass ratio $q = 0.064 - 0.82$ and 49 spin combinations. The "hang-up effect" was observed: minimum merging time was observed for systems with spin direction opposite to orbital angular momentum; the maximum merging time was observed for systems with the same spin direction and orbital angular momentum.

(ii) A system with beating in waveforms and spotted-like-signal at time-frequency picture should have non-zero individual spin $\chi_{1,2}$ at least for a more massive BH and non-symmetric mass ratio q. The transition point of the possibility to observe this effect is mass ratio $q \sim 0.2$. This effect can be possibly observed with the next 3G generation of gravitational wave telescopes.

(iii) A system with non-zero effective spin χ_{eff} should have non-zero spin $\chi_{1,2}$ at least for a more massive BH.

(iv) A system with zero spin for a more massive BH is most hardly specified from observations.

5. Acknowledgements

MS thanks the International Astronomical Union for the grant support to participate in the online video-conference IAU Symposium No. 362. The work of MS and PB was supported under the special program of the NRF of Ukraine "Leading and Young Scientists Research Support" - "Astrophysical Relativistic Galactic Objects (ARGO): life cycle of active nucleus", No. 2020.02/0346. The work of PB was supported by the Volkswagen Foundation under the special stipend No. 9B870 (2022). The work of MS and PB was also supported by the Volkswagen Foundation under the Trilateral Partnerships grant No. 97778. MG and KM were partially supported by the Polish National Science Center (NCN) through the grant UMO-2016/23/B/ST9/02732.

References

Abbott B. P., et al., 2017, Classical and Quantum Gravity, 34, 044001

Abbott B. P., et al., 2019, Physical Review X, 9, 031040

Abbott R., et al., 2021, Physical Review X, 11, 021053

Arca-Sedda M., Rizzuto F. P., Naab T., Ostriker J., Giersz M., Spurzem R., 2021, ApJ, 920, 128

Berczik P., et al., 2011, in International conference on High Performance Computing. pp 8–18

Blanchet L., 2006, Living Reviews in Relativity, 9, 4

Buonanno A., Chen Y., Vallisneri M., 2003, Phys. Rev. D, 67, 104025

Campanelli M., Lousto C. O., Zlochower Y., 2006, Phys. Rev. D, 74, 041501

Cutler C., 1998, Phys. Rev. D, 57, 7089

Cutler C., Flanagan É. E., 1994, Phys. Rev. D, 49, 2658

Damour T., 2001, Phys. Rev. D, 64, 124013

Faye G., Blanchet L., Buonanno A., 2006, Phys. Rev. D, 74, 104033

Kidder L. E., 1995, Phys. Rev. D, 52, 821

Maliszewski K., Giersz M., Gondek-Rosińska D., Askar A., Hypki A., 2021, arXiv e-prints, p. arXiv:2111.09223

Punturo M., et al., 2010, Classical and Quantum Gravity, 27, 194002

Romero-Shaw I., Lasky P. D., Thrane E., 2021, ApJ, 921, L31

Tagoshi H., Ohashi A., Owen B. J., 2001, Phys. Rev. D, 63, 044006

The LIGO Scientific Collaboration et al., 2021a, arXiv e-prints, p. arXiv:2108.01045

The LIGO Scientific Collaboration et al., 2021b arXiv e-prints, p. arXiv:2111.03606

The Predictive Power of Computational Astrophysics as a Discovery Tool
Proceedings IAU Symposium No. 362, 2023
D. Bisikalo, D. Wiebe & C. Boily, eds.
doi:10.1017/S1743921322001739

Predicting the Expansion of Supernova Shells Using Deep Learning toward Highly Resolved Galaxy Simulations

Keiya Hirashima[1]†, **Kana Moriwaki**[2], **Michiko Fujii**[1],
Yutaka Hirai[3,4,5], **Takayuki Saitoh**[6] and **Junichiro Makino**[6,5]

[1]Department of Astronomy, Graduate School of Science, The University of Tokyo, 7-3-1 Hongo, Bunkyo-ku, Tokyo 113-0033, Japan

[2]Department of Physics, Graduate School of Science, The University of Tokyo, 7-3-1 Hongo, Bunkyo-ku, Tokyo 113-0033, Japan

[3]Department of Physics, University of Notre Dame, 225 Nieuwland Science Hall, Notre Dame, IN 46556, USA

[4]Astronomical Institute, Tohoku University, 6-3, Aramaki, Aoba-ku, Sendai, Miyagi 980-8578, Japan

[5]RIKEN Center for Computational Science, 7-1-26 Minatojima-Minami-machi, Chuo-ku, Kobe, Hyogo 650-0047, Japan

[6]Department of Planetology, Graduate School of Science, Kobe University, 1-1 Rokkodai-cho, Nada-ku, Kobe, Hyogo 657-8501, Japan

Abstract. The load imbalance and communication overhead of parallel computing are crucial bottlenecks for galaxy simulations. A successful way to improve the scalability of astronomical simulations is a Hamiltonian splitting method, which needs to identify such regions integrated with smaller timesteps than the global timestep for integrating the entire galaxy. In the case of galaxy simulations, the regions inside supernova (SN) shells require the smallest steps. We developed the deep learning model to forecast the region affected by the SN shell's expansion during one global step. In addition, we identified the particles with small timesteps using image processing. We can identify target particles using our method with a higher identification rate (88 % to 98 % on average) and lower "non-target"-to-"target" fraction (6.4 to 5.5 on average) compared to the analytic approach with the Sedov-Taylor solution. Our method using Hamiltonian splitting and deep learning will improve the performance of extremely high-resolution galaxy simulations.

Keywords. Deep Learning, Supernova, Galaxy: formation

---

1. Introduction

Galaxies have formed via several physical processes such as gravitational and hydrodynamic forces, radiative cooling and heating, star formation, supernova explosions, and chemical evolution. Galaxy formation simulations have been performed using N-body/hydrodynamics simulations such as smoothed particle hydrodynamics (SPH; Gingold and Monaghan 1977; Lucy 1977), which is a particle-based method using gas particles smoothed with a kernel size depending on the local density.

† E-mail: `hirashima.keiya@astron.s.u-tokyo.ac.jp`

The resolution of N-body/SPH simulations depends on the number of star and gas particles. The number of stars in the Milky-Way galaxy exceeds 10^{10}. We need more than 10^{10} particles to resolve individual stars (star-by-star simulation).

As higher-performance supercomputers have been developed, a higher resolution has been achieved. The highest resolution for large-scale galaxy simulations, IllustrisTNG (Weinberger et al. 2017; Pillepich et al. 2018) with 10^{10} particles were now within reach using 25,000 CPU cores.

A crucial bottleneck of galaxy simulations is the communication overhead. In massively parallel computing using more than ~ 1000 CPU cores, the communication takes longer than the calculations (see Figure 63 in Springel et al. 2021) simply because the communication costs increase as the parallelization degree increases.

Another problem is the worsening scalability due to the load imbalance. With a better resolution, we can resolve smaller-scale phenomena, and therefore we have to integrate them with a smaller timestep. Unfortunately, this increases the number of integration steps necessary for one simulation since the intended simulation time does not change. Additionally, the particles with small timesteps are a tiny fraction of the entire system. Such a situation worsens the scalability in parallel computing. Thus, we need to solve these problems to achieve star-by-star galaxy simulations.

A successful way to improve the scalability of astronomical simulations is Hamiltonian splitting, which allows us to integrate small-step regions separately. For example, in N-body simulations of a cluster in the galaxy Fujii et al. 2007 used each integration scheme for a cluster and the host galaxy derived with the Hamiltonian splitting. This method can also be used for galaxy simulations. Through Hamiltonian splitting, we need to pick up particles in the small-step regions, which must be integrated with small timesteps compared to the global step (the integration timestep for the entire galaxy). In the case of galaxy simulations, the regions inside supernova (SN) shells require the smallest steps. We need to integrate these regions for the decrease of the communication overhead.

The time evolution of SN's shell is given by an analytical solution (Sedov 1959) in isotropic and uniform interstellar medium (ISM). Here, we derive the radius R of a shell with the released energy E in the uniform density ρ at a specific time t. Introducing the dimensionless similarity variable ξ, the radius R is written as the following;

$$R(t) = \xi \left(\frac{E}{\rho}\right)^{1/5} t^{2/5}. \tag{1.1}$$

However, the real ISM is neither isotropic nor uniform. One way to improve the analytical approach is by dividing the surrounding gas into several regions and applying the analytical solution (Equation 1.1) using each density.

To improve the selection of the small-step regions, we develop a method using a deep learning method. In this study, we propose a computer vision approach to forecast the expansion of SN shells in inhomogeneous ISM using deep learning and to identify "target" particles using image processing. Here, "target" particles are the particles that will have small timesteps in the subsequent global timestep compared to the global time step.

2. Data Preparation

Our objective was to develop a deep learning model to predict the shell expansion of SN. We made the training data from hundreds of simulations of a SN explosion in inhomogeneous (turbulent) gas distribution using our SPH code, ASURA-FDPS (Saitoh et al. 2008; Iwasawa et al. 2016). We use SPH particles of the mass of 10 $M_\odot$ and the temperature of 100 K, which is the target resolution of our galaxy simulation. We assume a point source of explosions. The thermal energy of 10^{51} erg is injected into the center of gas distributions. The softening parameter is set to be 3 pc.

Table 1. The dataset settings for our deep learning model. We made datasets from simulations of SN explosions in the gas cloud. We classify datasets into three types according to the mean density $\bar{\rho}$ and spacial symmetries of gas clouds.

Datasets	$\bar{\rho}$ [cm^{-3}]	Spatial Symmetries	t [Myr]
Fiducial	1.864×10^2	non-uniform & anisotropic	0.133
Spherical	1.864×10^2	uniform & isotropic	0.133
Sparse-Spherical	3.728×10	uniform & isotropic	0.133

Table 2. The setup for mathematical morphology operations: "Dilation", "Erosion", "Gradients" and a self-made operator "Majority" (see the text). In the column of iterations, D_3, D_5, and E are the numbers of iterations varied in the experiments.

Operators	kernel or threshold	iterations
Dilation	(3,3)	1
Gradients	(3,3)	1
Dilation	(3,3)	D_3 (e.g. 1)
Dilation	(5,5)	D_5 (e.g. 3)
Majority	≥ 2	-
Erosion	(3,3)	E (e.g. 1)
Dilation	(3,3)	$E + 1$ (e.g. 2)
Majority	≥ 2	-

Table 3. The settings for experiments. In each experiment, we use training data and test data of SN explosions in the gas cloud of the mean density $\bar{\rho}$. Results of simulations are predicted up to the result of forecast horizon t in Table 1. We used the same trained model in Experiment 1, 2, and 3. The names of datasets correspond to those in Table 1.

Experiment	Training Data	test data
Experiment 1	Fiducial	Fiducial
Experiment 2	Fiducial	Spherical
Experiment 3	Fiducial	Sparse-Spherical

We use a data format of three-dimensional (3D) volume images for deep learning. We perform 300 simulations of supernova explosions described above and obtain 20 snapshots with a timestep $dt = 7.0 \times 10^{-3}$ Myr (Fiducial). The initial conditions are density distributions time-evolved for a specific time, in which gas filamentary structures are formed after a core collapse, from a gas sphere with turbulence. By smoothing particles with SPH kernels of size depending on the local densities, we get 3D volume images composed of 32^3 voxels† with resolutions of 1.875 pc. A sequence data are converted from each simulation containing 20 volume images.

3. Computer Vision Approach

Our deep learning model is based on Memory In Memory network‡ by Wang et al. (2018), which utilizes differential signals effectively to archive high predictability in 2D video prediction. We improved the original MIM in the following two points. First, we increased the internal dimension from two to three to deal with the 3D volume images. Second, the original MIM is the many-to-many model with sequential inputs and outputs. We designed our model as the one-to-many model to have one input and sequential outputs. We call our model 3D-MIM.

We predict the regions affected by a SN explosion by applying image processing. The following processing is performed for individual 2D slices of the 3D volume image. We

† The *voxel* in a 3D data is equivalent to the pixel in a two-dimensional (2D) image.
‡ https://github.com/Yunbo426/MIM

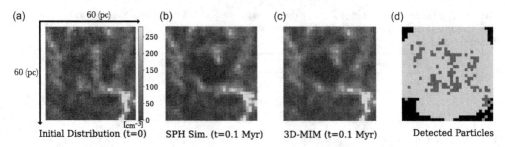

(a)   60 (pc)   (b)   (c)   (d)

60 (pc)

[cm⁻³]

Initial Distribution (t=0)   SPH Sim. (t=0.1 Myr)   3D-MIM (t=0.1 Myr)   Detected Particles

Figure 1. An example of forecast results by our deep learning model. One side of each panel corresponds to 60 pc. Color maps show the density distribution. The color bar and scale are the same in all panels. Panel (a) shows the initial distribution. Panel (b) shows the simulation result 0.1 Myrs after the supernova explosion (ground truth). Panel (c) shows the forecast result using our deep learning model. Panel (d) shows the forecast area and distribution of "target" particles. (Dark grey: "target" particles, Pale grey: "forecast area").

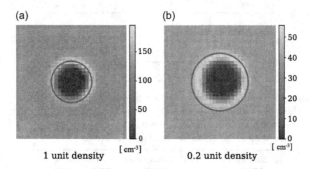

(a)   (b)

1 unit density   [cm⁻³]   0.2 unit density   [cm⁻³]

Figure 2. The test results of shell expansion 0.1 Myrs after an SN explosion with Experiment 2 (left) and Experiment 3 (right) in Table 3. The predictions are performed by the 3D-MIM. Red circles show the analytical solution (the Sedov-Taylor solution) of the expanding shell's radius.

calculate the pixel-by-pixel quotient of the densities just before the explosion and the densities predicted by the 3D-MIM. Then, we assign 1 to a pixel with a quotient less than 0.9 and 0 otherwise. After the binarization, we find several blobs¶ of pixels assigned 1. Among them, we give 1 to the blob whose barycenter is closest to the source of the explosion and 0 to the other pixels. We then apply three types of mathematical morphology operators: "Dilation", "Erosion", and "Gradients", and a self-made operator: "Majority". To optimize the order of operators applied to the image, we examine several different configurations with the order of operators shown in Table 2.

In practice, we perform the above process on 2D slices along each of the three axes and sum up the resulting three binarized volume images, obtaining a single volume image whose voxels have values from 0 to 3. We then process them using "Majority", which assigns 1 to each voxel with a value more than a threshold. Finally, we identify the particles inside them as those that require small timesteps in the future.

4. Results

4.1. *Forecasting the Expansion of a Supernova Shell*

Figure 1 shows an example of the forecast result using our deep learning model (Experiment 1; Table 3). With an initial distribution [Figure 1 (a)] is given as the input, our 3D-MIM forecasts the density distribution after 0.1 Myr [Figure 1 (c)].

¶ We call a chunk of four or more connected pixels a blob.

For further examination, we test our model with uniform density distributions for which analytical solutions are obtained using Equation (1.1). Figure 2 shows the test results of Experiments 2 and 3 at 0.1 Myr in Table 3. In these cases, the background gas is uniform and isotropic media. Experiment 2 has the unit density $(1.864 \times 10^2$ cm^{-3}, left), i.e., the same mean density as the training data, whereas Experiment 3 has 0.2 times as unit density $(3.728 \times 10$ cm^{-3}, right). The red circles show the Sedov solution calculated using Equation (1.1). The forecast result agrees well with the radius predicted by the Sedov solution when the unit density of the input is similar to the trained data. However, the model is not trained with such a uniform distribution. When the input density is smaller than the trained data, the shell expansion becomes smaller than the analytical solution. This indicates that the model has not learned the scaling law. However, it is practically not a problem for our purpose that is to find particles with small timesteps, and therefore we do not need very accurate predictions of the expansion of SN's shells.

4.2. *Identification of Target Particles*

To fit our method with particle simulations, we need to extract particles that will evolve with smaller timesteps under the effect of SN from forecast results by deep learning. We call the particles that will require smaller timesteps due to a SN explosion "target" particles and others "non-target" particles. We define the particles of which timesteps are smaller than the global timestep and temperatures are higher than 100 K as "target".

In the analytic approach, we calculate the radius of the shell at a subsequent global timestep in each direction using each mean density of the 20 tetrahedral domains and Equation (1.1) under the icosahedron domain decomposition. On the other hand, in the computer vision approach, we forecast the variation in the spatiotemporal sequence of density distribution due to SN shell using 3D-MIM and decide the region where the density will decrease more than a threshold at the subsequent global timestep. 2D slices made of 3D volume images representing density distributions are processed using some morphology operators to predict the distribution of "target" particles. Figure 1 (d) shows the forecast region and distribution of "target" particles. It can be found that the forecast region covers most "target" particles.

Figure 3 shows the comparison of the detectability of the "target" particles at global timesteps in the (a) analytic and (b) computer vision approaches, respectively. The vertical axis shows the ratio of the number of "non-target" particles to that of "target" particles inside the forecast region. The horizontal axis shows the ratio of the number of the successfully detected "target" particles to all "target" particles.

Compared to the analytic approach, the identification rate of the computer vision approach is better (88 % to 98 % on average and 3 % to 1 % on standard derivation), and the scattering is more negligible (6.4 to 5.5 on average and 0.93 to 0.64 on standard derivation), while the "non-target" ratio is similar for both approaches. This result shows that the computer vision approach can follow the complex change in the gas structure better than the analytic approach.

The computer vision approach itself costs a larger run-time than the analytic approach. However, when combined with the computer vision approach, consistent high-resolution galaxy simulations can be achieved in a realistic time.

5. Conclusion

We developed a new algorithm, composed of deep learning and image processing, that improves one step in the challenge of overcoming the difficulties of high parallel computing in high-resolution galaxy simulations. Our new deep learning model successfully forecasted the density distribution after an SN explosion in the non-uniform and

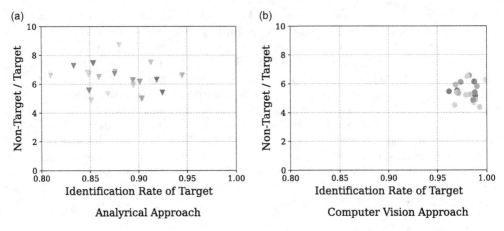

Figure 3. The comparison of the (a) analytical and (b) computer vision (CV) approach (Experiment 1) to identify "target" particles for the global timestep $T = 0.1$ Myr. The same color indicates the same simulation data. The vertical axis shows the number ratio of the "non-target" particles to the "target" particles inside the forecast region. The horizontal axis shows the number ratio of the successfully detected "target" particles to all "target" particles.

anisotropy gas cloud until 0.1 Myr. The forecast result for uniform distributions also agrees well with the analytic solution when the unit density of the input is similar to the training data, although the model is not trained using such a uniform distribution. In addition, by image processing, our new algorithm can identify "target" particles while excluding "non-target" particles better than the method based on the analytic solution. The average increases from around 87% to around 98% (as seen in Fig. 3). We are implementing the method under a parallel computing environment and trying to run it within a running time comparative to the analytic approach. We will include this method in our N-body/SPH code, ASURA-FDPS (Saitoh et al. (2008); Iwasawa et al. (2016)), and perform a star-by-star galaxy simulation using a massively parallel computer like Fugaku.

References

Fujii, M., Iwasawa, M., Funato, Y., & Makino, J. 2007, *PASJ*, 59, 1095.

Gingold, R. A. & Monaghan, J. J. 1977, *MNRAS*, 181, 375–389.

Iwasawa, M., Tanikawa, A., Hosono, N., Nitadori, K., Muranushi, T., & Makino, J. 2016, *PASJ*, 68(4), 54.

Lucy, L. B. 1977, *AJ*, 82, 1013–1024.

Pillepich, A., Springel, V., Nelson, D., Genel, S., Naiman, J., Pakmor, R., Hernquist, L., Torrey, P., Vogelsberger, M., Weinberger, R., & Marinacci, F. 2018, *MNRAS*, 473(3), 4077–4106.

Saitoh, T. R., Daisaka, H., Kokubo, E., Makino, J., Okamoto, T., Tomisaka, K., Wada, K., & Yoshida, N. 2008, *PASJ*, 60(4), 667–681.

Sedov, L. I. 1959,.

Springel, V., Pakmor, R., Zier, O., & Reinecke, M. 2021, *MNRAS*, 506(2), 2871–2949.

Wang, Y., Zhang, J., Zhu, H., Long, M., Wang, J., & Yu, P. S. 2018, *arXiv e-prints*, arXiv:1811.07490.

Weinberger, R., Springel, V., Hernquist, L., Pillepich, A., Marinacci, F., Pakmor, R., Nelson, D., Genel, S., Vogelsberger, M., Naiman, J., & Torrey, P. 2017, *MNRAS*, 465(3), 3291–3308.

The Predictive Power of Computational Astrophysics as a Discovery Tool
Proceedings IAU Symposium No. 362, 2023
D. Bisikalo, D. Wiebe & C. Boily, eds.
doi:10.1017/S1743921322001831

Toward Realistic Models of Core Collapse Supernovae: A Brief Review

Anthony Mezzacappa[iD]

Department of Physics and Astronomy University of Tennessee, Knoxville, USA
Nielsen Physics Building – 401 1408 Circle Drive Knoxville, TN 37996-1200
email: `mezz@utk.edu`

Abstract. Motivated by their role as the direct or indirect source of many of the elements in the Universe, numerical modeling of core collapse supernovae began more than five decades ago. Progress toward ascertaining the explosion mechanism(s) has been realized through increasingly sophisticated models, as physics and dimensionality have been added, as physics and numerical modeling have improved, and as the leading computational resources available to modelers have become far more capable. The past five to ten years have witnessed the emergence of a consensus across the core collapse supernova modeling community that had not existed in the four decades prior. For the majority of progenitors – i.e., slowly rotating progenitors – the efficacy of the delayed shock mechanism, where the stalled supernova shock wave is revived by neutrino heating by neutrinos emanating from the proto-neutron star, has been demonstrated by all core collapse supernova modeling groups, across progenitor mass and metallicity. With this momentum, and now with a far deeper understanding of the dynamics of these events, the path forward is clear. While much progress has been made, much work remains to be done, but at this time we have every reason to be optimistic we are on track to answer one of the most important outstanding questions in astrophysics: How do massive stars end their lives?

Keywords. supernovae: general, neutrinos, hydrodynamics, relativity

---

1. Introduction

Core collapse supernovae are directly or indirectly responsible for the lion's share of the elements in the Universe. As such, they are among the most important astrophysical phenomena to be studied and understood. There is a long history to core collapse supernova modeling, beginning with the work of Colgate and White 1966. In the fifty six years since, such modeling has progressed considerably, shedding light on the key phenomena responsible for these explosions. This progress has been very encouraging to the core collapse supernova modeling community, but much work remains, as we will discuss here.

Contemporary core collapse supernova theory (at least for the lion's share of such supernovae, which originate from slowly rotating progenitors) centers around the question of how the supernova shock wave, which forms as the result of stellar core collapse and bounce at super-nuclear densities and stalls as the result of the enervating processes of nuclear dissociation and neutrino losses, is revived. Oddly enough, the phenomena on which contemporary core collapse supernova theory rests entered the picture one at a time, about once every decade. While Colgate and White were the first to propose that core collapse supernovae could be neutrino driven, present-day efforts can be traced back to the work of Wilson in 1982, as documented in Wilson 1985 and further elaborated in Bethe and Wilson 1985. Wilson demonstrated that electron-neutrino and -antineutrino

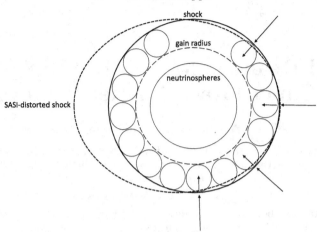

Figure 1. After stellar core bounce and shock formation, the core is stratified into regions defined by the proto-neutrino star surface (neutrinospheres), the gain radius, and the stalled shock. The region between the neutrinospheres and the gain radius is a net neutrino cooling region. The region between the gain radius and the shock is a net neutrino heating region, also known as the gain region. Shown also, schematically, are convection in the gain region resulting from neutrino heating by the proto-neutron star below it, and the distortion of the shock due to the Standing Accretion Shock Instability (SASI).

absorption on neutrons and protons below the shock, respectively, newly liberated from core nuclei by shock dissociation, can deposit sufficient energy to render the accretion shock a dynamical shock once again. Figure 1 illustrates the stratification of the stellar core region below the shock shortly after bounce. Neutrinos and antineutrinos of all three flavors emerge from the proto-neutron star. The region between the proto-neutron star surface and the shock divides into a net neutrino cooling region, due to the inverse of the weak interactions responsible for heating the material, and a net neutrino heating region above it. Neutrino heating and cooling balance at the gain radius. The gain region is defined as the region between the gain radius and the shock. A decade later, core collapse supernova models were freed of the constraints imposed by spherical symmetry when the first two-dimensional simulations of Herant et al. 1992 were performed. Enter neutrino-driven convection. Not surprisingly, the post-shock material heated from below by neutrinos emerging from the neutrinospheres defining the surface of the proto-neutron star becomes convectively unstable, with multiple benefits with regard to aiding neutrino shock reheating. Most important, now continued accretion, which fuels the neutrino luminosities that drive the explosion, can continue while explosion develops. This is not possible in spherical symmetry. In 2003, yet another instability entered the picture. In the context of core-collapse-supernova-informed axisymmetric hydrodynamics studies, Blondin et al. 2003 discovered that the supernova shock wave itself may become unstable to non-radial perturbations. This was confirmed in all subsequent axisymmetric core collapse supernova simulations. In axisymmetry, the instability is manifest in an $\ell = 1$ "sloshing" mode. In three dimensions, the sloshing mode is joined by an $m = 1$ "spiral" mode (Blondin and Mezzacappa 2007). As illustrated in Figure 1, the SASI can increase the size of the gain region and, with it, post-shock neutrino heating. Neutrino-driven convection in the gain region becomes turbulent. In 2013, Murphy et al. 2013 were the first to demonstrate quantitatively that the resulting turbulent ram pressure in this region assists the thermal pressure in the region in driving the shock outward. Rotation and magnetic field effects round out the list of phenomena that play a role in the explosion mechanism, though much of the progress we will discuss here was accomplished without them. Not surprisingly, the centrifugal effects of rotation, albeit slow rotation for the majority of massive stars, aid shock revival and explosion (e.g., see Summa et al. 2018). And sufficient progress has been made to include magnetic fields in core collapse supernova models that we now know that magnetic effects are non-negligible, as well (e.g., see Obergaulinger et al. 2015). Magnetic field strengths in the postshock region are amplified in the absence of rotation by collapse, convection, and turbulence, and by rotation when rotation is present, leading to magnetic stresses, like turbulent stresses, that aid in moving the shock outward to larger radii. The effects of magnetic fields become even more pronounced in the presence of significant core rotation, with outcomes changing qualitatively when magnetic fields are included (e.g., see Kuroda et al. (2020)).

2. Requirements

More than two decades ago, Bruenn et al. 2001 and Liebendörfer et al. 2001 demonstrated, by comparing Newtonian and general relativistic simulation outcomes directly, that core collapse supernovae are general relativistic phenomena. Newtonian approximations to gravity, hydrodynamics (or magnetohydrodynamics if magnetic fields are included), and neutrino kinetics are, generally speaking, not realistic. With general relativistic treatments of all three, the stratification shown in Figure 1 is much more compact, with significantly reduced radii for the neutrinospheres, and gain and shock radii, giving rise also to a significantly reduced gain region volume. In addition, infall velocities ahead of the shock are larger. On the plus side of the ledger, the neutrinospheres are hotter, increasing the luminosities of the emergent neutrinos, as well as their RMS energies, though gravitational redshift will downgrade the neutrino spectra as they propagate outward through the cooling and heating layers.

The neutrino heating rate per gram in the gain region is given by

$$\dot{\epsilon} = \frac{X_n}{\lambda_0^a} \frac{L_{\nu_e}}{4\pi r^2} \langle E_{\nu_e}^2 \rangle \left\langle \frac{1}{\mathcal{F}_{\nu_e}} \right\rangle + \frac{X_p}{\bar{\lambda}_0^a} \frac{L_{\bar{\nu}_e}}{4\pi r^2} \langle E_{\bar{\nu}_e}^2 \rangle \left\langle \frac{1}{\mathcal{F}_{\bar{\nu}_e}} \right\rangle, \qquad (2.1)$$

where ϵ is the internal energy of the stellar core fluid per gram, $X_{n,p}$ are the neutron and proton mass fractions, respectively, $L_{\nu_e,\bar{\nu}_e}$ are the electron-neutrino and -antineutrino luminosities, respectively, $\mathcal{F}_{\nu_e,\bar{\nu}_e}$ are the inverse flux factors for the electron-neutrinos and -antineutrinos, respectively, and $\lambda_0^a, \bar{\lambda}_0^a$ are constants related to the weak interaction coupling constants. Thus, knowledge of the neutrino luminosities, spectra, and angular distributions are needed to compute the neutrino heating rates. This requires knowledge of the neutrino distribution functions, $f_{\nu_e,\bar{\nu}_e}(r, \theta, \phi, E, \theta_p, \phi_p, t)$, from which these quantities can be calculated. The neutrino distribution functions are determined by solving their respective Boltzmann kinetic equations. Thus, the core-collapse supernova problem is a phase space problem, in the end involving 6 dimensions plus time: 3 spatial dimensions (e.g., r, θ, and ϕ) and 3 momentum-space dimensions (as we will see: neutrino energy, a direction cosine, and an additional momentum-space angle).

It is not feasible at present, even on today's leadership-class supercomputing architectures, to perform multi-second (later we will see what time scales must be considered) core collapse supernova simulations with Boltzmann neutrino kinetics. Instead, today's leading core collapse supernova models are based on solutions to the equations for the neutrino angular moments, defined in terms of the neutrino distribution function. The moment equations are obtained by integrating the Boltzmann kinetic equations over neutrino angle. A two-moment approach is the canonical implementation, in which one solves for the spectral neutrino number or energy density and the spectral neutrino number or energy flux in each of three dimensions – i.e., the lowest four angular moments of the neutrino distribution. Important angular information encoded by the neutrino distribution functions is kept, but not all of it. For example, the spectral number density and the three spectral number fluxes, one for each of the three spatial dimensions, are defined in terms of the distribution function as

$$\mathcal{N}(r, \theta, \phi, E, t) \equiv \int_0^{2\pi} d\phi_p \int_{-1}^{+1} d\mu f(r, \theta, \phi, \mu, \phi_p, E, t), \qquad (2.2)$$

$$\mathcal{F}^i(r, \theta, \phi, E, t) \equiv \int_0^{2\pi} d\phi_p \int_{-1}^{+1} d\mu n^i f(r, \theta, \phi, \mu, \phi_p, E, t), \qquad (2.3)$$

respectively, where $\mu \equiv \cos \theta_p$ is the neutrino direction cosine defined by θ_p, one of the angles of propagation defined in terms of the outward pointing radial vector defining the neutrino's position at time t. In three dimensions, two angles are needed to uniquely define a neutrino propagation direction. The angle ϕ_p provides the second. n^i is the component of the neutrino direction cosine in the ith direction, given as a function of μ and ϕ_p. E is the neutrino energy. E, θ_p, ϕ_p can be viewed as spherical momentum-space coordinates. Integration of the neutrino Boltzmann equation over the angles θ_p and ϕ_p, weighted by 1, n^i, $n^i n^j$, ... defines an infinite set of evolution equations for the infinite number of angular moments of the distribution function, which is obviously impossible to solve. In a moments approach, the infinite set of equations is rendered finite by truncation, after the equation for the zeroth moment in the case of one-moment closure (e.g., flux-limited diffusion) or after the equations for the first moments in the case of two-moment closure (e.g., M1 closure). Neutrinos are, of course, Fermions, and Fermions obey Fermi–Dirac statistics. As a result, closure of the system of moment equations must also obey Fermi–Dirac statistics. It must be *realizable*. But not all of the closures implemented in leading core-collapse supernova models to date are. Commonly used closures such as the Minerbo, Levermore, and Cernohorsky–Bludman closures obey Maxwell–Boltzmann, Bose–Einstein, and Fermi–Dirac statistics, respectively, with the Minerbo closure being the one most commonly deployed. This does not present an issue at low occupancies. Rather, it is in the important region deep within the proto-neutron star, where neutrino occupancies are large and their fluxes are small, where we can expect a breakdown of realizability (see Chu et al. (2019)).

The coupling of the neutrinos to the stellar core matter is mediated by what is now known to be an extensive set of weak interactions. Charged-current electron-neutrino and -antineutrino absorption on nucleons behind the shock powers the supernova, but the neutrino luminosities and RMS energies emerging from the neutrinospheres, which enter the heating rate, Equation (2.1), are defined by a much larger set of charged- and neutral-current interactions – critical among these: non-isoenergetic scattering on nucleons, electron–positron annihilation, and nucleon–nucleon bremsstrahlung. For a complete listing, the reader is referred to Mezzacappa et al. 2020. As we will discuss later, the set of relevant neutrino weak interactions continues to evolve. With regard to the the neutrino weak interactions, two things must be emphasized: (1) All of the relevant neutrino interactions in the above-cited list must be included. (2) State of the art implementations of each of the interactions, as described in Mezzacappa et al. 2020, must be used.

3. The Current State of the Art

Despite the challenges the core collapse supernova modeling community has faced and that, to a certain extent, still lie ahead, dictated by the rather austere requirements just discussed, we are arguably in a very encouraging period of core collapse supernova theory. At present, the efficacy of neutrino shock revival has been demonstrated by all leading core collapse supernova modeling groups worldwide, across progenitor characteristics such as mass, metallicity, and rotation (Hanke et al. 2013; Lentz et al. 2015; Melson et al. 2015b,a; Summa et al. 2018; Roberts et al. 2016; Kuroda et al. 2016; O'Connor and Couch

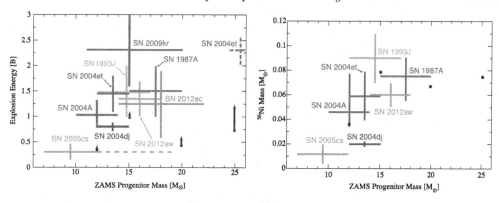

Figure 2. Explosion energies and ^{56}Ni mass produced (black dots) for four axisymmetric core collapse supernova models for progenitors of 12, 15, 20, and 25 M$_\odot$, plotted against observations as a function of ZAM mass (from Bruenn et al. 2016). The black arrows in the explosion energies plot indicate that the explosion energies are still increasing at the times the simulations were stopped. The length of the arrows is tied to the magnitude of the rates of change of the explosion energies at these times.

2018; Vartanyan et al. 2019; Burrows et al. 2019; Kuroda et al. 2020; Stockinger et al. 2020). This is a marked change relative to where the field was only ten years ago, let alone more than fifty years ago. The modeling community is now entering a new period, of quantitative rather than qualitative prediction. For example, modelers are now able to compare quantitatively predictions for observables such as explosion energies, ^{56}Ni production, and remnant neutron star masses and kicks, with observations. An example of this is shown in Figure 2, where the explosion energy predictions of Bruenn et al. 2013 and Bruenn et al. 2016 for axisymmetric simulations beginning with progenitors of mass 12, 15, 20, and 25 M$_\odot$ are plotted against observed explosion energies, as a function of progenitor mass. A more recent example of this encouraging agreement, across a large collection of progenitor masses, can be found in Burrows and Vartanyan 2021.

The above cited efforts provide a strong foundation on which to build future core collapse supernova models. Nonetheless, they suffer from one or more of three significant shortcomings: (1) General relativity is included only approximately. (2) Neutrino transport is not three-dimensional. (3) They do not include all of the necessary neutrino interactions and/or state of the art treatments of them. We consider each of these in turn.

The models of Hanke et al. 2013; Lentz et al. 2015; Melson et al. 2015b,a; Summa et al. 2018; O'Connor and Couch 2018; Vartanyan et al. 2019; Burrows et al. 2019; Stockinger et al. 2020 all deploy an effective gravitational potential in the context of Newtonian hydrodynamics. In this approach, the monopole contribution to the Newtonian gravitational potential, as defined through a decomposition of the potential in terms of spherical harmonics, is replaced by a corrected potential whose form is suggested by comparing the equations for Newtonian hydrostatic equilibrium and the Tolman–Oppenheimer–Volkov (TOV) equation for general relativistic hydrostatic equilibrium. At an instant of time in the simulation, the three-dimensional data is spherically averaged, and the resulting quantities are used to compute the corrected potential. This approximation to general relativistic gravity was first proposed by Rampp and Janka 2002. The goal is to capture the stronger gravitational fields that result in general relativity given that, along with rest mass, internal energy and pressure contribute to the gravitational field, as well. The fundamental problem with this approach is, of course, that the correction for general relativity is *ad hoc*. In the context of two-dimensional

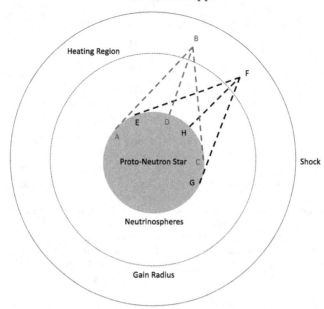

Figure 3. In the ray-by-ray approximation, a hot spot at D would correspond to a hot surface between A and C, thereby overestimating the neutrino heating at B. A (relatively) cold spot at H would correspond to a (relatively) cold surface between E and G, thereby underestimating the heating at F, to which the hot spot at D would in reality contribute.

models, Müller et al. 2012 obtained qualitatively different results (e.g., explosion versus no explosion) in two models of the same progenitor mass, with Newtonian hydrodynamics and an effective potential used in one model and general relativistic hydrodynamics and gravity used in the other. Explosion occurred in the general relativistic model.

The models of Hanke et al. 2013; Lentz et al. 2015; Melson et al. 2015b,a; Summa et al. 2018; Stockinger et al. 2020 deploy the so-called Ray-by-Ray (RbR) approximation to neutrino transport. The sophistication of neutrino transport implementations in spherically symmetric core collapse supernova models motivated this approach, which was first proposed by Rampp and Janka 2002. For each (θ, ϕ) on an angular grid in a numerical simulation, a solution of the neutrino transport equations is obtained for all r, assuming spherical symmetry. For a spherically symmetric source (in our case, given a spherically symmetric proto-neutron star), the RbR solution is exact. If the conditions at the neutrinospheres differ from spherical symmetry over a protracted period of time, the RbR approximation will, as a function of (θ, ϕ), lead to over- and under-estimation of the neutrino heating in the gain region (see Figure 3). Such protracted deviations from spherical symmetry can happen in, for example, axisymmetric simulations. Given the imposed symmetry, accretion funnels impinging on the neutrinospheres, thereby heating them, can be pinned to particular (θ, ϕ), leading to inaccurate neutrino heating, which in turn can alter the explosion outcome artificially (e.g., see Skinner et al. 2016). More recently, the RbR approximation was investigated by Glas et al. 2019 in the context of three-dimensional models and, given the absence of any imposed symmetry, yielded results in good agreement with three-dimensional transport, in the cases considered. Nonetheless, future simulations should endeavor to deploy three-dimensional neutrino transport.

The models of Roberts et al. 2016; Kuroda et al. 2016, 2020, while three-dimensional and general relativistic (although Roberts et al. 2016 neglect velocity-dependent terms in the neutrino moment equations), deploy a subset of the relevant neutrino interactions and, in some cases, not the most advanced implementations of them. Specifically, in all three cases, the neutrino opacities given by Bruenn 1985 are used. In the simulations conducted by Kuroda et al. 2016, 2020, the production of neutrino–antineutrino pairs via nucleon–nucleon bremsstrahlung, as given by Hannestad and Raffelt 1998, is used, as well. In the Bruenn 1985 opacities, electron capture on nuclei is treated using the Independent Particle Model (IPM) for nuclei. In this model the nucleons in the nuclei are assumed to be noninteracting. Given the IPM, electron capture on nuclei during stellar core collapse is blocked, due to neutron final-state blocking in this approximation, and electron capture is given by capture on free protons only. Consequently, the deleptonization of the core (the electron neutrinos produced by electron capture initially escape until neutrino trapping densities are reached) is underestimated, which in turn changes the initial shock location and strength. Simulations by Hix et al. (2003), which used the more advanced models of electron capture on nuclei of Langanke et al. 2003, in which interactions between nucleons in nuclei are taken into account, as well as thermal effects, demonstrated that the shock forms more deeply in the core, is less energetic when it does form, and has to propagate through more of the iron core before reaching the silicon layer and the density drop-off that accelerates explosion. Bruenn 1985 opacities also assume that scattering on nucleons is isoenergetic. While the large rest mass of the nucleons leads kinematically to small neutrino energy transfer, Müller et al. 2012 demonstrated that the high collision rate in the vicinity of the neutrinospheres leads to heating of the electron-neutrinospheres and, in turn, neutrino shock reheating itself. They deployed the more advanced neutrino–nucleon scattering rates of Burrows and Sawyer 1998 and Reddy et al. 1998, which account for the small but important energy exchange that occurs in these scattering events. Finally, Bruenn 1985 opacities include the production of neutrino–antineutrino pairs via electron–positron annihilation but neglect the additional source of neutrino pairs from nucleon–nucleon bremsstrahling, which can dominate neutrino pair production via electron–positron annihilations in certain regions of the core.

4. Additional Challenges

4.1. *Time Scales*

The significant rates of change in the explosion energies in some of the explosion models considered in Figure 2 are an indication that explosion energies in multi-dimensional models will evolve over much longer time scales than the canonical one-second of post-bounce evolution considered in the past. In fact, the compendium of two-dimensional models published to date indicate that several seconds of postbounce simulation will be required (e.g., see Burrows and Vartanyan 2021). While not so much a challenge for two-dimensional simulations, three-dimensional simulations are much more costly. Among the dozens of explosion simulations documented in the references cited in this review, the final explosion energies have been determined in only a few cases, for this reason. This illustrates the general problem we face in conducting a sufficient number of three-dimensional core collapse supernova models to span progenitor parameter space while at the same time conducting such simulations for sufficiently long periods of time in order to predict all relevant observables and compare these predictions with observations. At this time, advances in core collapse supernova theory are throttled more by the availability of computing time than by the necessary advances documented in Section 3.

4.2. *Progenitors*

Much of the progress in core collapse supernova theory has been achieved by considering spherical, non-rotating progenitors of various masses and metallicities. The reason for this is simple. There are at present no core collapse supernova progenitors that have been obtained through three-dimensional stellar evolution simulations, and there will not be any such progenitors for some time to come. Nonetheless, important progress has been made to determine the impact such simulations and the progenitors they will produce may have on core collapse supernova models. There are two primary considerations, based on considerations of single-star and binary-star evolution: (1) What deviations from spherical symmetry at the onset of collapse might we expect and what impact will those deviations have on the supernova mechanism? (2) What is the impact on the evolution of the progenitor of being in a binary system?

Couch et al. 2015 were the first to consider the impact of deviations from spherical symmetry for single massive stars. They followed the final minutes of silicon burning in three dimensions, to the onset of core collapse and, subsequently, followed collapse, bounce, shock formation, and shock propagation. The non-spherical progenitor structure in their model induced by convective silicon burning led to enhanced post-shock turbulence, which in turn aided explosion. That deviations from spherical symmetry in core collapse supernova progenitors aids explosion was corroborated in the later work of Müller et al. 2017 and Vartanyan et al. 2022. Going forward, models in three dimensions should begin, at least in some cases for a comparative analysis, with the late stages of stellar evolution rather than with the onset of collapse, until three-dimensional progenitors evolved through all stages of stellar evolution become available. Of course, stellar evolution simulations will face the same challenges. It will be difficult to perform the number of three-dimensional stellar evolution simulations needed to adequately span the full range of core collapse supernova progenitors.

SN1987A occupies a special place in core collapse supernova theory. The very recent work of Utrobin et al. 2021 determined that only one of the three-dimensional models considered, a binary progenitor model, satisfied (almost) all of the observational constraints. Also in the context of three-dimensional models, earlier efforts to understand the impact of binary stellar evolution on the progenitors of core collapse supernovae – specifically, the impact of mass loss in binary systems – were initiated by Müller et al. 2019. They considered both single-star progenitors between 9.6 and 12.5 $M_\odot$ and ultra-stripped progenitors with He core masses between 2.8 and 3.5 $M_\odot$. They concluded that the differences between core collapse supernova models initiated from single or binary-stripped stars *of the same He core mass* were no larger than the stochastic variations among models initiated from single stars. However, they did stress the importance of binary evolution on the determination of the He core mass itself. A second, complementary study by Vartanyan et al. 2021 investigated the impact of binary mass loss during the first Roche-lobe overflow phase, considering the impact on stripped but not ultra-stripped progenitors. They concluded that, for the same initial mass, binary-stripped progenitors are more "explodable" relative to their single-star counterparts. The results of these studies make clear that at least some future three-dimensional core collapse supernova models must factor in the effects of binary stellar evolution on the progenitors used.

4.3. *Weak Interactions*

The early history of core collapse supernova theory was intertwined with the development of the electroweak theory of the weak interactions (see Mezzacappa et al. 2020). As weak interaction theory progressed, so did core collapse supernova theory. This continues to this day. There are two things to consider: (1) The continued addition of new weak interaction channels of relevance to the core collapse supernova mechanism. (2) The uncertainties associated with the cross sections for all of the weak interactions involved. We discuss both here.

The work by Bollig et al. 2017 is the most recent example of the addition of a whole new class of weak interaction channels to core collapse supernova theory. They demonstrated that muons cannot be ignored as a constituent of the stellar core material and, with them, neutrino–muon interactions. Past simulations ignored muons given their large rest mass and, consequently, their small expected populations, but the work cited here makes clear this is not a good assumption. Bollig et al. 2017 found that the inclusion of muons could *qualitatively* alter the outcome of the supernova simulations they performed.

On the other hand, the work by Melson et al. 2015a considered the sensitivity of core collapse supernova models to *uncertainties* in the cross sections of the weak interactions already included in all leading core collapse supernova models. In particular, they considered variations of the neutrino–nucleon scattering cross section consistent with experimental uncertainties and found that variations of this cross section could again qualitatively change the outcome of the models they considered. Neutrino–nucleon scattering is one of the most important opacities in the proto-neutron star. This motivated the decision to vary this particular opacity. A more systematic and complete study will need to be undertaken in which all of the neutrino opacities are varied in a statistically meaningful way. Such a study would take into consideration the known feedbacks that occur when neutrino opacities are varied (e.g., see Lentz et al. 2012). Unfortunately, in the context of three-dimensional modeling, such sensitivity studies are at present prohibitive. Going forward, at least in the context of three-dimensional modeling, we will need to rely on more limited studies based on varying targeted opacities, as was done in the studies cited here.

5. Horizons

The core collapse supernova simulations cited here, which represent the culmination of decades of core collapse supernova modeling, assumed that neutrinos are massless. We now know this is not the case. Neutrinos have mass and, as a result, quantum mechanical effects such as vacuum flavor mixing, matter-enhanced flavor mixing, and finally, and perhaps most importantly, neutrino-enhanced flavor mixing can occur. Such effects can be expected to impact observations of core collapse supernova neutrinos from the next Galactic or near-extra-Galactc event, are likely to impact core collapse supernova nucleosynthesis, and may even impact the explosion mechanism itself. Here, we concern ourselves with the last possibility.

Electron neutrinos and antineutrinos interact via both charged and neutral currents, whereas muon and tau neutrinos and antineutrinos interact via neutral currents only. As a result, muon and tau neutrinos decouple from the stellar core material at higher densities and, consequently, higher temperatures, and emerge with harder spectra. On the other hand, neutrino shock revival is mediated by electron neutrinos and antineutrinos. Mixing between electron flavor and muon and tau flavor neutrinos, if it occurs below the gain region, could enhance neutrino shock reheating given its sensitive dependence on the electron flavor neutrino and antineutrino spectra [see Equation (2.1)]. Herein lies the importance of neutrino mixing to the explosion mechanism.

Core collapse supernovae are unique neutrino environments. Nowhere else in the Universe do we have trapped degenerate seas of neutrinos and antineutrinos of all flavors, diffusing through the proto-neutron star, emerging from the region in which the neutrinospheres are embedded. In such environments, neutrino–neutrino interactions become important. Of particular note here: Sawyer 2005 demonstrated that under certain conditions, which depend on the angular distributions of the electron neutrinos and antineutrinos in the core, so-called "fast" flavor transformations can occur in the neutrino decoupling region around the neutrinospheres. Given these would occur in regions below the gain region, they would impact neutrino shock reheating – i.e., they would factor into the core collapse supernova explosion mechanism. Indeed, in the context of three-dimensional models, Abbar et al. 2021 and Nagakura et al. 2021 found conditions for fast flavor transformations in the post shock region in all of the models they considered that exploded. Given that fast flavor transformations and explosion coexist in these models, the models' outcomes (e.g., the final explosion energies) may depend on fast-flavor-transformation physics. For a review of this rapidly evolving subject, the reader is referred to Duan et al. 2010, Mirizzi et al. 2016, and Tamborra and Shalgar 2021.

To include flavor transformations in core collapse supernova simulations, classical neutrino kinetics – specifically Boltzmann neutrino kinetics, given the need to keep track of the neutrino and antineutrino angular distributions – would need to be replaced by quantum neutrino kinetics. A three-dimensional, general relativistic treatment of quantum neutrino kinetics, with sufficient angular and energy resolution in neutrino momentum space and a complete set of neutrino weak interactions, remains a long-term goal.

6. Summary and Outlook

Efforts to ascertain the core collapse supernova explosion mechanism(s) have been ongoing for more than five decades. Recent, rapid progress is very promising. Based on the sophisticated three-dimensional simulations performed to date, which span progenitor mass and metallicity, the majority with no or slow rotation, there is consensus across core collapse supernova modeling groups that core collapse supernovae can be driven by neutrino shock reheating aided by turbulent neutrino-driven convection and, depending on the progenitor, the SASI. Nonetheless, much work remains to be done. First and foremost, all three-dimensional models with classical neutrino kinetics must be further developed to include (1) general relativity for gravity, hydrodynamics, and neutrino kinetics, (2) magnetohydrodynamics to capture the role magnetic fields play for both slowly and rapidly rotating progenitors, and (3) the full set of relevant neutrino weak interactions. In the long-term, efforts to extend classical neutrino kinetics to quantum neutrino kinetics must continue, to explore the role of fast flavor transformations. This quantum kinetics development must ultimately be founded on the extension of classical neutrino kinetics based on the moment formalism, used in most of the three-dimensional models cited here, to classical neutrino kinetics based on solutions of the neutrino Boltzmann equations. At the same time, all models must remain abreast of the evolving set of relevant neutrino weak interactions and the increasingly constrained set of viable nuclear equations of state (e.g., see Tews et al. 2017).

In the past, the core collapse supernova modeling community has contended with uncertainties associated with (a) approximations to critical components of core collapse supernova models, (b) differences in the numerical methods deployed by different groups, (c) the differing numerical resolutions used in the simulations carried out by these groups, (d) the neutrino interaction cross sections, and (e) the nuclear equation of state. Moreover, the past five decades has taught us that core collapse supernova models are sensitive to small changes in important quantities, such as the neutrino luminosities, RMS energies, cross sections, etc. While this may seem daunting at first, the core collapse supernova modeling community has made significant progress over the past fifty-plus years through systematic improvements that address all of the categories of uncertainty listed above, and no doubt will continue to do so in the future.

For additional information, we refer the reader to recent reviews by Janka et al. 2016; Müller 2016, 2020; Mezzacappa et al. 2020 and Burrows and Vartanyan 2021.

AM acknowledges support from the National Science Foundation through grants PHY 1806692 and 2110177, the Department of Energy through its Scientific Discovery through Advanced Computing Program through grant DE-SC0018232, and the Department of Energy's Exascale Computing Project.

References

Abbar, S., Capozzi, F., Glas, R., Janka, H. T., & Tamborra, I. 2021, On the characteristics of fast neutrino flavor instabilities in three-dimensional core-collapse supernova models. *Phys. Rev. D*, 103, 063033.

Bethe, H. A. & Wilson, J. R. 1985, Revival of a stalled supernova shock by neutrino heating. *Ap.J.*, 295, 14.

Blondin, J., Mezzacappa, A., & DeMarino, C. 2003, Stability of Standing Accretion Shocks, with an Eye toward Core-Collapse Supernovae. *Ap.J.*, 584, 971.

Blondin, J. M. & Mezzacappa, A. 2007, Pulsar spins from an instability in the accretion shock of supernovae. *Nature*, 445, 58.

Bollig, R., Janka, H. T., Lohs, A., Martínez-Pinedo, G., Horowitz, C. J., & Melson, T. 2017, Muon Creation in Supernova Matter Facilitates Neutrino-Driven Explosions. *Phys. Rev. Lett.*, 119, 242702.

Bruenn, S. W. 1985, Stellar core collapse - Numerical model and infall epoch. *Ap.J. Suppl.*, 58, 771.

Bruenn, S. W., De Nisco, K. R., & Mezzacappa, A. 2001, General Relativistic Effects in the Core Collapse Supernova Mechanism. *Ap.J.*, 560, 326.

Bruenn, S. W., Lentz, E. J., Hix, W. R., Mezzacappa, A., Harris, J. A., Messer, O. E. B., Endeve, E., Blondin, J. M., Chertkow, M. A., Lingerfelt, E. J., Marronetti, P., & Yakunin, K. N. 2016, The Development of Explosions in Axisymmetric Ab Initio Core-Collapse Supernova Simulations of 12-25 $M_\odot$ Stars. *Ap.J.*, 818, 123.

Bruenn, S. W., Mezzacappa, A., Hix, W. R., Lentz, E. J., Messer, O. E. B., Lingerfelt, E. J., Blondin, J. M., Endeve, E., Marronetti, P., & Yakunin, K. N. 2013, Axisymmetric Ab Initio Core-collapse Supernova Simulations of 12-25 $M_\odot$ Stars. *Ap.J.*, 767, L6.

Burrows, A., Radice, D., & Vartanyan, D. 2019, Three-dimensional supernova explosion simulations of 9-, 10-, 11-, 12-, and 13-$M_\odot$ stars. *Mon. Not. R. Astron. Soc*, 485, 3153.

Burrows, A. & Sawyer, R. F. 1998, Effects of correlations on neutrino opacities in nuclear matter. *Phys. Rev. C*, 58, 554.

Burrows, A. & Vartanyan, D. 2021, Core-collapse supernova explosion theory. *Nature*, 589, 29.

Chu, R., Endeve, E., Hauck, C. D., & Mezzacappa, A. 2019, Realizability-preserving dg-imex method for the two-moment model of fermion transport. *J. Comp. Phys.*, 389, 62.

Colgate, S. A. & White, R. H. 1966, The Hydrodynamic Behavior of Supernovae Explosions. *Ap.J.*, 143, 626.

Couch, S. M., Chatzopoulos, E., Arnett, W. D., & Timmes, F. X. 2015, The Three-dimensional Evolution to Core Collapse of a Massive Star. *Ap.J.*, 808, L21.

Duan, H., Fuller, G. M., & Qian, Y.-Z. 2010, Collective Neutrino Oscillations. *Annu. Rev. Nucl. Part. Sci.*, 60, 569.

Glas, R., Just, O., Janka, H.-T., & Obergaulinger, M. 2019, Three-dimensional core-collapse supernova simulations with multidimensional neutrino transport compared to the ray-by-ray-plus approximation. *Ap.J.*, 873, 45.

Hanke, F., Müller, B., Wongwathanarat, A., Marek, A., & Janka, H.-T. 2013, SASI Activity in Three-dimensional Neutrino-hydrodynamics Simulations of Supernova Cores. *Ap.J.*, 770, 66.

Hannestad, S. & Raffelt, G. 1998, Supernova Neutrino Opacity from Nucleon-Nucleon Bremsstrahlung and Related Processes. *Ap.J.*, 507, 339.

Herant, M., Benz, W., & Colgate, S. A. 1992, Postcollapse Hydrodynamics of SN 1987A: Two–Dimensional Simulations of the Early Evolution. *Ap.J.*, 395, 642.

Hix, W. R., Messer, O. E. B., Mezzacappa, A., Liebendörfer, M., Sampaio, J. M., Langanke, K., Dean, D. J., & Martinez-Pinedo, G. 2003, Consequences of Nuclear Electron Capture in Core Collapse Supernovae. *Phys. Rev. Lett.*, 91, 201102.

Janka, H.-T., Melson, T., & Summa, A. 2016, Physics of Core-Collapse Supernovae in Three Dimensions: A Sneak Preview. *Annu. Rev. Nucl. Part. Sci.*, 66, 341.

Kuroda, T., Arcones, A., Takiwaki, T., & Kotake, K. 2020, Magnetorotational Explosion of a Massive Star Supported by Neutrino Heating in General Relativistic Three-dimensional Simulations. *Ap.J.*, 896, 102.

Kuroda, T., Takiwaki, T., & Kotake, K. 2016, A New Multi-energy Neutrino Radiation-Hydrodynamics Code in Full General Relativity and Its Application to the Gravitational Collapse of Massive Stars. *Ap.J. Suppl.*, 222, 20.

Langanke, K., Martínez-Pinedo, G., Sampaio, J. M., Dean, D. J., Hix, W. R., Messer, O. E., Mezzacappa, A., Liebendörfer, M., Janka, H.-T., & Rampp, M. 2003, Electron Capture Rates on Nuclei and Implications for Stellar Core Collapse. *Phys. Rev. Lett.*, 90, 241102.

Lentz, E. J., Bruenn, S. W., Hix, W. R., Mezzacappa, A., Messer, O. E. B., Endeve, E., Blondin, J. M., Harris, J. A., Marronetti, P., & Yakunin, K. N. 2015, Three-dimensional core-collapse supernova simulated using a 15 $M_\odot$ progenitor. *Ap.J.*, 807, L31.

Lentz, E. J., Mezzacappa, A., Bronson Messer, O. E., Liebendörfer, M., Hix, W. R., & Bruenn, S. W. 2012, Interplay of Neutrino Opacities in Core-Collapse Supernova Simulations. *Ap.J.*, 760, 94.

Liebendörfer, M., Mezzacappa, A., Thielemann, F.-K., Messer, O. E. B., Hix, W. R., & Bruenn, S. W. 2001, Probing the gravitational well: No supernova explosion in spherical symmetry with general relativistic Boltzmann neutrino transport. *Phys. Rev. D*, 63, 103004.

Melson, T., Janka, H.-T., Bollig, R., Hanke, F., Marek, A., & Müller, B. 2015,a Neutrino-driven Explosion of a 20 Solar-mass Star in Three Dimensions Enabled by Strange-quark Contributions to Neutrino-Nucleon Scattering. *Ap.J.*, 808, L42.

Melson, T., Janka, H.-T., & Marek, A. 2015,b Neutrino-driven Supernova of a Low-mass Iron-core Progenitor Boosted by Three-dimensional Turbulent Convection. *Ap.J.*, 801, L24.

Mezzacappa, A., Endeve, E., Messer, O. E. B., & Bruenn, S. W. 2020, Physical, numerical, and computational challenges of modeling neutrino transport in core-collapse supernovae. *Living Reviews in Computational Astrophysics*, 6, 4.

Mirizzi, A., Tamborra, I., Janka, H. T., Saviano, N., Scholberg, K., Bollig, R., Hüdepohl, L., & Chakraborty, S. 2016, Supernova neutrinos: production, oscillations and detection. *Nuovo Cimento Rivista Serie*, 39, 1.

Müller, B. 2016, The Status of Multi-Dimensional Core-Collapse Supernova Models. *Proc. Astron. Soc. Aust.*, 33, e048.

Müller, B. 2020, Hydrodynamics of core-collapse supernovae and their progenitors. *Living Reviews in Computational Astrophysics*, 6, 3.

Müller, B., Janka, H.-T., & Marek, A. 2012, A New Multi-Dimensional General Relativistic Neutrino Hydrodynamics Code for Core-Collapse Supernovae II. Relativistic Explosion Models of Core-Collapse Supernovae. *Ap.J.*, 756, 84.

Müller, B., Melson, T., Heger, A., & Janka, H.-T. 2017, Supernova simulations from a 3D progenitor model - Impact of perturbations and evolution of explosion properties. *Mon. Not. R. Ast. Soc.*, 472, 491.

Müller, B., Tauris, T. M., Heger, A., Banerjee, P., Qian, Y.-Z., Powell, J., Chan, C., Gay, D. W., & Langer, N. 2019, Three-dimensional simulations of neutrino-driven core-collapse supernovae from low-mass single and binary star progenitors. *Mon. Not. R. Ast. Soc.*, 484, 3307.

Murphy, J. W., Dolence, J. C., & Burrows, A. 2013, The Dominance of Neutrino-driven Convection in Core-collapse Supernovae. *Ap.J.*, 771, 52.

Nagakura, H., Burrows, A., Johns, L., & Fuller, G. M. 2021, Where, when, and why: Occurrence of fast-pairwise collective neutrino oscillation in three-dimensional core-collapse supernova models. *Phys. Rev. D*, 104, 083025.

Obergaulinger, M., Janka, H. T., & Aloy, M. A. Magnetic Field Amplification in Non-Rotating Stellar Core Collapse. In Pogorelov, N. V., Audit, E., & Zank, G. P., editors, *Numerical Modeling of Space Plasma Flows ASTRONUM-2014* 2015, volume 498 of *Astronomical Society of the Pacific Conference Series*, 115.

O'Connor, E. & Couch, S. 2018, Exploring Fundamentally Three-dimensional Phenomena in High-fidelity Simulations of Core-collapse Supernovae. *Ap.J.*, 865, 81.

Rampp, M. & Janka, H.-T. 2002, Radiation hydrodynamics with neutrinos. Variable Eddington factor method for core-collapse supernova simulations. *Astron. Astrophys.*, 396, 361.

Reddy, S., Prakash, M., & Lattimer, J. M. 1998, Neutrino interactions in hot and dense matter. *Phys. Rev. D*, 58, 013009.

Roberts, L. F., Ott, C. D., Haas, R., O'Connor, E. P., Diener, P., & Schnetter, E. 2016, General-Relativistic Three-Dimensional Multi-group Neutrino Radiation-Hydrodynamics Simulations of Core-Collapse Supernovae. *Ap.J.*, 831, 98.

Sawyer, R. F. 2005, Speed-up of neutrino transformations in a supernova environment. *Phys. Rev. D*, 72, 045003.

Skinner, M. A., Burrows, A., & Dolence, J. C. 2016, Should One Use the Ray-by-Ray Approximation in Core-collapse Supernova Simulations? *Ap.J.*, 831, 81.

Stockinger, G., Janka, H. T., Kresse, D., Melson, T., Ertl, T., Gabler, M., Gessner, A., Wongwathanarat, A., Tolstov, A., Leung, S. C., Nomoto, K., & Heger, A. 2020, Three-dimensional models of core-collapse supernovae from low-mass progenitors with implications for Crab. *Mon. Not. R. Ast. Soc.*, 496, 2039.

Summa, A., Janka, H.-T., Melson, T., & Marek, A. 2018, Rotation-supported neutrino-driven supernova explosions in three dimensions and the critical luminosity condition. *Ap.J.*, 852, 28.

Tamborra, I. & Shalgar, S. 2021, New developments in flavor evolution of a dense neutrino gas. *Ann. Rev. Nucl. Part. Sci.*, 71, 165.

Tews, I., Lattimer, J. M., Ohnishi, A., & Kolomeitsev, E. E. 2017, Symmetry Parameter Constraints from a Lower Bound on Neutron-matter Energy. *Ap.J.*, 848, 105.

Utrobin, V. P., Wongwathanarat, A., Janka, H. T., Müller, E., Ertl, T., Menon, A., & Heger, A. 2021, Supernova 1987A: 3D Mixing and Light Curves for Explosion Models Based on Binary-merger Progenitors. *Ap.J.*, 914, 4.

Vartanyan, D., Burrows, A., Radice, D., Skinner, M. A., & Dolence, J. 2019, A successful 3D core-collapse supernova explosion model. *Mon. Not. R. Ast. Soc.*, 482, 351.

Vartanyan, D., Coleman, M. S. B., & Burrows, A. 2022, The collapse and three-dimensional explosion of three-dimensional massive-star supernova progenitor models. *Mon. Not. R. Ast. Soc.*, 510, 4689.

Vartanyan, D., Laplace, E., Renzo, M., Götberg, Y., Burrows, A., & de Mink, S. E. 2021, Binary-stripped Stars as Core-collapse Supernovae Progenitors. *Ap.J.*, 916, L5.

Wilson, J. R. Supernovae and Post–Collapse Behavior. In Centrella, J. M., LeBlanc, J. M., & Bowers, R. L., editors, *Numerical Astrophysics* 1985, pg. 422, Boston. Jones and Bartlett.

The Predictive Power of Computational Astrophysics as a Discovery Tool
Proceedings IAU Symposium No. 362, 2023
D. Bisikalo, D. Wiebe & C. Boily, eds.
doi:10.1017/S1743921322001417

Hot spot drift in synchronous and asynchronous polars: synthesis of light curves

Andrey Sobolev [ID], Dmitry Bisikalo and Andrey Zhilkin [ID]

Institute of astronomy of the Russian academy of sciences, 119017, Pyatnitskaya st., 48,
Moscow, Russia
email: asobolev@inasan.ru

Abstract. In this paper, the effect of hot spots movement by accretor surface on the appearance of bolometric light curves for two types of polars - synchronous V808 Aur and asynchronous CD Ind is studied. The analysis was carried out under the assumption of a dipole configuration of the magnetic field, in which the axis of the dipole passes through the accretor center. It is shown that a noticeable shift of the flow maximum at the light curve corresponding to the position of the spots in synchronous polars is determined by a change in the magnitude of mass transfer rate. At the same time, the maximum deviation of the spots from the magnetic poles was 30°. In asynchronous polars, assuming a constant of the mass transfer rate, the spots movement caused by a change in the orientation of the dipole axis relative to the donor has a significant effect on the appearance of light curve. The greatest displacement of the spots from the magnetic poles, which equals to 20°, was observed at the moments when the accretion jet switched from one pole to the other. It is concluded that the comparison of synthetic and observational light curves provides an opportunity to study the physical properties of polars.

Keywords. close binary star, polar, MHD, flow structure, donor, accretor, hot spot, temperature map

1. Introduction

In our previous work [Bisikalo (2021)] we already investigated the movement of hot spots in synchronous and asynchronous polars. The three-dimensional pictures of the flow structure constrained by the results of numerical calculations and maps of the temperature distribution over the accretor surface together provide quantitative estimates of the hot spots movement. Such estimates are almost impossible to obtain directly from the observations of real polars. Therefore, in this paper for the purposes of comparison with observational data, we construct light curves.

2. Statement of a problem

Numerical MHD modeling of the synchronous polar V808 Aur [Gabdeev (2016)] allowed us to obtain a general picture of the hot spots movement by the accretor surface for different states of the binary system, which correspond to various values of the mass transfer rate. The analysis has shown that variations in the mass transfer rate form different patterns of the matter flow from the donor to the accretor: its high value leads to an elongation of the ballistic part of the jet trajectory and its significant deviation from the direction to the accretor; at lower values of the rate, the ballistic part of the jet gradually decreases and the main influence on the movement of matter is exerted by the accretor magnetosphere.

The maps of temperature distribution over primary component surface have shown that two energy release zones are formed in the considered synchronous polar: the first one (main) is in the vicinity of north magnetic pole and the secondary one is near the south pole. Hot spots are characterized by different intensities: the northern zone, corresponding to the accretion of matter from the jet, has a temperature about 5 times higher than the southern one, and this ratio persists with a change in the mass transfer rate. The position of hot spots is determined by the nature of accretion. The southern spot formed by the matter from the polar common envelope does not change its position with variations in the mass transfer rate. At the same time, its area decreases proportionally with a diminution in the binary system state. The temperature, on the other hand, experiences a slight increase with this change in the polar state, since the accretion of matter from the common envelope occurs more concentrated. This is accompanied by the growth in spot luminosity.

The location of the northern hot spot, on the contrary, significantly depends on the accretion rate. At the very high polar state, the spot has a maximum longitude offset relative to the north magnetic pole by about 30°. Its latitude deviation in this case is 5–7°. At the high state, the longitude offset decreases to 15°, while the value of the latitude of spot practically does not change, but the area of the energy release zone decreases by a factor of 2. At the intermediate state of the polar, the northern spot, while maintaining its geometric dimensions, closely approaches the magnetic pole. At the same time, its distance from the pole in longitude and latitude is 5°. Finally, at the low state, the northern region of energy release practically coincides with the north magnetic pole, its area decreases by a factor of 4 times compared to the very high state.

For the asynchronous polar CD Ind [Schwope (1997)], MHD modeling, which was performed at a constant value of the mass transfer rate, revealed the presence of jet switching processes from one magnetic pole to the other. In these processes, there is a noticeable change in the flow structure: the formation of an arch of matter in the region of magnetosphere, as well as the local accumulation of matter in the jet. The formation of these elements causes a significant change in the parameters of hot spots, their luminosity and position relative to the magnetic poles.

The constructed temperature maps for the asynchronous polar showed a number of features of the spots drift caused by the switching of the jet between the magnetic poles at a fixed value of the mass transfer rate. At the beginning of switching, the hot spot is located close to the pole which the jet will be re-connected to, it has a maximum area and a significant elongation along the longitude of the accretor. The moment of formation of an arch of matter in the inner region of the magnetosphere is characterized by a significant displacement of both spots relative to the magnetic poles: when they approach each other, the maximum distance from the pole is 20°. Due to the orbital rotation of the polar, the temperatures and accretion rates of hot spots at this stage differ by almost 2 times. Upon completion of the switch, the formed hot spot at the currently active pole retains a removal value of about 15°, and its temperature and area decrease slightly. It is worth noting that both switching processes follow the same scenario.

3. Results of calculations

For the purposes of comparing the results of numerical modeling with observational data, we synthesized bolometric light curves that allow us to separate the radiation flux of the hot spots from the rest part of the polar, since the latter have 2-3 orders of magnitude lower luminosity. In this paper, the synthesis of the desired light curves was performed by the method described in detail in one of our previous works [Sobolev & Zhilkin (2019)].

From the previously calculated temperature maps, the values of the displacement of hot spots relative to the magnetic poles are known, as well as the position of the poles themselves, so it is not difficult to identify areas corresponding to energy release zones on the light curves.

In Fig. 1 the bolometric light curves for the synchronous polar V808 Aur are presented. The upper panel of the figure shows a plot for the value of the mass transfer rate of $10^{-7} M_\odot$/year, and the lower one — for $10^{-10} M_\odot$/year. As it follows from the obtained temperature maps, these curves correspond to the deviation of spot from the pole equal to $30°$ and close to $0°$, respectively.

In the figures shown, the apparent magnitude is plotted along the y axis, and the time scale in fractions of the orbital period (synchronous polar) and in orbital periods (asynchronous polar) is plotted along the x axis.

From Fig. 1, upper panel, it can be seen that the displacement of the maximum flow corresponding to the northern hot spot is 0.06 of the orbital period, which in the accepted angle reference system is $22°$. Note that since the hot spot is an extended object, the displacement angle indicated for it refers to its center, which does not necessarily have the highest temperature. For this reason, the displacement value estimated from the light curve may not coincide with the value obtained from the analysis of temperature maps. The position of southern hot spot on this curve cannot be clearly distinguished, however, it is known that it does not change for all calculated values of the mass transfer rate. Thus, it can be argued that on this curve the location of spot will be the same as in Fig. 1, lower panel. If the south magnetic pole is in the orbital phase of 0.53, and the southern spot is in the phase of 0.49 ($350°$), then the value of the displacement of the latter is 0.04 of the orbital period, or $15°$. The northern hot spot for the case of the minimum mass transfer rate practically coincides with the north pole, as follows from the temperature map.

In Fig. 2 the light curves for the asynchronous polar CD Ind are presented: the upper panel shows the moment of flow switching from the south magnetic pole to the north one, the lower panel — reverse switching from the north to the south pole.

For clarity, the plot shows in detail a fragment of the curve that coincides directly with the moment of switching (the 8th orbital period). When analyzing the asynchronous polar curves, it should be taken into account that the process of switching the jet between the magnetic poles proceeds quickly enough — within the fraction of the orbital period. This means that although it is possible to show all the phases of the process in detail during modeling, only the initial and final stages of the process are available to the terrestrial observer due to its location. On the observational light curve, the position of the hot spots will be fixed exactly at the specified time points.

As it can be seen from Fig. 2, the light curves for both accretion switching processes are almost identical and differ only in the value of the radiation flux. The position of the hot spots in this figure relative to the magnetic poles is shown as it will be seen by an Earth observer. The figure shows that the deviation of the northern energy release zone is $25°$, the southern zone is about $30°$, which is comparable with model calculations. Since the position of the spots in this polar is determined only by the dynamics of the switching process, in both cases it coincides. To display the motion of the spots accross the switching process, it is necessary to synthesize a separate family of light curves for the eighth orbital period, each of them will be shifted by a certain phase angle in tenths of a period.

4. Conclusions

The paper presents a method for analyzing the drift of hot spots in synchronous and asynchronous polars based on the construction of synthetic light curves. This method

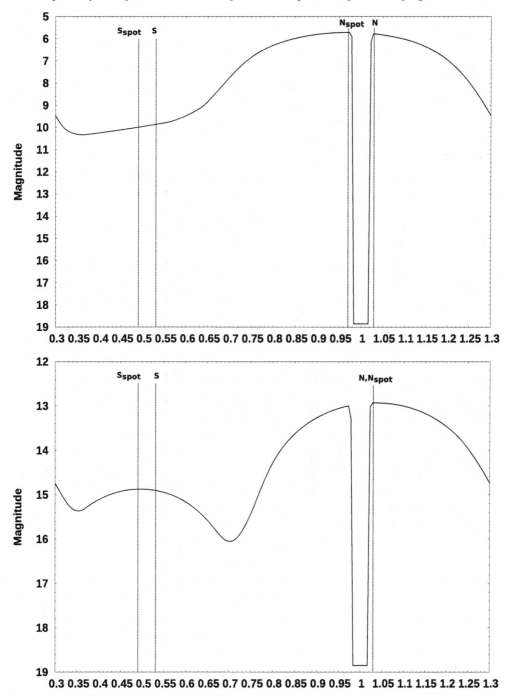

Figure 1. Synthetic bolometric light curves for the synchronous polar V808 Aur at a high state $(10^{-7} M_\odot/\text{year}$, upper panel) and at a low state $(10^{-10} M_\odot/\text{year}$, lower panel). The following values are indicated on the plot: N and S — north and south magnetic pole, N_spot and S_spot — northern and southern hot spot.

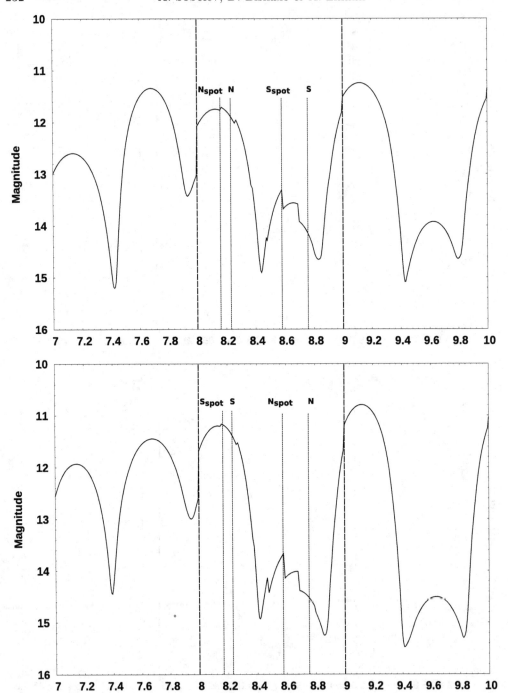

Figure 2. Synthetic bolometric light curves for the asynchronous polar CD Ind. It is shown a part of curve, corresponding to the moment of jet switching from the south pole to the north one (upper panel) and from the north pole to the south one (lower panel). Here the designations are the same as in the figure 1.

has the following features. Its advantage is the possibility for direct comparison both the observational data and simulation results. This comparison provides a qualitative picture of the hot spots distribution and thus allows us to investigate the physical properties of polars. At the same time, the quantitative assessment performed by this method is ambiguous.

For example, according to the light curve, it is impossible to state certainly how the hot spot is displaced along the accretor surface. The movement of the spot along the star longitude leads to a shift of the maximum on a light curve, this moment can be fixed, but the shift in latitude cannot be unambiguously determined, since it leads to a change in brightness of the spot, and brightness variations may be caused by other reasons. In addition, as it was shown, the maximum on the light curve does not necessarily correspond to the spot center, so the deviation value of hot spots from the magnetic poles found on the light curves are approximate.

To display fleeting processes in polars, the duration of which is significantly less than the selected time scale, e. g., the orbital period, the presented method gives rough estimates. Therefore, a good tool for studying physical processes in polars will be the joint use of comparison of observational and synthetic light curves and numerical modeling.

References

Bisikalo, D. Sobolev, A. Zhilkin, A. "Hot Spots Drift in Synchronous and Asynchronous Polars: Results of Three-Dimensional Numerical Simulation." Galaxies, 2021, Vol. 9, No. 4, p.110.

Gabdeev, M. *et al.* 2016, Photometric and Spectral Studies of the Eclipsing Polar CRTS CSS 081231 J071126+440405; Astrophys. Bull., 71, 101.

Schwope, A. Buckley, D. O'Donoghue, D. Hasinger, G. Astron. and Astrophys., 1997, 326, p. 195

Sobolev, A., Zhilkin, A. 2019, Method of constracting a synthetic light curve for eclipsed polars; INASAN Proceedings, 3, p. 231.

The Predictive Power of Computational Astrophysics as a Discovery Tool
Proceedings IAU Symposium No. 362, 2023
D. Bisikalo, D. Wiebe & C. Boily, eds.
doi:10.1017/S1743921322001776

The predictive power of numerical simulations to study accretion and outflow in T Tauri Stars

Ana I. Gómez de Castro[1,2]

[1]Joint Center for Ultraviolet Astronomy, Universidad Complutense de Madrid, Avda. Puerta de Hierro s/n, 28040 Madrid, Spain
email: aig@ucm.es

[2]S.D. Física de la Tierra y Astrofísica, Fac. CC Matemáticas, Plaza de Ciencias 3, 28040 Madrid, Spain

Abstract. T Tauri Stars (TTSs) offer a unique chance to study the physics of non-relativistic accretion engines. In this invited talk, the current status of the field is presented with special emphasis on the predictive power of the numerical simulation of magnetospheric accretion and close binary systems and its impact on astronomical observations.

Keywords. T Tauri Stars, protostellar jets, accretion physics

---

1. Introduction

Gravitational engines are widely spread in astronomical environments; they are constituted by a source a gravity (a star, a compact object or a supermassive black hole) and a surrounding mass repository in the form of an accretion disk that taps the mass flow onto the gravity source. Collimated bipolar flows are generated as a self-regulating mechanism to carry away from the system part of the angular momentum excess that needs to be removed for accretion to operate. Outflows reach terminal speeds roughly equal to the Keplerian velocity at the innermost border of the disk thus, in engines driven by black holes or supermassive blackholes the outflow velocity is very close to the speed of light making these systems the most powerful engines in nature. At the other end of the energy domain are the pre-main sequence (PMS) stars with jet speeds of several hundred kilometers per second (see Table 1). These non-relativistic sources allow studying the inner structure of the engine in detail, as well as the impact of stellar and disk magnetic fields in the process.

In low mass pre-main sequence (PMS) stars (also known as T Tauri stars or TTSs) a magnetized shear layer is generated between the inner border of the disk, rotating at Keplerian velocities, and the star. The Keplerian shear amplifies the field producing a strong toroidal component; an external dynamo sets in. This toroidal field and the associated magnetic pressure push the field lines outwards from the disk rotation axis, inflating and opening them up in a *butterfly-like pattern*, so producing a current layer between the stellar and the disk dominated regions as displayed in Figure 1. Magnetic field dissipation in the current layer produces high energy radiation and particles. The magnetic link between the star and the disk is broken and reestablished continuously by magnetic reconnection. The opening angle of the current layer, as well as its extent, depends on the stellar and disk fields, the accretion rate and the ratio between the inner disk radius and the stellar rotation frequencies. Hot, pressure driven outflows are

Table 1. Gravitational engines: PMS stars compared with supermassive black holes.

Type of source	Mass of the source ($M_\odot$)	Accretion rate ($M_\odot$ yr^{-1})	Outflow velocity (km s^{-1})	Outflow mechanical power (MW)
Protostar	1	1×10^{-8}	300	6×10^{20}
Supermassive Blackhole	10^8	1	0.98c	6×10^{32}

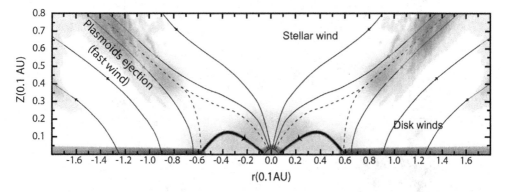

Figure 1. Sketch of the accretion engine in a PMS star. The star acts as a magnetic rotor that interacts with the plasma orbiting around it, in Keplerian orbits. The results of numerical simulations on the interaction between the stellar magnetosphere and the disk are shown; they are color coded from light green to brown and represent the emissivity of the C III] line at 191nm (Gómez de Castro & von Rekowski, 2011). The magnetic configuration is outlined as well as the reconnection layer where magnetic bubbles are thought to be generated (after Gómez de Castro et al. 2016).

produced from the star, in the region closer to the rotation axis whilst cool centrifugally driven flows are produced by the disk; plasmoids are ejected from the current layer generating a third outflowing component.

This article deals with the predictive power of the numerical simulations in this field; from the first works on protostellar jet formation (Goodson et al. 1997; Goodson & Winglee 1999) to the late developments on magnetospheric accretion (Romanova et al. 2021). In most cases, numerical simulations have just attempted to provide a physical explanation to observed phenomena but there are some few cases in which numerical simulations have been instrumental for the planning and analysis of the observations, as will be shown in the last section.

2. Numerical simulations of protostellar jets and accretion: predictions

Early in the 90's, the basic physics of the generation of bipolar collimated outflows from young stars was laid down and the two main branches of models were outlined: disk winds (Pudritz et al. 1991) and magneto-centrifugally winds from stars (Shu et al. 1994). Disk winds were first proposed by Blandford & Payne (1982) in the context of extragalactic jets from an extrapolation of the hydromagnetic solution for the fast component of the solar wind (Weber & Davis 1967). This theory was later adapted to the physics of pre-main sequence (PMS) stars and reproduced successfully some of the properties of the large scale jets (Pudritz et al. 2007). Otherwise, X-winds were based on the magnetic interaction between the star and the disk, and later evolved into the current magnetospheric accretion/launching theories (von Rekowski & Branderburg 2004, 2006). The current paradigma (as shown in Figure 1) includes both in a natural manner.

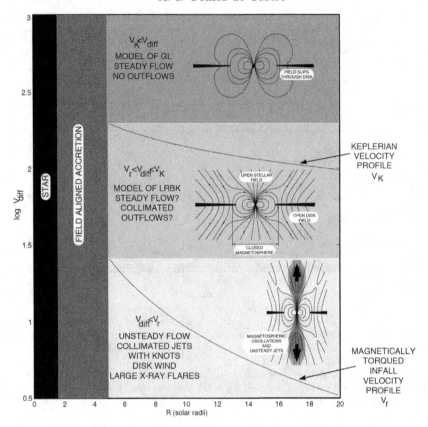

Figure 2. Summary of three classes of magnetically dominated accretion (after Goodson & Winglee 1999); v_K, v_r and v_{diff} are the keplerian, radial and diffusion velocities in the inner disk.

Early numerical models of the whole system date back to the works by Goodson et al. (1997) and Goodson & Winglee (1999). The key parameter regulating the engine is the magnetic diffusivity (ill constrained from the observations). In the high diffusivity limit, the stellar field can continuously slip through the inner disk (Ghosh & Lamb, 1979), and in the low diffusivity limit, the radial velocity is larger than the diffusion velocity and an unsteady, strong flow is initiated (Goodson et al. 1997). For intermediate diffusivities, the field wraps up until the associated magnetic pressure opens it; steady solutions are possible provided the diffusion velocity keeps being higher than the radial velocity at the inner edge of the disk (see sketch in Figure 2). In the numerical implementation of these models, often cylindrical symmetry (2.5 D) is imposed to solve the equations of the magnetohydrodynamics (MHD) and transport is hypothesized to follow the α prescription, sometimes implementing a soft dependence of α on the radius. In Goodson et al. (1997), the stellar magnetosphere is assumed to be a magnetic dipole aligned with the disk axis and the inner border of the disk is set at 8.5 R$_*$ stellar radii. *The most important prediction from these simulations is the oscillation of the inner disk border* (being the time scale depending on the diffusivity). The time-scale and the amplitude of the oscillation of the inner border of the disk also depends on the disk density, *i.e.* on the accretion rate and the evolutionary state of the disk (Matt et al. 2002). *The prediction is that they should become more pronounced in low gas-density disks such as the transitional disks.*

Let us focus now, on the core of the engine: the interaction between the stellar magnetosphere and the disk. A major step forward was the implementation of the "cubed

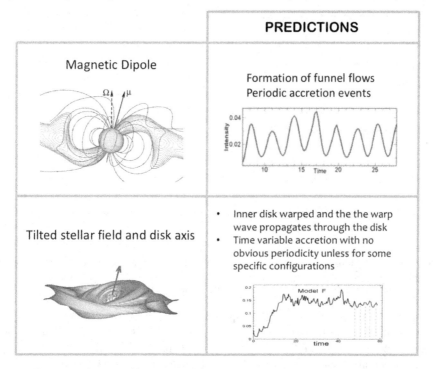

Figure 3. Some predictions of the numerical simulations of magnetospheric accretion. Top, formation of funnel flows and periodic accretion events (Romanova et al. 2004). Bottom, generation of a warp wave in the accretion disk and disapearence of the periodic accretion pattern for most of the configurations (Romanova et al. 2021).

sphere" mathematical set-up into the simulations which enabled for the first time to track the accretion flow from the inner disk onto the star (Romanova et al. 2002). The simulation was based on a clever selection of the quasi equilibrium initial conditions and solved for the accretion flow in a 2.5D, MHD framework (Romanova et al. 2003, 2004). These simple models have evolved over the years to include multipolar components in the stellar magnetic field (Long et al. 2008; Romanova et al. 2012) or allow for a misalignment between the disk axis and the magnetic rotator (Romanova et al. 2021). *A major prediction from the models is the formation of funnel flows* that channel the disk material over the stellar surface; these funnels are destroyed if the multipolar components become important. Also, accretion onto dipolar magnets produces a periodic signal due to the rotation of the star but this periodicity gets lost when multipolar components become important. As, the higher moments of the stellar magnetic field (cuadrupoles, octupoles...) decrease more pronouncedly with distance than the lower ones, it should be expected that at the distance of the inner border of the disk, the dipolar moment is the dominant. *Another interesting prediction is that tilted rotators result in the formation of a warp wave that propagates through the disk. Also accretion periodicity gets lost in most of the possible configurations* (see Figure 3).

3. Comparison between numerical predictions and observations

3.1. *Magnetospheric accretion*

Magnetospheric accretion is the most accepted theory for describing the observed characteristics of the TTSs (e.g, Hartmann et al., 2016, Gómez de Castro et al., 2013,

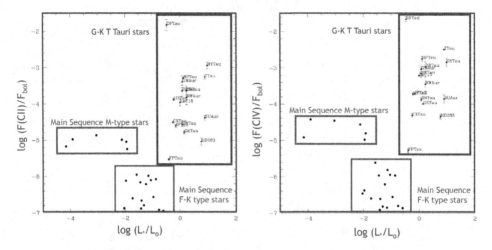

Figure 4. The normalised surface fluxes versus stellar luminosities for spectral tracers of chromospheric-like plasma ($T_e \sim 10^4$K) and plasma at $T_e \sim 4 \times 10^4$K. The location of the TTSs and the main sequence cool stars in the diagram is indicated.

Bouvier et al., 2007). Most numerical simulations hypothesize dipolar stellar magnetospheres extending up to 3-10 R_* with field strengths of ~ 1 kG. *The numerical work came after* the realization that the TTSs have extended magnetospheres and that mass infall was being channelled onto specific locations attached at the stellar surface. Actual field measurements came even later (Johns-Krull et al. 1999; Johns-Krull 2007), as well as the observation of multipolar fields (Donati et al. 2007, 2008, 2019, Gregori et al. 2012). Key observations in shaping the magnetospheric paradigma were:

• *The ultraviolet excess*: the first ultraviolet observations of the TTSs back in 80's already showed that the line fluxes produced by chromospheric and transition region tracers are 50-100 times stronger than those of their main sequence analogues (see Figure 4). This excess was used in the first estimates of an equivalent emitting volume ($2R_* - 4R_*$) and the associated magnetospheric radius driving at the first concepts of "extended magnetospheres".

• *The rotational modulation of the UV flux.* Firm evidence of the rotational modulation of the UV flux came in the early 90's when several well-known TTSs were monitored to find that the light curves of some of them were rotationally modulated at UV wavelengths indicating the presence of hot spots (8,000 K - 40,000 K) on the stellar surface (Simon et al. 1990; Gómez de Castro & Fernández, 1996; Gómez de Castro & Franqueira, 1997). The detection of hot plasma localized on specific areas of the stellar surface was the first clear evidence of the existence of accretion shocks: shocks on the stellar surface where the infalling material releases its kinetic energy into heating. The stability of these locations over several rotation periods provided the first hints on the magnetospheres channelling the accretion flow. However, the actual extension of the accretion funnel has been measured only very recently, through the time lag between optical and ultraviolet observations (Espaillat et al. 2021).

• *The peculiar ultraviolet spectrum*: the atmospheres of cool stars are stratified in three main regions: the warm (T$\sim 10^4$ K) chromosphere, the hot (T$\geq 10^6$ K) corona and the transition region (TR) between them. There are well characterized correlations between the flux radiated in the various spectral tracers of these regions. These so-called fluxflux relations are used to model energy transport in cool stars and call for a universal mechanism operating in them. Flux-flux relations were also been measured in the TTSs (Yang et al. 2012; Gómez de Castro & Marcos-Arenal 2012). When compared with their main sequence analogues, it becomes evident that there is excess radiation from low ionization species (T$\sim 10^4$ K) with respect to the highly ionized ones (T$\sim 10^5$ K). This indicates that radiation is released by a different mechanism. The most successful models propose that the excess gravitational energy of the accreting matter is released into heating at accretion shocks where the temperature reaches 0.3-1 MK, i.e., coronal-like temperatures, driving a photoionization cascade that results in the observed scalings (Calvet & Gullbring 1998; Gómez de Castro & Lamzin 1999).

Currently, the field has reached maturity and simulations and observations have entered into the standard feedback loop to narrow down the parameter space and study other details of the system such as the dusty environment in the inner disk (see i.e. Li et al. 2022), the development of the stellar wind and the role that the combined action of the stellar wind and the magnetosphere torques may have in the acceleration of stellar rotation from the classical TTS or CTTS stage to the mildly or non-accreting sources in the weak TTS or WTTS stage (see i.e. Ireland et al. 2020 or Pantolmos et al. 2020). In most cases, numerical simulations are used rather as an analysis tool than a predictive tool.

3.2. *Enhanced velocity dispersion at high accretion rates*

A different example on the predictive value of numerical simulations comes from the expectations of the velocity dispersion at the base of the outflow; this effect is illustrated in Figure 5 for some representative engines (von Rekowski & Branderburg 2006; Gómez de Castro & von Rekowski, 2011). The simulations shown assumed cylindrical symmetry and solved the full set of resistive MHD equations to study the mass infall/outflow from an accretion disk around a star with a dipolar magnetic field aligned with the disk. Line profiles were computed for some selected spectral tracers and the basic parameters: dispersion and centroid were evaluated and represented. These calculations were made for a broad grid of models ranging from stellar fields of 1kG to 5kG, and from various assumptions on the disk magnetization (passive disk or disk with a dynamo built in). As shown in the Figure 5, the dispersion (the line broadening) grows with the field strength and with the magnetic activity of the disk. Moreover, as the spectral lines are formed in the outflow inclination effects are noticeable. However, this prediction is not fulfilled by the data; though the observed line broadenings are as high as expected for outflows from 1 kG - 2kG stars, no inclination effects are apparent implying that most of the line flux is not produced in the wind. Later on, it astronomical observations showed that such large broadenings can be produced by the magnetosphere itself.

Winds, accreting matter and the stellar atmosphere share spectral tracers in the TTSs making it difficult to disentangle the various components. However, some specific spectral lines (He II line at 164 nm or the N V resonance multiplet) are known to be radiated by the star (either by the atmosphere or by the accretion shocks). The high sensitivity of the Cosmic Origins Spectrograph (COS) in Hubble enabled to resolve the profiles of these tracers in a fair sample of TTSs and it was found that the profiles are very broad and that the broadening decreases as the stars approach the main sequence (see Figure 6).

3.3. *PMS close binaries*

Numerical simulations of the evolution of disks around PMS binaries point out that an inner gap develops within the circumbinary (CB) disk; the gap radius is about 2-3 times the semi-major axis of the orbit. The characteristics of the gap depend on the binary mass ratio, for instance, for secondary to primary mass ratios $q \ll 1$, the gap becomes an annular ring around the primary through which the secondary travels. It also depends on the relative mass of the secondary to the protostellar disk; if the disk mass is large compared to the mass of the secondary, inward orbital migration of the secondary may

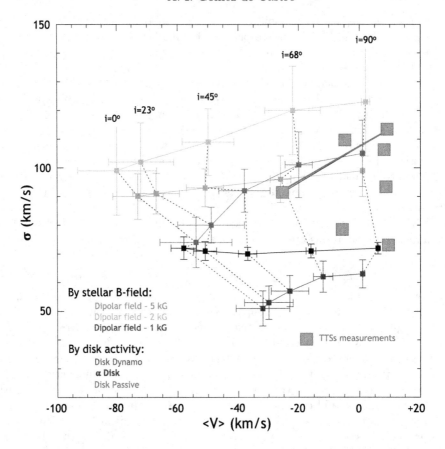

Figure 5. Velocity dispersion versus velocity centroid at the base of the jet (see Gómez de Castro & von Rekowski, 2011 for further details). The time-averaged values $\tilde{\sigma}$ and $<\tilde{V}>$ are represented as squares, while the error bars indicate the time variability during quiescence. Dark blue, black, green and light blue are used to represent the models: passive disk, matter falling onto a stellar dipolar field of 1 kG (Reference), 2kG (Mag-2kG) and 5 kG (Mag-5kG), respectively. Red is used for a magnetized disk with a dynamo built in. The *observed values* for all the TTSs observed with the HST/STIS (DE Tau, AK Sco, RY Tau, RW Aur, T Tau and RU Lup) are plotted as big brown squares. The two observations of RY Tau are plotted.).

occur on time scales comparable to the disk accretion time scale, eventually leading to collision (Lin & Papaloizou, 1993). There have been extensive theoretical studies of this situation, using both analytic and numerical methods (e.g., Goldreich & Tremaine, 1980; Lin & Papaloizou, 1986; Nelson & Papaloizou, 2003).

The inner hole excavated by the binary orbit is filled with transient dusty structures that transport the material from the inner border of the disk onto the stars. Angular momentum transport along the streamers dominates the dynamical evolution; according to hydrodynamical (HD) numerical simulations, streamers may carry as much as a 30% of the total accretion flow (Shi et al., 2012). In close PMS systems, accretion disks can either take up or release angular momentum and the details of evolution depend on the mass ratio between the two stars and on the orbit eccentricity (Artymowicz & Lubow, 1994; Bate & Bonnell, 1997, Hanawa et al., 2010, de Val Borro et al., 2011; Shi et al., 2012). Highly eccentric orbits favour the formation of spiral waves within the inner disk that do channel the flow as the accreting gas streams onto each star.

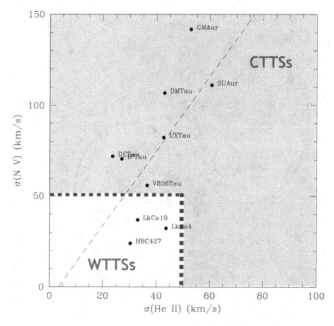

Figure 6. Broadening of the N V line (dispersion) compared with the broadening of the He II line.

For some specific orbital parameters circumstellar (CS) disks are formed around each component within the Roche Lobe; this configuration is shown in Figure 7 for the AK Sco system ($q \simeq 1$). HD simulations show that the gas flow consists of a circumbinary disk, a gap, circumstellar accretion disks, and a system of shock waves and tangential discontinuities; these elements are outlined in Figure 7. Note that the velocity distribution is non-Keplerian in the inner region of the circumbinary disk and that gas motion is governed by the bow shocks, one per star, and the gravity of the stellar components. The size and the shape of the gap are substantially determined by the bow shocks.

The study of AK Sco is a good example of the predictive power of numerical simulations. The orbit is elliptical ($e \simeq 0.47$) and the outer boundaries of the circumstellar disks (and the accretion streams passing by) get close enough at periastron passage to effectively lose the angular momentum leading to an increase of the accretion rate. Numerical results predict variations in the accretion rate *but they are not expected to be evenly distributed between the two components*; see the pronounced asymmetry in some cycles (*i.e.* cycle 12 in Figure 8). Moreover, inter-cycle variations occur in the total accretion rate, in its distribution between the components of the system, and also, in the details of the temporal evolution of the infall (see also Sytov & Fateeva, 2019 for a study of accretion under a different configuration in terms of stellar masses and eccentricity). These predictions were confirmed during the monitoring carried out with the Hubble Space Telescope. AK Sco was tracked during periastron passage in three consecutive orbits (Gómez de Castro et al. 2016; Gómez de Castro et al., 2020). Hubble UV observations show that during the first and the last cycles, the accretion rate increases at the periastron passage and the electron density becomes higher. However, not significant variations were observed during the intermediate cycle passage though the bulk UV radiation from the system increased by 20%; the different state of the system between the first two cycles at periastron passage is shown in Figure 9.

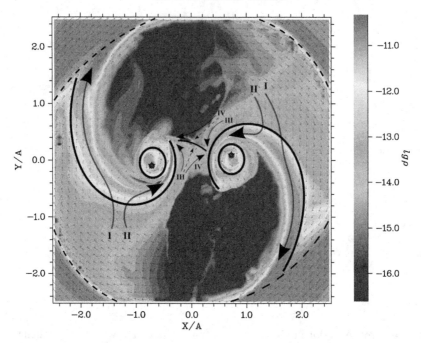

Figure 7. Generic configuration of matter distribution in a PMS close binary with q ≃ 1. The stellar orbit creates a large gap within the CB disk and matter is channelled onto the stars by the variable gravitational field generated by the binary orbit. Most of the matter is transferred through streamers that connect the inner border of the CB disk to the stars. CS disks are accommodated within the Roche lobe. If the mass of the components is significantly different, only the CS disk of the primary remains estable. The main dynamical features within the gap are outlined: (I) - the outflow to the circumbinary envelope; (II) - the accretion stream; (III, IV) - parts of accretion streams that form the inner stationary shock wave and contribute to the accretion rates. The dashed line indicates the boundary of the gap. Note that the arrows mark the direction of the flow in the bow shock (from Gómez de Castro et al., 2013).

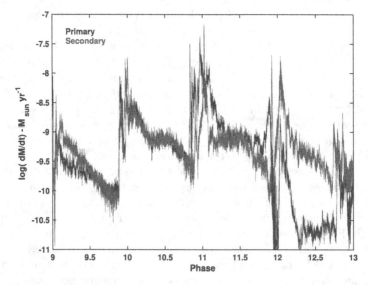

Figure 8. Predicted accretion rate onto both components of the system by HD numerical simulations of the evolution of the system (Gómez de Castro et al. 2016).

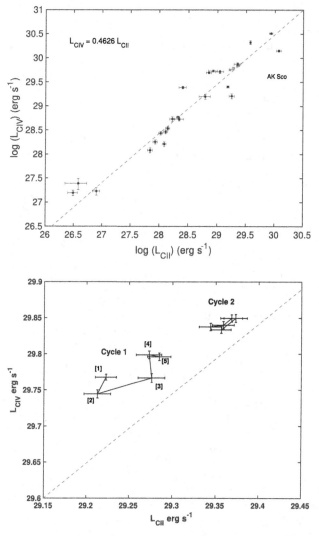

Figure 9. C IV - C II flux-flux diagram. Top, diagram for all the T Tauri stars observed with Hubble with the same configuration (STIS/G140L). AK Sco observations are plotted in blue (cycle 1) and red (cycle 2). The regression line is indicated. Bottom, zoom on the location of AK Sco observations. The order of the observations in cycle 1 is marked as [1], [2].... 1-σ error bars are plotted; note the different behaviour of the star in the two cycles (Gómez de Castro et al., 2020).

4. Summary: the predictive power of numerical simulations

Numerical simulations are a very powerful analysis tool and they are crucial for the interpretation of the data however, the physical framework needs to be clear beforehand and this comes from previous understanding, unavoidably linked to data acquisition, of the system. For instance, numerical simulation of magnetospheric accretion came roughly 10 years after the detection of hot spots on the surface of the TTSs and the evidence of the accretion shocks. However, numerical simulations are needed to predict the distribution of matter in the magnetosphere, the formation of funnel flows and to study the instabilities in the shock fronts. Also, they have contributed to disentangle the complex environment around the TTSs where the signatures of the magnetosphere, the outflow, and the infalling gas are observed in the same tracers. In some specific cases, such as the observation of close binary systems, numerical simulations have made detailed predictions instrumental to define the observing campaigns to test them, successfully.

References

Artymowicz, P., Lubow, S.H. 1994, *ApJ*, 421, 651

Bate, M.R., Bonnell, I.A. 1997, *MNRAS*, 288, 1041

Blandford, R. D., Payne, D.G. 1982, *MNRAS*, 199, 883

Bouvier, J., Alencar, S. H. P., Harries, T. J., Johns-Krull, C. M. et al. 2007, *Protostars and Planets V, B. Reipurth, D. Jewitt, and K. Keil (eds.), University of Arizona Press, Tucson*, 951

Calvet, N., Gullbring, E. 1998, *ApJ*, 509, 802

Donati, J. -F., Jardine, M. M., Gregory, S. G., Petit, P., et al. 2007, *MNRAS*, 380, 1297

Donati, J. -F., Jardine, M. M., Gregory, S. G., Petit, P., et al. 2008, *MNRAS*, 386, 1234

Donati, J. -F., Bouvier, J., Alencar, S. H., Hill, C., et al. 2019, *MNRAS*, 483, L1

de Val-Borro, M., Gahm, G. F., Stempels, H. C, et al. 2011, *A&A*, 535, 6

Ghosh, P., Lamb, F.K. 1979, *ApJ*, 232, 259

Goldreich, P.,S., Tremaine, S. 1980, *ApJ*, 241, 425

Gómez de Castro, A. I., Fernández, M. 1996, *MNRAS*, 283, 55

Gómez de Castro, A. I., Franqueira, M. 1997, *A&A*, 323, 541

Gómez de Castro, A. I., Lamzin, S. 1999, *MNRAS*, 304, L41

Gómez de Castro, A. I., von Rekowski, B. 2011, *MNRAS*, 411, 849

Gómez de Castro, A. I., Marcos-Arenal, P. 2012, *ApJ*, 749, 190

Gómez de Castro, A. I. 2012, *ApJ*, 775, 131

Gómez de Castro, A. I. 2013, *Planets, Stars and Stellar Systems Vol. 4, by Oswalt, Terry D.; Barstow, Martin A.*, 279

Gómez de Castro, A.I., Gaensicke, B., Neiner, C. & Barstow, M.A. 2016a, *JATIS*, 2, id. 041215

Gómez de Castro, A.I., Loyd, R. O. P., France, K., Sytov, A. et al. 2016b, *ApJ*, 818, L17

Gómez de Castro, A. I., Vallejo, J. C., Canet, A., Loyd, P. et al. 2020, *ApJ*, 904, 120

Goodson, A. P., Winglee, R. M., Bohm, K.-H. 1997, *ApJ*, 489, 199

Goodson, A. P., Winglee, R. M. 1999, *ApJ*, 524, 159

Gregory, S. G., Donati, J. -F., Morin, J., Hussain, G. A. J., et al. 2012, *ApJ*, 755, 97

Hanawa, T., Ochi, Y., Ando, K. 2010, *MNRAS*, 708, 485

Hartmann, L., Herczeg, G., Calvet, N. 2016, *ARAA*, 54, 135

Ireland, L. G., Zanni, C., Matt, S. P., Pantolmos, G. 2021, *ApJ*, 906, 4

Johns-Krull, C. M., Valenti, J. A.; Koresko, C. 1999, *ApJ*, 516, 900

Johns-Krull, C. M. 2007, *ApJ*, 664, 975

Li, R., Chen, Y.-X. L., Douglas N. C. 2022, *MNRAS*, 510, 5246

Lin, D.N.C., Papaloizou, J.C.B. 1992, *Protostars and planets III (A93-42937 17 90)*, 749

Long, M., Romanova, M. M., Lovelace, R. V. E. 2008, *MNRAS*, 386, 1274

Matt, S., Goodson, A. P., Winglee, R. M. 2002, *ApJ*, 574, 232

Nelson, R.P., Papaloizou, J.C.B. 2003, *Proceedings of the Conference on Towards Other Earths: DARWIN/TPF and the Search for Extrasolar Terrestrial Planets, Edited by M. Fridlund, T. Henning, compiled by H. Lacoste*, ESA SP-539, 175

Pantolmos, G., Zanni, C., Bouvier, J. 2020, *A&A*, 643, A129

Pudritz, R. E., Pelletier, G., Gómez de Castro, A. I. 1991, *The Physics of Star Formation and Early Stellar Evolution, NATO Advanced Study Institute (ASI) Series C*, 342, 539

Pudritz, R. E., Ouyed, R., Fendt, Ch., Brandenburg, A. 2007, *Protostars and Planets V, B. Reipurth, D. Jewitt, and K. Keil (eds.), University of Arizona Press, Tucson*, 277

Romanova, M. M., Ustyugova, G. V., Koldoba, A. V., Lovelace, R. V. E. 2002, *ApJ*, 578, 420

Romanova, M. M., Ustyugova, G. V., Koldoba, A. V., Wick, J.V., et al. 2003, *ApJ*, 595, 1009

Romanova, M. M., Ustyugova, G. V., Koldoba, A. V., Lovelace, R. V. E. 2004, *ApJ*, 610, 910

Romanova, M.M., Kulkarni, A.K., Long, M., & Lovelace, R.V.E. 2008, *AIP Conference Proceedings*, 1068, 87

Romanova, M. M., Ustyugova, G. V., Koldoba, A. V., Lovelace, R. V. E. 2012, *MNRAS*, 421, 63

Romanova, M. M., Koldoba, A. V., Ustyugova, G. V., Blinova, A. A. et al. 2021, *MNRAS*, 506, 372

Simon, T., Vrba, F. J., Herbst, W. 1990, *AJ*, 100, 1957

Shi, J.-M., Krolik, J.H., Lubow, S.H. et al. 2012, *ApJ*, 749, 118

Shu, F., Najita, J., Ostriker, E., Wilkin, F. et al. 1994, *ApJ*, 429, 781

Sytov, A. Yu., Fateeva, A.M. 2021, *ARep*, 63, 1045

von Rekowski, B., Brandenburg, A. 2006, *AN*, 327, 53

von Rekowski, B., Brandenburg, A. 2004, *A&A*, 420, 17

Weber, E.J., Davis, L. 1967, *ApJ*, 148, 217

Yang, H., Herczeg, G. J., Linsky, J.L., Brown, A. et al. 2012, *ApJ*, 744, 121

Discussion

QUESTION 1: I have a question concerning the topology of the magnetic field. What sets the footpoints of the outermost closed magnetic field lines, is it only the rotation or spin of the T Tauri star or is it also the relative amplitude of the momentum of the magnetic field of the star? How far is it with respect to the corotation radius?

ANSWER: The location depends both on the stellar magnetic field (including the details of the field topology and orientation) but also on the rotation velocity and the diffusivity of the material in the inner disk. Its location is ill determined from observations.

QUESTION 2: I am interested in the hot spot from accretion. Some people claim that the structure can be viewed by some recent observations not by light cap but by others, some method like obscuration or others. Do you have comments on that?

ANSWER: You see the spot basically at any time; it is very difficult not to see the accretion area. What it is difficult to separate the contribution from the hot spot from the rest of the atmosphere, since it shares similar spectral tracers. Monitoring and rotational modulation provide the best chances to pick out the contribution from the spot.

The Predictive Power of Computational Astrophysics as a Discovery Tool
Proceedings IAU Symposium No. 362, 2023
D. Bisikalo, D. Wiebe & C. Boily, eds.
doi:10.1017/S1743921322001545

Feedback from the Vicinity of Massive Protostars in the First Star Formation

Kazutaka Kimura[iD]

Center for Gravitational Physics and Quantum Information, Yukawa Institute for
Theoretical Physics, Kyoto University, Kyoto 606-8502, Japan
email: kazutaka.kimura@yukawa.kyoto-u.ac.jp

Abstract. Many simulations have been performed to elucidate the formation process of first stars. In first star formation, radiative feedback is a key process in determining stellar masses. However, previous simulations which follow the feedback process don't resolve the small scale ($\lesssim 10$ AU) to realize long-term calculation, and the structure near massive protostars is still unknown. To clarify how the radiation from the protostar works, we need to resolve small scale and calculate the interaction between the radiation and the dense gas in such a region. As a first step towards understanding the phenomenon in this region, we perform the high-resolution simulation around the massive protostar without radiative transfer. We find that dense gas covers the protostar even in the polar direction and the HII region cannot expand. Solving the radiative transfer for getting accurate results is our future work. We are currently developing the new radiation hydrodynamics code for that.

Keywords. stars: formation, accretion, accretion disks, HII regions

---

1. Introduction

First stars, also called Pop III stars, are formed in the early universe just after the Big Bang (Greif 2015; Haemmerlé *et al.* 2020). Their radiation heats and ionizes the interstellar medium and their supernova explosions spread heavy elements into the surrounding gas. First stars are thus critical to subsequent structure formation. Especially, the initial mass function (IMF) determines their roles and much research has been devoted to the study of the IMF.

In first star formation, radiative feedback is a key process which determines stellar masses (McKee & Tan 2008), as well as turbulence and magnetic fields (e.g. Sharda *et al.* 2021). In multi-dimensional simulations (Hosokawa *et al.* 2011; Sugimura *et al.* 2020), stellar radiation ionizes surrounding gas and creates HII region in polar direction. That region expands, and finally blows off the surrounding gas, which would otherwise accrete onto the central star. As a result, the final stellar masses are larger than present-day stellar ones and range from $10 - 1000$ $M_\odot$ (Hirano *et al.* 2014).

However, to reduce the computational cost for long-term calculations, previous simulations which follow the above feedback process don't resolve the structure within ~ 10 AU of the central protostar. Radiation from the protostar interacts with the dense gas near the protostar. Hence, to clarify how much the radiation can escape to a larger scale and whether the HII region can expand, we need to perform a high-resolution calculation near the massive protostar.

Here, we perform the simulation around the massive protostar with the resolution of 0.015 AU without radiative transfer. To solve radiative transfer is future work. We are currently developing the new radiation hydrodynamics code based on SFUMATO-RT

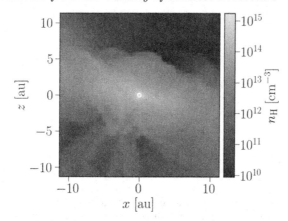

Figure 1. Simulation snapshots of the gas number density seen from the edge-on view. The central white circle represents the sink particle corresponding to the protostar. The sink radius is 0.25 AU, the stellar mass is 24 $M_\odot$, and the accretion rate is $\sim 10^{-4} - 10^{-3}$ $M_\odot$/yr.

which can follow the radiative transfer accurately in the very optically thick region near the massive protostar. SFUMATO-RT is the radiation hydrodynamics code developed by Sugimura *et al.* (2020) and includes adaptive mesh refinement, sink particle method, primordial gas chemistry network and cooling processes. Sink particle method is often used in star formation simulations to save the computational cost by simplifying the small scale structure (c.f. Federrath *et al.* 2010) and we also use it in this work.

Our simulation method is as follows. Here, we use the original SFUMATO-RT. As a initial condition, we pick up a typical primordial star-forming cloud from the cosmological simulations (Hirano *et al.* 2014, 2015). We can't simulate from this initial condition to the formation of massive protostar with high resolution all the time due to expensive computational cost. Thus, we first simulate with low resolution. At this time, the sink radius is 64 AU. Then, after the massive star is formed, we gradually shrink the sink radius to 0.25 AU.

2. Structure near the massive protostar without radiative transfer

Fig. 1 shows the gas number density in the simulation seen from the edge-on view. At this time, the protostellar mass is 24 $M_\odot$ and the accretion rate is $\sim 10^{-4} - 10^{-3}$ $M_\odot$/yr. As we can see, the protostar is covered with the dense gas even in the polar direction. Moreover, Fig. 2 shows the one-dimensional profile of the gas density and temperature in the polar direction. The blue line represents the simulation result and the orange dashed line represents the analytical solution of the static equilibrium of the gas pressure and the stellar gravity. The vertical dashed line corresponds to the sink radius. We can see the bounded structure (< 6 AU) by the protostellar gravity with the high density and temperature and fit the simulation result well with the analytical solution by tuning the mass and entropy in the bounded structure. The density is high up to 10^{14} cm^{-3} near the central protostar. In this bounded structure, the gas is fully ionized. Such a hot and ionized disk has been found in many previous one-dimensional disk models (e.g. Takahashi & Omukai 2017; Matsukoba *et al.* 2019; Kimura *et al.* 2021).

Note that in our simulation, we estimate the cooling rate by using the jeans length derived from the density and temperature at each cell. This method is accurate under the condition in which the self-gravity of the gas is dominant. However, near the massive protostar, the stellar gravity is strong and the self-gravity of the gas is subdominant. Then, this method is not accurate, and we overestimate the temperature. If we solve radiative transfer with the new code we are currently developing, this bounded structure

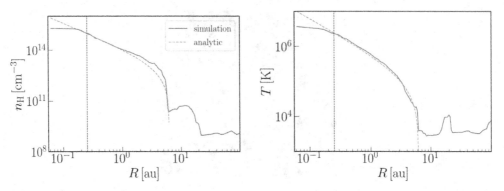

Figure 2. One-dimensional profile of gas density and temperature in the polar direction at the snapshot corresponding to Fig. 1. The blue line represents the simulation result, and the orange dashed line show the analytic solution of static equilibrium of the gas pressure and the stellar gravity.

would become cooler and smaller with lower gas pressure. Here, we estimate whether the HII region can expand or not, using this profile in the next section.

3. Can HII region expand?

Here, we estimate whether the HII region can expand when the gas structure in our simulation is realized. As shown in Fig. 2, in the bounded structure ($< 6\ AU$), the gas is very hot and fully ionized. Then, we judge whether the radiation from the surface can create the expanding HII region. The radiation from the bounded structure has an effective temperature of 10^4 K because the gas is very optically thick in the bounded structure and the photosphere lies almost at the surface. The density just outside the bounded structure is about 10^{10} cm^{-3}. At this time, the Strömgren radius is

$$r_{St} = \left(\frac{3\dot{Q}}{4\pi\alpha_B n_H^2}\right)^{1/3} \simeq 2\ \text{AU} \left(\frac{\dot{Q}}{10^{49}\ \text{s}^{-1}}\right)^{1/3} \left(\frac{n_H}{10^{10}\ \text{cm}^{-3}}\right)^{-2/3}, \qquad (3.1)$$

where $\dot{Q}$ is the emissivity of EUV photons and α_B is the case B recombination coefficient. This radius is within the bounded structure. Moreover, the escape velocity is

$$v_{\text{esc}} = \sqrt{\frac{2GM_*}{r}} \simeq 134\ \text{km/s} \left(\frac{M_*}{20\ M_\odot}\right)^{1/2} \left(\frac{r}{2\ \text{AU}}\right)^{1/2} \qquad (3.2)$$

where G is the gravitational constant. This value is larger than the typical sound velocity in the HII region and we conclude that IIII region cannot expand.

As said in the previous section, in our simulation we overestimate the temperature by the method using the jeans length. Additionally, we don't treat the radiation pressure by diffusive photons. Near the massive protostar, the diffusion process of the radiation is effective because the gas is very dense and optically thick. If we include these effects, there is the possibility that the bounded structure would change, that the radiation pressure can push off the gas in the polar direction, and that the HII region can expand. We plan to perform such a simulation in the future with the new code we are currently developing.

4. Summary

In first star formation, the radiation from the protostar creates the expanding HII region and this determines the stellar mass. However, previous simulations don't resolve the structure in the vicinity of the massive protostar, and how the gas and radiation interact with each other in this region is still unknown. To understand the gas structure in this region, here as a test calculation we perform the high-resolution simulation around the primordial massive protostar without radiative transfer.

As a result, the hot and dense bounded structure is formed around the protostar. If this gas structure is realized, the HII region cannot expand and the radiative feedback is not effective, which leads to very massive first stars. However, if we include the effects of the diffusive process of radiation and the radiation pressure, there is the possibility that HII region would expand.

Now, we are developing the new radiation hydrodynamics code which can calculate the radiative transfer in the optically thick region based on the SFUMATO-RT. In the future, we will perform the simulation with this new code to elucidate the structure near the massive protostar.

References

Federrath, C., Banerjee, R., Clark, P. C., Klessen R. S., 2010, *ApJ*, 713, 269

Greif, T. H. 2015, *Computational Astrophysics and Cosmology*, 2, 3

Haemmerlé, Mayer, Klessen, Hosokawa, Madau, & Bromm, 2020, *SSRv*, 216, 48

Hirano, S., Hosokawa, T., Yoshida, N., et al. 2014, *ApJ*, 781, 60,

Hirano, S., Hosokawa, T., Yoshida, N., Omukai, K., & Yorke, H. W. 2015, *MNRAS*, 448, 568

Hosokawa, T., Omukai, K., Yoshida, N., & Yorke, H. W. 2011, *Science*, 334, 1250

Kimura, K., Hosokawa, T., Sugimura, K. 2021, *ApJ*, 911, 52

Matsukoba, R., Takahashi, S. Z., Sugimura, K., & Omukai, K. 2019, *MNRAS*, 484, 2605

McKee, C. F., & Tan, J. C. 2008, *ApJ*, 681, 771

Sharda, P., Federrath, C., Krumholz, M. R., Schleicher, D. R. G., 2021, *MNRAS* (Letters), 503, 2014

Sugimura, K., Matsumoto, T., Hosokawa, T., Hirano, S., & Omukai, K. 2020, *ApJ* (Letters), 892, L14

Takahashi, S. Z., & Omukai, K. 2017, *MNRAS*, 472, 532

Discussion

AUDIENCE: You increased the resolution to the protostar and you try to get a kind of asymptotic regime which does not depend on the size of the sink. You showed the profile. Where does this analytic model come from?

KIMURA: This is a static equilibrium of the gas pressure and the gravity of the protostar.

AUDIENCE: Just a quick question about the initial configuration. How would you expect more angular momentum, for example, a few increased angular momentum? Do you think that it would make it easier to blow off the polar gas? Because it would be less dense, for example. Would you like to comment on that?

KIMURA: I have no exact answer. But I think this bounded structure is determined from the mass and entropy.

AUDIENCE: I'm cuious about the velocity structure of the disk. Whether it's Keplerian or maybe wouldn't at some radius?

KIMURA: Within 1 AU, there is a difference from the Keplerian motion.

The Predictive Power of Computational Astrophysics as a Discovery Tool
Proceedings IAU Symposium No. 362, 2023
D. Bisikalo, D. Wiebe & C. Boily, eds.
doi:10.1017/S1743921322001442

A cloud-cloud collision in Sgr B2? 3D simulations meet SiO observations

Wladimir E. Banda-Barragán[1][iD], Jairo Armijos-Abendaño[2] and Helga Dénes[3]

[1]Hamburger Sternwarte, Universität Hamburg, Gojenbergsweg 112, D-21029 Hamburg, Germany
email: wbanda@hs.uni-hamburg.de

[2]Observatorio Astronómico de Quito, Escuela Politécnica Nacional, Interior del Parque La Alameda, 170136, Quito, Ecuador

[3]ASTRON - The Netherlands Institute for Radio Astronomy, NL-7991 PD Dwingeloo, The Netherlands

Abstract. We compare the properties of shocked gas in Sgr B2 with maps obtained from 3D simulations of a collision between two fractal clouds. In agreement with ^{13}CO(1-0) observations, our simulations show that a cloud-cloud collision produces a region with a highly turbulent density substructure with an average $N_{\rm H2} \gtrsim 5 \times 10^{22}$ cm^{-2}. Similarly, our numerical multi-channel shock study shows that colliding clouds are efficient at producing internal shocks with velocities of $5 - 50$ km s^{-1} and Mach numbers of $\sim 4 - 40$, which are needed to explain the $\sim 10^{-9}$ SiO abundances inferred from our SiO(2-1) IRAM observations of Sgr B2. Overall, we find that both the density structure and the shocked gas morphology in Sgr B2 are consistent with a $\lesssim 0.5$ Myr-old cloud-cloud collision. High-velocity shocks are produced during the early stages of the collision and can ignite star formation, while moderate- and low-velocity shocks are important over longer time-scales and can explain the extended SiO emission in Sgr B2.

Keywords. hydrodynamics, radio lines: ISM, Galaxy: centre, ISM: clouds

---

1. Introduction

The Galactic centre, at a distance of ~ 8 kpc (Boehle et al. 2016; Gravity Collaboration et al. 2019), contains $\sim 3 \times 10^7$ M$_\odot$ of molecular gas with most of it lying inside a ring of clouds known as the Central Molecular Zone (CMZ). As a result, this zone harbours several regions of active star formation. One of such regions is Sgr B2, which contains $\sim 10^6 - 10^7$ M$_\odot$ of molecular gas (Molinari et al. 2011; Santa-Maria et al. 2021). Sgr B2 is located at a projected distance of ~ 100 pc away from the very centre of the Galaxy, which hosts Sgr A*, a supermassive black hole with an estimated mass of 4×10^6 M$_\odot$ (Gravity Collaboration et al. 2019). Given its location, the CMZ is subjected to strong tidal forces, which makes the local interstellar medium (ISM) highly pressurised when compared to the ISM in the disc of the Galaxy. Typical thermal pressures in the CMZ are $\sim 10^6 - 10^7$ K cm^{-3} (see Spergel and Blitz 1992; Santa-Maria et al. 2021), which are $\sim 1 - 2$ dex higher (see Crocker 2012) than the typical values of $\sim 10^4$ K cm^{-3} found in the Galactic disc.

Dynamical models of the CMZ propose that the cloud complexes in this region follow either a closed ∞-shaped orbit (see Molinari et al. 2011) or open orbits with at least four gas streams (see Kruijssen et al. 2015). Sgr B2 is located at the Eastern end of the CMZ (de Pree et al. 1995), at the intersection between the so-called x$_1$ and x$_2$ orbits of the inner Galaxy, which are produced by the gravitational effects of the Galactic bar (Binney et al. 1991). In the Molinari et al. scenario, which is also supported by previous authors (e.g. Hasegawa et al. 1994), Sgr B2 has an orbital speed of $\sim 80 \, \mathrm{km \, s^{-1}}$ and is the result of collisions of gas travelling along these two orbits. On the other hand, in the Kruijssen et al. scenario, Sgr B2 has an orbital speed of $\sim 130 \, \mathrm{km \, s^{-1}}$ and the star formation observed in it is due to gas compression caused by this cloud having passed the pericentre of the Galaxy $\sim 0.7 \, \mathrm{Myr}$ ago. In order to assess whether the cloud-cloud collision or the pericentre-passage scenarios or both can explain the formation and evolution of Sgr B2, combining information from observations and numerical simulations is essential.

In this report, we briefly summarise the findings of our recent paper, Armijos-Abendaño et al. (2020), in which we report and explain new observations of SiO emission in Sgr B2, together with new hydrodynamical simulations of cloud-cloud collisions. Since this report only provides a summary of our findings, we encourage the readers to read our full paper for a more thorough discussion.

2. Shocked gas structure and kinematics in Sgr B2

As in all the other molecular clouds in the CMZ, the kinematics of Sgr B2 is complex. Observations reveal that this region contains hot cores with number densities $> 10^6 \, \mathrm{cm^{-3}}$ embedded in a dense envelope with number densities $\sim 10^5 \, \mathrm{cm^{-3}}$ covering a few parsecs, and a more diffuse envelope with number densities $\sim 10^3 \, \mathrm{cm^{-3}}$ covering a diameter of $\sim 40 \, \mathrm{pc}$ (Schmiedeke et al. 2016). In this paper, we study the large-scale structure of Sgr B2 on scales covering $(15' \times 15')$, equivalent to an area of $(36 \times 36) \, \mathrm{pc^2}$ (at the distance of the Galactic centre).

Using the IRAM 30-m telescope, we have observed Sgr B2 and detected several molecular lines in emission, including SiO $J = 2\text{-}1$ at 86.8 GHz, $C^{18}O$ $J = 1\text{-}0$ at 109.8 GHz, and ^{13}CO $J = 1\text{-}0$ at 110.2 GHz (see the full sample in Table 1 of Armijos-Abendaño et al. 2020). The first molecular line, SiO $J = 2\text{-}1$, is a shock tracer (Schilke et al. 1997; Louvet et al. 2016), which we use to study the structure and kinematics of shocked gas in this region, the second one, $C^{18}O$ $J = 1\text{-}0$, allows us to estimate the SiO abundances with respect to molecular hydrogen (H_2), and the third molecular line, ^{13}CO $J = 1\text{-}0$, provides information on the typical hydrogen column densities. From the latter two, we find SiO relative abundances $N_{SiO}/N_{H_2} \sim 10^{-9}$, and a mean column number density of $\bar{N}_{H_2} \gtrsim 5 \times 10^{22} \, \mathrm{cm^{-2}}$, respectively.

We find that the SiO emission is very extended in Sgr B2, covering the full $(36 \times 36) \, \mathrm{pc^2}$ surveyed area and displaying a turbulent substructure with several arcs, cavities, and cores. Our observations also unveil a complex shocked gas kinematics as SiO emission covers a wide range of velocities, $[-5, +115] \, \mathrm{km \, s^{-1}}$. The spatial distribution of gas in different velocity channels reveal important properties about shocked gas in this region. Our maps indicate that shocked gas is turbulent in all velocity channels, and spatially anti-correlated at low and high velocities. Figure 1 shows two sample maps of integrated SiO emission at high velocities $[70, 85] \, \mathrm{km \, s^{-1}}$ and low velocities $[10, 25] \, \mathrm{km \, s^{-1}}$. These maps are spatially complementary, in agreement with expectations from cloud-cloud collision scenarios proposed earlier by e.g. Sato et al. (2000) based on a similar complementarity found based on ^{13}CO maps.

This spatial complementarity could also be caused by the superposition of clouds disconnected in 3D along the line of sight, but our position-velocity maps (such as the one displayed in Figure 1) show that the structures at low and high velocities are connected by bridges and V-shaped features, which are also characteristic of cloud-cloud collisions (see Torii et al. 2017; Enokiya et al. 2019). Therefore, we find that these

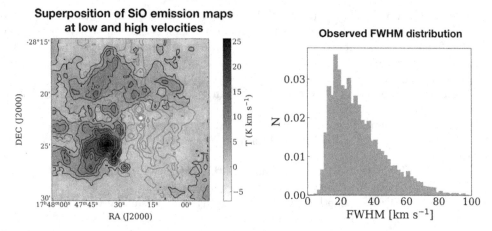

Figure 1. Left: Superimposed maps of SiO emission at low (black) and high (magenta) velocities, showing spatial anti-correlation. Black contour levels ([10, 25] km s^{-1}) are: 2.5, 5, 7.5, 10, 12.5, 15, 17.5, 20 K km s^{-1}; magenta contour levels ([70, 85] km s^{-1}) are: 3, 6, 9, 12, 15 K km s^{-1}. Right: Distribution of the FWHM of SiO J = 2-1 line emission. These panels have been adapted from Armijos-Abendaño et al. (2020), the reader is referred to that paper for further details.

structures are kinematically connected and likely belong to parcels of gas from the colliding clouds or gas streams, which would have been moving at different initial speeds. Similarly, we find that stellar feedback is unlikely to create the large-scale structure of Sgr B2 as we find no clear connection between broad SiO components and the local star-forming regions as expected in regions where stellar feedback plays a more important role (Jiménez-Serra et al. 2010). The small sizes < 0.3 pc of the local H II regions (Mehringer et al. 1993) also suggest stellar feedback would only be important in shaping sub-pc to pc structures, and that star formation, rather than the cause, is a product of the strong supersonic turbulence driven by e.g. the collision of gas clouds.

3. A cloud-cloud collision in Sgr B2

Cloud-cloud collisions are ubiquitous in ISM interactions (e.g. in shock-multicloud models, see Banda-Barragán et al. 2020, 2021), but can a cloud-cloud collision really explain the properties of shocked gas in Sgr B2? To answer this question and confirm whether or not a cloud-cloud collision can produce Sgr B2 and its star formation, we carry out two analyses: 1) we decompose the observed SiO spectra using GaussPy+ (Riener et al. 2019) in order to study the typical SiO integrated line intensities and FWHM line widths in this region, and 2) we compare these line widths, which we use as shock velocity proxies (in fact we assume that $v_{\text{shock}} \lesssim$ FWHM), to shock speeds measured directly from numerical simulations of cloud-cloud collisions (see Section 7.2 of Armijos-Abendaño et al. 2020).

From the first analysis, we find SiO integrated line intensities of the order of 11 ± 3 K km s^{-1}, peaking at ~ 5 K km s^{-1}, and SiO line widths of 31 ± 5 km s^{-1}, slightly higher than the background gas ^{13}CO, peaking at 21 km s^{-1}. The fact that SiO emission line widths are in the range of $5 - 50$ km s^{-1} suggests that shocks in Sgr B2 are predominantly moderate- and low-velocity shocks. Since we do not find a correlation between star-forming regions and zones with broad SiO lines, we argue that shocks in Sgr B2 emerge in supersonically-turbulent gas that has been produced by stirring after a cloud-cloud collision.

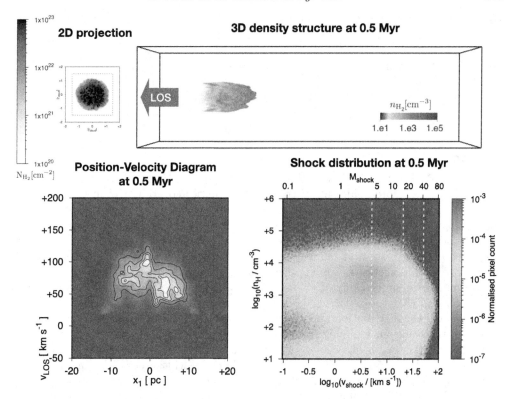

Figure 2. Cloud-cloud collision simulation at 0.5 Myr showing the 3D density structure (top right corner), the 2D column numbers density projection (top left corner), and the corresponding position-velocity diagram and shock distribution at the same time. These panels have been adapted from Armijos-Abendaño et al. (2020), the reader is referred to that paper and the simulation movies at `https://tinyurl.com/y5bc3smn` for further details.

From the second analysis regarding numerical simulations (see Figure 2), we find that colliding fractal clouds are efficient at producing internal shocks with velocities $\sim 5 - 50\,\mathrm{km\,s^{-1}}$ and typical shock Mach numbers of $\sim 4 - 40$ (akin to those reported by Henshaw et al. 2016 for the CMZ). In our models, clouds are initially moving at a relative velocity of $120\,\mathrm{km\,s^{-1}}$ as we find that speed is needed to produce shocks in high-velocity channels. Similarly, our simulations indicate that shocked gas (i.e. SiO emission) is efficiently produced during the early stages of the collision for $t < 0.5\,\mathrm{Myr}$ in all velocity channels. Most of the emission is concentrated at velocities in the range of $[25, 70]\,\mathrm{km\,s^{-1}}$. As gas decelerates following the collision, we find that shocked gas with lower and higher velocities than those limits quickly disappears at later stages of the evolution, allowing us to constrain the age of the collision. Position-velocity diagrams for SiO, obtained from the simulations, also show that V-shaped structures are short-lived and disappear at later stages, which further confirms the $t < 0.5\,\mathrm{Myr}$ time-scale. This time-scale is consistent with the estimated age of the shell-like structures in Sgr B2 reported by Tsuboi et al. (2015).

During the collision we find that the distribution of shocks also varies with time. At the very early stages of the collision for $t < 0.2\,\mathrm{Myr}$, the dominant shocks are high-velocity shocks with $v_{\mathrm{shock}} > 50\,\mathrm{km\,s^{-1}}$, which can easily ignite star formation on time-scales of $\sim 0.1\,\mathrm{Myr}$. Between $t = 0.2\,\mathrm{Myr}$ and $0.5\,\mathrm{Myr}$, the dominant shocks are moderate- and low-velocity shocks with speeds in the range of $v_{\mathrm{shock}} = 5 - 50\,\mathrm{km\,s^{-1}}$, which can maintain the widespread SiO emission. This range of shock velocities is also consistent with other shock models that reproduce the observed mid-J CO emission in Sgr B2 (see Santa-Maria et al. 2021). On the other hand, for $t > 0.5$ Myr, very-low-velocity shocks with $v_{\text{shock}} = 2 - 5 \, \text{km s}^{-1}$ and subsonic waves dominate the shock distribution. Thus, in this scenario the source of SiO emission is replenishment by a population of moderate- and low-velocity shocks as they have the appropriate velocities to trigger grain mantle sputtering (Gusdorf et al. 2008), which generally needs shocks speeds $> 7 \, \text{km s}^{-1}$, within the reasonably-long timescales of $\lesssim 10^5$ yr (Harada et al. 2015), which are needed to explain the observed SiO abundance in this region.

4. Conclusion

Our IRAM observations of SiO J=2-1 emission in Sgr B2 show that shocked gas in this region is widespread and has a complex kinematics covering a wide velocity range, $[-5, +115] \, \text{km s}^{-1}$. Shocked gas in this region has a turbulent substructure with a fractal morphology characterised by arcs, cavities, and cores. The spatial anti-correlation of gas at low and high velocities, and the presence of V-shaped features on position-velocity maps strongly suggests that Sgr B2 is a product of a cloud-cloud collision between gas likely travelling along the x_1 and x_2 orbits of the inner Galaxy.

Our numerical simulations of collisions between fractal clouds suggest that a cloud-cloud collision that took place $\lesssim 0.5$ Myr ago can readily explain both the gas density structure and the distribution and kinematics of shocked gas in Sgr B2. During the early stages of the collision, high-velocity shocks are produced with the ability to ignite star formation. Later on, shocks evolve into moderate- and low-velocity shocks with speeds in the range of 5 to $50 \, \text{km s}^{-1}$, which can explain the widespread SiO emission. Thus, in this scenario, Sgr B2 can be interpreted as a structure produced by turbulent stirring associated with colliding clouds or colliding gas streams.

The authors gratefully acknowledge the Gauss Centre for Supercomputing e.V. (www.gauss-centre.eu) for funding this project by providing computing time (via grant pn34qu) on the GCS Supercomputer SuperMUC-NG at the Leibniz Supercomputing Centre (www.lrz.de). WEBB is supported by the Deutsche Forschungsgemeinschaft (DFG) via grant BR2026/25, and by the National Secretariat of Higher Education, Science, Technology, and Innovation of Ecuador, SENESCYT, via grant 1711298438. We also thank the referee for their very helpful comments and suggestions, which helped us improve this paper.

References

Armijos-Abendaño, J., Banda-Barragán, W. E., Martín-Pintado, J., Dénes, H., Federrath, C., & Requena-Torres, M. A. 2020, Structure and kinematics of shocked gas in Sgr B2: further evidence of a cloud-cloud collision from SiO emission maps. *MNRAS*, 499(4), 4918–4939.

Banda-Barragán, W. E., Brüggen, M., Federrath, C., Wagner, A. Y., Scannapieco, E., & Cottle, J. 2020, Shock-multicloud interactions in galactic outflows - I. Cloud layers with lognormal density distributions. *MNRAS*, 499(2), 2173–2195.

Banda-Barragán, W. E., Brüggen, M., Heesen, V., Scannapieco, E., Cottle, J., Federrath, C., & Wagner, A. Y. 2021, Shock-multicloud interactions in galactic outflows - II. Radiative fractal clouds and cold gas thermodynamics. *MNRAS*, 506(4), 5658–5680.

Binney, J., Gerhard, O. E., Stark, A. A., Bally, J., & Uchida, K. I. 1991, Understanding the kinematics of Galactic Centre gas. *MNRAS*, 252, 210.

Boehle, A., Ghez, A. M., Schödel, R., Meyer, L., Yelda, S., Albers, S., Martinez, G. F., Becklin, E. E., & Do, T. 2016, . *ApJ*, 830, 17.

Crocker, R. M. 2012, Non-thermal insights on mass and energy flows through the Galactic Centre and into the Fermi bubbles. *MNRAS*, 423(4), 3512–3539.

de Pree, C. G., Gaume, R. A., Goss, W. M., & Claussen, M. J. 1995, The Sagittarius B2 Star-forming Region. II. High-Resolution H66 alpha Observations of Sagittarius B2 North. *ApJ*,

451, 284.

Enokiya, R., Torii, K., & Fukui, Y. 2019, . *Publ. Astron. Soc. Japan*, 00, 1–16.

Gravity Collaboration, Abuter, R., Amorim, A., Baubóck, M., Berger, J. P., Bonnet, H., Brandner, W., Clénet, Y., Coudé Du Foresto, V., de Zeeuw, P. T., Dexter, J., Duvert, G., Eckart, A., Eisenhauer, F., Förster Schreiber, N. M., Garcia, P., Gao, F., Gendron, E., Genzel, R., Gerhard, O., Gillessen, S., Habibi, M., Haubois, X., Henning, T., Hippler, S., Horrobin, M., Jiménez-Rosales, A., Jocou, L., Kervella, P., Lacour, S., Lapeyrère, V., Le Bouquin, J. B., Léna, P., Ott, T., Paumard, T., Perraut, K., Perrin, G., Pfuhl, O., Rabien, S., Rodriguez Coira, G., Rousset, G., Scheithauer, S., Sternberg, A., Straub, O., Straubmeier, C., Sturm, E., Tacconi, L. J., Vincent, F., von Fellenberg, S., Waisberg, I., Widmann, F., Wieprecht, E., Wiezorrek, E., Woillez, J., & Yazici, S. 2019, A geometric distance measurement to the Galactic center black hole with 0.3% uncertainty. *A&A*, 625, L10.

Gusdorf, A., Pineau des Forêts, G., Cabrit, S., & Flower, D. R. 2008, . *A&A*, 490, 695–706.

Harada, N., Riquelme, D., Viti, S., Jiménez-Serra, I., Requena-Torres, M. A., Menten, K. M., Martín, S., Aladro, R., Martín-Pintado, J., & Hochgürtel, S. 2015, . *A&A*, 584, A102.

Hasegawa, T., Sato, F., Whiteoak, J. B., & Miyawaki, R. 1994, . *ApJ*, 429, L77–L80.

Henshaw, J. D., Longmore, S. N., Kruijssen, J. M. D., Davies, B., Bally, J., Barnes, A., Battersby, C., Burton, M., Cunningham, M. R., Dale, J. E., Ginsburg, A., Immer, K., Jones, P. A., Kendrew, S., Mills, E. A. C., Molinari, S., Moore, T. J. T., Ott, J., Pillai, T., Rathborne, J., Schilke, P., Schmiedeke, A., Testi, L., Walker, D., Walsh, A., & Zhang, Q. 2016, Molecular gas kinematics within the central 250 pc of the Milky Way. *MNRAS*, 457(3), 2675–2702.

Jiménez-Serra, I., Caselli, P., Tan, J. C., Hernand ez, A. K., Fontani, F., Butler, M. J., & van Loo, S. 2010, Parsec-scale SiO emission in an infrared dark cloud. *MNRAS*, 406(1), 187–196.

Kruijssen, J. M. D., Dale, J. E., & Longmore, S. N. 2015, The dynamical evolution of molecular clouds near the Galactic Centre - I. Orbital structure and evolutionary timeline. *MNRAS*, 447(2), 1059–1079.

Louvet, F., Motte, F., Gusdorf, A., Nguyên Luong, Q., Lesaffre, P., Duarte-Cabral, A., Maury, A., Schneider, N., Hill, T., Schilke, P., & Gueth, F. 2016, Tracing extended low-velocity shocks through SiO emission. Case study of the W43-MM1 ridge. *A&A*, 595, A122.

Mehringer, D. M., Palmer, P., Goss, W. M., & Yusef-Zadeh, F. 1993, . *ApJ*, 412, 684–695.

Molinari, S., Bally, J., Noriega-Crespo, A., Compiègne, M., Bernard, J. P., Paradis, D., Martin, P., Testi, L., & Barlow, M. 2011, . *ApJ*, 735, L33.

Riener, M., Kainulainen, J., Henshaw, J. D., Orkisz, J. H., Murray, C. E., & Beuther, H. 2019, GAUSSPY+: A fully automated Gaussian decomposition package for emission line spectra. *A&A*, 628, A78.

Santa-Maria, M. G., Goicoechea, J. R., Etxaluze, M., Cernicharo, J., & Cuadrado, S. 2021, Submillimeter imaging of the Galactic Center starburst Sgr B2. Warm molecular, atomic, and ionized gas far from massive star-forming cores. *A&A*, 649, A32.

Sato, F., Hasegawa, T., Whiteoak, J. B., & Miyawaki, R. 2000, . *ApJ*, 535, 857–868.

Schilke, P., Walmsley, C. M., Pineau des Forets, G., & Flower, D. R. 1997, SiO production in interstellar shocks. *A&A*, 321, 293–304.

Schmiedeke, A., Schilke, P., Möller, T., Sánchez-Monge, A., Bergin, E., Comito, C., Csengeri, T., Lis, D. C., & Molinari, S. 2016, . *A&A*, 588, A143.

Spergel, D. N. & Blitz, L. 1992, Extreme gas pressures in the galactic bulge. *Nat*, 357(6380), 665–667.

Torii, K., Hattori, Y., Hasegawa, K., Ohama, A., Haworth, T. J., Shima, K., Habe, A., Tachihara, K., Mizuno, N., Onishi, T., Mizuno, A., & Fukui, Y. 2017, Triggered O Star Formation in M20 via Cloud-Cloud Collision: Comparisons between High-resolution CO Observations and Simulations. *ApJ*, 835(2), 142.

Tsuboi, M., Miyazaki, A., & Uehara, K. 2015, . *Publ. Astron. Soc. Japan*, 67, 90.

Discussion

Question 1: How do you do this morphological comparison between the simulations and the observations? It sounds like a very complicated task to make sense out of it.

Answer 1: We basically carry out two comparisons. One-to-one comparisons are always very hard, but what we did was to take the SiO emission maps from the observations in different velocity channels and study the shock distributions in our simulations within the same velocity channels. In that way we compared the shock velocities that we got from the simulations with the line widths of SiO emission from the observations, and we found that they are consistent if the time-scale of the collision evolution is < 0.5 Myr. The second comparison involves position-velocity diagrams, which display zig-zag and V-shaped features that we can also reproduce with our simulations within a similar time-scale. Both are qualitative comparisons.

Question 2: Would machine learning help here to do such a comparison in all dimensions at the same time?

Answer 2: Possibly yes. We actually did incorporate a bit of machine learning in the analysis. The spectral decomposition of the observed emission lines was done using GaussPy+, which uses machine learning for the spectral decomposition. Machine learning would also very likely help to do multi-dimensional analyses on the data in the future.

Question 3: Did you compare your simulations/observations with other molecular line emissions, like CO, e.g. ^{13}CO?

Answer 3: Yes, in the paper we also included analyses for $C^{18}O$ and ^{13}CO, which also show similar kinematics and in the latter case also a spatial anti-correlation at low- and high- velocities similar to that seen in SiO emission. We also decomposed the ^{13}CO spectra and found that it has a similar FWHM distribution than SiO, but it is a little bit shifted towards lower velocities. In general, SiO has larger FWHM than CO because gas is supersonically turbulent.

Question 4: How about the spatial distribution of CO? I guess you can reproduce the main features.

Answer 4: Yes, we can reproduce the main (fractal-like) density features too.

Question 5: It seems like with these models where you have two colliding clouds, you get these two-velocity features with the bridge between them. It has never been clear to me if this is a unique feature of cloud-cloud collisions although it is always interpreted as a sign of it. I guess with your models, you could actually do different kinds of models that are not necessarily cloud-cloud collisions and see if you get any similar looking features. Have you thought about this?

Answer 5: I think these bridges are quite unique features of cloud-cloud collisions because if we would have just two clouds moving at different velocities, not colliding but disconnected along the line of sight, these features would be absent. In the paper we also discussed how the initial density distribution of the colliding clouds affect the density maps and for example if we assume that the clouds do not have this turbulent substructure, but they are just spherical clouds with uniform densities, then it would be harder to reproduce the observations.

Question 6: I was thinking more of the idea of for example if you have a cloud that has been torn apart by shear then you might see a feature like that.

Answer 6: I agree. I think shear could also cause a feature like the one seen in cloud-cloud collisions.

Question 7: In the movie of the colliding clouds you showed, on the left one you see these streamers at the back of this cloud. What are they?

Answer 7: The simulation is idealised and I set up one of the clouds at rest while the other one is moving on the grid. Because we have one cloud initially moving across the domain, this cloud is going to suffer from stripping by Kelvin-Helmholtz instabilities. Therefore, these streamers are sort of tails that form from mass stripping via dynamical instabilities. The simulation movies are available at `https://tinyurl.com/y5bc3smn`.

The Predictive Power of Computational Astrophysics as a Discovery Tool
Proceedings IAU Symposium No. 362, 2023
D. Bisikalo, D. Wiebe & C. Boily, eds.
doi:10.1017/S1743921322001508

Formation process of the Orion Nebula Cluster

Michiko S. Fujii[1], **Long Wang[2]**, **Takayuki R. Saitoh[3]**,
Yutaka Hirai[4,5] and Yoshito Shimajiri[6]

[1]The University of Tokyo

[2]Sun Yat-sen University

[3]Kobe University

[4]University of Notre Dame

[5]Tohoku University

[6]National Astronomical Observatory of Japan

Abstract. The Orion Nebula Cluster (ONC) is one of the nearest open clusters, which we can directly compare to numerical simulations. We performed a simulation of star cluster formation similar to the ONC using our new N-body/smoothed particle hydrodynamics code, ASURA+BRIDGE. We found that the hierarchical formation of star clusters via clump mergers can explain the observed three peaks in the stellar age distribution as well as the dynamically anisotropic structures of the ONC.

Keywords. methods: n-body simulations, methods: numerical, open clusters and associations: individual (Orion Nebula Cluster)

1. Introduction

The Orion Nebula Cluster (ONC) is one of the nearest open clusters. It is $\sim 400\,\mathrm{pc}$ away from us (Kounkel et al. 2017), and the age is estimated to be 1–3 Myr (Hillenbrand and Hartmann 1998). Thanks to the distance, we can observe the detailed structures of this cluster. Recent observations have shown that it has three distinct age populations (Beccari et al. 2017). One possible scenario explaining it is the complete ejection of O-stars during the formation of the ONC. Kroupa et al. (2018) suggested that massive stars formed in star clusters suspend star formation until they are completely ejected from the clusters due to binary-single encounters. Once all massive stars are ejected, star formation starts again. The complete ejection of massive stars was examined using N-body simulations (Wang et al. 2019), but the formation of new stars was not included in their study.

2. Simulation

We performed an N-body/smoothed-particle hydrodynamics simulation for the formation of an ONC-like cluster. We used our new code ASURA+BRIDGE (Fujii et al. 2021b,a). In this code, star formation is assumed using a probabilistic formation following a given initial mass function (Hirai et al. 2021). Once massive stars formed more massive than $10\,M_\odot$, stellar feedback is switched on. We assumed that individual massive stars form HII regions with a Strömgren radius calculated using local gas density and the photon emission rate. The details of the simulation are described in Fujii et al. (2021a).

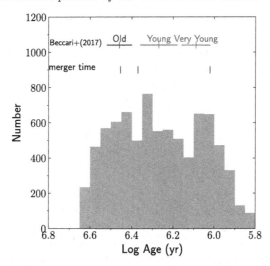

Figure 1. Age distribution of stars in the star cluster at 8.7 Myr. Blue, green, and red bars indicate the three age distributions identified in Beccari et al. (2017). Black lines indicate the merger timings of clumps.

One of the main advantages of ASURA+BRIDGE is that we can integrate stellar orbits without any gravitational softening. In ASURA+BRIDGE, stars are integrated using a direct-tree hybrid integrator, Particle-Particle Particle-Tree (P^{3}T; Oshino et al. 2011; Iwasawa et al. 2015). We adopted an open-source code, PETAR (Wang et al. 2020a), for the integration of stars. PETAR includes a Slow-Down Algorithmic Regularization (SDAR) scheme for the integration of binaries and multiples (Wang et al. 2020b), which enables us fast and stable integration of star clusters including hard binaries.

For the initial condition, we adopted a uniform spherical molecular cloud with a turbulent velocity field with a mass of $2 \times 10^4 M_\odot$ and a radius of 12 pc. The initial SPH particle mass is 0.01 $M_\odot$, and the softening length for the gas particle is 0.07 pc. We did not use any softening for the star-star interactions to follow the dynamical evolution of star clusters.

3. Results

3.1. *Three Populations in the ONC*

Star formation started after about one initial free-fall time of the cloud (~ 5 Myr). Once massive stars are formed, they begin ionizing the surrounding gas. However, not the entire region is immediately ionized. Dense regions continue star formation until enough massive stars are formed. In this simulation, the star formation was terminated at ~ 9 Myr.

We found that the star cluster evolved via mergers of smaller clumps and that these mergers enhanced the star formation and made the three populations in age. In Figure 1, we present the age distribution of the cluster at 8.7 Myr from the beginning of the simulation, which is the time we assume that the simulated cluster is the closest to the current ONC. We found three peaks in the age distribution and that each peak corresponds to a clump merger. Clump mergers bring dense gas to the clump center with stars, and it enhances the star formation inside the merged cluster. We confirmed this process by analyzing the amount of dense gas in the cluster.

3.2. *Dynamical Structures of the ONC*

We investigated the dynamical structures of our simulated cluster. We found that we observe anisotropic structures in the velocity space only within $\sim 0.5\,$Myr after clump mergers. We followed the three-dimensional velocity structure (velocity dispersions) and virial ratio every 0.05 Myr. We found that the velocity anisotropy and virial ratio increase after clump mergers and drop in $\sim 0.5\,$Myr. Observed high velocity anisotropy ($\sigma_{v_r}/\sigma_{v_\alpha} = 1.44^{+0.21}_{-0.18}$ and $\sigma_{v_\delta}/\sigma_{v_\alpha} = 1.23^{+0.17}_{-0.16}$; Theissen et al. 2021) and virial ratio (0.7 ± 0.3; Kim et al. 2019) of the ONC suggest that the ONC experienced a clump merger in the past 0.5 Myr.

4. Summary

We performed an N-body/SPH simulation of an ONC-like star cluster. In this simulation, we can follow the orbits of individual stars, including their strong dynamical interactions. In our simulation, stellar clump mergers brought dense molecular gas into the forming cluster and enhanced the star formation in the cluster. This process formed three peaks in the age distribution of stars in the cluster at the time similar to the age of the ONC. We also found that clump mergers cause velocity anisotropy similar to that observed in the ONC. The virial ratio also increased after clump mergers.

References

Beccari, G., Petr-Gotzens, M. G., Boffin, H. M. J., Romaniello, M., Fedele, D., Carraro, G., De Marchi, G., de Wit, W. J., Drew, J. E., Kalari, V. M., Manara, C. F., Martin, E. L., Mieske, S., Panagia, N., Testi, L., Vink, J. S., Walsh, J. R., & Wright, N. J. 2017, A tale of three cities. OmegaCAM discovers multiple sequences in the color-magnitude diagram of the Orion Nebula Cluster. *A&A*, 604, A22.

Fujii, M. S., Saitoh, T. R., Hirai, Y., & Wang, L. 2021,a SIRIUS project. III. Star-by-star simulations of star cluster formation using a direct N-body integrator with stellar feedback. *Publ. Astron. Soc. Japan*, 73a(4), 1074–1099.

Fujii, M. S., Saitoh, T. R., Wang, L., & Hirai, Y. 2021,b SIRIUS project. II. A new tree-direct hybrid code for smoothed particle hydrodynamics/N-body simulations of star clusters. *Publ. Astron. Soc. Japan*, 73b(4), 1057–1073.

Hillenbrand, L. A. & Hartmann, L. W. 1998, A Preliminary Study of the Orion Nebula Cluster Structure and Dynamics. *ApJ*, 492(2), 540–553.

Hirai, Y., Fujii, M. S., & Saitoh, T. R. 2021, SIRIUS project. I. Star formation models for star-by-star simulations of star clusters and galaxy formation. *Publ. Astron. Soc. Japan*, 73(4), 1036–1056.

Iwasawa, M., Portegies Zwart, S., & Makino, J. 2015, GPU-enabled particle-particle particle-tree scheme for simulating dense stellar cluster system. *Computational Astrophysics and Cosmology*, 2, 6.

Kim, D., Lu, J. R., Konopacky, Q., Chu, L., Toller, E., Anderson, J., Theissen, C. A., & Morris, M. R. 2019, Stellar Proper Motions in the Orion Nebula Cluster. *AJ*, 157(3), 109.

Kounkel, M., Hartmann, L., Loinard, L., Ortiz-León, G. N., Mioduszewski, A. J., Rodríguez, L. F., Dzib, S. A., Torres, R. M., Pech, G., Galli, P. A. B., Rivera, J. L., Boden, A. F., Evans, Neal J., I., Briceño, C., & Tobin, J. J. 2017, The Gould's Belt Distances Survey (GOBELINS) II. Distances and Structure toward the Orion Molecular Clouds. *ApJ*, 834(2), 142.

Kroupa, P., Jeřábková, T., Dinnbier, F., Beccari, G., & Yan, Z. 2018, Evidence for feedback and stellar-dynamically regulated bursty star cluster formation: the case of the Orion Nebula Cluster. *A&A*, 612, A74.

Oshino, S., Funato, Y., & Makino, J. 2011, Particle-Particle Particle-Tree: A Direct-Tree Hybrid Scheme for Collisional N-Body Simulations. *Publ. Astron. Soc. Japan*, 63, 881.

Theissen, C. A., Konopacky, Q. M., Lu, J. R., Kim, D., Zhang, S. Y., Hsu, C.-C., Chu, L., & Wei, L. 2021, The 3-D Kinematics of the Orion Nebula Cluster: NIRSPEC-AO Radial Velocities of the Core Population. *arXiv e-prints*, arXiv:2105.05871.

Wang, L., Iwasawa, M., Nitadori, K., & Makino, J. 2020,a PETAR: a high-performance N-body code for modelling massive collisional stellar systems. *MNRAS*, 497a(1), 536–555.

Wang, L., Kroupa, P., & Jerabkova, T. 2019, Complete ejection of OB stars from very young star clusters and the formation of multiple populations. *MNRAS*, 484(2), 1843–1851.

Wang, L., Nitadori, K., & Makino, J. 2020,b A slow-down time-transformed symplectic integrator for solving the few-body problem. *MNRAS*, 493b(3), 3398–3411.

The Predictive Power of Computational Astrophysics as a Discovery Tool
Proceedings IAU Symposium No. 362, 2023
D. Bisikalo, D. Wiebe & C. Boily, eds.
doi:10.1017/S1743921322001399

PION: Simulations of Wind-Blown Nebulae

Jonathan Mackey[1,2] , **Samuel Green**[1,2], **Maria Moutzouri**[1,2],
Thomas J. Haworth[3], **Robert D. Kavanagh**[4] , **Maggie Celeste**[1,5],
Robert Brose[1,2], **Davit Zargaryan**[1,2] and **Ciarán O'Rourke**[6]

[1]Dublin Institute for Advanced Studies, 31 Fitzwilliam Place, Dublin 2, Ireland

[2]DIAS Dunsink Observatory, Dunsink Lane, D15 XR2R, Ireland

[3]Astronomy Unit, School of Physics and Astronomy, Queen Mary University of London, London E1 4NS, UK

[4]Leiden Observatory, Leiden University, PO Box 9513, 2300 RA, Leiden, The Netherlands

[5]School of Physics, Trinity College Dublin, The University of Dublin, Dublin 2, Ireland

[6]Irish Centre for High-End Computing (ICHEC), NUI Galway, Galway City, Ireland

Abstract. We present an overview of PION, an open-source software project for solving radiation-magnetohydrodynamics equations on a nested grid, aimed at modelling asymmetric nebulae around massive stars. A new implementation of hybrid OpenMP/MPI parallel algorithms is briefly introduced, and improved scaling is demonstrated compared with the current release version. Three-dimensional simulations of an expanding nebula around a Wolf-Rayet star are then presented and analysed, similar to previous 2D simulations in the literature. The evolution of the emission measure of the gas and the X-ray surface brightness are calculated as a function of time, and some qualitative comparison with observations is made.

Keywords. (magnetohydrodynamics:) MHD shock waves, methods: numerical, stars: mass loss, stars: Wolf-Rayet, ISM: bubbles

1. Introduction

The late stages of massive-star evolution, after the main sequence, remain very uncertain (Langer 2012; Smith 2014), despite significant advances in recent years. In particular, uncertainties in the mechanisms of mass loss of Red Supergiants (RSG) and Luminous Blue Variables (LBV) make it difficult to predict whether a given star can lose its envelope through mass loss during its lifetime and become a Wolf-Rayet (WR) star. Mass transfer and mass loss induced by interaction with a binary companion is common in the evolution of massive stars (Sana et al. 2012), and so the relative importance of mass loss through winds and eruptions, versus mass-loss through binary interaction, is an active field of research. This lack of knowledge translates to a large systematic uncertainty when making predictions for evolutionary tracks of massive stars and for connecting observed properties of supernova progenitors with the zero-age main-sequence progenitor properties (e.g. mass).

Many post-main-sequence massive stars are surrounded by nebulae that can be composed of mass lost during previous evolution and/or interstellar gas swept up by the expanding stellar wind (García-Segura et al. 1996). The dynamical timescale of these nebulae is $\sim 10^4 - 10^5$ years, not much shorter than the $\sim 10^5 - 10^6$ year timescales of post-main-sequence nuclear burning in massive stars. Constraints on the previous evolution of stars can therefore be obtained by comparing predictions of stellar evolution calculations with observed nebulae (Mackey et al. 2012; Meyer et al. 2020). This is not a trivial undertaking because the shocks driven by expanding wind bubbles are often radiative and subject to dynamical instabilities (Garcia-Segura & Mac Low 1995), and effects of thermal conduction (Comerón & Kaper 1998; Meyer et al. 2014) and magnetic fields (van Marle et al. 2015; Meyer et al. 2017) can be significant. Multi-dimensional simulations are often required to capture the effects of these processes on the evolution of a circumstellar nebula.

2. Methods

In Mackey et al. (2021) we presented the first public release of PION, a radiation-magnetohydrodynamics (R-MHD) code for simulating the circumstellar medium (CSM) around massive stars and, more generally, feedback of massive stars on the interstellar medium (ISM). PION uses the finite-volume method on a rectilinear grid of cubic cells to solve the equations of hydrodynamics or ideal-MHD following the time-integration scheme of Falle et al. (1998). A short-characteristics raytracing module is used to calculate photon fluxes at each cell, that can be used to obtain non-equilibrium ionization, heating and cooling rates (Mackey 2012). Nested grids (i.e. static mesh-refinement) focus resolution on certain points in the domain, usually on the star for the case of wind-blown nebulae (cf. Yorke & Kaisig 1995; Freyer et al. 2003). PION also has stellar-wind boundary conditions for non-rotating and rotating, evolving stars, following methods developed for Heliosphere modelling (Pogorelov et al. 2004). The source code and tutorials for PION v2.0 are available from a website† and a VCS repository‡ with a BSD 3-clause license.

PION is written in C++ and is parallelised using domain decomposition into sub-domains with the Message Passing Interface (MPI). Since v2.0 was released we have worked on improving parallel scaling to larger number of cores, by upgrading the MPI communication algorithms and implementing hybrid OpenMP/MPI parallelisation. The idea behind this is that the ratio of (duplicated) boundary data to domain data increases with the number of sub-domains (PION allocates one sub-domain per MPI process on each level of the nested grid), and this is the fundamental limit to the parallel scaling. If we reduce the number of MPI processes for a given core count, by using OpenMP threads, then the parallel scaling should improve.

It was relatively straightforward to rewrite the loops over cells within each sub-domain so that they are loops over 1D columns of cells. Each OpenMP thread is then given a 1D column of cells to calculate on for the various tasks (calculation of timestep, hydrodynamic fluxes in each direction, radiative heating and cooling, ionization and recombination, and cell updates). The most difficult work involved making various function calls threadsafe for OpenMP execution, which required some re-design of the data structures. Code optimization work is still ongoing, but there is already signficiant improvement.

Fig. 1 shows preliminary results of the strong scaling using upgraded MPI communication and either 1 or 5 OpenMP threads per MPI process. The speedup achieved with respect to a simulation using 32 cores is compared with the ideal scaling. The scaling that was achieved using the same simulation in Mackey et al. (2021) is shown for comparison, and it is clear that we have made significant progress. Using 5 threads per rank we are able to achieve good scaling up to at least 2560 cores, 5× as many cores as previously. The speedup has still not saturated at this number of cores, for a test calculation with 256^3 cells per level and 3 refinement levels. This gives us confidence that we could achieve good scaling to 10 000 cores or more on larger grids of, e.g., 512^3 or 1024^3 cells per refinement level. These scaling improvements are enabled by our participation in the

† https://www.pion.ie
‡ https://git.dias.ie/massive-stars-software

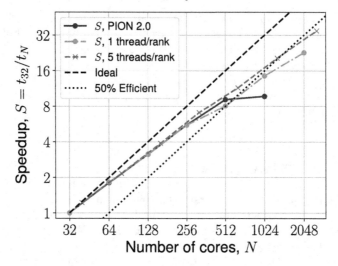

Figure 1. Strong scaling of PION for a 3D MHD simulation of a bow shock around a runaway massive star. The simulation has 256^3 grid cells per level and 3 levels of refinement. Results from PION v2.0 from Mackey et al. (2021) (blue solid line) are compared with an upgraded version of PION run with 1 (cyan dot-dashed line) and 5 (magenta dashed line) OpenMP threads per MPI process. The speedup, S, is defined as the run duration using N cores, t_N, divided by the run duration using 32 cores, t_{32}.

EuroCC-Ireland Academic Flagship Programme, and will be included in the next release of PION.

Synthetic emission maps for some observational tracers (emission measure, broadband X-ray emission, Hα) can be generated with a projection code distributed with PION, which stores the projected maps in binary VTK format. A python library is provided that can read both PION snapshots and the VTK projected-data files and generate simple plots. For calculating infrared dust emission, PION snapshots are exported to the TORUS Monte-Carlo radiative transfer code (Harries et al. 2019). This then determines the spatially varying dust temperature via a radiative equilibrium calcuation, and produces emission maps of the thermally emitting dust grains at specified frequencies (Green et al. 2019).

3. Wolf-Rayet Nebulae

Nebulae around WR stars are often spectacular objects, as the intense ionizing radiation and dense, fast winds of the stellar core sweep through the remnants of the extended stellar envelope, lost either through binary stripping or winds/eruptions. Good examples are the nebula M1-67 around WR 124 (Grosdidier et al. 1998) and NGC 3199 around WR 18 (Toalá et al. 2017). The first multidimensional calculations of this process by García-Segura et al. (1996) followed the CSM around a 35 M$_\odot$ star as it evolved from a O star $\to$ RSG $\to$ WR, using 1D stellar-evolution calculations as a time-dependent inner boundary condition to 2D hydrodynamical simulations in spherical geometry. Later the same evolutionary calculation was used in radiation-hydrodynamical (R-HD) simulations using a 2D nested grid in cylindrical coordinates (Freyer et al. 2006).

In Mackey et al. (2021) we again used the same evolutionary calculation on a 3D Cartesian nested grid to study the performance of PION on a calculation where there are 2D reference results in the literature. The grid has 256^3 cells on each level and 4 levels, the coarsest level being a cube with domain $\{x, y, z\} \in [-30, 30] \times 10^{18}$ cm and each finer level 2$\times$ smaller in each dimension and centred on the star located at the

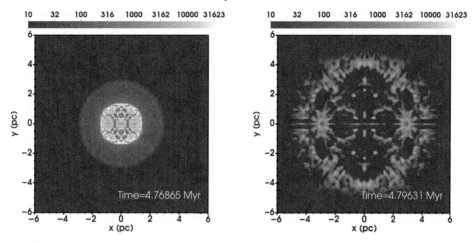

Figure 2. Emission measure from the expanding nebula 14 000 years (left) and 42 000 years (right) after the RSG→WR transition, plotted on a logarithmic scale with units cm^{-6} pc. Time is shown since the beginning of the stellar evolution calculation; the RSG phase ended at $t \approx$ **4.755** Myr.

origin. The wind boundary region has a radius of 20 cells, $\approx 5.86 \times 10^{17}$ cm. The time evolution of the stellar wind and radiation, as well as snapshots showing slices through the 3D simulation domain, are presented in Mackey et al. (2021).

Here we present synthetic observations of the simulation, calculating the projected Emission Measure (EM) and X-ray surface brightness using a raytracing method, neglecting internal absorption. The EM is plotted in Fig. 2 as the fast wind from the WR star sweeps up the wind bubble of the previous RSG phase. The slow and dense wind from the RSG expanded to distance from the star (located at the origin) of $r \approx 3$ pc, and the WR wind-bubble expands through this dense gas at about 120 km s^{-1}. The wind termination shock is adiabatic, but the forward shock is strongly radiative leading to dynamical instability in the swept-up shell. 14 000 years after the RSG→WR transition (at $t \approx 4.755$ Myr), the shell has expanded to 1.7 pc and remains relatively regular and symmetric. Once the swept-up shell breaks out of the RSG wind region it breaks up into clumps and filaments, and 42 000 years after the transition it is very fragmented and qualitatively similar in appearance to Galactic WR nebulae like M1-67.

X-ray emission from 0.3-10 keV is shown in Fig. 3, calculated using the method in Green et al. (2019), plotted at the same times as for the EM. When the expanding WR wind is still contained within the RSG wind region the X-ray emission is bright and fills the WR wind bubble, slightly limb-brightened. At later times it is about 10× fainter and associated with the boundary layers where hot coronal gas is dynamically mixing with photoionized (and much denser) nebular gas. For both plots (EM and X-ray) the clumps and filaments have a very regular and symmetric appearance. This is because the instabilities are seeded by the computational grid, and so the resulting structures that develop reflect this cubic Cartesian grid.

In reality RSG winds are clumped, asymmetric and variable (O'Gorman et al. 2015; Humphreys et al. 2021), reflecting their turbulent outer convective layers, and so we should not expect detailed agreement between our simple model and any observed nebula. Also, rotation was not included in this evolutionary calculation and this would introduce an asymmetry in the WR and RSG winds. In future work we will investigate WR nebulae with higher resolution, 3D R-MHD simulations that are based on new stellar evolution sequences for single-star progenitors of classical WR stars.

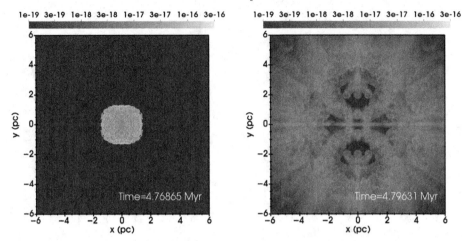

Figure 3. X-ray emission (0.3-10 keV) from the WR nebula 14 000 years (left) and 42 000 years (right) after the RSG→WR transition, plotted on a logarithmic scale with units $\mathrm{erg\,cm^{-2}\,s^{-1}\,arcsec^{-2}}$.

4. Conclusions

PION is a 1D-3D R-MHD code for simulating expanding nebulae around stars – the current public release is v2.0, available from https://www.pion.ie. It has demonstrated parallel scaling to 1024 MPI processes on production simulations, as presented in Mackey et al. (2021), although the scaling for problems with radiative transfer is not so good. Newly implemented hybrid OpenMP/MPI algorithms show much better scaling to at least 2560 cores on the same test problems, and these improvements will be included in the next public release of PION. A 3D R-HD simulation of an expanding WR nebula was presented, including the time-evolution of projected Emission Measure maps and X-ray surface-brightness maps. These show many features that are qualitatively similar to Galactic WR nebulae such as M1-67 and NGC 3199, and motivate more detailed 3D simulations with the goal of determining which (if any) of the Galactic WR nebulae are produced by mass loss through stellar winds as predicted by single-star evolution.

Acknowledgements

JM is grateful to N. Langer and L. Grassitelli for stimulating discussions and for providing evolutionary calculations. JM acknowledges funding from a Royal Society-Science Foundation Ireland University Research Fellowship (20/RS-URF-R/3712). DZ, RB acknowledge funding from an Irish Research Council (IRC) Starting Laureate Award (IRCLA\2017\83). MM acknowledges funding from a Royal Society Research Fellows Enhancement Award (RGF\EA\180214). TJH is funded by a Royal Society Dorothy Hodgkin Fellowship. The authors wish to acknowledge the DJEI/DES/SFI/HEA Irish Centre for High-End Computing (ICHEC) for the provision of computational facilities and support (project eurocc-af-2). This work is supported by the EuroCC project funded by the European High-Performance Computing Joint Undertaking (JU) under grant agreement No 951732 and the Irish Department of Further and Higher Education, Research, Innovation and Science.

References

Comerón, F., & Kaper, L. 1998, A&A, 338, 273
Falle, S., Komissarov, S., & Joarder, P. 1998, MNRAS, 297, 265
Freyer, T., Hensler, G., & Yorke, H. W. 2003, ApJ, 594, 888

—. 2006, ApJ, 638, 262

García-Segura, G., Langer, N., & Mac Low, M. 1996, A&A, 316, 133

Garcia-Segura, G., & Mac Low, M.-M. 1995, ApJ, 455, 160

Green, S., Mackey, J., Haworth, T. J., Gvaramadze, V. V., & Duffy, P. 2019, A&A, 625, A4

Grosdidier, Y., Moffat, A. F. J., Joncas, G., & Acker, A. 1998, ApJL, 506, L127

Harries, T. J., Haworth, T. J., Acreman, D., Ali, A., & Douglas, T. 2019, Astronomy and Computing, 27, 63

Humphreys, R. M., Davidson, K., Richards, A. M. S., Ziurys, L. M., Jones, T. J., & Ishibashi, K. 2021, AJ, 161, 98

Langer, N. 2012, ARA&A, 50, 107

Mackey, J. 2012, A&A, 539, A147

Mackey, J., Green, S., Moutzouri, M., Haworth, T. J., Kavanagh, R. D., Zargaryan, D., & Celeste, M. 2021, MNRAS, 504, 983

Mackey, J., Mohamed, S., Neilson, H. R., Langer, N., & Meyer, D. M.-A. 2012, ApJL, 751, L10

Meyer, D. M.-A., Mackey, J., Langer, N., Gvaramadze, V. V., Mignone, A., Izzard, R. G., & Kaper, L. 2014, MNRAS, 444, 2754

Meyer, D. M.-A., Mignone, A., Kuiper, R., Raga, A. C., & Kley, W. 2017, MNRAS, 464, 3229

Meyer, D. M. A., Petrov, M., & Pohl, M. 2020, MNRAS, 493, 3548

O'Gorman, E., et al. 2015, A&A, 573, L1

Pogorelov, N. V., Zank, G. P., & Ogino, T. 2004, ApJ, 614, 1007

Sana, H., et al. 2012, Science, 337, 444

Smith, N. 2014, ARA&A, 52, 487

Toalá, J. A., Marston, A. P., Guerrero, M. A., Chu, Y. H., & Gruendl, R. A. 2017, ApJ, 846, 76

van Marle, A. J., Meliani, Z., & Marcowith, A. 2015, A&A, 584, A49

Yorke, H. W., & Kaisig, M. 1995, Computer Physics Communications, 89, 29

Discussion

PORTEGIES ZWART: The symmetries are beautiful but not very realistic. One way to get rid of them is by introducing some random noise, but would they disappear if you increase the resolution and you get automatically randomness?

MACKEY: It shouldn't – it depends how symmetric the Riemann solver is. If it is perfectly symmetric then all of the octants should be identical. At the moment I am looking into adding a stochastic model for clumpiness to the RSG wind, because we know that the winds of cool stars are very clumped. We expect these non-linear clumps in the wind will then seed all of the structure that we see in the WR nebulae.

MOHAMED: Could you say something about the computational cost of the TORUS post-processing of PION simulations.

MACKEY: At the beginning the TORUS calculations were taking longer than the hydro simulations, but Tom and Sam did a lot of work on this. In the end with MPI and multi-threading, the TORUS calculations took about 10 000 core hours for a 3D simulation that required about 50 000 core hours.

The Predictive Power of Computational Astrophysics as a Discovery Tool
Proceedings IAU Symposium No. 362, 2023
D. Bisikalo, D. Wiebe & C. Boily, eds.
doi:10.1017/S1743921322001880

Infrared appearance of wind-blown bubbles around young massive stars

Maria S. Kirsanova[iD] and Yaroslav N. Pavlyuchenkov

Institute of Astronomy, Russian Academy of Sciences,
119017, 48 Pyatnitskaya str., Moscow, Russia
email: kirsanova@inasan.ru

Abstract. Thousands of ring-like bubbles appear on infrared images of the Galaxy plane. Most of these infrared bubbles form during expansion of H II regions around massive stars. However, the physical effects that determine their morphology are still under debate. Namely, the absence of the infrared emission toward the centres of the bubbles can be explained by pushing the dust grains by stellar radiation pressure. At the same time, small graphite grains and PAHs are not strongly affected by the radiation pressure and must be removed by another process. Stellar ultraviolet emission can destroy the smallest PAHs but the photodestruction is ineffective for the large PAHs. Meanwhile, the stellar wind can evacuate all types of grains from H II regions. In the frame of our chemo-dynamical model we vary parameters of the stellar wind and illustrate their influence on the morphology and synthetic infrared images of the bubbles.

Keywords. ISM: bubbles, ISM: dust, extinction, ISM: HII regions, stars: winds, outflows

---

1. Introduction

H II regions around young OB-type stars represent natural laboratories to study properties of cosmic dust particles. Observations made with *Spitzer* and *WISE* telescope revealed different spatial morphology of the middle and far-infrared (IR) emission in H II regions, see e.g. Churchwell *et al.* (2006); Deharveng *et al.* (2010); Anderson *et al.* (2014). The mid-IR emission of these objects looks like a compact arc or a small ring at $24\,\mu m$, surrounded by a larger ring of emission at $8\,\mu m$, see e.g. Simpson *et al.* (2012). The larger ring is observed toward the dense envelope of neutral gas, swept up by a shock front from an expanding H II region. Therefore, these objects are often called infrared bubbles. Subsequent observations with *Herschel* have demonstrated that the ring-like appearance is also typical for far-IR emission, see e.g. Molinari *et al.* (2010); Anderson *et al.* (2012). Series of papers by Pavlyuchenkov *et al.* (2013); Akimkin *et al.* (2015, 2017) aims to study different factors which determine quantitatively and qualitatively spatial distribution and intensity of the mid and far-IR emission from the H II regions and surrounding neutral material. They found that none of the following factors: photo-destruction by ultraviolet photons, radiation pressure and overall expansion of H II regions can explain all the observed IR features by itself. Recent observations of H II regions by *SOFIA* telescope revealed the importance of stellar wind in dynamics of ionised gas around O-type stars, see e. g. Pabst *et al.* (2019); Luisi *et al.* (2021). Following these results, we numerically explore the influence of the wind on dust dynamics and simulate mid and far-IR emission from H II regions. Our main aim is to find out if the inclusion of wind allows us to reproduce all the observed IR features simultaneously. Moreover, apart from the H II regions, our calculations can be useful to interpret mid-IR observations of old stellar populations:

such as luminous blue variables or Wolf-Rayet stars, see e.g. Gvaramadze *et al.* (2010), Wachter *et al.* (2010).

2. Simulations

We considered three models with different parameters of stellar wind in order to demonstrate its crucial role in dynamics of dust grains. The three models have common parameters such as spectral type of ionising star, O7 V with effective temperature 37000 K, and initial uniformly-distributed gas number density $n_0 = 10^3$ cm^{-3}. The dust-to-gas mass ratio has standard value 1/100, and the dust is uniformly mixed with gas in the beginning of the simulations. First model has no stellar wind. It is needed to compare results of two other models with different parameters of the wind. For this model, we estimated thermal energy E_{therm} of the HII region at a particular moment of model time when radius of the ionized bubble reached 0.7 pc. In the second model the wind is "weak", where energy that brings stellar wind is comparable with E_{therm}, i. e. $E_{\text{wind}} \approx E_{\text{therm}}$. The wind mass-loss rate in this model is to be $dM/dt = 4 \times 10^{-8}$ M$\odot$ yr^{-1} and the wind terminal velocity $v_\infty = 500$ km s^{-1}. The third model has "strong wind", where $E_{\text{wind}} \approx 100 E_{\text{therm}}$, $dM/dt = 1.5 \times 10^{-7}$ M$\odot$ yr^{-1} and $v_\infty = 2000$ km s^{-1}. To calculate the radiation intensity distributions we adopt the same radiative transfer model as in Pavlyuchenkov *et al.* (2013). This model takes into account the stochastic (transient) heating of PAHs and small grains by single photons. The thermal state of a grain of particular type is described by the probability density distribution over the temperature, $P(T)$, which represents a fraction of the grains having temperature in the range $(T, T + dT)$.

3. Results

A general physical structure of the HII region and surrounding gas at the moment when its radius is ≈ 0.7 pc is shown in Fig. 1. An ionised gas region is surrounded by a dense envelope of neutral hydrogen (atomic and molecular), which has been shovelled by the shock preceding the ionization front. The ionised region is rarefied and appears as a hot bubble in the strong wind model, while the empty cavity occupies less than half of the ionised volume in the weak wind model. Stellar wind makes expansion of the ionised region faster, therefore the age of the HII region is 59, 58 and 31 thousand yr in the models without wind, with the weak and strong wind, respectively. The gas velocity is up to an order of magnitude higher (10 vs 1 km s^{-1}) in the strong wind model at the ionisation front and in the dense neutral envelope. Radial distributions of dust grains with different sizes have different radii of the empty inner cavity in the model without wind, namely the larger the grain the larger the cavity. PAH particles, having the largest surface-to-mass ratio, are well coupled to the gas in contrast with the big grains, as we discussed in Akimkin *et al.* (2015). There is no segregation of the grain ensemble in the models with the wind, and the radii of the inner cavities are determined by the momentum transfer from the wind to the gas and dust medium.

We illustrate the importance of stochastic heating in Fig. 2 which shows how $P(T)$ depends on the distance from the star for selected dust components which correspond to PAHs, very small graphite grains (VSGs), and big graphite grains (BGs). The distribution of $P(T)$ for PAHs is broad over the entire cloud. In the very centre of the H II region, the VSGs are heated up to 1000 K, and the temperature distribution is quite narrow, because under the strong radiation from the ionising star these grains tend to have equilibrium temperatures despite their relatively small size. Further from the star, the temperature distribution becomes wider, so that in the middle of the ionised region the temperature of VSGs fluctuates approximately from 50 K to 200 K. The $P(T)$ for big grains is represented

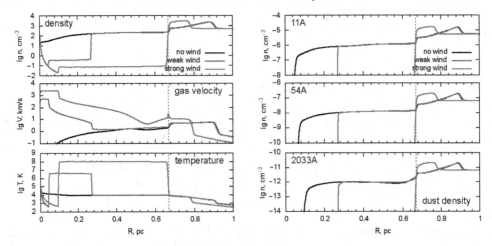

Figure 1. Summary of the model physical structure of the HII region at the time, when its radius is ≈ 0.7 pc. Left: distributions of gas density (top), velocity (middle), and gas temperature (bottom). Right: distributions of dust number densities for the considered dust components: PAH (11Å, top), graphite (54Å, middle), and silicate grain (0.2μm bottom).

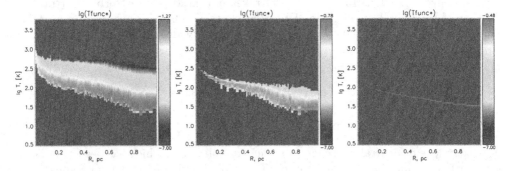

Figure 2. Radial dependence of temperature probability, $P(T)$, for the selected dust grain types: PAHs (11Å), small graphites (middle, 50Å), and big graphites (right, 0.2μm).

by delta-function everywhere in the considered region, so the BGs temperature can be reliably described in the framework of the thermal equilibrium.

Comparing simulated distributions of infrared emission in three considered models, we find the most prominent differences at 8 and 24 μm, see Fig. 3. There is a bright peak of the 8 μm emission near the centre of H II region in the model without wind, related to small radial dust drift under the effect of radiation pressure. This effect alone does not clear the H II region, therefore simulations with no wind can not explain typical observed distributions of the 8 μm even qualitatively. The weak wind model also produces a peak at 8 μm inside of the bubble, while the strong wind model evacuates all PAHs from the H II region and carries out only the ring-like 8 μm emission.

On the contrary, the simulated 24 μm distribution is qualitatively consistent with the observed morphology in the models without and with weak wind. They both produce the bright inner ring around the ionising star and the less bright outer ring related with the dense shovelled envelope. The strong wind model produces a nonobservable one ring-like structure at the dense envelope because the bubble is empty.

All three considered models produce ring-like emission at 70 μm and longer wavelengths with approximately flat intensity distribution within H II region. Strong wind makes the shovelled neutral envelope thin and dense, therefore the far-infrared emission is two times

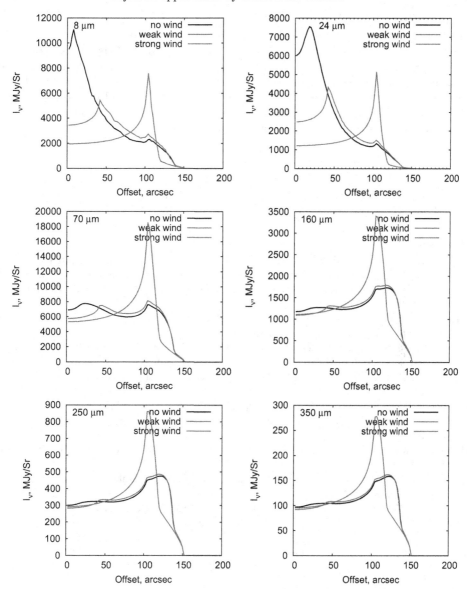

Figure 3. Radial intensity profiles at 8 μm, 24 μm, 70 μm, 160 μm, 250 μm and 350 μm calculated for the models with and without stellar wind.

brighter than in the other two models. We conclude that the far-infrared dust emission is not sensitive to the stellar wind parameters.

4. Conclusion

Different distributions of the 8 and 24 μm emission around young massive stars can be attributed to the specific parameters of stellar wind. There is a bright 8 μm emission from H II regions in the models without or with the weak stellar wind. Therefore, PAHs should be removed from the H II regions by some other way to produce the commonly observed 8 μm ring. Ultraviolet photons can destroy PAHs, and we consider this way as one of the possibilities to reduce brightness at 8 μm inside of the H II regions. For example, Pavlyuchenkov *et al.* (2013) proposed phenomenological model of the PAH destruction by stellar UV photons and found that this mechanism can be effective to create the 8 μm rings. However, subsequent detailed modelling by Murga *et al.* (2022) confirmed the importance of stellar UV photons only for small PAHs with sizes less than $\approx 6 - 7$ Å. Extreme UV or ever soft X-ray photons could be effective to destroy larger PAHs, therefore we consider this direction as promising continue of our numerical study. Central peaks at 24 μm can be produced in the models without and with the weak wind. All dust types are removed from H II regions in the strong wind model. Far-IR dust emission is not sensitive to the stellar wind parameters.

5. Acknowledgements

We are thankful to S. Yu. Parfenov and J. Mackey for fruitful discussions. We also thank referee for valuable comments and suggestions. The study is supported by Russian Science Foundation (project 21-12-00373).

References

Akimkin V. V., Kirsanova M. S., Pavlyuchenkov Ya. N., Wiebe D. S. 2015, *MNRAS*, 449, 440
Akimkin V. V., Kirsanova M. S., Pavlyuchenkov Ya. N., Wiebe D. S. 2017, *MNRAS*, 469, 630
Anderson L. D., Zavagno A., Deharveng L., Abergel A., Motte F., André P., Bernard J.-P., et al., 2012, *A&A*, 542, A10.
Anderson L. D., Bania T. M., Balser D. S., Cunningham V., Wenger T. V., Johnstone B. M., Armentrout W. P. 2014, *ApJS*, 212, 1.
Churchwell E., Povich M. S., Allen D., Taylor M. G., Meade M. R., Babler B. L., Indebetouw R., et al. 2006, *ApJ*, 649, 759
Deharveng L., Schuller F., Anderson L. D., Zavagno A., Wyrowski F., Menten K. M., Bronfman L., et al. 2010, A&A, 523, A6
Gvaramadze V. V., Kniazev A. Y., Fabrika S., 2010, *MNRAS*, 405, 1047.
Luisi M., Anderson L. D., Schneider N., Simon R., Kabanovic S., Güsten R., Zavagno A., et al., 2021, *SciA*, 7, eabe9511
Molinari S., Swinyard B., Bally J., Barlow M., Bernard J.-P., Martin P., Moore T., et al., 2010, *A&A*, 518, L100
Pabst C., Higgins R., Goicoechea J. R., Teyssier D., Berne O., Chambers E., Wolfire M., et al., 2019, *Nature*, 565, 618
Pavlyuchenkov Ya. N., Kirsanova M. S., Wiebe D. S. 2013, *ARep*, 57, 573
Simpson R. J., Povich M. S., Kendrew S., Lintott C. J., Bressert E., Arvidsson K., Cyganowski C., et al., 2012, *MNRAS*, 424, 2442.
Murga M. S., Kirsanova M. S., Wiebe D. S., Boley, P. A. 2022, *MNRAS*, 509, 800
Wachter S., Mauerhan J. C., Van Dyk S. D., Hoard D. W., Kafka S., Morris P. W., 2010, *AJ*, 139, 2330.

Discussion

MACKEY: It seems like in your models the dust and the gas are very well coupled. Maybe this is because of the quite high density. Did you look at some cases where there is quite significant drift between the gas and the dust? Have you explored that?

KIRSANOVA: Yes, they are coupled, but the dust is not frozen to the gas. We obtain empty H II regions in the models with stellar wind. If we don't include the wind, we see significant differences. We calculate models with quite high density in order to obtain quantitative agreement between our synthetic observations in the mid and far-IR and the observed values.

The Predictive Power of Computational Astrophysics as a Discovery Tool
Proceedings IAU Symposium No. 362, 2023
D. Bisikalo, D. Wiebe & C. Boily, eds.
doi:10.1017/S1743921322001302

Numerical 2D MHD simulations of the collapse of magnetic rotating protostellar clouds with the Enlil code

Sergey Khaibrakhmanov[1,2] , **Sergey Zamozdra**[2],
Natalya Kargaltseva[1,2], **Andrey Zhilkin**[3] **and** **Alexander Dudorov**[2†]

[1]Ural Federal University, 51 Lenina str., Ekaterinburg, 620000, Russia
email: khaibrakhmanov@csu.ru

[2]Chelyabinsk State University, 129 Br. Kashirinykh str., Chelyabinsk, 454001, Russia

[3]Institute of Astronomy of the Russian Academy of Sciences (INASAN), Moscow, 119017, Russia

Abstract. We numerically investigate the gravitational collapse of rotating magnetic protostellar clouds. The simulations are performed using 2D MHD code 'Enlil'. The code is based on TVD scheme of increased order of accuracy. We developed a model of the initially non-uniform cloud, which self-consistently treats gas density and large-scale magnetic field distribution. Simulation results for the typical parameters of a solar mass cloud are presented. In agreement with our previous results for the uniform cloud, the isothermal collapse of the non-uniform cloud results in formation of hierarchical structure of the cloud, consisting of flattened envelope and thin quasi-magnetostatic primary disk near its equatorial plane. The non-uniform cloud collapses longer than the uniform one, since the magnetic field is dynamically stronger at the periphery of the cloud in the former case.

Keywords. magnetic fields, magnetohydrodynamics (MHD), numerical simulation, star formation, interstellar medium

1. Introduction

Contemporary star formation takes place in gravitationally bound cores of molecular clouds, which are called protostellar clouds (PSCs hereafter) or prestellar cores. PSCs rotate with typical ratio of the rotational energy to the gravitational energy of few percents (Goodman et al. 1993; Caselli et al. 2002). Large-scale magnetic field is ubiquitous in the PSCs with typical magnetic field strength of $10^{-5} - 10^{-4}$ G and non-dimensional mass-to-flux ratio of 2–3 (Li et al. 2009; Crutcher 2012).

The first numerical simulations of the gravitational collapse of PSCs have shown that electromagnetic and centrifugal forces cause the flattening of the cloud along the rotation axis and magnetic field direction and further formation of disks around protostars during the collapse (Nakano 1979; Black & Scott 1982; Dorfi 1982). This result has been confirmed by numerous simulations later on (see, e.g., Dudorov et al. 1999b, 2000; Zhao et al. 2020).

Modern observations have revealed large flattened envelopes around very young protostars in so-called class 0 young stellar objects. High-angular resolution observations with The Atacama Large Millimeter/Submillimeter Array (ALMA) indicate to the presence

---

†Professor Alexander Dudorov has passed away.

Table 1. Models of initial non-uniform large-scale magnetic field in PSCs.

N	Approach	References
1	Magneto-hydrostatic equilibrium	Stodólkiewicz (1963), Mouschovias (1976), Tomisaka et al. (1988), Carry & Stahler (2001), Allen et al. (2003)
2	Relaxation to magneto-hydrostatic equilibrium	Habe et al. (1991), Leao et al. (2013)
3	Straight magnetic lines with prescribed ratio of gas and magnetic pressures P/P_{m}	Mouschovias & Morton (1991), Banerjee & Pudritz (2006)
4	Straight magnetic lines with $B \propto N$ (column density)	Hennebelle & Ciardi (2009)
5	Straight magnetic lines with prescribed radial profile $B(R)$	Seifried et al. (2012), Myers et al. (2013)
6	Straight magnetic lines with constant mass-to-flux ratio	Gray et al. (2018)
7	Curved magnetic lines with prescribed vector potential $A_\phi(R)$	Tsukamoto et al. (2020)

of small possibly keplerian disks around protostars inside those flattened envelopes (see, e.g., Tobin et al. 2020). It is of great interest to study the conditions for the formation of rotationally supported protostellar disks (RSD), since they further evolve into accretion disks, which ultimately become protoplanetary disks of young stars. The main problem in this field is so-called magnetic braking catastrophe, i. e. too efficient transport of the cloud/disk angular momentum preventing the formation of RSD (see recent review Zhao et al. 2020). In order to tackle this problem, it is important to simulate initial stages of the cloud's collapse taking into account dissipative magneto-gas-dynamics (MHD) effects.

One of the problem in the simulations of the collapse of PSCs is a choice of the initial conditions. According to observations (see review in Gomez et al. 2021), PSCs have non-uniform density distribution. Usually, the radial profile of the cloud's density is approximated by power-law functions, although power-law indexes may vary in a wide range. Large-scale magnetic field of PSCs is also non-uniform. Typically PSCs have magnetic field with hour-glass geometry (Li et al. 2009), which is in agreement with theoretical predictions (Dudorov & Zhilkin 2008).

In Table 1, we summarize theoretical approaches used to set initial non-uniform large-scale magnetic field in the simulations of the PSC's collapse. Solution of the magneto-hydrostatic equilibrium equations (N1) is the difficult problem, therefore the problem of cloud's relaxation to the magneto-hydrostatic equilibrium can be considered (N2). The approaches (N3–N6) are more simple: straight magnetic lines with prescribed intensity. The straight lines guarantee a divergence free magnetic field. The approach N7 admits curved magnetic lines. All the approaches N3–N7 give more realistic field than the uniform field, but remain non-physical one.

In our previous works, we investigated the isothermal collapse of initially uniform magnetic rotating PSC (Khaibrakhmanov et al. 2021; Kargaltseva et al. 2021). We analyzed the hierarchical structure of the collapsing cloud, consisting of a flattened envelope, which contains quasi-magnetostatic primary disk (PD) inside with the first hydrostatic core in its center. The simulations have shown that the fast shock MHD wave moves outward from the PD boundary into the envelope, and a region of magnetic braking is formed behind the wave front. The PD play a key role in the evolution of the angular momentum of the system. In this work, we further develop our model and propose new self-consistent approach to model the initial state of the PSC with non-uniform magnetic field. The collapse of such a non-uniform PSC with typical parameters is numerically simulated and compared with our results for the uniform PSC.

2. Problem statement

We model the collapse of non-uniform spherical rotating cloud with non-uniform magnetic field. The density distribution is Plummer-like (Whitworth & Ward-Thompson 2001):

$$\rho = \frac{\rho_c}{1 + (r/r_1)^2}, \tag{2.1}$$

where ρ_c is the central density, $r_1 = \pi^{-1/2}(M_0/\rho_c)^{1/3}$ is the characteristic radius of the cloud's dense part, M_0 is the mass of the cloud. The function (2.1) has the asymptotic power-law index -2 that fits the PSC observations (Gomez et al. 2021).

The initial magnetic field is poloidal. It is calculated under the assumption of the PSC formation via spherically symmetric contraction of uniform medium with density ρ_0 penetrated by uniform magnetic field $\mathbf{B}_0$. If the magnetic field is frozen into gas then $\mathbf{B} \propto \rho \delta l$, where δl is the length element. In spherical coordinates (r, θ, ϕ)

$$\delta l_r \propto dR, \quad \delta l_\theta \propto R, \tag{2.2}$$

where R is the radius of spherical layer, dR is the width of the layer. Since the mass of the layer $dm = 4\pi\rho R^2 dR$ is conserved then relations (2.2) yields

$$B_r \propto R^{-2}, \quad B_\theta \propto \rho R. \tag{2.3}$$

Let $\rho = \rho_c f(r)$. The mass of the matter inside a sphere with initial radius R_i is conserved, therefore

$$\rho_c V_\star = \rho_0 R_i^3/3, \quad V_\star \equiv \int_0^R f(r) r^2 dr. \tag{2.4}$$

Let's express R_i from eq. (2.4) and substitute into relations (2.3) to obtain the law of frozen magnetic field evolution during the cloud formation

$$\frac{B_r}{B_{r0}} = R^{-2} \left(3 V_\star \frac{\rho_c}{\rho_0} \right)^{2/3}, \quad \frac{B_\theta}{B_{\theta 0}} = R f(R) \left(\frac{\rho_c}{\rho_0} \right)^{2/3} (3 V_\star)^{-1/3}. \tag{2.5}$$

In the case of uniform contraction $f(r) = 1$ and $V_\star = R^3/3$, so the law (2.5) yields $B_r, B_\theta \propto \rho^{2/3}$ as expected.

The collapse of the cloud is investigated using the equations of gravitational MHD. Numerical modeling is performed with the help of the 2D numerical code 'Enlil' based on the TVD scheme of the increased order of accuracy (Dudorov et al. 1999a,b; Zhilkin et al. 2009). Thermal evolution of the cloud is simulated using the equation of state with variable adiabatic index (see Kargaltseva et al. 2021).

3. Results

In this section we compare two simulations of initially uniform (run I) and non-uniform cloud (run II). We consider the PSC with mass of $M_0 = 1 \, M_\odot$ and temperature of 10 K. Uniform cloud has density $\rho_c = 1.5 \cdot 10^{-18}$ g cm^{-3} and radius $R_0 = 0.022$ pc. Non-uniform cloud has central density $\rho_c = 3.5 \cdot 10^{-18}$ g cm^{-3}, radius $R_0 = 0.021$ pc and characteristic radius $r_1 = 0.015$ pc. The ratios of thermal, rotation and magnetic energies of the cloud to the modulus of its gravitational energy are $\varepsilon_t = 0.3$, $\varepsilon_m = 0.2$, $\varepsilon_w = 0.01$, respectively.

In Figure 1, we show the structure of the clouds in runs I and II at the onset of the collapse (left panels), and at the moment of the first hydrostatic core formation (right panels).

Our simulations show that the general picture of the collapse in run II is similar to that in run I. In both cases, the cloud acquires a hierarchical structure at the end of

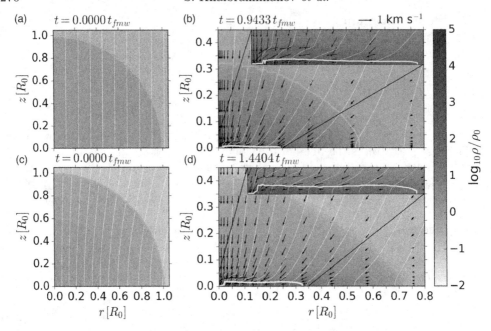

Figure 1. Two-dimensional structure of collapsing PSCs with initially uniform (upper panels) and non-uniform (lower panels) distributions of density and magnetic field. Left panels: initial state, right panels: moment of the first hydrostatic core formation. Initial parameters of the clouds: $\varepsilon_t = 0.3$, $\varepsilon_m = 0.2$, $\varepsilon_w = 0.01$. A quarter of the cloud in the region of positive r and z is considered. Color filling shows logarithm of non-dimensional density, arrows show the velocity field, and white lines are the poloidal magnetic field lines. The green line shows the border of the PD, the blue line is the boundary of the cloud. Insets in the upper right corners of panels (b) and (d) show the zoomed-in region near the primary disk.

the isothermal collapse. The hierarchy consists of a flattened envelope with thin quasi-magnetostatic PD near its equatorial plane. Further the first core forms in the center of the PD.

Figure 1(d) shows that flattened envelope has radius $R \approx 0.75\,R_0$ and half-thickness $Z \approx 0.4\,R_0$, while the PD is characterized by $R \approx 0.35\,R_0$ and half-thickness $Z \approx 0.02\,R_0$. These characteristics are larger than in run I, according to Figures 1(b) and (d). The degrees of flattening of each structure, Z/R, are similar in both runs: $Z/R \approx 0.6$ for the envelope and $Z/R \approx 0.04 - 0.06$ for the PD.

The main difference between runs I and II is that initially non-uniform cloud evolves slower. The first core forms at $t = 1.44\,t_{\rm fmw}$ in run II, while it happens at $t = 0.94\,t_{\rm fmw}$ in run I. Here $t_{\rm fmw}$ is the typical dynamical time of the collapse taking into account the effects of rotation and magnetic field (Dudorov & Sazonov (1982), see also Khaibrakhmanov et al. (2021)). This is explained by the fact that initial magnetic field is dynamically stronger at the periphery of the cloud in run II as compared to run I.

In order to describe the hierarchical structure of the cloud in run II, we plot density profiles $\rho(r, 0)$ and $\rho(0, z)$, as well as vertical velocity profiles at several radii in Figure 2. Figure 2(a) clearly demonstrates the flattening of the cloud. There are three pronounced density jumps in the vertical density profile. Analysis of this dependence together with the velocity profile at $r = 0.05\,R_0$ shows that the first jump at $z \approx 0.02\,R_0 \approx 70$ au corresponds to the surface of the quasi-magnetostatic ($v_z \to 0$) PD. The second jump, $z \approx 0.05\,R_0 \approx 200$ au, lies at the front of fast MHD shock wave propagating from the PD into the envelope. The third jump at $z \approx 0.5\,R_0 \approx 2000$ au is the contact boundary of the cloud.

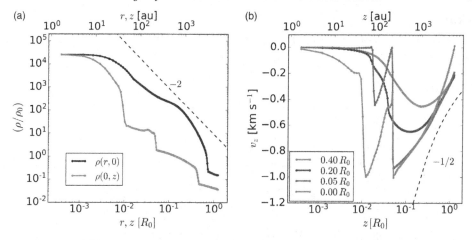

Figure 2. Panel (a): the profiles of gas density along the equatorial plane of the cloud (blue line) and along the rotation axis (orange line) in run II. Moment of time corresponds to Figure 1(d). Panel (b): corresponding profiles of vertical velocity v_z along the rotation axis at $r = 0$ (orange line), $r = 0.05\,R_0$ (green line), $r = 0.2\,R_0$ (purple line) and $r = 0.4\,R_0$ (blue line). Dashed lines with labels show typical slopes.

Analogous jumps are seen in profiles $v_z(z)$ at $r = 0$ and $r = 0.2\,R_0$. At further distances, $r > 0.2\,R_0$, the quasi-magnetostatic equilibrium is not established.

Figures 1(d) and 2(b) show that the half-thickness of the primary disk is minimal, $Z_{\min} \approx 40$ au, near the rotation axis, $r = 0$. Ten computational cells fit into the primary disk in this region, and maximum cell size Δz is of 3 au, i.e. $\Delta z \ll Z_{\min}$. At further radii r, number of cells that fit into the primary disk is even more. This means that the adopted grid resolution is sufficient to resolve the internal structure of the primary disk.

4. Conclusions and discussion

In this work, we presented a numerical MHD model of the collapse of initially non-uniform PSC. Distinctive feature of the model is self-consistent treatment of both initial density and large-scale magnetic field distribution in the cloud. Initial density profile has the asymptotic power-law index -2 that fits the PSC observations. Initial magnetic field distribution is determined considering that the cloud forms as a result of spherically symmetric contraction under condition of frozen-in magnetic field.

Numerical simulation of the collapse of such a non-uniform PSC of solar mass under typical parameters show that general picture of the collapse is similar to that of the initially uniform magnetic field. The cloud acquires a hierarchical structure during the isothermal collapse, consisting of flattened envelope and thin quasi-magnetostatic PD near its equatorial plane. The main difference between initially uniform and non-uniform clouds is that the isothermal collapse lasts longer in the latter case. This is explained by the following. For a fixed total magnetic energy of the cloud, the magnetic field is dynamically stronger at the periphery of non-uniform cloud, since the gas pressure in this case is smaller than in the case of uniform cloud, while the magnetic pressures are nearly the same.

In future, we plan to apply the developed model to study angular momentum and magnetic flux evolution of PSCs for various initial parameters in order to investigate efficiency of magnetic braking at the initial stages of the collapse. Results of the simulations will be used to construct synthetic maps of collapsing PSCs in sub-mm range and to investigate observational appearance of PD.

Acknowledgments

This work is supported by the Russian Science Foundation, project 19-72-10012. The authors thank anonymous referee for useful comments.

References

Allen, A., Li, Z.-Y., & Shu, F. H. 2003. *Astrophys. J.*, 599, 363.

Banerjee, R. & Pudritz, R. E. 2006. *Astrophys. J.*, 641, 949.

Black, D. C., & Scott, E. H. 1982. *Astrophys. J.*, 263, 696.

Carry, C. L. & Stahler, S. W. 2001. *Astrophys. J.*, 555, 160.

Caselli, P., Benson, P. J., Myers, P. C., & Tafalla, M. 2002. *Astrophys. J.*, 572, 1, 238.

Crutcher, R. M. 2012. *Annu. Rev. Astron. Astr.*, 50, 29.

Dorfi, E. 1982. *Astron. Astrophys.*, 114, 151.

Dudorov, A. E., & Sazonov, Yu. V. 1982. *Nauchnye Informatsii*, 50, 98.

Dudorov, A. E., Zhilkin, A. G., & Kuznetsov, O. A. 1999. *Matem. Modelir.*, 11, 101.

Dudorov, A. E., Zhilkin, A. G., & Kuznetsov, O. A. 1999. *Matem. Modelir.*, 11, 110.

Dudorov, A. E., Zhilkin, A. G., Lazareva, N. Y., & Kuznetsov, O. A. 2000. *Astronomical and Astrophysical Transactions*, 19, 515.

Dudorov, A. E., & Zhilkin, A. G. 2008. *Astron. Rep.*, 52, 790.

Gomez, G. C., Vázquez-Semadeni, E., & Palau, A. 2021. *Mon. Not. R. Astron. Soc.*, 502, 4, 4963.

Goodman, A. A., Benson, P. J., Fuller, G. A., & Myers, P. C. 1993. *Astrophys. J.*, 406, 528.

Gray, W. J., McKee, C. F., & Klein, R. I. 2018. *Mon. Not. R. Astron. Soc.*, 473, 2124.

Habe, A., Uchida, Y., Ikeuchi, S., & Pudritz, R. E. 1991. *Publ. Astron. Soc. Japan*, 43, 703.

Hennebelle, P. & Ciardi, A. 2009. *Astron. Astrophys.*, 506, L29.

Kargaltseva, N. S., Khaibrakhmanov, S. A., Dudorov, A. E., & Zhilkin A. G. 2021, *Bulletin of the Lebedev Physics Institute*, 48, 268.

Khaibrakhmanov, S. A., Dudorov, A. E., Kargaltseva, N. S., & Zhilkin A. G. 2021, *Astron. Rep.*, 65, 693.

Leao, M. R. M., de Gouveia Dal Pino, E. M., Santos-Lima, R., & Lazarian, A. 2013. *Astrophys. J.*, 777, 46.

Li, H.-b., Dowell, C. D., Goodman, A., Hildebrand , R., & Novak, G. 2009, *Astrophys. J.*, 704, 891.

Mouschovias, T. C. 1976. *Astrophys. J.*, 206, 753.

Mouschovias, T. C. & Morton, S. A. 1991. *Astrophys. J.*, 371, 296.

Myers, A. T., McKee, C. F., Cunningham, A. J., Klein, R. I., & Krumholz, M. R. 2013. *Astrophys. J.*, 766, 97.

Nakano, T. 1979. *Publ. Astron. Soc. Japan*, 31, 697.

Seifried, D., Pudritz, R. E., Banerjee, R., Duffin D., & Klessen, R. S. 2012. *Mon. Not. R. Astron. Soc.*, 422, 347.

Stodólkiewicz, J. S. 1963. *Acta Astron.*, 13, 30.

Tobin, J. J., Sheehan, P. D., Megeath, S. T., et al. 2020. *Astrophys. J.*, 890, 2, 130.

Tomisaka, K., Ikeuchi, S., & Nakamura, T. 1988. *Astrophys. J.*, 335, 239.

Tsukamoto, Y., Machida, M. N., Susa, H., Nomura, H., & Inutsuka, S. 2020. *Astrophys. J.*, 896, 158.

Whitworth, A. P. & Ward-Thompson, D. 2001. *Astrophys. J.*, 547, 317.

Zhao, B., Tomida, K., Hennebelle, P., et al. 2020. *Solar System Research*, 216, 43.

Zhilkin, A. G., Pavlyuchenkov, Y. N., & Zamozdra, S. N. 2009. *Astron. Rep.*, 53, 590.

The Predictive Power of Computational Astrophysics as a Discovery Tool
Proceedings IAU Symposium No. 362, 2023
D. Bisikalo, D. Wiebe & C. Boily, eds.
doi:10.1017/S174392132200134X

Two-dimensional MHD model of gas flow dynamics near a young star with a jet and a protoplanetary disk

V. V. Grigoryev⬡, D. V. Dmitriev and T. V. Demidova⬡

Crimean astrophysical observatory

Abstract. A numerical nonstationary two-dimensional MHD model of a protoplanetary disk near T Tauri star with a jet is developed. The model assumes consideration of pure gas ionization, optically thin cooling, anisotropic thermal conductivity and viscosity. The relaxation of gas-dynamic flows is analyzed. Based on the evolution of plasma flows, profiles of hydrogen spectral lines have been obtained.

Keywords. T Tau stars, protoplanetary disks, MHD, Radiative transfer

---

1. The model

The two-dimensional model is implemented using the PLUTO package (Mignone et al. 2007), which allows solving the non-stationary MHD equations by the Godunov method. The equations are: mass continuity equation, motion equation (magnetic pressure, gravitational force, external forces and viscosity are included), energy equation (the same processes as in motion equation with optically thin cooling and thermal conductivity are included).

This equation system is closed by ideal gas equation, taking into account recombination processes by optically thin cooling, described in module SNEq (includes 16 different line emissions coming from some of the most common elements and the hydrogen ionization and recombination rate coefficients, see details in PLUTO documentation).

Viscosity is taken into account isotropically based on Shakura and Sunyaev (1976) model. Anisotropic thermal conductivity is taken into account in accordance with analytical formulas Balbus (1986), Cowie and McKee (1977).

Two-dimenstional grid ($(R, \theta) \in [0.35; 8]L_0 \times (0; \pi)$, 64×128 cells) is uniform on θ and logarithmic on R-direction, $L_0 = 4.2 \times 10^{11}$ cm $= 6R_\odot$, where $R_\odot$ — solar radius.

The initial conditions are the same as in Romanova et al. (2002), Type I. The outer boundary condition on $R \approx 0.23$ a.u. is outflow, in the border cells closest to the star a stellar atmosphere is modelled with a small thickness. The boundary conditions on θ satisfy the axisymmetry ones, but in the angle no more than $3.5°$ jet is modelled: matter is less dense than the corona of the disk, and it flies away from the star at a speed of no less than local escape velocity. The value $3.5°$ is chosen empirically and can be changed later.

2. Results

The plot Fig. 1 (left) shows the density at time $t = 38t_0$ ($t_0 \approx 2.5 \times 10^4$ s, $2\pi t_0$ is one orbital period at the distance L_0 from the center of the star). The matter of the disk has formed two accretion columns. Initially, they were formed by the corona material during about one period of rotation. These accretion columns enclose two hot bubbles that are

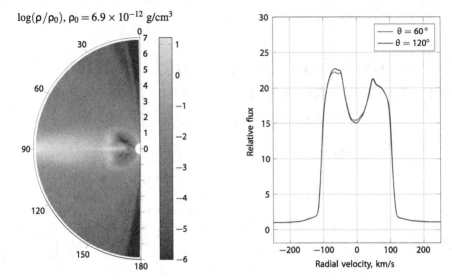

$\log(\rho/\rho_0)$, $\rho_0 = 6.9 \times 10^{-12}$ g/cm^3

Figure 1. Left — density distribution at $t = 38t_0$. Right — emission in Br_γ at $t = 49.5t_0$.

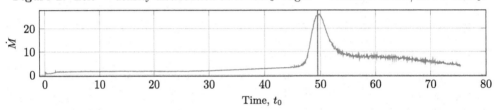

Figure 2. Accretion rate $\dot{M}$ in units 10^{-9} $M_\odot$/year during the simulation. $M_\odot = 2 \times 10^{33}$ g is solar mass.

part of the magnetosphere. Analysis of the dynamics shows unstable accretion rate (see Fig. 2) and variable equatorial radius of the magnetosphere, so there is no equilibrium yet.

To calculate hydrogen emission spectra we solve system of statistical equilibrium equations for a 15-level hydrogen atom (Grinin and Katysheva 1980). Sobolev (1960) approximation is used to solve radiation transfer in lines and find mean intensities. After solving statistical equilibrium equations for level populations we use ray-by-ray integration with Doppler line absorbtion core to find emergent transition line profile (Dmitriev et al. 2019). The spectral line Br_γ at the moment of maximum accretion rate is shown in Fig. 1 right.

Acknowledgements

The simulations were performed using the resources of the Joint SuperComputer Center of the Russian Academy of Sciences (Savin et al. 2019): *https://www.jscc.ru/*. The work was partially supported by the RScF grant No. 19-72-10063.

References

Balbus, S. A. 1986, *ApJ*, 304, 787.
Cowie, L. L. & McKee, C. F. 1977, *ApJ*, 211, 135–146.
Dmitriev, D. V., Grinin, V. P., & Katysheva, N. A. 2019, *Astron. Lett.*, 45(6), 371–383.
Grinin, V. P. & Katysheva, N. A. 1980, *Bulletin Crimean Astrophysical Observatory*, 62, 52.
Mignone, A., Bodo, G., Massaglia, S., Matsakos, T., Tesileanu, O., Zanni, C., & Ferrari, A. 2007, *ApJS*, 170(1), 228–242.

Romanova, M. M., Ustyugova, G. V., Koldoba, A. V., & Lovelace, R. V. E. 2002, *ApJ*, 578(1), 420–438.

Savin, G., Shabanov, B., Telegin, P., & Baranov, A. 2019, *Lobachevskii J Math*, 40(1), 1853–1862.

Shakura, N. I. & Sunyaev, R. A. 1976, *MNRAS*, 175, 613–632.

Sobolev, V. V. 1960, *Moving envelopes of stars*. Cambridge: Harvard University Press.

The Predictive Power of Computational Astrophysics as a Discovery Tool
Proceedings IAU Symposium No. 362, 2023
D. Bisikalo, D. Wiebe & C. Boily, eds.
doi:10.1017/S1743921322001375

PRESTALINE: A package for analysis and simulation of star forming regions

G. Van Looveren[1], O. V. Kochina[2] and D. S. Wiebe[2,3]

[1]Institute for Astrophysics, University of Vienna,
email: `gwenael.van.looveren@univie.ac.at`

[2]Institute of Astronomy, Russian Academy of Science,
email: `okochina@inasan.ru, dwiebe@inasan.ru`

[3]Lebedev Physical Institute, Samara 443011, Russia; Institute of Astronomy, Russian
Academy of Sciences

Abstract. PRESTALINE is a package allowing a user to simulate and analyse spectra of various astrophysical objects. The package is based on the numerical models PRESTA (Kochina & Wiebe (2017)) and RADEX (van der Tak et al. (2007)). PRESTALINE provides the direct comparison of theoretical models with observations and allows estimating physical conditions in a studied object, such as kinetic temperature and chemical composition. Here we present the results of applying PRESTALINE to the test object DR21(OH) and discuss possible applications and future extensions of the project.

Keywords. astrochemistry, line: profiles, ISM: molecules

1. Introduction

Astrochemical modelling is a standard approach for the investigation of star-forming regions. The problem in its application is in the need of making a proper comparison of theoretical (molecular abundances calculated with a chemical model) and observational (molecular spectra) results. Usually it are the observers who make the step towards the theory: observed spectral lines are identified, and their intensities are converted into column densities of corresponding species. These observationally inferred column densities can then be compared to the column densities obtained as an output from astrochemical numerical models. But this approach is not free of complications. Generally, as all solutions of the inverse problem, it suffers from degeneracy problems. Also, to make the problem of minimisation tractable, we need many simplifying assumptions on the physical structure of the object. Conversion of line intensities into column densities is often made under the LTE assumption, which is not always adequate for realistic physical conditions. Theoreticians on the other side blindly rely on the values provided by observers, which may lead to confusion instead of discovery.

A more correct approach can be found in solving the direct problem: converting column densities obtained in numerical astrochemical models into line intensities and creating a synthetic spectrum, suitable for direct comparison with the observed spectrum. In this work we present the package PRESTALINE (Van Looveren, Kochina & Wiebe 2021) designed for that purpose. PRESTALINE simulates the chemical evolution, uses the column densities and physical properties of the object to calculate synthetic line intensities and then provides the visual comparison of the observed and calculated spectra. As a separate feature, the package allows using the observed spectrum for a rough estimation of parameters of the studied object, such as temperature and chemical composition, in a quantitative way.

A description of the package is given in Section 2. Section 3 demonstrates capabilities of the estimation tool and its application to the spectrum of the DR21(OH) star-forming region. The conclusions and plans for further development and applications of the package are described in Section 4.

2. Synthetic spectra

PRESTA. The astrochemical module of PRESTALINE is based on PRESTA, a 1D model of the chemical evolution of prestellar and protostellar objects, which is being developed at the Institute of Astronomy of the Russian Academy of Sciences. PRESTA simulates a static or collapsing spherically symmetric protostellar core, irradiated by both diffuse external radiation and a central source (protostar) with assumed parameters. The modelled object is characterized by gas and dust temperature radial profiles (these temperatures can be different), density radial profile, and extinction for external and internal radiation. PRESTA currently uses the chemical network from the work of Albertsson et al. (2014) that includes deuterated species. The model may contain several dust populations and also may take into account evolving distributions of temperature and density (warm-up and collapse). The outputs of PRESTA are radial profiles of relative abundances as a function of time and/or column densities of each species as a function of time.

At the first step, PRESTA prepares basic files containing the information on the modelled object needed for further processing and creating the synthetic spectra. These data are kinetic temperature, density of collision partners (we consider only H_2 here), and column densities of the considered species. In order to be more consistent with the properties of the real object and allow for a more detailed study, the computational domain is divided into several zones, each with its own physical conditions. Due to limitations of RADEX on the next processing step, the number of zones cannot be too large, as the column densities of less abundant species within the zone may become smaller than the minimum that RADEX can handle. In this work we assumed that the object is divided into three zones: a hot dense inner zone, a warm dense intermediate zone, and a cold and less dense outer zone. For each of the zones separate output files are produced with average temperature, gas density, and the column densities of the species. These files are passed to the second module of the package as the input ones.

RADEX. In order to relate the chemical model of a molecular cloud to a spectrum, the column densities need to be converted into spectral intensities or similar quantities (several options for conversion are included in PRESTALINE). PRESTALINE uses RADEX to achieve this purpose. RADEX is a non-LTE, statistical equilibrium radiative transfer code developed by van der Tak et al. (2007). It provides information on each transition such as central wavelength, excitation temperature, and opacity.

Visualisation. The next part of PRESTALINE is a visualisation module. As thermal broadening is usually the dominant broadening mechanism in molecular clouds, a Gaussian line shape is a good approximation, so the excitation temperatures calculated by RADEX are converted into the amplitudes of Gaussians. To ensure that the lines and only the lines are modeled in sufficient detail, a window around the central wavelength is created. If a new line overlaps with an earlier modelled line, their intensities are added. This allows for line blending and more realistic spectra. Additionally, the visualisation takes into account the optical depth of a transition to reproduce the broadening of saturated lines.

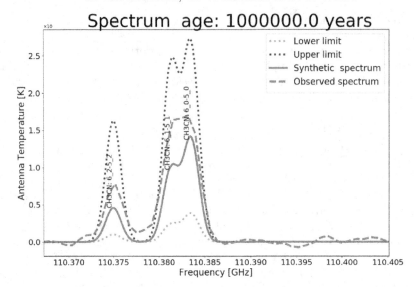

Figure 1. An example of a synthetic spectrum compared with the observed spectrum of DR(21)OH. The dotted lines represent the upper (calculated with column density times 5) and lower (calculated with column density divided by 5) limits.

All lines originating within a zone are combined into a single spectrum. This spectrum is then used as the background radiation of the next, more outward zone. The background reduced in intensity according to the opacity of the zone closer to the observer. This process is repeated for all the three zones to get the emergent spectrum. This approach allows for more complex spectra as outer zones can diminish and even extinguish the contribution of the inner ones.

In order to demonstrate the capabilities of PRESTALINE, we used the spectrum of DR21(OH), a well-known region of massive star formation belonging to the Cygnus-X complex. The spectrum was kindly provided by S. Kalenskii (Kalenskii & Johansson 2010). The region has already been studied with PRESTA (Kochina et al. (2013)), so for the test modelling we could use information on the physical properties we already have. The comparison of the synthetic spectrum created by PRESTALINE with the observed spectrum is shown in Figure 1. To provide a quick way to assess how close a synthetic spectrum is to an observed one, PRESTALINE produces error margins. This is done by dividing the column density of each considered species by a factor of 5 for the lower margin and multiplying it by 5 for an upper margin.

3. Estimation tool

The PRESTALINE package also has the capability to estimate the parameters of an object by analysing the observed spectrum. The first of these parameters is the temperature of the molecular cloud. The temperature can then be used to fit the column densities of various species.

Temperature estimation. The temperature is the first parameter which PRESTALINE fits, as further parameters depend on the temperature. The temperature is estimated through a rotation diagram, a method discussed in detail in Kalenskii & Johansson (2010). In this method the line intensity is related to the energy of the transition, which depends on the gas temperature. The energy of the transition can be calculated from molecular data. The line intensity is found by fitting a Gaussian to the observed spectrum

CN, Density estimates

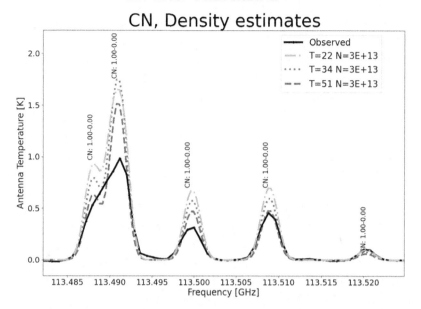

Figure 2. Estimation of CN column density for different temperatures using a 1-zone approach. A 1-zone approach is unable to reproduce the ratios of transitions of the observed spectrum.

and calculating the area under it. In the case of two overlapping lines, two Gaussians are fitted simultaneously to get the best result. As the lines selected for this purpose are either single or a blended pair, a double Gaussian is sufficient. The slope described by these points is the inverse of the temperature. It is important to note that this method assumes optically thin lines, making the selection of lines important.

For this work, CH_3CCH and CH_3CN are chosen. The first of these is associated with the warm dense intermediate region, whereas the latter is most commonly found in the hot dense core region. When this method is applied to the observed spectrum, it results in $34 \pm 12K$ and $51 \pm 9K$ for transitions of CH_3CCH and CH_3CCN, respectively. Figure 2 shows a synthetic spectrum for a 1-zone model of 22 K (the lower error bound), 34 K (the CH_3CCH temperature), 51 K (the CH_3CN temperature) and the observed spectrum. From this graph it is clear that the temperature strongly influences both the intensity of the lines and the ratio between intensities of the same molecule's lines. The 1-zone models cannot reproduce the observed spectrum satisfactorily. By combining multiple zones with different parameters (e.g. different temperatures), better agreement with the observation can be achieved. This multiple zone model is also more representative of real protostellar objects.

Density estimation. With the temperature determined, the next parameter is the column density. The column density of each species is fitted separately and can be fitted either with a 1-zone or a 3-zone model. At this step, RADEX is used again to calculate the line intensities for each transition. This requires the previously determined temperature and a column density, the parameter we wish to fit. As RADEX can handle 20 orders of magnitude in number density, it can be time consuming to fully cover all possible combinations especially when using multiple zones. For this reason PRESTALINE currently uses a random search algorithm to probe the parameter space.

Each set of column densities is turned into a synthetic spectrum as described above. The spectra are then compared to the observation to determine the uncertainty of the

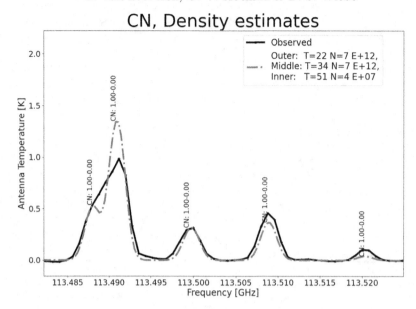

Figure 3. Estimation of CN column density using a 3-zone approach. A 3-zone approach allows much better agreement with the observations.

fit. Using the 3-zone model allows the fitting procedure to find better fits for some complex lines as shown in figure 3, but it also adds uncertainty to the results. When an outer zone is very dense and opaque, different parameters for the inner zone result in the same spectrum and uncertainty. Additionally, an error calculation cannot see the difference between overestimation or a comparable underestimation, though this can have significant physical meaning. For this reason PRESTALINE will then store the three best fits for the user to compare.

4. Conclusion

At the symposium we presented PRESTALINE, a package for the numerical analysis of molecular spectra. The tool intends to bridge a gap between theoretical and observational astrochemistry, both by deriving synthetic spectra from astrochemical models and by deriving parameters from observed spectra. The synthetic spectra allow a chemical model to be compared directly to observations through statistical equilibrium, which avoids the assumption of LTE. On the other hand, the estimation tool allows the user to quickly assess some of the object parameters.

In the section on temperature estimation, we alluded to the importance of temperature tracers for different regions. Adding molecules to the temperature estimation and grouping them according to their formation region will further improve the derived temperature profile and consequently the density profiles. To further improve the ease of use, the estimation part will be expanded with a tool for the analysis of the radial velocity.

One of the future applications of PRESTALINE is to build an atlas of spectra for a grid of objects with the given parameters and facilitate estimating specifics of studied objects.

PRESTALINE can be used for studies of various objects, different in geometry and physical conditions from the star-forming regions presented in the current work, such as planetary nebulae or protoplanetary disks. Extensions of the chemical network will also allow calculating the synthetic spectra for hotter regions: galactic nuclei and others.

Acknowledgements

The studies at Lebedev Physical Institute were supported by the Ministry of Science and Higher Education of the Russian Federation by the grant No. 075-15-2021-597.

We thank Sergei Kalenskii for valuable support and for the spectral data provided.

References

T. Albertsson, D. Semenov and T. Henning 2014, *ApJ* 784,39

S. V. Kalenskii & L. E.B. Johansson 2010, *Astron. Rep.*, 54, 295

O. V. Kochina, D. S. Wiebe, S. V. Kalenskii and A. I. Vasyunin 2013, *Astron. Rep.* 57, 818

O. V. Kochina & D. S. Wiebe 2017, *Astron. Rep.*, 61, 103

F. F. S. van der Tak, J. H. Black, F. L. Schöier & D. J. Jansen and E. F. van Dishoeck 2007, *A&A*, 468, 627

G. Van Looveren, O. V. Kochina & D. S. Wiebe 2021, *Open Astronomy* vol. 30, no. 1

Discussion

BANDA BARRAGAN: Can PRESTALINE be adapted for use, with outputs from grid-based simulations. For example, we have a temperature map from the simulation or a column density map, can I adapt the code for a use with the simulation output as well to produce some synthetic spectra?

VAN LOOVEREN: Yes, at the moment the visualisation part and RADEX part take as an input: the column density, temperature and collision partner column density. If these are provided the different sections can work on their own. In a similar way if you have a profile of the temperature, the density and radiation field, this can be used as an input for PRESTA.

The Predictive Power of Computational Astrophysics as a Discovery Tool
Proceedings IAU Symposium No. 362, 2023
D. Bisikalo, D. Wiebe & C. Boily, eds.
doi:10.1017/S1743921322001223

Episodic accretion onto a protostar

Tomoyuki Hanawa[1], Nami Sakai[2] and Satoshi Yamamoto[3]

[1] Center for Frontier Science, Chiba University, 1-33 Yayoi-cho, Inage-ku, Chiba 263-8522,
Japan
email: `hanawa@faculty.chiba-u.jp`

[2] RIKEN Cluster for Pioneering Research, 2-1 Hirosawa, Wako-shi, Saitama 351-0198, Japan
email: `nami.sakai@riken.jp`

[3] Department of Physics, The University of Tokyo, Bunkyo-ku, Tokyo 113-0033, Japan
email: `yamamoto@taurus.phys.s.u-tokyo.ac.jp`

Abstract. We introduce hydrodynamic simulations in which a protostar captures a cloudlet with a relatively small angular momentum. The cloudlet accretes onto the protostar and perturbs the gas disk rotating around the protostar. This cloudlet capture can reproduce some features observed in the molecular emission lines from TMC-1A. First, the cloudlet can reproduce the blue asymmetry observed in the CS emission. Second, the cloudlet can explain the slow infall observed in the $C^{18}O$ emission. Third, the impact of the cloudlet can explain the offset of the SO emission from the disk center. We also argue that a warm gas should confine the cloudlet through pressure. A protostar may obtain substantial mass by capturing cloudlets.

Keywords. Young stellar objects – Interstellar medium – Interstellar molecules – Protostars – Star formation

1. Introduction

Gas accretion is one of the important characteristics of young stellar objects. It supplies mass and angular momentum to the disks rotating around the central stars. Though the accretion rate tends to decline with age, it is variable and may be sporadic. We often assume that the gas accretion is more or less symmetric unless the system does not contain binary or multiple stars. However, some protostars show a highly asymmetric Doppler shit in the molecular emission lines observed with ALMA (see, e.g., Yen et al. 2014 for L1489 IRS; Sakai et al. 2016 for TMC-1A). More recently Garufi et al. (2022) have detected SO and SO_2 emissions in the region of disk impacted by gas streamer in DG Tau and HL Tau. These observations suggest the possibility that the gas accretion is intrinsically asymmetric in these systems.

Gas accretion onto more evolved objects can also be asymmetric. Dullemond et al. (2019) have demonstrated that capture of a cloudlet forms an arc-like feature observed in AB Aur. The cloudlet capture model can explain the misalignment of inner and outer disks observed in some transitional systems (Küffmieir et al. 2021). It can also explain variability in the accretion rate because the capture should be episodic.

The cloudlet capture model assumes that the accreting gas has a relatively small angular momentum and hence the centrifugal radius is small. Then, the accreting gas approaches the central star without losing the angular momentum. This mode of gas accretion is quite different from the classical picture in which the angular momentum transfer drives gas accretion. The cloudlet capture is similar to the ballistic model of an infalling–rotating envelope of Sakai et al. (2016) in which the centrifugal radius is

100 au. The ballistic model, however, does not take into account hydrodynamic effects and asymmetry. Here we introduce our hydrodynamic simulations for TMC-1A in which a protostar associated with a rotating gas disk captures a cloudlet. The cloudlet collides with the disk and destroys a part of it. The model can explain the observed asymmetry in the molecular emission lines from TMC-1A.

2. Model and Methods

We consider the three components of gas, cloudlet and gas disk, as well as the surrounding warm medium, in our simulations. Both the cloudlet and gas disk consist of cold molecular gas, while the warm gas consists of warm atomic gas. All these components should remain nearly isothermal because the thermal timescale is much shorter than the dynamical timescale. We employ an artificially low specific heat ratio, $\gamma = 1.05$, and do not take into account heating and cooling explicitly to mimic the nearly isothermal state. We assume that the mass of the central star is $M = 0.53$ M$_\odot$, though it is highly uncertain in the literature. We do not consider the magnetic fields and the self-gravity for simplicity. Hence, the basic equations are expressed as

$$\frac{\partial \rho}{\partial t} + \boldsymbol{\nabla} \cdot (\rho \boldsymbol{v}) = 0, \quad \frac{\partial}{\partial t}(\rho \boldsymbol{v}) + \boldsymbol{\nabla}\,(\rho \boldsymbol{v}\boldsymbol{v} + P) = \rho \boldsymbol{g}, \quad \boldsymbol{g} = -\frac{GM}{\max(|\boldsymbol{r}|, a)^3}\boldsymbol{r}, \tag{2.1}$$

$$\frac{\partial}{\partial t}(\rho E) + \boldsymbol{\nabla} \cdot [(\rho E + P)\boldsymbol{v}] = \rho \boldsymbol{v} \cdot \boldsymbol{g}, \quad E = \frac{\boldsymbol{v}^2}{2} + \frac{P}{(\gamma - 1)\rho}, \tag{2.2}$$

where ρ, P, $\boldsymbol{v}$, and G denote the density, pressure, velocity, and gravitational constant, respectively. We reduce the gravitational acceleration, $\boldsymbol{g}$, artificially in the region of $r \leq a = 50$ au to avoid too steep pressure gradient.

We use the cylindrical coordinates, (r, φ, z), in our hydrodynamic simulations. The central star is located at the origin of the cylindrical coordinates. We integrate the hydrodynamic equations according to Hanawa & Matsumoto (2021) so that we can achieve both the angular momentum conservation and free-stream preservation. Our numerical code does not take account of the increase in the protostar mass. The spatial resolution is $\Delta r = \Delta z \simeq r\Delta\varphi = 1.0$ au in the region of $|r| \leq 64$ au and $|z| \leq 64.5$ au, while it is $\Delta r/r \simeq \Delta z/|z| \simeq \Delta\varphi = 1/64$.

The warm gas has mean molecular weight of 1.27 $m_{\rm H}$, respectively, at the initial stage, where $m_{\rm H}$ denotes the mass of hydrogen atom. The warm gas is in hydrostatic equilibrium. The gas disk has a radius of 100 au and is supported mainly by the centrifugal force against gravity. The gas disk is in hydro static balance in the vertical direction with the surrounding warm gas. The cloudlet has the same pressure as the surrounding warm gas and a uniform velocity corresponding to the hyperbolic orbit at the cloudlet center. The periastron of the hyperbolic orbit, i.e., the centrifugal barrier, is set to be 50 au. Both the cloudlet and disk have an initial temperature of 78 K and a mean molecular weight of 2.3 $m_{\rm H}$.

We introduce the color (scalar) field, c, to trace the gas. It is assigned to be $c = 1$ for the cloudlet, $c = -1$ for the disk, and $c = 0$ for the warm gas. We follow the change in the color by solving

$$\frac{\partial(c\rho)}{\partial t} + \boldsymbol{\nabla} \cdot (c\rho \boldsymbol{v}) = 0, \tag{2.3}$$

simultaneously.

We denote the time, t, in yr and the distance in au. For further details, see Hanawa et al. (2022), on which this talk is based.

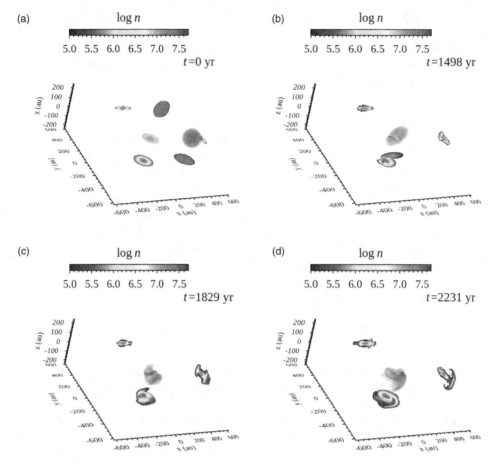

Figure 1. Evolution of the density distribution in model A in which the orbit of the cloudlet is coplanar to the rotating disk. Each panel shows the density by volume rendering in the center. The density distributions on the planes of $x = 0$, $y = 0$, and $z = 0$ are projected on the right, deeper and lower sides, respectively.

3. Results

Figure 1 shows the density evolution in model A by a series of snapshots. Each panel displays the number density by volume rendering and cross sections. Note that our numerical model is scalable, i.e., still valid if both the density and pressure are magnified by an arbitrary factor. Thus, the color bar shows not the absolute density but a measure. Figure 1(a) shows the initial stage at $t = 0$. The blue sphere on the right hand side denotes the cloudlet while the red disk surrounded by blue ring denotes the disk. The cloudlet is located at 500 au away from the central star and has the radius of 100 au. The disk radius is 100 au at the initial stage. At $t = 1498$ yr, the head of the cloudlet collides with the outer edge of the disk. The cloudlet rotates around the protostar while disturbing the pre-existing gas disk. The impact of the cloudlet forms spiral arms in the disk as shown in the cross sections of $z = 0$ in Figure 1 (c) and (d). Part of the cloudlet accretes onto the disk, whereas the rest leaves.

The cloudlet approaches the protostar from one side. Thus, it appears as a blue-shifted component to the left of the protostar if we observe it from the lefthand side in Figure 1, i.e., from the direction of $x = -\infty$. After the passage of the periastron, the cloudlet should appear as a red-shifted component to the right of the protostar. Hence, model A

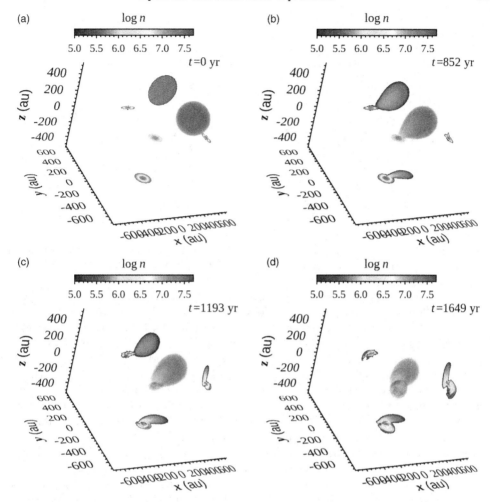

Figure 2. The same as Figure 1 but for model C.

can reproduce the blue asymmetry observed in TMC-1A qualitatively if we choose the appropriate stage.

Model A, however, mismatches the observations in some respects. First, the velocity of the infall is much higher than the observed one, about 1.6 km s^{-1}. The free-fall velocity reaches 4.34 km s^{-1} at the distance of 50 au from the star for the assumed stellar mass, 0.53 M$_\odot$. Aso et al. (2015) suggested that the infall is decelerated by magnetic force from the observation of C^{18}O emission. Second, the cloudlet is too compact to explain the extension of the blue-shifted emission.

We have constructed model C to overcome the shortcomings mentioned. First, the orbital plane of the cloudlet is inclined to the disk plane so that they have different inclinations to the line of sight. If the orbital plane is nearly face-on, the line of sight velocity is much lower than the velocity in 3D. Thus far, the infall velocity has been derived under the assumption that the infalling gas shares the same orbital plane with the disk. The inclination is estimated from the oblateness of the disk shape measured from the radio continuum emission. The cloudlet is assumed to be spherical at the initial stage and to have the mass of 1.02×10^{-4} M$_\odot$. If such a cloudlet accretes onto the protostar every thousand years, the average mass accretion rate amounts to $\dot{M} = 10^{-7}$ M$_\odot$ yr^{-1}.

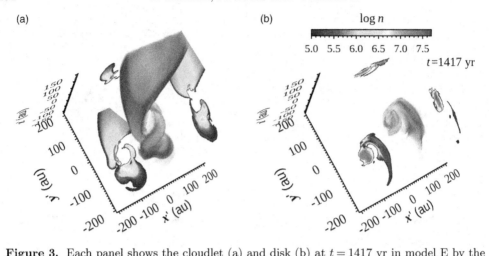

Figure 3. Each panel shows the cloudlet (a) and disk (b) at $t = 1417$ yr in model E by the volume rendering and cross sections. The color scale is common.

Figure 2 is the same as Figure 1 but for model C. The orbital plane of the cloudlet is inclined by 30° to the disk plane in model C. The cloudlet has a radius of 200 au, and its center is located 500 au away from the protostar at the initial stage. The cloudlet reaches the gas disk earlier in model C than in model A [see Figure 2(b)]. As shown in Figure 2(c), the cloudlet hits the upper side of the disk to form spiral arms. The head of the cloudlet goes through the disk midplane at a later stage shown in Figure 2(d).

The cloudlet changes its shape and size while approaching the protostar. The tidal force of the protostar elongates the cloudlet in the direction of the orbit and the warm gas confines the cloudlet through the pressure. Though the elongation and confinement are seen in model A, they are more prominent in model C.

Our numerical model is similar to those of Dullemond et al. (2019) and Küffmieir et al. (2020, 2021) except for the presence of the warm gas. The cloudlet expands to form an extended second-generation disk in their simulations, while it evolves into an elongated stream in our simulations. The difference is that the confinement of cloudlet is ascribed to the warm gas.

Figure 3 shows the cloudlet and disk at $r = 1417$ yr separately. The viewing angles are different from that of Figure 2, and the horizontal coordinates are rotated by 37.5°,

$$(x', y') = (x \cos 37.°5 + y \sin 37.°5, -x \sin 37.°5 + y \cos 37.°5). \qquad (3.1)$$

A part of the cloudlet already passed the periastron, though the main part is still approaching the periastron. The former can be seen as a minor red-shifted component if we observe it in this viewing angle. Figure 3(a) shows that the collision of the cloudlet with the disk produces a shock wave. The shock compression should increase the temperature temporally and changes its chemical composition accordingly.

The outer part of the disk is detached from the inner part to form a large spiral arm. The impact on the disk depends on the ratio of the cloudlet mass to the disk mass. Because the disk has the same mass in models A and C, the impact is larger in model C.

4. Comparison with Observations

Our model C can reproduce several features observed in the molecular emission lines from TMC-1A. First, the model can explain the blue asymmetry observed in CS ($J = 2 - 1$) line (Sakai et al. 2016) if the viewing angle is set appropriately. The cloudlet is chemically fresh, i.e., its chemical composition is close to that of a molecular cloud.

Second, it can resolve the issue of apparently slow infall estimated from the $C^{18}O$ line emission by Aso et al. (2015) under the assumption that the orbital plane is inclined to the disk plane and nearly faced on. Third, it can explain the asymmetry that the SO emission is offset to the south from the protostar. The impact of the cloudlet onto the disk induces a shock wave and releases SO from the dust through sputtering and shattering. To match the observation, the cloudlet should approach the protostar from the Northwest and the orbital plane should be close to the plane, though the Northwest side should be far from us.

A warm atomic gas is an important constituent in our model, though it is neither visible in our model nor in observations because of the low density. However, the warm gas confines the cloudlet by the pressure and assists its elongation. If the gas pressure is much lower, the cloudlet disperses to form an extended arc or a disk as shown by Dullemond et al. (2019) and Küffmieir et al. (2020, 2021). The shape of newly accreted molecular gas can serve as a probe of the environment around the young stellar objects.

This work was supported by JSPS KAKENHI Grant Nos. JP18H05222, JP19K03906, JP20H05845, JP20H05847, JP20H00182.

References

Aso, Y., Ohashi, N., Saigo, K. et al. 2015, *ApJ*, 812, 27
Dullemond, C.P., Küffmeier, M., Goikovic, F. et al. 2019, *A&Ap*, 628. A20
Garufi, A., Podio, L. Codella, C. et al. 2022, *A&Ap*, 658, A104.
Hanawa, T., Matsumoto, Y. 2021, *ApJ*, 907, 43
Hanawa, T., Sakai, N., Yamamoto, S. 2022, *ApJ*, 932, 122.
Küffmeier, M., Goicovic, F.G., Dullemond, C.P. 2020, *A&Ap*, 633, 3
Küffmeier, M., Dullemond, C.P., Reissi, S., Goicovic, F.G 2020, *A&Ap*, 656, 161
Sakai, N., Oya, Y., Lópezz Sepulcre, A. et al. 2016, *ApJ* (Letters), 820, L34
Yen, H.-W., Takakuwa, S., Ohashi, N. et al. 2014, *ApJ*, 793, 1

Discussion

BISIKALO: How high is the temperature of the cold gas?

HANAWA: The temperature of the cold gas is a little high and about 80 K.

BISIKALO: How did you take into account the viscosity?

HANAWA: The gas is assumed to be inviscid because it accretes onto the disk on a dynamic timescale. The viscosity has little effect on the accretion flow even if we take into account the viscosity of Shakura–Sunyaev type of $\alpha \simeq 0.1$.

BISIKALO: Did you take into account the self-gravity?

HANAWA: No.

The Predictive Power of Computational Astrophysics as a Discovery Tool
Proceedings IAU Symposium No. 362, 2023
D. Bisikalo, D. Wiebe & C. Boily, eds.
doi:10.1017/S1743921322001296

Photoevaporation of Protoplanetary Disks

Ayano Komaki[1], Riouhei Nakatani[2] and Naoki Yoshida[1,3,4]

[1]Department of Physics, The University of Tokyo, 7-3-1 Hongo, Bunkyo, Tokyo 113-0033,
Japan
email: `ayano.komaki@phys.s.u-tokyo.ac.jp`

[2]RIKEN Cluster for Pioneering Research, 2-1 Hirosawa, Wako-shi, Saitama 351-0198, Japan

[3]Kavli Institute for the Physics and Mathematics of the Universe (WPI), UT Institute for
Advanced Study, The University of Tokyo, Kashiwa, Chiba 277-8583, Japan

[4]Research Center for the Early Universe (RESCEU), School of Science, The University of
Tokyo, 7-3-1 Hongo, Bunkyo, Tokyo 113-0033, Japan

Abstract. Photoelectric effect of dust grains by UV radiation is an important process for disk
heating, but as a disk evolves, the amount of dust grains decreases. Photoeaporation is a disk
dispersal process, which is caused by high-energy radiation. We perform a set of photoevaporation simulations solving hydrodynamics with radiative transfer and non-equilibrium chemistry
in a self-consistent way. We run a series of simulations with varying the dust-to-gas mass ratio
in a range $\mathcal{D} = 10^{-1}$–10^{-8}. We show that H_2 pumping and X-ray heating mainly contribute to
the disk heating in case of $\mathcal{D} \leq 10^{-3}$ and photoelectric effect mainly heats the gas in $\mathcal{D} \geq 10^{-3}$
cases. The mass-loss profile changes significantly with respect to the main heating process. The
outer disk is more efficiently dispersed when photoelectric effect is the main heating source.

Keywords. protoplanetary disks, formation,hydrodynamics,pre–main-sequence

---

1. Introduction

Planets are formed inside protoplanetary disks (PPDs). Recent observations suggest
that the protoplanetary disks disperse in a few Myr (Haisch 2001; Cieza 2007; Richert
(2018)). The disk lifetime could be the time limit for planet formation. Planets move in
the radial direction by interaction with the disk. This process is called migration. Clearly,
the disk evolution and dispersal timescale determine the resulting planetary systems.

Accretion, magnetohydrodynamics (MHD) winds and photoevaporation are suggested
to be main disk dispersal processes. Accretion is mass-loss onto the star by viscous
evolution. Magnetorotational instability contributes to mass-loss by generating magnetorotational winds. MHD simulations show that this process is effective around
a low-mass star. Photoevaporation is a process that the high-energy radiation (far-
ultraviolet; FUV; $6\,\mathrm{eV} \leq h\nu < 13.6\,\mathrm{eV}$, extreme-ultraviolet; EUV; $13.6\,\mathrm{eV} \leq h\nu \leq 100\,\mathrm{eV}$,
X-rays; $100\,\mathrm{eV} \leq h\nu$) from the central star heats the disk gas. The gas with sufficient
kinetic energy eventually escapes from the system. EUV photons are absorbed by atomic
hydrogen to form HII region and they do not reach the dense region. Since FUV and
X-ray photons are absorbed by dust grains and various elements, they penetrate into a
dense region of a disk and generate dense winds.

The dust properties are characterized by dust-to-gas mass ratio and size distribution.
The dust size distribution is well explained as $\mathrm{d}n \propto a^{-3.5}\,\mathrm{d}a$, where a represents dust
size. This is called MRN distribution (Mathis 1977). Dust grains in protoplanetary disks
go through physical process such as dust sedimentation, dust growth and radial drift.

Recent SED observations by ALMA suggest that these process deplete small dust grains and change the dust size distribution (Espaillat 2014). Since small dust grains are the main cause of photoelectric heating, the heating rate of photoelectric effect is expected to decrease and temperature profile changes as the disk evolves. We characterize the amount of small dust grains with dust-to-gas mass ratio, $\mathcal{D}$ and perform a series of photoevaporation simulations varying $\mathcal{D}$ as a parameter.

The luminosity of FUV radiation is also known to decrease with disk evolution. We consider the effect on photoevaporation in this paper. The high-energy radiation is generated by chromosphere and accretion. The accretion rate is known to decrease with a timescale of $\sim 1\,\mathrm{Myr}$, and the lunimosity also decreases (Calvet & Gullbring 1998; Gullbring 1998). We perform a set of simulations with varying luminosity.

2. Methods

We solve hydrodynamics, radiative transfer and non-equilibrium chemistry in a a self-consistent way. We use PLUTO to solve hydrodynamics (Mignone 2007). We assume the disk is axisymmetric around the rotational axis, and symmetric with respect to the disk midplane. We perform 2D photoevaporation simulations with spherical coordinates (r, θ). In order to calculate the angular momentum, we consider 3D gas velocity, $v = (v_r, v_\theta, v_\phi)$. The equations we solve are

$$\frac{\partial \rho}{\partial t} + \nabla \cdot (\rho v) = 0, \tag{2.1}$$

$$\frac{\partial (\rho v_r)}{\partial t} + \nabla \cdot (\rho v_r v) = -\frac{\partial P}{\partial r} - \rho \frac{GM_*}{r^2} + \rho \frac{v_\theta^2 + v_\phi^2}{r}, \tag{2.2}$$

$$\frac{\partial (\rho v_\theta)}{\partial t} + \nabla \cdot (\rho v_\theta v) = -\frac{1}{r}\frac{\partial P}{\partial \theta} - \rho \frac{v_r v_\theta}{r} + \rho \frac{v_\phi^2}{r} \cot \theta, \tag{2.3}$$

$$\frac{\partial (\rho v_\phi)}{\partial t} + \nabla^l \cdot (\rho v_\phi v) = 0, \tag{2.4}$$

$$\frac{\partial E}{\partial t} + \nabla \cdot H v = -\rho v_r \frac{GM_*}{r^2} + \rho(\Gamma - \Lambda), \tag{2.5}$$

$$\frac{\partial n_{\mathrm{H}} y_i}{\partial t} + \nabla \cdot (n_{\mathrm{H}} y_i v) = n_{\mathrm{H}} R_i, \tag{2.6}$$

where ρ, P, M_*, y_i, R_i represent gas density, pressure, central stellar mass, abundance of each chemical species and reaction rate, respectively. Each of E, H, Γ, Λ expresses the energy, enthalpy, heating rate and cooling rate per unit volume, and G, n_{H} are the gravitational constant and the number density of elemental hydrogen.

We incorporate FUV, EUV, X-ray photons as radiation from the central star. We assume that the system is $\sim 1\,\mathrm{Myr}$ old, and determine each luminosity following Gorti & Hollenbach (2009). We incorporate EUV/X-ray photoionization heating, FUV photoelectric heating (Bakes & Tielens 1994), and heating by H_2 photodissociation and by H_2 pumping (Kuiper 2020; Nakatani 2021). We also incorporate dust-gas collisional cooling (Yorke & Welz 1996), fine-structure cooling of CII and OI (Hollenbach & McKee 1989; Osterbrock 1989; Santoro & Shull 2006), molecular line cooling of H_2 and CO (Galli & Palla 1998; Omukai *et al.* 2010), hydrogen Lyman α line cooling (Anninos 1997), and radiative recombination cooling (Spitzer 1978) as cooling sources. We take into account 10 chemical species, HI, HII, H^-, HeI, H_2, H_2^+, CO, OI, CII and electrons.

We calculate the dust temperature in the following manner. We incorporate both the direct and diffuse radiation to solve the dust temperature. We perform ray-tracing for the direct component and adopt the flux-limited-diffusion (FLD) approximation for the diffused component (Kuiper 2010; Kuiper & Klessen 2013; Kuiper 2020).

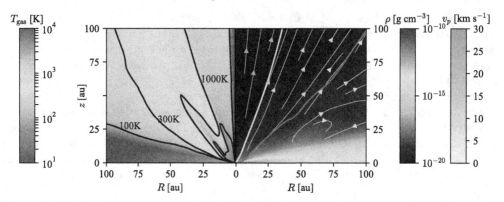

Figure 1. The disk structure of the simulation with $\mathcal{D} = 10^{-6}$. The color map on the left portion shows the gas temperature, T_{gas}, and the right portion shows the density distribution, ρ. The right portion also shows the distribution of H-bearing species. The yellow line on the right side represents the ionization front where the abundance of HII is 0.5, and the brown line indicates the dissociation front where the abundance of H_2 is 0.25. The navy contour lines on the left side represent where the gas temperature is 100 K, 300 K, 1000 K, and 3000 K. The arrows represent the poloidal velocity of the gas, which is defined as $v_p = \sqrt{v_r^2 + v_\theta^2}$. For clarity, we do not plot the arrows where the velocity is less than $0.1 \,\text{km}\,\text{s}^{-1}$, which are mostly in the optically thick disk region.

We define the computational domain of the polar angle to $[0, \pi/2]$, and the radial coordinate to $[0.1 r_g, 20 r_g]$. The gravitational radius r_g is defined as

$$r_g = \frac{GM_*}{(10\text{km s}^{-1})^2} \simeq 8.87 \text{ au} \left(\frac{M_*}{1\,M_\odot} \right).$$

for a fully ionized gas with $T_{\text{gas}} = 10^4$ K. We assume that the disk mass is 3% of the central stellar mass i.e., $M_{\text{disk}} = 0.03 M_*$ (Andrews & Williams 2005). See Nakatani (2018a,b) and Komaki (2021) for more details.

3. Results

Fig. 1 shows the time-averaged simulation snapshot with $\mathcal{D} = 10^{-6}$. We average from 840 yr to 5000 yr to avoid the effect of the initial condition and transient fluctuations. Averaging over time smoothes out short-period fluctuations of the mass-loss rates. EUV radiation heats the gas near the central star and at high latitudes. The gas forms the HII region there. The gas temperature is lowered via adiabatic cooling and becomes $T_{\text{gas}} \sim 3000$ K. The HI region and H_2 region are formed at the lower latitudes. FUV photoelectric effect effectively heats the HI region up to $\gtrsim 3000$ K for $\mathcal{D} \geq 10^{-2}$. In the case of $\mathcal{D} = 10^{-3}$, the main heating source is also FUV photoelectric effect, but the gas is cooled by OI line emission down to $\lesssim 3000$ K. In the runs with $\mathcal{D} = 10^{-4} - 10^{-8}$, the photoelectric heating by FUV radiation is weak, therefore the thermochemical structure is different from the high $\mathcal{D}$ cases. X-ray radiation is the main heating source instead of FUV photoelectric effect in the HI region. The gas temperature is efficiently lowered by adiabatic cooling and OI line emission down to 300 K.

The FUV photoelectric heating rate decreases as the dust-to-gas mass ratio becomes smaller, following $\Gamma_{\text{FUV}} \propto \mathcal{D}$. In the runs with $\mathcal{D} = 10^{-1} - 10^{-2}$, the disk gas is dominantly heated by FUV photoelectric effect, while in the run with $\mathcal{D} = 10^{-3}$, the heating rate for FUV photoelectric effect becomes almost same as the heating rates for X-ray radiation and H_2 pumping. In case of $\mathcal{D} < 10^{-3}$, the disk gas is heated mainly by X-ray radiation and H_2 pumping. The heating rate by H_2 pumping is highest at the H_2 dissociation front.

The pumped H_2 molecules heat the gas by collisions with HI or H_2. H_2 pumping heats the gas effectively in a dense region whose density is higher than the critical density, $\rho \sim 5.0 \times 10^{-20}$ g cm^{-3}. H_2 pumping heating becomes relatively important at the inner region. H_2 pumping has a few times higher heating rate than X-ray heating at the H_2 dissociation front. In the inner region with $r \sim 10 r_g$, H_2 pumping heating rate is higher than X-ray heating. The ratio becomes ~ 300 at the innermost region. In contrast, in the outer regions, H_2 pumping has a heating rate a few times lower than X-ray heating. The whole thermochemical structure is determined by both H_2 pumping and X-ray heating. The disk gas temperature is 200–300 K at $r = 10 r_g$, and this is lower by a factor of ~ 1.5–2 than the fiducial $\mathcal{D}$ case.

We calculate the mass-loss rate from the whole disk based on the simulation results using

$$\dot{M} = \int_{S, \eta > 0} \rho v \cdot dS, \tag{3.1}$$

where dS represents the surface unit area at $r = 100$ au. We define η as the total enthalpy and it is calculated by

$$\eta = \frac{1}{2} v_p^2 + \frac{1}{2} v_\phi^2 + \frac{\gamma}{\gamma - 1} c_s^2 - \frac{GM_*}{r}.$$

We define $v_p = \sqrt{v_r^2 + v_\theta^2}$ as the poloidal velocity, γ as the specific heat ratio, and c_s as the sound velocity. We assume that only the gas satisfying $\eta > 0$ escapes from the system eventually.

Based on Fig. 2a, we classify the disk dispersal by the difference of the disk heating profile.

$$\dot{M} = \begin{cases} 1.0 \times 10^{-9} \times 10^{-0.15\mathcal{D}^2 - 0.49\mathcal{D}} \ M_\odot \ \mathrm{yr}^{-1} & (\mathcal{D} \geq 10^{-3}) \\ 5.6 \times 10^{-10} \ M_\odot \ \mathrm{yr}^{-1} & (\mathcal{D} \leq 10^{-5}) \end{cases}.$$

In the run with $\mathcal{D} = 10^{-1}$, the disk is optically thick because of abundant dust grains. The photoevaporative flows are launched from the region with lower density, which result in the lower mass-loss rate compared to $\mathcal{D} = 10^{-2}$ case. When the dust-to-gas mass ratio is $\mathcal{D} \leq 10^{-4}$, the gas temperature is raised by H_2 pumping and X-ray heating. Since the abundance of H_2 does not depend on the dust amount, we expect that the mass-loss rate with even lower dust-to-gas mass ratio would stay constant. In the cases of $\mathcal{D} = 10^{-6}$–10^{-8}, the mass-loss rate converges to $\dot{M} \sim 5$–$6 \times 10^{-10} \ M_\odot \ \mathrm{yr}^{-1}$.

4. Implications

Observations suggest that the luminosity of high-energy radiation differs among the same spectral type stars (Gullbring 1998; Vidotto 2014). Accretion rate decreases as the disk evolves (Hartmann 1998). Since high-energy radiation is generated by accretion shock, the luminosity also decreases with disk evolution. The FUV luminosity decreases with the accretion rate following $L_{\mathrm{FUV}} \propto \dot{M}_{\mathrm{acc}}$ (Calvet & Gullbring 1998; Gullbring 1998). Considering the accretion rate decays following $t^{-3/2}$, L_{FUV} drops to $0.1 L_{\mathrm{FUV,f}}$ in ~ 3 Myr, where $L_{\mathrm{FUV,f}} = 10^{31.7}$ erg s^{-1} is a fiducial value of a $1 \ M_\odot$ star. Komaki (2021) suggest that among FUV and X-ray photons, the photons with higher luminosity become the main heating source of the disk gas and generate photoevaporative flows. The change of the luminosity could affect the main heating process and mass-loss by photoevaporation. In $\mathcal{D} = 10^{-2}$–10^{-6} cases, we perform a set of photoevaporation simulations varying FUV luminosity as a parameter with $L_{\mathrm{FUV}} = 0.1 L_{\mathrm{FUV,f}}, 10 L_{\mathrm{FUV,f}}$. We set X-ray luminosity as a constant value, $L_{\mathrm{X\text{-}ray}} = 10^{30.4}$ erg s^{-1}.

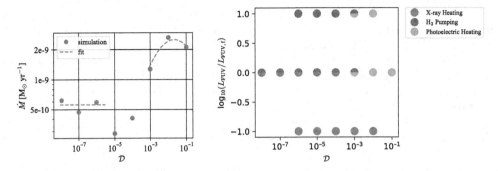

(a) The time-averaged mass-loss rate (b) The main heating process is plotted for each run
for each dust-to-gas mass ratio.   varying $\mathcal{D}$ and L_{FUV}.

Figure 2. (a) shows the time-averaged mass-loss rate for each dust-to-gas mass ratio. The
blue dots represent the simulation results. The dotted line is a fit. (b) shows the main heating
process, plotted for each run varying $\mathcal{D}$ and L_{FUV}. The orange, brown and magenta represent
photoelectric heating, heating by H_2 pumping and X-ray heating. $L_{\mathrm{FUV,f}}$ expresses the fiducial
FUV luminosity.

Fig. 2b shows the main heating process in our simulation set. When the luminosity
satisfies $L_{\mathrm{FUV}} < L_{\mathrm{X\text{-}ray}}$, X-ray radiation mainly heats the disk gas regardless of the dust-
to-gas mass ratio. When the luminosity satisfies $L_{\mathrm{FUV}} > L_{\mathrm{X\text{-}ray}}$, the gas temperature is
heated mainly by FUV radiation. Especially in case of $\mathcal{D} \leq 10^{-3}$, H_2 pumping becomes
the main heating process while in case of $\mathcal{D} \geq 10^{-3}$, the gas is heated by photoelectric
effect. As a disk evolves, small dust grains and FUV luminosity both decreases. As shown
in the previous section, the temperature structure depends on the disk heating process.
The mass-loss profile also changes with the temperature structure. The disk surface
mass-loss rate is also suggested to vary with a disk evolution.

5. Discussion

Q1: Can a photoevaporation blow away a dust?

A: Hutchison (2016) performed a simulation and show that dust grains smaller than
0.3 μm are blown up by photoevaporative winds.

Q2: Does dust photoevaporation correlate with gas mass-loss photoevaporation?

A: Yes. Since the dust photoevaporation is dependent on the gas velocity, I think dust
photoevaporation rate depends on the gas photoevaporation rate.

Q3: What is the roles of EUV and X-ray radiation?

A: The H_{II} region is formed in the low-polar-angle region. EUV photons heat the gas
effectively only in this region and do not contribute to the disk heating. As for X-ray radiation, X-ray is not the main heating source, but contribute to mass-loss. In $M_* = 3\,M_\odot$
case, the mass-loss rate is lower than other cases because of the low X-ray luminosity.
This shows that X-ray radiation contribute to the photoevaporation to some extent.

References

Andrews, Sean M. & Williams, Jonathan P. 2005, *ApJ*, 631, 1134
Anninos, Peter and Zhang, Yu and Abel, Tom and Norman, Michael L. 1997, *NewA*, 2, 209

Bakes, E. L. O. and Tielens, A. G. G. M. 1994, *ApJ*, 427, 822

Calvet, Nuria and Gullbring, Erik 1998, *ApJ*, 509, 802

Cieza, Lucas and Padgett, Deborah L. and Stapelfeldt, Karl R. and Augereau, Jean-Charles and Harvey, Paul and Evans, Neal J., II and Merín, Bruno and Koerner, David and Sargent, Anneila and van Dishoeck, Ewine F. and Allen, Lori and Blake, Geoffrey and Brooke, Timothy and Chapman, Nicholas and Huard, Tracy and Lai, Shih-Ping and Mundy, Lee and Myers, Philip C. and Spiesman, William and Wahhaj, Zahed 2007, *ApJ*, 667, 308

Espaillat, C. and Muzerolle, J. and Najita, J. and Andrews, S. and Zhu, Z. and Calvet, N. and Kraus, S. and Hashimoto, J. and Kraus, A. and D'Alessio, P. 2014, *Protostars and Planets VI*

Galli, Daniele and Palla, Francesco 1998, *AAP*, 335, 403

Gorti, U. and Hollenbach, D. 2009, *ApJ*, 690, 1539

Gullbring, Erik and Hartmann, Lee and Briceño, Cesar and Calvet, Nuria 1993, *ApJ*, 492, 323

Haisch, Karl E., Jr. and Lada, Elizabeth A. and Lada, Charles J. 2001, *ApJL*, 553, L153

Hartmann, Lee and Calvet, Nuria and Gullbring, Erik and D'Alessio, Paola 1998, *ApJ*, 495, 385

Hollenbach, David and McKee, Christopher F. 1989, *ApJ*, 342, 306

Komaki, Ayano and Nakatani, Riouhei and Yoshida, Naoki 2021, *ApJ*, 910, 51

Kuiper, R. and Klahr, H. and Dullemond, C. and Kley, W. and Henning, T. 2010, *AAP*, 511, A81

Kuiper, R. and Klessen, R. S. 2013, *AAP*, 555, A7

Kuiper, Rolf and Yorke, Harold W. and Mignone, Andrea 2020, *ApJS*, 250, 13

Mathis, J. S. and Rumpl, W. and Nordsieck, K. H. 1977, *ApJ*, 217, 425M

Mignone, A. and Bodo, G. and Massaglia, S. and Matsakos, T. and Tesileanu, O. and Zanni, C. and Ferrari, A. 2007, *ApJS*, 170, 228

Nakatani, Riouhei and Hosokawa, Takashi and Yoshida, Naoki and Nomura, Hideko and Kuiper, Rolf 2018, *ApJ*, 857, 57

Nakatani, Riouhei and Hosokawa, Takashi and Yoshida, Naoki and Nomura, Hideko and Kuiper, Rolf 2018, *ApJ*, 865, 75

Nakatani, Riouhei and Kobayashi, Hiroshi and Kuiper, Rolf and Nomura, Hideko and Aikawa, Yuri 2021, *ApJ*, 915, 90

Osterbrock, Donald E. 1989, *University Science Books*

Omukai, Kazuyuki and Hosokawa, Takashi and Yoshida, Naoki 2010, *ApJ*, 722, 1793

Ribas, Álvaro and Bouy, Hervé and Merín, Bruno 2015, *AAP*, 576, A52

Richert, A. J. W. and Getman, K. V. and Feigelson, E. D. and Kuhn, M. A. and Broos, P. S. and Povich, M. S. and Bate, M. R. and Garmire, G. P. 2018, *MNRAS*, 477, 5191R

Santoro, Fernando and Shull, J. Michael 2006, *ApJ*, 643, 26

Spitzer, Lyman 1978, *Physical processes in the interstellar medium (Wiley-Interscience)*

Vidotto, A. A. and Gregory, S. G. and Jardine, M. and Donati, J. F. and Petit, P. and Morin, J. and Folsom, C. P. and Bouvier, J. and Cameron, A. C. and Hussain, G. and Marsden, S. and Waite, I. A. and Fares, R. and Jeffers, S. and do Nascimento, J. D. 2014, *MNRAS*, 441, 2361

Yorke, H. W. and Welz, A. 1996, *AAP*, 315, 555

The Predictive Power of Computational Astrophysics as a Discovery Tool
Proceedings IAU Symposium No. 362, 2023
D. Bisikalo, D. Wiebe & C. Boily, eds.
doi:10.1017/S1743921322001727

Modeling protoplanetary disk evolution in young star forming regions

Martijn J. C. Wilhelm[1], Simon Portegies Zwart[1], Claude Cournoyer-Cloutier[2], Sean Lewis[3], Brooke Polak[4], Aaron Tran[5], Mordecai-Mark Mac Low[6,5] and Stephen L. W. McMillan[3]

[1]Leiden Observatory, Leiden University,
P.O. Box 9513, NL-2300 RA, Leiden, the Netherlands
email: `wilhelm@strw.leidenuniv.nl`

[2]Department of Physics and Astronomy, McMaster University, Hamilton, Canada

[3]Department of Physics, Drexel University, Philadelphia, USA

[4]Institut für Theoretische Astrophysik, Zentrum für Astronomie, Universität Heidelberg, Heidelberg, Germany

[5]Department of Astronomy, Columbia University, New York, USA

[6]Department of Astrophysics, American Museum of Natural History, New York, USA

Abstract. Stars form in clusters, while planets form in gaseous disks around young stars. Cluster dissolution occurs on longer time scales than disk dispersal. Planet formation thus typically takes place while the host star is still inside the cluster. We explore how the presence of other stars affects the evolution of circumstellar disks. Our numerical approach requires multi-scale and multi-physics simulations where the relevant components and their interactions are resolved. The simulations start with the collapse of a turbulent cloud, from which stars with disks form, which are able to influence each other. We focus on the effect of extinction due to residual cloud gas on the early evolution of circumstellar disks. We find that this extinction protects circumstellar disks against external photoevaporation, but these disks then become vulnerable to dynamic truncation by passing stars. We conclude that circumstellar disk evolution is heavily affected by the early evolution of the cluster.

Keywords. methods: numerical, stars: formation, planetary systems: protoplanetary disks, ISM: clouds

1. Introduction

Stars form in clusters through the gravitational collapse of a giant molecular cloud. Gaseous circumstellar disks (often also referred to as protoplanetary disks) are left over from this process, which are dispersed on a time scale of about 10 Myr (Ribas et al. 2014; Michel et al. 2021). Planets form in these disks on a shorter time scale (Tychoniec et al. 2020). The star formation process lasts a few megayears but it takes some 100 Myr for the cluster to dissolve in the Galactic tidal field (Krumholz et al. 2019). The question arises how the dense stellar and gaseous environment affects the evolution of the circumstellar disks, and therewith the planet formation process.

Evidence for a strong interaction between stars and disks is visible in the Orion Nebula Cluster, where several circumstellar disks show comet-like tails directed away from the cluster's most massive star θ^1C Ori. The material in these tails is being stripped by the radiation of θ^1C Ori, a process termed external photoevaporation (O'Dell et al. 1993;

Johnstone et al. 1998; Haworth et al. 2018). In a dense stellar environment circumstellar disks can also be stripped during close encounters, a process we'll refer to as dynamic truncation.

Several studies have simulated populations of circumstellar disks in star-forming regions that include the processes of dynamic truncations (Rosotti et al. 2014; Portegies Zwart 2016; Vincke & Pfalzner 2016; Concha-Ramírez et al. 2019a), external photoevaporation (Winter et al. 2019; Nicholson et al. 2019; Parker et al. 2021a,b), or both (Concha-Ramírez et al. 2019b, 2021a,b). In these studies, the initial masses, positions, and velocities of the stars were generally adopted from different observationally motivated parametrizations. The overlap in time scales of the various formation and evolutionary processes, however, requires a more subtle approach in which gas dynamics, star formation, disk evolution, and the various radiative processes are resolved together.

We perform multi-physics simulations based on the Torch model (Wall et al. 2019, 2020), which is assembled using the Astrophysics Multipurpose Software Environment (AMUSE; Portegies Zwart et al. 2009; Pelupessy et al. 2013; Portegies Zwart et al. 2013). Our simulations start with the gravitationally driven hydrodynamical collapse of a giant molecular cloud in which stars form with disks. These disks are subsequently affected by internal and external processes, such as viscous growth, internal and external photoevaporation, and dynamic trunction.

Here we report on a series of ten calculations in which we explore the ecology of circumstellar disks in their star-forming environments. We focus on the effect of extinction by residual gas of the giant molecular cloud (which we'll refer to as the intracluster medium from here on) on the evolution of circumstellar disks. For this reason our runs consist of pairs with identical realizations and star formation procedures, but where we include intracluster extinction in one and neglect it in the other.

2. Method

Our simulations start with the hydrodynamical collapse of a turbulent molecular cloud, in which stars form with disks that are affected by internal and external processes.

2.1. *Simulation environment*

The collapse of the molecular cloud and star formation are simulated using Torch (Wall et al. 2019, 2020). Torch couples the hydrodynamical adaptive mesh refinement code FLASH ((Fryxell et al. 2000); with additional implementations of radiative transfer (Baczynski et al. 2015), stellar winds, and supernovae) with the stellar dynamics code ph4 (McMillan et al. 2012) and the stellar evolution code SeBa (Portegies Zwart & Verbunt 1996; Toonen et al. 2012), using the AMUSE framework. Gas and stellar dynamics are coupled using the Bridge method (Fujii et al. 2007; Portegies Zwart et al. 2020), adapted to couple between (star) particles and a grid-based distribution of gas.

The circumstellar disk population is simulated using the model of Concha-Ramírez et al. (2021b). Each disk is simulated as a 1D viscous accretion disk using VADER (Krumholz & Forbes 2015), which models accretion onto the host star ($\sim 10^{-8}$ $M_\odot$ yr^{-1} in our model) and viscous spreading of the disk. The model implements internal and external photoevaporation as extra module. The mass loss through external photoevaporation due to far-UV (FUV) radiation is obtained by interpolation on the FRIED grid (Haworth et al. 2018). In contrast with Concha-Ramírez et al. (2021b) we neglect external photoevaporation due to extreme-UV radiation because their method is difficult to implement when extinction is in play.

Dynamic truncations are implemented with an event-based approach. When two stars pass each other within 0.02 pc, the closest approach between them is estimated as the

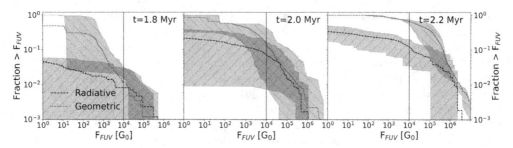

Figure 1. The fraction of circumstellar disks exposed to an FUV radiation field greater than some value, at different moments in time. The dashed curves give runs with the radiative method (including extinction). The dotted curves give runs with the geometric method (without extinction). The vertical lines denote a radiation field of 10^4 G_0, which is the maximum radiation field on the FRIED grid. The results are aggregated over five simulation runs starting from the same cloud initial conditions, but with different stellar initial mass function realizations. Shaded regions indicate typical run-to-run variation.

periastron of their two-body Keplerian orbit. We subsequently calculate the truncation radius for the disks of both stars following Portegies Zwart (2016). Any disk material beyond the truncation radius is removed.

Each simulation is run twice. In the *radiative* models, in which we account for intracluster extinction, the FUV radiation flux on a specific disk is directly taken from FERVENT, the radiative transfer solver within FLASH. In the *geometric* models, we calculate the local FUV flux by superposing the contribution of each star using the inverse square law. This allows us to study the importance of extinction due to the intracluster medium on the photoevaporation of circumstellar disks.

For efficiency, only > 7 $M_\odot$ stars exert feedback by emitting far (5.6-13.6 eV) and extreme (13.6+ eV) UV radiation, and stellar winds. We refer to these stars as massive stars.

2.2. *Initial conditions*

The calculations start with a 10^4 $M_\odot$ spherical cloud with a Gaussian density profile with a radius of 7 pc. To mediate collapse, the initial cloud is turbulent with a virial ratio (the absolute ratio of kinetic to potential energy) of 0.13. The cloud is embedded in a uniform neutral medium with a number density of hydrogen of 1.25 cm^{-3}. Stars form according to the Kroupa (2001) initial mass function, with masses from 0.08 $M_\odot$ to 150 $M_\odot$. Each star with mass $M_* < 1.9$ $M_\odot$ receives a disk upon formation with an initial structure following Concha-Ramírez et al. (2021b), but with a mass $M_d = 0.1$ $M_\odot$ $(M_*/M_\odot)^{0.73}$, and radius $R_d = 117$ au $(M_*/M_\odot)^{0.45}$ (rescaled from Wilhelm et al. 2022).

3. Results

Our simulation results are strongly affected by the time of birth, the location, and the velocity of massive stars. Their relative rarity compared to low-mass stars introduces a high degree of stochasticity in our results. To control this, we performed a total of five models (each duplicated) with identical initial hydrodynamical state (i.e., density, temperature, and velocity structure) but with a different random sequence of choices from the stellar initial mass function.

In Fig. 1 we present the fraction of circumstellar disks exposed to an FUV radiation field greater than some value, at 1.8 Myr (just after the massive star has formed in three of five runs), 2.0 Myr (when a massive star has formed in all runs), and 2.2 Myr (which

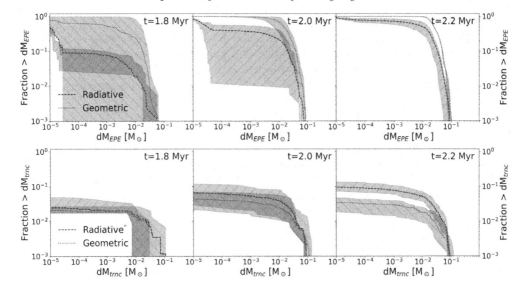

Figure 2. The fraction of circumstellar disks that lost mass through external photoevaporation (top) and dynamic truncation (bottom) greater than some value, for the radiative and geometric method, at different moments in time. The dashed curves give runs with the radiative method (including extinction). The dotted curves give runs with the geometric method (without extinction). The results are aggregated over five simulation runs starting from the same cloud initial conditions, but with different stellar initial mass function realizations. Shaded regions indicate typical run-to-run variation.

four runs have reached). The results are aggregated over all runs where data was present but split by radiation field method. The radiation fields increase with time as the number of massive stars increases. At any time, the radiation field perceived by the stars in the radiative runs is smaller than that in the geometric runs. At 2.2 Myr, on average about 0.4 Myr after the first massive star has formed, only $\sim$30% of disks in the radiative runs are exposed to a FUV radiation field greater than 1 G_0 ($1.6 \cdot 10^{-3}$ erg s^{-1} cm^{-2}, comparable to the mean interstellar level (Habing 1968)), but $\sim$15% of disks are exposed to a radiation field in excess of 10^4 G_0. Compare this with the geometric runs, where $\sim$70% of disks are exposed to radiation fields greater than 10^4 G_0. This implies that the intracluster medium in the parent cloud effectively shields circumstellar disks for at least $\sim$0.5 Myr after the formation of the first massive star.

The external radiation field leads to mass loss in the circumstellar disks. The total amount of mass lost in these disks is presented in the top row of panels in Fig. 2. The disks in the geometric runs lose more mass than those in the radiative runs. In the geometric runs at 2.2 Myr, almost every disk has lost $\gtrsim 0.01$ M$_\odot$ (which is close to all the mass in the disk for the lower-mass stars). At that same time, only half the disks in the radiative runs have lost that amount of material. This demonstrates that the shielding of the intracluster medium effectively protects the circumstellar disks.

The amount of mass lost by dynamic truncation, presented at three moments in time in Fig. 2 (bottom row), shows a reversed trend. More mass is lost by dynamic truncation in the radiative runs than in the geometric runs. At 2.2 Myr, $\sim$10% of disks in the radiative runs have lost 10^{-3} M$_\odot$ (or about 1 M$_{\rm Jup}$), against $\sim$3% in the geometric runs. These findings imply that if external photoevaporation is less effective in evaporating disks, the relative importance of dynamic truncation increases.

4. Discussion & conclusion

We have run simulations coupling the formation of stars from a collapsing cloud including massive star feedback, with the evolution of a population of circumstellar disks. We have studied the effect of extinction by residual cloud material on the external photoevaporation of circumstellar disks, and compared the efficiency of this mass loss channel to dynamic truncation due to stellar encounters.

Intracluster extinction effectively shields circumstellar disks from FUV radiation and hence reduces mass loss from external photoevaporation. For example, by $\sim$0.5 Myr after the formation of the first massive star, virtually all disks in the geometric runs are exposed to radiation fields in excess of the mean interstellar level, as opposed to $\sim$30% of disks in the radiative runs. Due to disks retaining larger radii, the mass lost through dynamic truncations increases, compared to the case without extinction. However, for the majority of disks the mass loss through external photoevaporation is still greater than through dynamic truncation.

In our simulations, which include feedback by stars more massive than 7 $M_\odot$, this gas has not been cleared for at least $\sim$0.5 Myr after the formation of the first massive star. Our simulation does not include protostellar outflows, which are produced by stars of all masses. While less energetic than feedback from massive stars, this can clear out the neighborhood of stars with disks prior to when massive star feedback becomes effective, decreasing the effectiveness of shielding.

In the radiative runs, $\sim$10% of disks lose $> 1 M_{\rm Jup}$ through dynamic truncations, compared to $\sim$3% of disks in the geometric runs. However, this depends on the dynamics of newly formed stars, which are spawned from their parent sink particle with a random uniform position offset of up to 0.17 pc and a normally distributed velocity offset proportional to the sink's sound speed ($1.9 \cdot 10^4$ cm s^{-1}). Resolving this numerical issue would require the formation of individual stars, which requires increased numerical resolution.

The relative time and distance with respect to massive stars considerably affect circumstellar disk evolution, even on the short time scale of planet formation. It would be interesting to investigate if and how the formation of planets is affected by these environmental variations in disk evolution.

Acknowledgements

The simulations in this work have been carried out on the Cartesius supercomputer, hosted by the Dutch national high performance computing center SURFsara.

This work could not have been done without the help of the AMUSE community and the entire Torch team. We thank Steven Rieder and Inti Pelupessy for their continuing work on AMUSE, and the rest of the Torch team (including Sabrina Appel, William Farner, Joe Glaser, Ralf Klessen, and Alison Sills) for many interesting discussions and a very pleasant collaboration.

References

Baczynski, C., Glover, S. C. O. & Klessen, R. S. 2015, *MNRAS*, 454, 1
Concha-Ramírez, F., Vaher, E. & Portegies Zwart 2019a, *MNRAS*, 482, 1
Concha-Ramírez, F., Wilhelm, M. J. C., Portegies Zwart, S. & Haworth, T. 2019b, *MNRAS*, 490, 4
Concha-Ramírez, F., Wilhelm, M. J. C., Portegies Zwart, S., van Terwisga, S. E. & Hacar, A. 2021a, *MNRAS*, 501, 2
Concha-Ramírez, F., Portegies Zwart, S. & Wilhelm, M. J. C. 2021b, *ArXiV* 2101.07826
Fryxell, B., Olson, K., Ricker, P., Timmes, F. X., Zingale, M., Lamb, D. Q., MacNeice, P., Rosner, R., Truran, J. W. & Tufo, H. 2000, *ApJS* 131, 1
Fujii, M., Iwasawa, M., Funato, Y. & Makino, J. 2007, *PASJ* 59, 6

Habing, H.-J. 1968, Bulletin of the Astronomical Institutes of the Netherlands 19

Haworth, T. J., Clarke, C. J., Rahman, W., Winter, A. J. & Facchini, S. 2018, *MNRAS* 481, 1

Johnstone, D., Hollenbach, D. & Bally, J. 1998, *ApJ* 499, 2

Kroupa, P. 2001, *MNRAS* 322, 2

Krumholz, M. R. & Forbes, J. C. 2015, Astronomy and Computing, 11

Krumholz, M. R., McKee, C. F. & Bland-Hawthorn, J. 2019, *ARAA* 57

McMillan, S., Portegies Zwart, S., van Elteren, A. & Whitehead, A. 2012, *ASP-CS* 453

Michel, A., van der Marel, N. & Matthews, B. C. 2021, *ApJ* 921, 1

Nicholson, R. B., Parker, R. J., Church, R. P., Davies, M. B., Fearon, N. M. & Walton, S. R. J. 2019, *MNRAS* 485, 4

O'Dell, C. R., Wen, Z. & Hu, X. 1993, *ApJ* 410

Parker, R. J., Nicholson, R. B. & Alcock, H. L. 2021a, *MNRAS* 502, 2

Parker, R. J., Alcock, H. L., Nicholson, R. B., Panić, O. & Goodwin, S. P. 2021b, *ApJ* 913, 2

Pelupessy, F. I., van Elteren, A., de Vries, N., McMillan, S. L. W., Drost, N. & Portegies Zwart, S. F. 2013, *A&A* 557

Portegies Zwart, S. F., Verbunt, F. 1996, *A&A* 309

Portegies Zwart, S. et al. 2009, *New Astron.* 14, 4

Portegies Zwart, S., McMillan, S. L. W., van Elteren, A., Pelupessy, I. & de Vries, N. 2013, Computer Physics Communications, 184, 3

Portegies Zwart, S. F. 2016, *MNRAS* 457, 1

Portegies Zwart, S., Pelupessy, I., Martínez-Barbosa, C., van Elteren, A. & McMillan, S. 2020, Communications in Nonlinear Science and Numerical Simulations 85

Ribas, A., Merin, B., Bouy, H. & Maud, L. T. 2014, *A&A* 561

Rosotti, G. P., Dale, J. E., de Juan Ovelar, M., Hubber, D. A., Kruijssen, J. M. D., Ercolano, B. & Walch, S. 2014, *MNRAS* 441, 3

Toonen, S., Nelemans, G., Portegies Zwart, S. 2012, *A&A* 546

Tychoniec, L, Manara, C. F., Rosotti, G. P., van Dishoeck, E. F., Cridland, A. J., Hsieh, T.-H., Murillo, N. M., Segura-Cox, D., van Terwisga, S. E. & Tobin, J. J. 2020, *A&A* 640

Vincke, K. & Pfalzner, S. 2016, *ApJ* 828, 1

Wall, J. E., McMillan, S. L. W., Mac Low, M.-M., Klessen, R. S. & Portegies Zwart, S. 2019, *ApJ* 887, 1

Wall, J. E., Mac Low, M.-M., McMillan, S. L. W., Klessen, R. S., Portegies Zwart, S., Pellegrino, A. 2020, *ApJ* 904, 2

Wilhelm, M. J. C. & Portegies Zwart, S. 2022, *MNRAS* 509, 1

Winter, A. J., Clarke, C. J., Rosotti, G. P., Hacar, A. & Alexander, R. 2019, *MNRAS* 490, 4

Discussion

PORTEGIES ZWART: If you go to [the figure showing the evolution of the radiation field], on the far left you see that the [line indicating no extinction] crosses the [line indicating extinction] near a G_0 of about 10^4 or so. Why do they cross?

WILHELM: That is a result of the resolution. In the case with extinction I get the radiation from different cells in the AMR grid whereas with no extinction I calculate it from the actual distance between the stars. When you get stars that are very close together they might be within the same cell and you get some inaccuracies. It's not going to have an impact on the final results because radiation fields greater than 10^4 [G_0] are nearest neighbor extrapolated to the [FRIED] grid which is at 10^4 [G_0], so all of these disks actually effectively experience radiation fields of 10^4 [G_0].

The Predictive Power of Computational Astrophysics as a Discovery Tool
Proceedings IAU Symposium No. 362, 2023
D. Bisikalo, D. Wiebe & C. Boily, eds.
doi:10.1017/S1743921322001478

Long-Term Evolution of Convectively Unstable Disk

Lomara Maksimova* and Yaroslav Pavlyuchenkov

Institute of Astronomy, Russian Academy of Sciences
Pyatnitskaya str., 48, Moscow 119017,
*email: lomara.maksimova@gmail.com

Abstract. We continue studying convection as a possible factor of episodic accretion in proto-planetary disks. Within the model of a viscous disk, the accretion history is analyzed at different rates and regions of matter inflow from the envelope onto the disk. It is shown that the burst-like regime occurs in a wide range of parameters. The long-term evolution of the disk is modeled, including the decreasing-with-time matter inflow from the envelope. It is demonstrated that the disk becomes convectively unstable and maintains burst-like accretion onto the star for several million years. The general conclusion of the study is that convection can serve as one of the mechanisms of episodic accretion in protostellar disks, but this conclusion needs to be verified using more consistent hydrodynamic models.

Keywords. protoplanetary disk, numerical model, convective instability, viscous disk

---

We simulate an axially symmetric, geometrically thin viscous Keplerian disk without a radial pressure gradient. The surface density of the disk and its evolution is modeled according to the equation by Pringle (1981). We also include the accretion of matter from the envelope onto the disk:

$$\frac{\partial \Sigma}{\partial t} = \frac{3}{R}\frac{\partial}{\partial R}\left[\sqrt{R}\frac{\partial}{\partial R}\left(\nu\sqrt{R}\Sigma\right)\right] + W(R,t), \qquad (0.1)$$

where $\Sigma(R,t)$ is the surface viscosity; R is the distance to the star; t is time; $\nu(R,t)$ is the kinematic viscosity coefficient; and $W(R,t)$ is the matter inflow from the envelope under the assumption that the specific angular momentum of the settling matter coincides with that of the disk. Radial evolution is modelled in parallel with reconstructing the disks vertical structure to identify convectively unstable regions in the disk. The density and temperature distributions are calculated in the polar direction under the approximation of a hydrostatic-equilibrium disk. The main factor governing the disk evolution within this model is the dependence of the viscosity coefficient $\nu(R,t)$ on the radius. Detailed description of the model is presented in Pavlyuchenkov *et al.* (2020), Maksimova *et al.* (2020).

Due to this model, we obtained the evolution of the accretion rate from the disk onto the star (see Fig 1). The filled area in the accretion rate distribution at 0.17–3.7 Myr indicates a burst-like accretion regime at this figure scale, the numerous bursts merge into a single continuous band. Characteristic forms of accretion bursts at times around 0.4, 1.5, and 3.5 Myr are shown in Fig 2. After 3.7 Myr, the bursts cease to occur, and the accretion rate smoothly decreases with time. A comparison between the accretion rate onto the star and the rate of matter inflow from the envelope (the dashed line in Fig. 1) leads to a conclusion about the importance of the disk mass accumulation process during

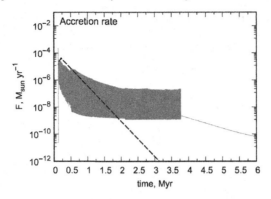

Figure 1. Evolution of accretion rate from the disk onto the star (the dashed black line indicates an adopted matter inflow from the envelope onto the disk.

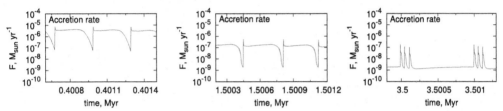

Figure 2. The accretion rate of matter from the disk onto the star for three time intervals around 0.4 Myr (left), 1.5 Myr (center), and 3.5 Myr (right).

the first million years. The subsequent evolution of disk is defined by the redistribution of accumulated mass while the matter inflow from the envelope becomes negligibly small.

In conclusion, we note that:

• The mass accumulation process in the disk through matter inflow from the envelope has an important effect on the disk evolution and its periodic accretion events.

• The disk soon becomes convectively unstable and remains so for almost 4 Myr. Meanwhile, the instability captures an area of several tens of astronomical units and then gradually shrinks.

• Burst parameters (intensity, duration, and frequency), as well as their shape, change with time, which is associated with a change in the disk mass and the integral flow of matter through it. The bursts may take very bizarre shapes.

Finally, it should be recalled that the presented model is rather illustrative because of the many underlying physical assumptions. Its main purpose is to demonstrate the possible role of convection as a driver of episodic accretion in protostellar disks. We believe that further investigation of the convection role should rely on more consistent models, which will consider hydrodynamic effects, dust evaporation, and gas dissociation and ionization processes.

Acknowledgements

The research by Maksimova Lomara was carried out in the framework of the project "Study of stars with exoplanets" under a grant from the Government of the Russian Federation for scientific research conducted under the guidance of leading scientists (agreement № 075-15-2019-1875)

References

Pringle, J. E. 1981, Annu. Rev. Astron. Astrophys., 19, 117

Pavlyuchenkov, Ya. N., Tutukov, A. V., Maksimova, L. A., Vorobyov, E. I. 2020, Astron. Rep., 64, 1

Maksimova, L. A., Pavlyuchenkov, Ya. N., Tutukov, A. V. 2020, Astron. Rep., 64, 10

The Predictive Power of Computational Astrophysics as a Discovery Tool
Proceedings IAU Symposium No. 362, 2023
D. Bisikalo, D. Wiebe & C. Boily, eds.
doi:10.1017/S1743921322001612

Heliosphere in the Local Interstellar Medium

Nikolai V. Pogorelov⊙

Department of Space Science and Center for Space Plasma and Aeronomic Research,
University of Alabama in Huntsville, Huntsville, AL 35805, USA
email: np0002@uah.edu

Abstract. The Sun moves with respect to the local interstellar medium (LISM) and modifies its properties to heliocentric distances as large as 1 pc. The solar wind (SW) is affected by penetration of the LISM neutral particles, especially H and He atoms. Charge exchange between the LISM atoms and SW ions creates pickup ions (PUIs) and secondary neutral atoms that can propagate deep into the LISM. Neutral atoms measured at 1 au can provide us with valuable information on the properties of pristine LISM. *Voyager* 1 and 2 spacecraft perform in situ measurements of the LISM perturbed by the presence of the heliosphere and relate them to the unperturbed region. We discuss observational data and numerical simulations that shed light onto the mutual influence of the SW and LISM. Physical phenomena accompanying the SW–LISM interaction are discussed, including the coupling of the heliospheric and interstellar magnetic field at the heliopause.

Keywords. Sun: solar wind, ISM: kinematics and dynamics, ISM: magnetic field

1. Introduction

The interaction of the solar wind (SW) with the local interstellar medium (LISM) is a natural laboratory that allows the space physics and astrophysics communities to investigate a number of interesting physical phenomena in partially ionized plasma. Although the interaction of two plasma streams seems to be a trivial problem in the MHD sense, this is not so because the density of neutral hydrogen (H) atoms in the LISM surrounding the heliosphere is maybe three times higher than that of protons. While a tangential discontinuity, called the heliopause (HP), is formed in the ionized component, the interstellar neutral (ISN) atoms are able to penetrate deep into the heliosphere. The interaction of the ISN atoms with the SW ions occurs predominantly through the resonant charge exchange. As a result of such interaction, new neutral atoms are born with the properties of the parent SW ions and new ions with the properties of the parent ISN atoms. Newly born (secondary) neutral atoms and ions have properties strongly dependent on where in the heliosphere they were created. The flow of SW ions at distances exceeding 10-15 solar radii is super-fast magnetosonic, so its deceleration by the HP and LISM counter-pressure in the heliotail creates a heliospheric termination shock (TS). Thus, the secondary neutral atoms born inside the TS will be cool but have high radial velocity components. They make a so-called neutral SW. The secondary atoms born in the heliosheath (the SW region between the TS and HP, HS) are hot, but have low bulk speed. Because of the large charge-exchange mean free path, both populations of secondary neutral atoms can easily propagate back into the LISM and affect its properties to distances of 500-1000 AU in the upwind direction (Gruntman 1982). Therefore, the pristine LISM becomes heated and decelerated by these secondary atoms via charge exchange. This is why, even if it is superfast magnetosonic at very large distances, there may be no bow shock in the LISM in front of the HP (see Fig. 1), at least in certain radial directions. The presence of

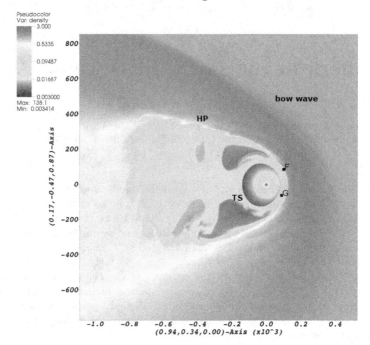

Figure 1. The picture of the SW–LISM interaction is shown through the plasma density distribution in the plane formed by the *V1* and *V2* trajectories. Letters F and G show the spacecraft positions in 2015. While the HP crossing distance at *V1* is closely reproduced, we also predicted that *V2* may cross the HP at a close distance. [From Pogorelov et al. (2015) with permission of the American Astronomical Society.]

the heliosphere affects the LISM plasma to much larger distances in the heliotail region. Moreover, ions of different energies are affected differently (e.g., PeV galactic comic rays, GCRs, may be affected to distances of the order of 1 pc), which makes it difficult to establish the part of the LISM modified by the heliosphere exactly. It is for this reason, Fraternale & Pogorelov 2021 proposed to extend the term Very Local Interstellar Medium (VLISM) to the LISM affected by the presence of the heliosphere, regardless of what physical processes are responsible for such modification and which physical quantities are affected. The space filled by the VLISM is sometimes called the outer heliosheath (OHS).

The secondary ions born in the SW due to charge exchange are quickly isotropized, but never reach the state of thermodynamic equilibrium (Vasyliunas & Siscoe 1976). These non-thermal ions are also called pickup ions (PUIs). They carry the majority of SW thermal energy starting from distances of the order of 10 AU and extending to the HP itself. We loosely distinguish the inner and outer heliosphere (IH and OH) by assuming that the former starts at some critical, Sun-centered sphere with radial velocity component exceeding the fast magnetosonic speed and ends when the effect of PUIs on the SW flow becomes noticeable. The OH extends to the HP itself. Pickup protons co-move with the thermal protons in the direction of the HP (Parker 1963). The effect of ion, especially PUI, streaming along magnetic field lines cannot be excluded, but it is likely limited by scattering in the turbulent magnetic field. It is worthwhile to mention that the mere applicability of the MHD treatment of collisionless SW plasma relies upon such scattering. Since the thermal energy of PUIs is high and they can experience charge exchange themselves, newly born atoms are called energetic neutral atoms (ENAs). PUIs can also be born in the OHS behind the HP, where they ultimately give birth to new ENAs. Those ENAs which are born in the HS and OHS can propagate back to Earth, where they are measured by the *Interstellar Boundary Explorer* (McComas et al. 2017b), *Cassini*/INCA (Krimigis et al. 2009), and *Solar Heliospheric Observatory* (*SOHO*/HSTOF) (Hilchenbach et al. 1998; Czechowski et al. 2006). Since ENAs carry information about PUIs throughout the SW–LISM interaction region, they are used as a tool to describe the global structure of the heliosphere (Reisenfeld et al. 2021). This information is enhanced by in situ measurements performed by near-Earth spacecraft and *Voyager* 1 and 2, which crossed the TS and HP and are performing measurements in the VLISM (Stone et al. 2005, 2008, 2013, 2019). More recent, *New Horizons* (*NH*) mission is measuring the PUI and thermal SW properties at distances now exceeding 40 AU (McComas et al. 2017a), where PUIs are energetically dominant. *Interstellar Mapping Probe* (*IMAP*) is a new NASA mission, to be launched in 2024. It will perform ENA and PUI measurements with even higher accuracy (McComas et al. 2018).

More details of what has been described in this brief introduction can be found in the review papers (e.g., Bzowski et al. 2009; Izmodenov et al. 2009; Pogorelov et al. 2009, 2017b; Zank 1999, 2015).

The abundance of in situ data and remote observations makes numerical simulations challenging. Their predictive power is strongly constrained by observations and mostly applies to the regions lacking in situ measurements. For example, both *Voyagers* move into the nose of the SW–LISM interaction region. On the other hand, *IBEX* measurements of ENA fluxes are lacking those created at distances exceeding 500 AU into the heliotail. It is therefore virtually impossible to constrain the heliotail length and structure by data only. Although all numerical simulations based on identical models and physical boundary conditions give agreeable results, controversies are still possible on the border of model applicability. As far as in situ measurements are concerned, they are performed at one point per time and remain therefore incomplete as far as the global time-dependent picture is concerned.

In the following sections, a number of challenges are described that affect predictive capabilities of numerical modeling. Physical phenomena accompanying the SW–LISM interaction are so broad that many of them are applicable to various astrophysical objects, especially those involving wind-wind interactions.

2. Challenges in numerical modeling of the SW–LISM interaction

The challenges to be discussed can be separated into two categories. Firstly, it is important to choose a proper physical model. Secondly, any model should be accompanied by appropriate, typically time-dependent, boundary conditions. While ideal MHD models are commonly used to describe the SW and LISM plasma flow, the former is collisionless. The applicability of fluid equations is justified by ion scattering on magnetic field fluctuations abundant in the SW. On the other hand, p-H charge exchange mean free path is at least of the order of characteristic distances in the interaction region (e.g., the measured distance between the TS and HP is about 35 AU), so the transport of neutral atoms should be modeled kinetically (Baranov & Malama 1993). This does not mean that simpler models, where different populations of neutral atoms are treated gas dynamically, cannot be used (see a comparison of these models in Pogorelov et al. 2009).

2.1. *Data-driven boundary conditions*

To simulate the SW behavior, it is necessary to identify a set of boundary conditions (b.c.'s). Remote observations of the solar magnetic field are made routinely in the photosphere. These data can be used as boundary conditions for solar coronal models that

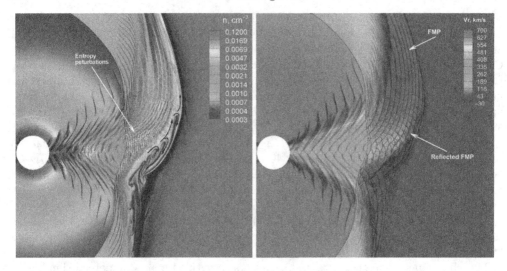

Figure 2. The only data-driven simulation of the CIR effect on the HS. Density (left panel) and radial component of the velocity (right panel) distributions in the meridional plane for the Solar Cycle 22 minimum. The entropy and fast magnetosonic perturbations (FMPs) are shown. The black arrows show the slow wind flow directions. [From Borovikov et al. (2012) with permission of the American Astronomical Union.]

propagate the SW beyond the critical sphere. In other words, solutions to the SW–LISM ineraction can be driven and validated by data.

To perform meaningful comparison with data at least the following observational issues should be addressed:

• Better shock identification algorithms to be developed especially for the HS data.

• A systematic search for the sectors and sector boundaries should be performed beyond the TS. Theoretical explanations should be provided for their presence and apparent occasional destruction possibly through local magnetic reconnection.

• SW and OHS observations should be used for theoretical studies of the formation of pressure fronts and their breaking.

• Relatively high-frequency, quasi-periodic structures have been observed in the HS. Their nature and origin should be revealed.

• The SW, including the one in the HS, is a driven, open dynamical system, and the appropriate statistical mechanics is that of Tsallis (2009). The q-triplet seems to be a universal characteristic of the system. Since the Tsallis statistics is associated with the Lorentzian distribution suitable for the mixture of thermal ions and PUIs, much could be learned from this approach, particularly on relatively small scales.

• Many corotating interaction regions (CIRs, Fig. 2), merged interaction regions (MIRs), and global MIRs (GMIRs, Fig. 3), associated with a decrease in the Galactic cosmic ray (GCR) intensity, have been observed in the HS. This is why, cosmic ray data should be involved in the data analysis.

• The question to be answered by combining observations and simulations is about why MIRs and GMIRs survive out to near the HP. In addition, no numerical model is able to reproduce $V2$ data in the HS satisfactorily (e.g., Fig. 4). Models that extend from the Sun through the HS are suitable for addressing these questions.

Our own Multi-Scale Fluid-Kinetic Simulation Suite (MS-FLUKSS, Pogorelov et al. 2014) is used to perform simulations of the SW flow and its interaction with the LISM with the b.c.'s at 20-25 solar radii provided to us by the well-developed POT3D-WSA coronal model constrained by remote and in situ SW observations (Arge et al. 2013;

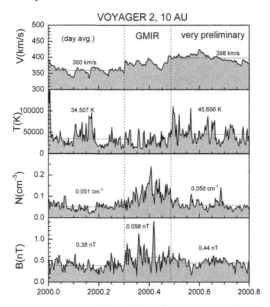

Figure 3. Quantity distributions in a typical GMIR observed by *V2*. The data is shifted by 234 days towards 10 AU.

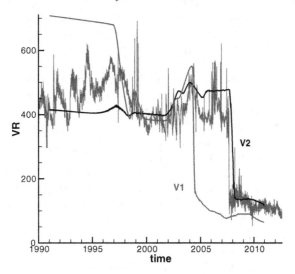

Figure 4. Radial component of the SW velocity vector along the *V2* (black line) and *V1* (red line) trajectories. *V2* observations are shown with the blue line. [From Pogorelov et al. (2013) with permission of the American Astronomical Society.]

Caplan et al. 2021). The inner b.c.'s for the ambient SW are being transferred with MHD simulations to the Earth's orbit with the uncertainty quantification (UQ) based on the WSA predictive metrics. The properties of a selected few *Voyager-* and *NH*-directed coronal mass ejections (CMEs) are derived from publicly available remote observations made by *SDO* AIA and HMI, *STEREO* A & B COR2 and H1/H2, and *SOHO* C2/C3 instruments. An additional UQ analysis is performed of the solution dependence on the observationally-derived CME properties at injection sites. *Parker Solar Probe* (*PSP*), *OMNI*, and *Ulysses* (SWOOPS and SWICS) data can be used for validation in the IH. In the OH, Voyager (MAG and PLS) and *NH* SWAP data can be engaged.

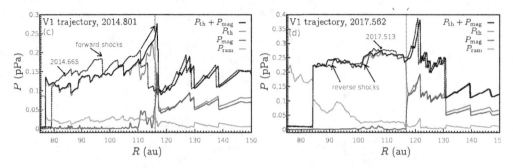

Figure 5. Two time frames with the pressure distributions along the *V1* trajectory show the forward and reverse shocks propagating through the HS. The dashed lines correspond to slightly shifted moments of time. One can also see shocks merging beyond the HP, the position of which is shown with vertical lines.

If PUIs are treated as a separate fluid, the mixture of thermal and non-thermal ions in the SW can modeled in the assumption of different kappa distribution functions, which makes it possible, to perform a UQ based on this important parameter. The UQ analysis can be done on the basis of error propagation from 1 AU through the heliosphere.

We build our research approach on the historically proven idea that theory and numerical simulations are of immense importance for the interpretation of observational data. The latter, while providing us with the ground truth about the physical processes occurring in the heliosphere, highly benefit from the capability of modeling to experiment with the boundary conditions, switch on and off different processes, and provide global solutions. Such global solutions are especially important if they are time-dependent and account for the realistic space dimension of the problem.

The b.c.'s in the unperturbed LISM are typically derived from the He atom measurements (McComas et al. 2015), especially those atoms that experience no charge exchange on the way to 1 AU. They give us the velocity components and temperature of the LISM. The prior estimates are currently being re-analyzed because the distribution function of pristine He atoms becomes anisotropic at 1 AU. (Wood et al. 2019; Swaczyna et al. 2020; Fraternale et al. 2021). The atom (H and He) and proton densities remain uncertain and are partially derived from the SW–LISM simulation results.

In situ measurements are typically made at each individual point and time and therefore may hide the actual complexity of a phenomenon. As shown in Fig. 5a, the numerical simulation from (Pogorelov et al. 2021) makes it possible not only to identify forward and reverse shocks in the HS, but also see their evolution in time. In addition, one can see overtaking of shocks propagating upstream in the LISM – the phenomenon so infrequent that it can hardly be observed by a single spacecraft.

Another surprising observation (Stone et al. 2008) was that *V2* crossed the TS at a much smaller heliocentric distance than *V1* (84 AU against 94 AU). While a number of possible explanation were proposed, some entirely unrealistic, the simulation of Pogorelov et al. (2013) driven by *Ulysses* measurement reproduced both the crossing time and the corresponding stand-off distances (see Fig. 4).

Pogorelov et al. (2021) show that time-dependent and data-driven numerical simulations can reproduce a surprising absence of the magnetic field rotation across the HP observed both by *V1* and *V2* (Burlaga et al. 2019).

2.1.1. An empirically driven solar wind model

The outer atmosphere of the Sun, the solar corona, is a plasma that expands to become the supersonic SW, which envelopes the planets and forms the heliosphere. The SW structure and dynamics are strongly influenced by the solar magnetic field, including the locations of fast and slow SW, and the position of the heliospheric current sheet (HCS). The magnetized SW is also the primary medium by which solar activity is transmitted to Earth and beyond, in the form of CMEs, which evolve and propagate in the wind, and energetic particles, which are transported with the magnetic field. The solar magnetic field is therefore a key component in any predictive SW model.

Empirical models are able to predict SW structure with reasonable success from a magnetic field model based on photospheric magnetic field maps (Riley et al. 2021). The OH evolves over long time periods compared to the solar rotation time scale. To model the global heliosphere over many years, we need to capture the large-scale evolution of the SW which is driven by changes in the solar magnetic field. We require a physically consistent description of the photospheric flux evolution that is continually updated. The processes by which the magnetic flux on the Sun evolves have been studied for many years. Assimilative Surface Flux Transport (SFT) models incorporate these processes (primarily differential rotation, meridional flow, supergranular diffusion, and random flux emergence) and have been successful in predicting the evolution of photospheric magnetic fields. Presently available models include the LMSAL Evolving Surface-Flux Assimilation Model (Schrijver & De Rosa 2003), the ADAPT model (Arge et al. 2013), and the Advective Flux Transport model (AFT) (Upton & Hathaway 2014). The map sequences provided by this models can be used to drive time-dependent, empirically driven MHD SW models that approximate its long-term evolution.

2.1.2. Coronal mass ejections and their uncertainty quantification

We have implemented two flux rope models in MS-FLUKSS: a modified spheromak model (Singh et al. 2020a,b) and a constant turn model based on FRiED model geometry (Isavnin 2016). Both can be initiated with desired speed, direction, orientation, mass, poloidal and toroidal magnetic fluxes, and a helicity sign. The initial kinematic and magnetic properties of CMEs are derived from various observations. We use the graduated cylindrical shell (GCS) model (Thernisien et al. 2009) to estimate CME speed, direction, and orientation using multiple viewpoint images from *STEREO* and *SOHO* coronagraphs. CME mass is calculated using the bright light coronagraph observations (Colaninno & Vourlidas 2009). The poloidal flux of CMEs is calculated from the reconnected flux under the post-eruption arcades (PEAs) (Gopalswamy et al. 2018; Singh et al. 2019). The toroidal flux of CMEs is calculated from coronal dimming at the CME footprints (Dissauer et al. 2019; Singh et al. 2020a). If there is no coronal dimming during a CME eruption, the toroidal flux can be estimated using the empirical relation between the toroidal and poloidal fluxes (Qiu et al. 2007). The helicity sign of CME flux ropes can be determined from magnetic field distribution in the source active regions (Luoni et al. 2011). All these parameters have observational errors associated with their estimates. This means that ensemble modeling of CMEs can be performed to get estimates about the uncertainties in CME simulations by creating ensemble members according to the expected errors in the initial CME model parameters.

2.2. Pick up ions crossing the termination shock. The width of the heliosheath

Although the SW plasma consists of the thermal SW ions and PUIs, we choose to solve the MHD system for the plasma mixture in the conservation-law form because

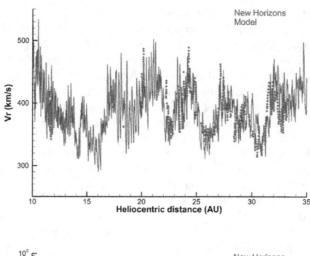

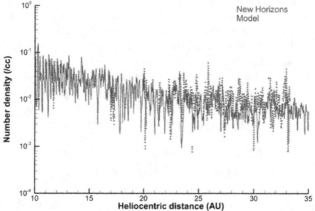

Figure 6. Radial velocity component and number density compared to the daily averaged *NH* SWAP data. [From Kim et al. (2016) with permission of the American Astronomical Society].

this allows us to satisfy the conservation laws of mass, momentum, energy, and magnetic flux at the TS efficiently. Clearly, some assumptions should be made about the distribution functions of protons and PUIs. It is usually assumed that the SW protons are Maxwellian. This is not true for PUIs. Zank et al. (2010) showed that assuming a kappa distribution for the mixture makes it possible to approximate the realistic distribution function. Heerikhuisen et al. (2019); DeStefano & Heerikhuisen (2017, 2020) studied the effect of the distribution function on the charge exchange source terms. The kappa distribution decreases the contribution of the most probable state while increasing the number of ions in the so-called "energetic tails." The important conclusion derived from DeStefano & Heerikhuisen (2017, 2020) is that the dependence of charge-exchange cross-sections on energy should be preserved when the integration is performed. This becomes crucial for κ approaching the minimum allowed value of 3/2. If this procedure is not followed, the effect of PUIs on the heliosphere is substantially exaggerated (Heerikhuisen et al. 2015).

We are using two approaches to take into account PUIs: (1) PUIs are not treated as a separate plasma component, but the conservation of mass, momentum, and energy for the mixture is preserved, while the solution dependence on the value of κ is investigated; (2) some sort of an isotropic distribution function for PUIs away from the TS is chosen, they are assumed to be co-moving with the SW ions, and the continuity and pressure equations are solved to describe the PUI flow (Pogorelov et al. 2016). The latter approach is easier to implement in the supersonic SW inside the TS (see, e.g., Kim et al. 2016). Figure 6 shows the comparison of our simulation based on model 2 with *NH* observations.

The description of PUIs crossing collisionless shocks, such as the TS, is impossible with the MHD, ideal or dissipative, approaches, because the details of kinetic shock structure, like the overshoot, are responsible for PUI reflections into the upstream region. Some kind of modified Rankine–Hugoniot-type boundary conditions describing the PUI transition across the TS are required. The fluid descriptions of the TS crossings by PUIs have been made so far either with simplified shock conditions (Pogorelov et al. 2016; Wu et al. 2016; Kornbleuth et al. 2020) or no conditions at all (Usmanov et al. 2016). Gedalin et al. (2020, 2021a,b) performed an extensive, probabilistic, test-particle simulations to derive such b.c.'s. The results have been validated with full-PIC simulations. It has been shown that the fraction of initially reflected PUIs is larger than that derived from the estimates involving only the cross-shock potential effect. In addition, the downstream perpendicular temperature of reflected PUIs appeared to be an order of magnitude lower than it was proposed by Chalov & Fahr (2000) and Zank et al. (2010).

The discussion of the HS width is complicated by the absence of in situ measurements for the TS and HP distances from the Sun at the same moment of time. Pogorelov et al. (2015) predicted that *V2* should cross the HP at a distance very close to that of *V1*. A separate treatment of PUIs indeed decreases the HS width (Pogorelov et al. 2016). However, data-driven numerical simulations reported by Kim et al. (2017) and Pogorelov et al. (2021), where PUIs were not treated separately, showed this width is variable: it has been smaller than 40 au since about 2014, reached 30 au in 2017, and now remains almost constant (about 35 au) in the *V1* direction. This is rather close to the difference between the observed crossings of the TS and HP.

2.3. *The effect of magnetic field dissipation on the thermodynamic properties and velocity of the SW in the IHS*

The HMF becomes almost radial at 0.1 au from the Sun. However, the regions of positive and negative polarity are separated by a current-carrying surface, which is called the heliospheric current sheet (HCS). Due to the Sun's rotation, the HMF lines acquire spiral shape. However, the Sun's magnetic and rotation axes do not coincide. For this reason, the idealized HCS, for the tilt between the Sun's magnetic and rotation axes (equal to $35°$) acquires a complex shape shown in Fig. 7 (*left panel*). In reality, this shape is not preserved beyond ~ 10 au because of the SW flow asymmetries and stream interaction. However, the width of each magnetic field sector decreases with decrease of the SW speed and vanishes near the SW stagnation point on the HP surface. In practice, this means that the sector structure cannot be resolved in its entirety for any chosen grid resolution. While one would expect some sort of numerical dissipation of the HMF in this case, the HMF strength starts to oscillate exhibiting the features of stochastic behavior, if the grid resolution is sufficiently high, (Fig. 7, *right panel*). *Ulysses* data driven simulations presented by Pogorelov et al. (2013) show that the calculated value of the (dominant) transverse HMF component is of the order of 1 μG, which is close to the average value of the same, strongly oscillating component in observations. In principle, magnetic field dissipation should result in the SW heating and its slower deceleration. Unfortunately, identification of such features in the turbulent SW behind the TS is a challenging task, which is still to be undertaken.

An approach proposed by Czechowski et al. (2010) and Borovikov et al. (2011) was based on the idea that the HMF can be assumed unipolar in numerical simulations while

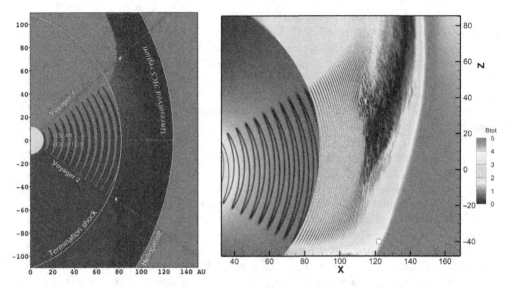

Figure 7. (*Left panel*) The HCS is shown using the boundary between the positive (red color) and negative (blue color) polarities tracked by the level-set method in the assumption of unipolar HMF. (*Right panel*) The distribution of the magnetic field magnitude with the Parker HMF distribution at the inner boundary. Transition to stochastic behavior occurs in the region where the HCS is no longer resolved. [From Borovikov et al. (2012) and Pogorelov et al. (2015) with permission of the American Astronomical Society.]

the magnetic field polarity is assigned after each time step by tracing the HCS shape with a level-set method. However, this approach turned out to have two major deficiencies. Firstly, even the level-set equation sooner or later stops to resolve the sector structure. Secondly, the HMF strength becomes unrealistically strong (Izmodenov & Alexashov 2020; Pogorelov et al. 2017a, 2021), which disagrees with *Voyager* data by a factor of ~ 7. Moreover, the plasma beta becomes less than 1 in the HS, so the SW flow gets under control of magnetic pressure, which can collimate the flow in the heliotail (Yu 1974). The collimation occurs inside the Parker field branches spiraling into the tail region. In the extreme case of unipolar HMF, such spiraling can result into the heliotail splitting into two branches (Opher et al. 2015; Pogorelov et al. 2015). Such splitting disappears even in the assumption of a flat HCS, which happens when the Sun's magnetic and rotation axes coincide. Moreover, the reasons for the collimation disappear because of the "kinking" or "sausage" instabilities. Moreover, the solar cycle effects, especially the presence of the dense, slow and rarefied, fast SW regions with the variable latitudinal extent of the boundary between them completely destroys the artificial collimation, in this way removing the possibility of split-tail structures.

2.4. *Magnetic field behavior at the heliopause*

Pogorelov et al. (2021) show that time-dependent and data-driven numerical simulations can reproduce a surprising absence of the magnetic field rotation across the HP observed both by *V1* and *V2* (Burlaga et al. 2019).

It is worth noting, however, that the behavior of magnetic field vectors across the HP along *V1* and *V2* trajectories cannot be reproduced with the boundary conditions assuming the unipolar HMF. The reasons are on the surface. As seen from Pogorelov et al. (2015, 2021); Izmodenov & Alexashov (2020), and Opher et al. (2020), the magnetic field strength is practically continuous across the HP in the assumption of unipolar HMF.

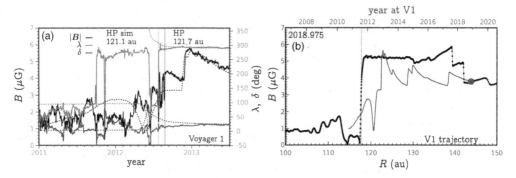

Figure 8. (Panel (a)) The magnetic field strength, and its elevation and azimuthal angles in the vicinity of the HP crossing measured by *Voyager* 1 (solid lines) and simulated in Kim et al. (2017) (dashed lines). The vertical lines show the HP position at the moment of *V1* entering the LISM. (Panel (b)) Two shocks approaching each other before merging in the OHS. The blue line shows the magnetic field distribution at a virtual *V1* propagating through the solution. [Adapted from Pogorelov et al. (2021) with permission of the American Astronomical Society.]

Since the HP is a tangential discontinuity, the equality of magnetic pressures across it can be satisfied only if the magnetic pressure dominates over the thermal pressure. This is in accord with our explanation of the exaggerated SW collimation inside the Parker field spirals bent into the heliotail.

2.5. *Propagation of waves and shocks through the VLISM, and their interaction. The structure of collisional shocks in the LISM*

Data-driven simulation of the SW-LISM interaction (Kim et al. 2017; Pogorelov et al. 2021) show that some of the shocks observed by *V1* can be well reproduced. It was also shown that each shock created at the HP and propagating upstream into the LISM decreases in intensity and speed as it propagates radially outward, and ultimately disappears at some distance from the HP. A shock can become stronger only when one shock overtakes another. Such shock interaction is shown in Fig. 8b of the magnetic field distribution along the *V1* trajectory, where we show both the time frames and the quantities taken by a virtual probe (indicated at each moment of time with a blue circle) co-moving with the *V1* along its trajectory.

In Fig. 8a, we present the distribution of $|\mathbf{B}|$, and its elevation and azimuthal angles, δ and λ, as a function of time along the *V1* trajectory. The observational data and simulation results are shown with solid and dashed lines, respectively. Although the simulated distributions are not in full agreement with the observations, one can see that δ and λ are continuous across the HP, which is shown with the vertical lines. Their values asymptotically approach those observed by the spacecraft.

2.6. *The nature of the density increase beyond the heliopause*

The increase in the plasma wave frequency observed by the *Voyager* Plasma Wave Instrument (PWS) has been of substantial interest for a long time. This increase implies that the plasma density is also increasing, on the average, with distance from the HP, i.e., the density maximum is not at the surface of the HP itself. The first two explanations for that phenomenon were proposed by Gurnett et al. (2015) and Cairns & Fuselier (2017). The former explanation is simply a misinterpretation of numerical simulations. The early, low-accuracy, discontinuity-capturing simulations resulted in a very wide dissipative structure of the HP (Steinolfson et al. 1994). This means that the density profile looked rather wide and smooth. If one identifies the position of the HP in a chosen direction as the point where the density profile has maximum gradient, the maximium of the density itself reveals itself as some distance from the HP. However, the numerical width of the HP cannot be as large as 40-50 au, which is the observed distance of the continuing increase in the plasma density. The current analysis of numerical simulations where the surface of the HP was fitted exactly, like those in Baranov & Malama (1993), showed that the density increase behind the HP should be separated from the sharp increase across the HP as a discontinuity. This phenomenon was also reconfirmed by the high-resolution, discontinuity-capturing simulations of Pogorelov et al. (2017a). The second explanation appealed to the magnetic-field induced anisotropies in the VLISM plasma. While the absence of such anisotropies cannot be excluded or confirmed by *Voyager* data, it is clear that the phenomenon exists even in the absence of magnetic field (Baranov & Malama 1993). Pogorelov et al. (2017a) suggested that this density increase is somehow related to the process of charge exchange in the VLISM. The argument in favor of such explanation was the similarity in the model-derived density-increase width with the proton-H atom mean free path. However, Pogorelov et al. (2021) pointed out that the density should not be expected to have maximum at the HP surface. By analogy with the supersonic blunt body problem, it is clear that this never happens, except for the critical streamline directed into the stagnation point on the HP surface.

2.7. *Plasma wave generation in the VLISM and its relationship to the Alfvén velocity distributions along the Voyager trajectories newly derived from the spacecraft data)*

The subject discussed in the previous section is intimately related with the distribution of the Alfvénic velocity along the *V1* and *V2* trajectories in the VLISM. There is a close association between the electron plasma oscillations (Gurnett et al. 2013) and the jumps in the magnetic field strength Burlaga et al. (2013) observed in the OHS. According to PWS measurements (Gurnett et al. 2015), *V1* observed radio emission in the 2–3 kHz range, which is thought to be excited by shocks propagating through plasma regions primed with nonthermal electrons resonantly accelerated by lower hybrid (LH) waves driven by a ring-beam instability of PUIs (Gurnett et al. 1993; Cairns & Zank 2002). Further acceleration of these electrons by a propagating shock may create electron beams moving away from it. These beams produce Langmuir waves, via the "bump-on-tail" instability (Filbert & Kellogg 1979; Cairns 1987). For a shock front convex outward with respect to the incoming plasma flow, the bump-on-tail velocity distribution is due to the existence of a threshold velocity below which electrons cannot reach a given point upstream of the shock. The region accessible to such beams is called the electron foreshock. The presence of nonthermal electrons is insufficient for the development LH instability. The instability growth rate and energy transfer to electrons should be sufficiently large, which occurs (see Cairns & Zank 2002 and references therein) if $\alpha_r = V_r/V_A < 5$, where V_r and V_A are the PUI ring-beam and Alfvén velocities, respectively. Magnetic field draping around the HP creates conditions for larger V_A. For PUIs born in the OHS by charge exchange of the VLISM ions with the neutral SW, which actually have a ring-beam distribution, V_r should be ~ 400 km/s.

Voyager data provide us with a new perspective on the plasma wave and radio emission generation. The analysis of these data in Pogorelov et al. (2021) showed that V_A decreases with distance from the HP, and is below 45 km/s at *V1* and 85 km/s at *V2* immediately after the HP. The Alfvénic velocity increases across the shocks traversing the OHS, but not substantially, since all shocks observed in that region so far have been rather weak. It is therefore unlikely that α_r would be smaller than 5 for ring-beam velocities corresponding to the neutral SW. For this reason, there remains a question about the physical mechanisms responsible for the plasma wave and radio emission generation on the OHS, since Roytershteyn et al. (2019) reported no substantial LH instability for a realistic, three-component distribution of PUIs beyond the HP.

3. Conclusions

This paper presented a number of observational phenomena that benefited from numerical simulations. Such simulations help understand in situ data obtained by an individual, or even two individual spacecraft, such as *V1* and *V2*. This is because one point in space per time observations may and do hide the details of stream and shock interaction. Of interest for large-scale simulations is the data analysis the VLISM turbulence performed by Fraternale & Pogorelov (2021). It was found, in particular, that the dissipative width of observed shock structures is too narrow to be attributed to Coulomb collisions alone. Moreover, high variability of the shock width structures suggests that that no actual shocks are observed in the collisional VLISM. Instead, *Voyagers* are crossed by sharp gradients which have not sufficient time to break in the turbulent plasma accompanied by transient wave activity.

Acknowledgement

This work was supported in part by NASA grants 80NSSC19K0260, 80NSSC18K1649, 80NSSC18K1212, and NSF-BSF grant PHY-2010450. It was also partially supported by the *IBEX* mission as a part of NASA's Explorer program. The authors acknowledge the Texas Advanced Computing Center (TACC) at The University of Texas at Austin for providing HPC resources on Frontera supported by NSF LRAC award 2031611. Supercomputer time allocations were also provided on SGI Pleiades by NASA High-End Computing Program award SMD-17-1537 and Stampede2 by NSF XSEDE project MCA07S033.

References

Arge, C. N., Henney, C. J., Hernandez, I. G., et al. 2013, in AIP Conference Series, Vol. 1539, Solar Wind 13, ed. G. P. Zank & et al., 11–14
Baranov, V. B., & Malama, Y. G. 1993, JGR, 98, 15157
Borovikov, S. N., Pogorelov, N. V., Burlaga, L. F., & Richardson, J. D. 2011, ApJL, 728, L21
Borovikov, S. N., Pogorelov, N. V., & Ebert, R. W. 2012, ApJ, 750, 42
Burlaga, L., Ness, N. F., Berdichevsky, D., et al. 2019, Nature Ast., 3, 1007
Burlaga, L. F., Ness, N. F., Gurnett, D. A., & Kurth, W. S. 2013, ApJL, 778
Bzowski, M., Mobius, E., Tarnopolski, S., Izmodenov, V., & Gloeckler, G. 2009, SSR, 143, 177
Cairns, I. H. 1987, JGR, 92, 2329
Cairns, I. H., & Fuselier, S. A. 2017, ApJ, 834, 197
Cairns, I. H., & Zank, G. P. 2002, Geophys. Res. Lett., 29, 47
Caplan, R. M., Downs, C., Linker, J. A., & Mikic, Z. 2021, ApJ, 915, 44
Chalov, S. V., & Fahr, H. J. 2000, A&A, 360, 381
Colaninno, R. C., & Vourlidas, A. 2009, ApJ, 698, 852
Czechowski, A., Hilchenbach, M., Hsieh, K. C., Kallenbach, R., & Kóta, J. 2006, ApJL, 647, L69
Czechowski, A., Strumik, M., Grygorczuk, J., et al. 2010, A&A, 516, A17
DeStefano, A. M., & Heerikhuisen, J. 2017, in J. Phys. Conf. Ser., Vol. 837, 012013
DeStefano, A. M., & Heerikhuisen, J. 2020, Physics of Plasmas, 27, 032901
Dissauer, K., Veronig, A. M., Temmer, M., & Podladchikova, T. 2019, ApJ, 874, 123
Filbert, P. C., & Kellogg, P. J. 1979, JGR, 84, 1369
Fraternale, F., & Pogorelov, N. V. 2021, ApJ, 906, 75
Fraternale, F., Pogorelov, N. V., & Heerikhuisen, J. 2021, ApJL, 921, L24
Gedalin, M., Pogorelov, N. V., & Roytershteyn, V. 2020, ApJ, 889, 116

—. 2021a, ApJ, 910, 107

—. 2021b, ApJ, 916, 57

Gopalswamy, N., Akiyama, S., Yashiro, S., & Xie, H. 2018, JASTP, 180, 35

Gruntman, M. A. 1982, Soviet Astronomy Letters, 8, 24

Gurnett, D. A., Kurth, W. S., Allendorf, S. C., & Poynter, R. L. 1993, Science, 262, 199

Gurnett, D. A., Kurth, W. S., Burlaga, L. F., & Ness, N. F. 2013, Science, 341, 1489

Gurnett, D. A., Kurth, W. S., Stone, E. C., et al. 2015, ApJ, 809

Heerikhuisen, J., Zirnstein, E., & Pogorelov, N. 2015, JGR, 120, 1516

Heerikhuisen, J., Zirnstein, E. J., Pogorelov, N. V., Zank, G. P., & Desai, M. 2019, ApJ, 874, 76

Hilchenbach, M., Hsieh, K. C., Hovestadt, D., et al. 1998, ApJ, 503, 916

Isavnin, A. 2016, ApJ, 833, 267

Izmodenov, V. V., & Alexashov, D. B. 2020, A&A, 633, L12

Izmodenov, V. V., Malama, Y. G., Ruderman, M. S., et al. 2009, SSR, 146, 329

Kim, T. K., Pogorelov, N. V., & Burlaga, L. F. 2017, ApJL, 843

Kim, T. K., Pogorelov, N. V., Zank, G. P., Elliott, H. A., & McComas, D. J. 2016, ApJL, 832, 72

Kornbleuth, M., Opher, M., Michael, A. T., et al. 2020, ApJL, 895, L26

Krimigis, S. M., Mitchell, D. G., Roelof, E. C., Hsieh, K. C., & McComas, D. J. 2009, Science, 326, 971

Luoni, M. L., Démoulin, P., Mandrini, C. H., & van Driel-Gesztelyi, L. 2011, Solar Physics, 270, 45

McComas, D. J., Bzowski, M., Fuselier, S. A., et al. 2015, ApJS, 220, 22

McComas, D. J., Zirnstein, E. J., Bzowski, M., et al. 2017a, ApJS, 233, 8

—. 2017b, ApJS, 229, 41

McComas, D. J., Christian, E. R., Schwadron, N. A., et al. 2018, SSR, 214, 116

Opher, M., Drake, J. F., Zieger, B., & Gombosi, T. I. 2015, ApJL, 800

Opher, M., Loeb, A., Drake, J., & Toth, G. 2020, Nature Ast., 4, 675

Parker, E. N. 1963, Interplanetary dynamical processes (Interscience Publishers)

Pogorelov, N. V., Bedford, M. C., Kryukov, I. A., & Zank, G. P. 2016, J. Phys. Conf. Series, 767

Pogorelov, N. V., Borovikov, S., Heerikhuisen, J., et al. 2014, in Proc. 2014 Ann. Conf. on Extreme Science and Engineering Discovery Environment, XSEDE '14, 22:1–22:8

Pogorelov, N. V., Borovikov, S. N., Heerikhuisen, J., & Zhang, M. 2015, ApJL, 812

Pogorelov, N. V., Fraternale, F., Kim, T. K., Burlaga, L. F., & Gurnett, D. A. 2021, ApJ, 917, L20

Pogorelov, N. V., Heerikhuisen, J., Roytershteyn, V., et al. 2017a, ApJ, 845

Pogorelov, N. V., Heerikhuisen, J., Zank, G. P., & Borovikov, S. N. 2009, SSR, 143, 31

Pogorelov, N. V., Suess, S. T., Borovikov, S. N., et al. 2013, ApJ, 772

Pogorelov, N. V., Fichtner, H., Czechowski, A., et al. 2017b, SSR, 212, 193

Qiu, J., Hu, Q., Howard, T. A., & Yurchyshyn, V. B. 2007, ApJ, 659, 758

Reisenfeld, D. B., Bzowski, M., Funsten, H. O., et al. 2021, ApJS, 254, 40

Riley, P., Lionello, R., Caplan, R. M., et al. 2021, A&A, 650, A19

Roytershteyn, V., Pogorelov, N. V., & Heerikhuisen, J. 2019, ApJ, 881, 65

Schrijver, C. J., & De Rosa, M. L. 2003, Solar Physics, 212, 165

Singh, T., Kim, T. K., Pogorelov, N. V., & Arge, C. N. 2020a, Space Weather, 18, e02405

Singh, T., Yalim, M. S., Pogorelov, N. V., & Gopalswamy, N. 2019, ApJL, 875, L17

—. 2020b, ApJ, 894, 49

Steinolfson, R. S., Pizzo, V. J., & Holzer, T. 1994, Geophys. Res. Lett., 21, 245

Stone, E. C., Cummings, A. C., Heikkila, B. C., & Lal, N. 2019, Nature Ast., 3, 1013

Stone, E. C., Cummings, A. C., McDonald, F. B., et al. 2005, Science, 309, 2017

—. 2008, Nature, 454, 71

—. 2013, Science, 341, 150

Swaczyna, P., McComas, D. J., Zirnstein, E. J., et al. 2020, Astrophys. J., 903, 48

Thernisien, A., Vourlidas, A., & Howard, R. A. 2009, Solar Physics, 256, 111

Tsallis, C. 2009, Introduction to nonextensive statistical mechanics: approaching a complex world (Springer Science & Business Media)

Upton, L., & Hathaway, D. H. 2014, ApJ, 780, 5

Usmanov, A. V., Goldstein, M. L., & Matthaeus, W. H. 2016, ApJ, 820, 17

Vasyliunas, V. M., & Siscoe, G. L. 1976, JGR, 81, 1247

Wood, B. E., Müller, H.-R., & Möbius, E. 2019, ApJ, 881, 55

Wu, Y., Florinski, V., & Guo, X. 2016, ApJ, 832, 61

Yu, G. 1974, ApJ, 194, 187

Zank, G. P. 1999, SSR, 89, 413

—. 2015, ARAA, 53, 449

Zank, G. P., Heerikhuisen, J., Pogorelov, N. V., Burrows, R., & McComas, D. 2010, ApJ, 708, 1092

The Predictive Power of Computational Astrophysics as a Discovery Tool
Proceedings IAU Symposium No. 362, 2023
D. Bisikalo, D. Wiebe & C. Boily, eds.
doi:10.1017/S1743921322002319

3D Realistic Modeling of Solar-Type Stars to Characterize the Stellar Jitter

Irina Kitiashvili[1] , **Samuel Granovsky[1,2,3]** , **Alan Wray[1]**

[1]NASA Ames Research Center
Moffett Field, MS 258-6, Mountain View, CA, USA
email: irina.n.kitiashvili@nasa.gov, alan.a.wray@nasa.gov

[2]New Jersey Institute of Technology
323 Dr Martin Luther King Jr Blvd, Newark, NJ, USA
email: sg2249@njit.edu

[3]Universities Space Research Association
7178 Columbia Gateway Drive, Columbia, MD, USA

Abstract. Detection of Earth-mass planets with the radial velocity method requires a precision of about 10cm/s to identify a signal caused by such a planet. At the same time, noise originating in the photospheric and subphotospheric layers of the parent star is of the order of meters per second. Understanding the physical nature of the photospheric noise (so-called stellar jitter) and characterizing it are critical for developing techniques to filter out these unwanted signals. We take advantage of current computational and technological capabilities to create 3D realistic models of stellar subsurface convection and atmospheres to characterize the photospheric jitter. We present 3D radiative hydrodynamic models of several solar-type target stars of various masses and metallicities, discuss how the turbulent surface dynamics and spectral line characteristics depend on stellar properties, and provide stellar jitter estimates for these stars.

Keywords. stars: individual (HD209458); line: profiles; techniques: radial velocities, spectroscopic; methods: numerical

1. Introduction

Since the discovery of the first exoplanetary system, thousands of planets have been detected near other stars. However, the detection of Earth-mass planets remains a challenging task because the radial velocity variations due to the planetary motion are about ten cm/s, whereas the stellar disturbances are often order m/s (e.g., Plavchan et al. 2015; Fischer et al. 2016). The problem is complicated because stellar surface disturbances (so-called stellar jitter) arise from multiple physical processes. Several approaches to estimate stellar jitter have been developed, such as least-squares deconvolution (Bellotti et al. 2022), scaling relations (Luhn et al. 2020), and others. These methods allow us to investigate the dependence of stellar perturbations on metallicity, age, mass, magnetic activity, and evolutionary stage (e.g., Bastien et al. 2016; Meunier et al. 2017; Oshagh et al. 2017; Collier Cameron et al. 2021; Sowmya et al. 2021).

To characterize stellar jitter, we use existing computational capabilities to perform 3D radiative models of stellar convection and use the resulting models of atmospheres to compute synthetic spectra. This approach has been used to investigate stellar surface

Table 1. Stellar parameters used to generate 3D radiative models.

Star	Mass	Teff	log(g)	FeH
HD25171	1.08	6166	4.329	−0.032
HD49674	1.00	5598	4.408	0.126
HD69830	0.87	5383	4.469	0.038
HD121504	1.18	6071	4.355	0.236
HD209458	1.05	6095	4.314	−0.144

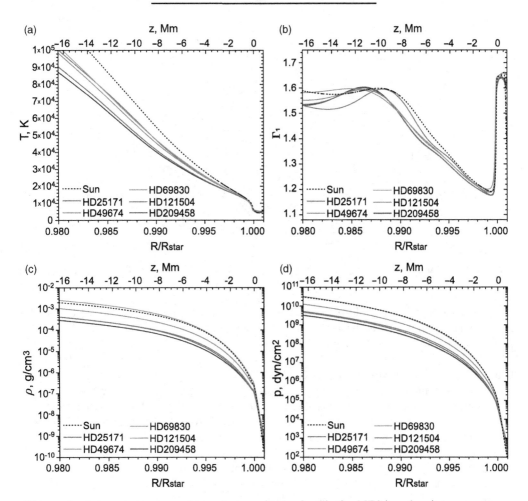

Figure 1. Internal structure of target stars obtained with the MESA code: a) temperature, b) adiabatic exponent, c) density, and gas pressure.

effects by various authors (e.g., Cegla et al. 2012, 2013; Dravins et al. 2017, 2021a,b). The present work presents initial results of modeling several target stars and computing spectral line syntheses over the stellar disks to investigate the jitter properties.

2. Computational setup to model stellar convection

To understand the nature of stellar jitter, we select five solar-type stars to characterize photospheric disturbances (Table 1). All these stars host Jupiter-size planets detected with the radial velocity method and are used as a testbed to investigate how different approaches to filtering out jitter signals can improve the signal quality of the radial

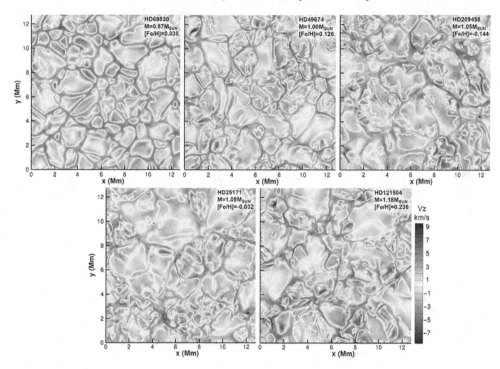

Figure 2. Granular structures at the photosphere reveal increased radial velocity amplitudes for more massive stars.

velocity residuals. As initial conditions, we use 1D mixing-length models of the stellar interiors (Fig. 1) generated with the MESA code (Paxton et al. 2011) in agreement with previously published stellar properties.

The 3D radiative hydrodynamics models of stellar convection are calculated using the StellarBox code (Wray et al. 2015, 2018). The radiative transfer calculations in this code are performed in the LTE approximation for four spectral bins; ray-tracing for 18 directional rays (Feautrier 1964) is implemented using the long-characteristics method. In this paper, we mostly consider time series obtained from simulations with 50 km grid resolution in the horizontal directions and, in the radial direction, varying from 20 km at the photosphere to 95 km near the bottom boundary of the computational domain. All models are computed for a 12.8 Mm-wide computational domain and cover 12 Mm of the convection zone and 1 Mm of atmosphere. The boundary conditions are periodic in the horizontal directions. No rotation is imposed.

3. Dynamics of the stellar convection

As stated above, we selected five target stars to to examine the properties of star-induced disturbances. Figure 1 shows the distribution of the radial (or vertical) component of velocity at the stellar photosphere for the target stars and the Sun. The photospheric structure for all stars reveals granulation patterns, including granules and mini-granules. These two populations of granular convection structures have been previously found in solar observations (Abramenko et al. 2012). However, the existence of the two populations is more prominent for more massive stars (Wray et al. 2015, 2018; Kitiashvili et al. 2016).

To investigate photosphere-originated disturbances, we synthesize six neutral iron lines (Fe I) at different locations on the stellar disk. The line profiles are used to examine the

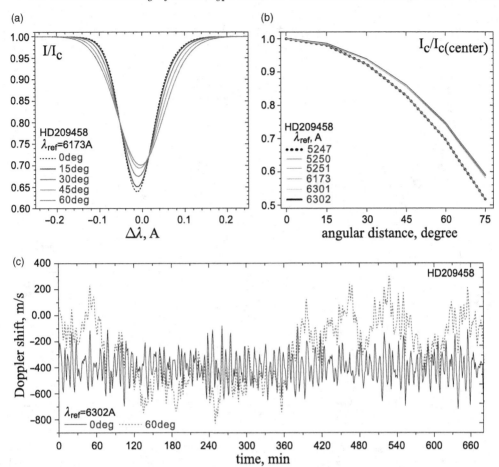

Figure 3. Center-to-limb effects: a) changes in the spectral line ($\lambda_{ref} = 6173$Å) at different distances from the disk center; b) limb darkening profiles for six FeI lines; c) Doppler shift variations as a function of time at the stellar disk center (black solid curve) and at 60 degrees to limb (red dotted curve).

dependence of the Doppler shift disturbances on wavelength and distance from the disc center. The spectral line synthesis is performed in the LTE approximation with the Spinor code (Frutiger et al. 2000) using stellar atmosphere models from the 3D simulations.

Synthesis of the spectral lines for different distances from the stellar disk center allows us to investigate the dependence of line properties on their location relative to the disk center. Increasing the distance from the disk center causes an increase in line width and a decrease in the line depth and continuum intensity (Fig. 3 a, b). The resulting limb darkening profiles of modeled spectral lines are clustered in two groups, shorter and longer wavelengths (panel b). Because of the difference in the line formation height, each line probes a specific range of the atmospheric layers. For instance, in the case of a patch location closer to the limb, the light is integrated over a thicker column of the atmosphere and thus is more strongly affected by flows above the photosphere. Figure 3c shows time-variations of the Doppler shift, integrated over the computational domain at the disk center (black solid curve) and at 60 degrees longitude (red dotted curve). Because of the effect of horizontal flows, the Doppler shift can also vary significantly with time.

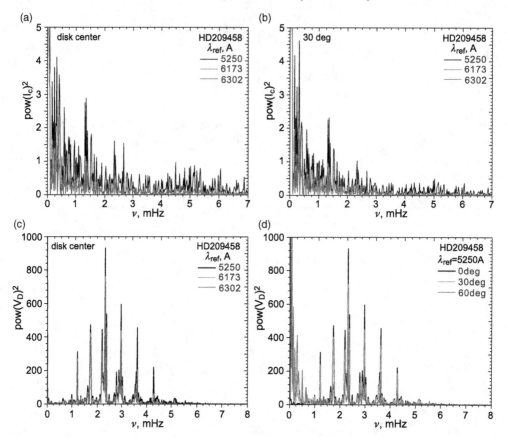

Figure 4. The power spectral density obtained from synthetic spectra of star HD209458. The power spectral density of the continuum intensity is shown for the disc center (panel a) and at 30 degrees longitude (panel b). The power spectral density of the Doppler-shift corresponds to the disc center of the simulated continuum intensity (panels a and b) and the Doppler-shift (panels c and d) at the disk center computed from three spectral lines (panel c), and for three distances from the disk center (0, 30, and 60 degrees) for one line (5250Å). In panels a-c, the power spectral density was obtained from synthesized data of three FeI lines: 5250Å (black curve), 6173Å (green), and 6302Å (red).

It is essential to note that the oscillatory properties depend on characteristics of the stellar convection, such as the size and lifetime of granulations, magnetic field flux and topology, etc. In the case of the hydrodynamic simulations presented in this paper, the properties of the stellar oscillations primarily depend on the distance from the disc center. Figure 4 shows a comparison of the power spectral density for the continuum intensity and the Doppler shift for three FeI lines. In particular, the power spectra of the continuum intensity show that the power decreases for longer wavelengths (Fig. 4a,b), which likely corresponds to a difference in the line formation heights. It is also confirmed by a decrease in the spectral density power of the Doppler shift at the disk center (panel c). Because radial motions are the main contributors to the signal, the continuum intensity (panel b) and Doppler shift oscillations are significantly reduced in the regions closer to the stellar limb.

4. Discussion and conclusions

We presented initial results of modeling solar-type stellar dynamics to investigate the nature of stellar jitter and to develop data-characterization and filtering techniques for robust detection of Earth-mass exoplanets. We performed 3D radiative hydrodynamic simulations of stellar convection and computed synthetic spectroscopic observables of target stars. We used the model of planet-hosting star HD209458 for a detailed study. We demonstrated the capabilities of this approach to capture a variety of known effects, such as limb darkening and the dependence of spectral-power density on wavelength and location on the stellar disk.

Our current effort is to produce a series of high-resolution hydrodynamic and MHD models and develop a pipeline for efficient generation of time-dependent synthetic observables, data analysis, and investigation of different techniques for filtering the photospheric disturbances in spectroscopic observations of stellar jitter.

Acknowledgment

The work is supported by the NASA Extreme Precision Radial Velocity Foundation Science Program.

References

Abramenko, V. I., Yurchyshyn, V. B., Goode, P. R., Kitiashvili, I. N., & Kosovichev, A. G. 2012, ApJ, 756, L27

Bastien, F. A., Stassun, K. G., Basri, G., & Pepper, J. 2016, ApJ, 818, 43

Bellotti, S., Petit, P., Morin, J., et al. 2022, A&A, 657, A107

Cegla, H. M., Shelyag, S., Watson, C. A., & Mathioudakis, M. 2013, ApJ, 763, 95

Cegla, H. M., Watson, C. A., Marsh, T. R., et al. 2012, MNRAS, 421, L54

Collier Cameron, A., Ford, E. B., Shahaf, S., et al. 2021, MNRAS, 505, 1699

Dravins, D., Ludwig, H.-G., Dahlén, E., & Pazira, H. 2017, A&A, 605, A91

Dravins, D., Ludwig, H.-G., & Freytag, B. 2021a, A&A, 649, A16

Dravins, D., Ludwig, H.-G., & Freytag, B. 2021b, A&A, 649, A17

Feautrier, P. 1964, Comptes Rendus Academie des Sciences (serie non specifiee), 258, 3189

Fischer, D. A., Anglada-Escude, G., Arriagada, P., et al. 2016, PASP, 128, 066001

Frutiger, C., Solanki, S. K., Fligge, M., & Bruls, J. H. M. J. 2000, A&A, 358, 1109

Kitiashvili, I. N., Kosovichev, A. G., Mansour, N. N., & Wray, A. A. 2016, ApJ, 821, L17

Luhn, J. K., Wright, J. T., Howard, A. W., & Isaacson, H. 2020, AJ, 159, 235

Meunier, N., Mignon, L., & Lagrange, A. M. 2017, A&A, 607, A124

Oshagh, M., Santos, N. C., Figueira, P., et al. 2017, A&A, 606, A107

Paxton, B., Bildsten, L., Dotter, A., et al. 2011, ApJS, 192, 3

Plavchan, P., Latham, D., Gaudi, S., et al. 2015, arXiv e-prints, arXiv:1503.01770

Sowmya, K., Nèmec, N. E., Shapiro, A. I., et al. 2021, ApJ, 919, 94

Wray, A. A., Bensassi, K., Kitiashvili, I. N., Mansour, N. N., & Kosovichev, A. G. 2015, arXiv e-prints, arXiv:1507.07999

Wray, A. A., Bensassiy, K., Kitiashvili, I. N., Mansour, N. N., & Kosovichev, A. G. 2018, Realistic Simulations of Stellar Radiative MHD, ed. J. P. Rozelot & E. S. Babayev, 39

The Predictive Power of Computational Astrophysics as a Discovery Tool
Proceedings IAU Symposium No. 362, 2023
D. Bisikalo, D. Wiebe & C. Boily, eds.
doi:10.1017/S1743921322001491

Dynamic complexity based analysis on the relationship between solar activity and cosmic ray intensity

Vipindas V.[ID]**, Sumesh Gopinath*, Vinod Kumar R. and Girish T.E.**

Department of Physics, University College, Thiruvananthapuram - 695034, Kerala, India

*Department of Physics, Sri Sathya Sai Arts and Science College, Saigramam, Thiruvananthapuram - 695104, Kerala, India
emails: vpndasv@gmail.com, sumeshgopinath@gmail.com, vinodkopto@gmail.com, tegirish5@yahoo.com

Abstract. The Earth's atmosphere is incessantly bombarded by energetic charged particles called cosmic rays (CR) which are having either solar or non-solar origin. Analysis based on information theoretic estimators can be effectively employed as a potential technique to analyze the dynamical changes in cosmic ray intensity during different solar cycles. In the present study, dynamical complexity based analysis using Jensen-Shannon divergence (JSD) has been employed which reveals the existence of some peculiar fluctuation properties in CRI flux at Jung neutron monitor station. JSD based dynamical complexity analyses confirm the existence of difference in dynamical properties of CR flux during solar cycles 20-21 and 22-23.

Keywords. Dynamical complexity, Jensen-Shannon divergence

1. Introduction

The variations in cosmic ray intensity (CRI) are continuously measured by ground station neutron monitors (NMs), operated at different geomagnetic cutoff rigidities, which are sensitive to energies of CRs. The flux rate of cosmic rays incident on the upper layers of Earth's atmosphere is modulated mainly by the solar wind. The 11-year solar cycle variations in CRI is anticorrelated with SSN (sunspot number) (Shuai Fu et al. 2021).

2. Data and Method

JSD can be considered as an information theoretic divergence estimator that measures the similarity between two distributions (Yin et al. 2020). When compared with the famous divergence estimator called Kullback-Leibler (KL) divergence, JSD is symmetric and is able to handle non-overlapped distributions thus leading to a smoother manifold. It is possible to determine the dynamical complexity derived from divergence measures of distributions (Schiepek and Strunk 2010). The dynamical complexity combines a 'fluctuation measure' which is sensitive to variations in amplitude or frequency of the time series and a 'distribution measure' which identifies the deviation of the data values from an ideal identical distribution over the range scale. The two measures are computed within a sliding window moving over the time series (for details regarding the method authors may refer to Kaiser 2017). Figure 1(a)-(d) shows the variation of JSD, dynamical complexity, SSN and CRI from which JSD has been calculated and the average dynamical complexity for different lags. While JSD varies between 0 and 1, the

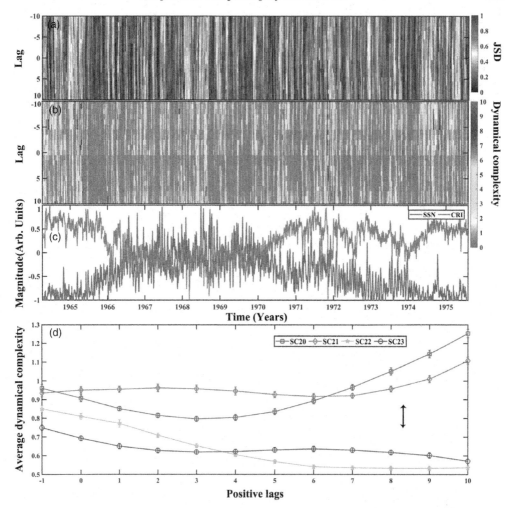

Figure 1. JS divergence analysis of solar cycle 20 for time lagged data at Jung: (a) JSD for lags -10 to 10, (b) Dynamical complexity for lags -10 to 10, (c) SSN and CRI variations for solar cycle 20, & (d) The variation of average dynamical complexity for solar cycles 20-23.

dynamical complexity varies between 0 and 10 with lower values being less complex and higher values show high degree of complexity.

3. Results and Conclusion

The variations in spectra during the periods of 1965, 1972, 1975 are due to higher divergence of associated distributions of SSN and CRI which in turn reflect in the dynamical complexity also. A peculiar bipartite pattern is seen for dynamical complexity between SC 20-21 and SC 22-23. Considering the 22-year Hale cycle, SC 20 and 21 mark the fall and rise periods with 1976 mark the minimum magnetic activity of the 22-year cycle. Similarly, SC 22 and 23 also characterize the descension and ascension periods where 1996 mark another magnetic minimum. This can also be interpreted on the grounds of earlier studies by Logachev et al. (2015). It is seen that there exists dynamical coupling in complexity of cosmic ray flux between 20-21 and 22-23 which is similar to the one reported for the proton data.

References

Shuai Fu et al. 2021, *ApJS*, 254, 37.

Yin. Y., Wang, W., Li, Q., Ren, Z., and Shang, P. 2020, *Fluct.Noise Lett.*, 20, 2150013.

Schiepek, G. and Strunk, G. 2010, *Biol. Cybern*, 102, 197.

Logachev, Y.I., Bazilevskaya, G.A., Vashenyuk, E.V. et al. 2015, *Geomagn. Aeron.*, 55, 277.

Kaiser, T. 2017, *dyncomp: an R package for Estimating the Complexity of Short Time Series.*

The Predictive Power of Computational Astrophysics as a Discovery Tool
Proceedings IAU Symposium No. 362, 2023
D. Bisikalo, D. Wiebe & C. Boily, eds.
doi:10.1017/S1743921322001466

Advances and Challenges in Observations and Modeling of the Global-Sun Dynamics and Dynamo

Alexander Kosovichev[1,2] ⓘ, **Gustavo Guerrero**[1,3], **Andrey Stejko**[1],
Valery Pipin[4] **and Alexander Getling**[5]

[1]Center for Computational Heliophysics, New Jersey Institute of Technology,
University Heights, Newark, NJ 07102, U.S.A.
email: alexander.g.kosovichev@njit.edu

[2]NASA Ames Research Center, Moffett Field, CA 94035 U.S.A.

[3]Physics Department, Universidade Federal de Minas Gerais
Av. Antonio Carlos, 6627, Belo Horizonte, MG 31270-901, Brazil
email: guerrero@fisica.ufmg.br

[4]Institute of Solar-Terrestrial Physics, Russian Academy of Sciences
Irkutsk, 664033, Russia
email: pip@iszf.irk.ru

[5]Skobeltsyn Institute of Nuclear Physics, Lomonosov Moscow State University,
Moscow, 119991 Russia
email: A.Getling@mail.ru

Abstract. Computational heliophysics has shed light on the fundamental physical processes inside the Sun, such as the differential rotation, meridional circulation, and dynamo-generation of magnetic fields. However, despite the substantial advances, the current results of 3D MHD simulations are still far from reproducing helioseismic inferences and surface observations. The reason is the multi-scale nature of the solar dynamics, covering a vast range of scales, which cannot be solved with the current computational resources. In such a situation, significant progress has been achieved by the mean-field approach, based on the separation of small-scale turbulence and large-scale dynamics. The mean-field simulations can reproduce solar observations, qualitatively and quantitatively, and uncover new phenomena. However, they do not reveal the complex physics of large-scale convection, solar magnetic cycles, and the magnetic self-organization that causes sunspots and solar eruptions. Thus, developing a synergy of these approaches seems to be a necessary but very challenging task.

Keywords. Sun: activity, Sun: helioseismology, Sun: magnetic fields, Sun: interior, MHD, turbulence

---

1. Introduction

The 11-year sunspots cycles have been regularly observed for more than 400 years. Tremendous observational material has been accumulated, but a basic understanding of the underlying physical processes is still lacking. There is no doubt that the magnetic cycles are caused by dynamo processes in the deep solar interior. These processes are associated with the interaction of turbulent convection with solar rotation beneath the solar surface, understanding of which is challenging. The observational data provided by helioseismology have limited spatial and temporal resolutions and provide only information about large-scale flows. The internal magnetic field has not been determined directly. Helioseismic inferences suffer from substantial systematic uncertainties because of so-called realization noise due to the random excitation of solar oscillation and complicated interaction between oscillations and convection. Helioseismic techniques and results must be tested by using numerical models and simulations of solar oscillations. A satisfactory theoretical understanding can be achieved only through realistic modeling and numerical simulations.

The realistic computational modeling of the solar interior and atmosphere requires resolving a wide range of scales from small-scale flows of granulation to global-scale zonal and meridional flows. The flows are characterized by extreme Reynolds numbers so that the full range of dynamical scales can never be resolved. However, high-resolution simulations of the near-surface convection have shown that the observed phenomena can be modeled with satisfactory accuracy using subgrid-scale turbulence models, such as the Large-Eddy Simulation (LES) methods. This method aims to resolve all dynamically essential scales and approximate the dissipation scales by applying turbulence scaling laws. The LES method requires that the simulations are performed with a grid resolution of 25-100 km. Therefore, the computational models are limited to small 'box' regions covering $\lesssim 100$ Mm horizontally and even less in depth (e.g. Rempel et al. 2009; Stein and Nordlund 2012; Kitiashvili et al. 2015). Nevertheless, these simulations reproduce the solar subsurface dynamics with a high degree of realism and have their prediction power. In particular, they predicted the formation of small-scale vortex structures in the solar granulation, uncovered mechanisms of acoustic wave excitation by turbulent convection, explained magnetic self-organization of solar magnetic fields, plasma flows in sunspots (the Evershed effects), local MHD dynamo, etc.

The global-Sun modeling is currently performed using two basic approaches: 1) by solving the 3D MHD equation in an anelastic approximation (Gough 1969) for suppressing the fast wave motions and excluding the surface and near-surface layers, typically, at $r > 0.95 R_\odot$; 2) by separating turbulent and large-scale flows by applying a mean-field approximation.

The 3D MHD computational models provide great insight into the global Sun dynamics and dynamo mechanisms (e.g. Ghizaru et al. 2010; Charbonneau and Smolarkiewicz 2013; Miesch et al. 2011; Simitev et al. 2015; Guerrero et al. 2019). However, they have been unable to reproduce the observations. The reason is that currently, the 3D MHD models cannot resolve the turbulent convective motion with the resolution necessary for describing the non-linear interaction of these motions with rotation and magnetic field – such interaction results in highly anisotropic turbulent transport affecting the large-scale momentum and energy balance.

A theoretical description found in the mean-field approximation separates the small-scale turbulent motions from large-scale flows and describes their interaction in turbulent transport coefficients depending on large-scale properties (e.g. Krause and Rädler 1980; Brandenburg et al. 1992; Kitchatinov et al. 1994; Pipin and Kitchatinov 2000). In addition, to turbulent diffusion, this approach predicts non-diffusive turbulent effects, which appear as additional terms in the mean-field equations. This theory has been developed to a high degree of sophistication. However, it cannot predict the strength and spectrum of turbulence and, in solar-stellar applications, uses parameters of convective velocity distribution from the mixing-length ansatz of the stellar evolution theory. Nevertheless, the mean-field theory provides a reasonable qualitative and quantitative description of the differential rotation, meridional circulation, solar-cycle evolution of the axisymmetric magnetic field and makes predictions that can be verified in observations. Traditionally, the dynamical and dynamo processes were considered separately. But, recently developed combined full-MHD mean-field models explained many observed phenomena, such as the

Solar differential rotation

Meridional circulation

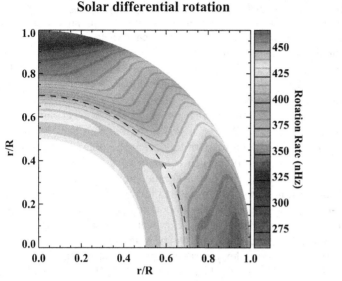
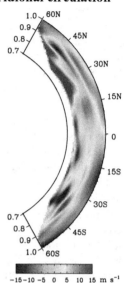

Figure 1. a) Mean solar rotation determined by global helioseismology from SDO/HMI JSOC data in 2010-2022. b) The distribution of the meridional flow velocity in the solar convection zone (Zhao et al. 2013).

differential rotation, migrating zonal flows ('torsional oscillations'), extended solar cycle phenomenon for zonal and meridional flows (Pipin and Kosovichev 2019, 2020). Despite the success, this theory cannot describe the observed 3D structure of the solar convection and magnetic fields. Although, initial steps towards a 3D mean-field description have been made.

This paper presents recent attempts to develop a synergy of the global 3D MHD modeling, mean-field theory, and helioseismic analysis by our group. By no means is this a comprehensive review of the current advances to develop computational models of the global-Sun dynamics and magnetism, to which many other groups have contributed.

2. Helioseismic Observations of the Solar Dynamo and Interior Dynamics

The helioseismic observations are generally divided into two complementary approaches: global and local helioseismology. Global helioseismology measures frequencies and frequency splitting of resonant acoustic oscillations and surface gravity waves, inversion of which provides estimates of the internal rotation (Fig. 1a) and axisymmetrical sound-speed variations in the whole spherical Sun. The measurements are typically performed using uninterrupted 72-day series of solar 5-minute oscillations. Only azimuthally averaged flows and structures can be measured by this technique.

Local helioseismology measures either local variations of the waves dispersion relation by using 3D Fourier transform (Ring-Diagram Analysis, Gough and Toomre 1983; Hill 1989) or acoustic travel times extracted from cross-correlations of solar oscillations (Time-Distance Helioseismology, Duvall et al. 1993; Kosovichev and Duvall 1997). These techniques provide maps of horizontal flows in shallow subsurface layers, albeit with different resolutions. In addition, time-distance helioseismology has been used to measure large-scale meridional flows through the whole of the solar convection zone.

Helioseismic measurements of the internal differential rotation, meridional circulation, and their variations with the solar cycle are essential for understanding solar dynamics and magnetism. It has been firmly established (Schou et al. 1998) that the solar differential rotation extends through the whole convection zone and has strong shearing flows: the tachocline at the bottom and the Near-Surface Shear Layer (NSSL) the top boundaries (Fig. 1a). The shearing flows play key roles in the solar dynamo processes. The tachocline is a likely place of magnetic field generation, and the NSSL 'shapes' the evolution of magnetic fields observed on the surface, in particular, forming the equator-ward migration of emerging magnetic flux - the butterfly diagram.

The distribution of the meridional circulation with depth in the convection zone is much less certain. The initial measurements (Zhao et al. 2013) showed that the circulation might consist of two circulation cells along the solar radius, with the polar-ward flows at the top and bottom of the convection zone and return equator-ward flow in the middle, contrary to the expectations that the circulation has a simple single-cell structure. This result caused significant attention and new measurement efforts because, if correct, it effectively rules out the flux-transport solar-cycle models, which predict the polar-ward meridional flow is responsible for the butterfly diagram. However, the exact flow structure has not been reliably established. Some studies show that the single-cell structure is more likely, while some others find even more complex multi-cell dynamics (Böning et al. 2017; Chen and Zhao 2017, 2018; Gizon et al. 2020a; Jackiewicz et al. 2015; Kholikov et al. 2014; Lin and Chou 2018; Schad et al. 2013). Additional uncertainty comes from potential variations of the meridional flows with the Sun's activity cycle. Because the measurements require a very long time series of solar oscillations, the results may depend on the time interval chosen for the analysis.

In such a situation, forward 3D acoustic modeling of helioseismic data is essential for resolving the discrepancies and determining the observational limits on the flow dynamics (Hartlep et al. 2013, 2008; Khomenko et al. 2009; Parchevsky and Kosovichev 2009; Parchevsky et al. 2014). Some recent advances in this direction are discussed in the next section.

3. Computational Solar Acoustics and Forward Modeling of Helioseismic Inferences

Helioseismic inferences of the internal structures and flows are under simplified assumptions about wave propagation. Relationships between the observed quantities, such as frequency shifts and travel-time anomalies, and the physical properties are obtained using a perturbation theory and expressed in the form of linear integral equations. The inversion procedures for solving these equations seek for the smoothest solution satisfying the relationships within the observational errors. Thus, the real observational constraints on complex flow structures remains unknown.

For getting insight into uncertainties of helioseismic measurements and obtain observational constraints the GALE (Global Acoustic Linearized Euler) code (Stejko et al. 2021a) had been developed. The code solves the conservation form of the linearized compressible Euler equations on a full global 3-dimensional grid: $0 \leq \phi \leq 2\pi$, $0 < \theta < \pi$, $0 < r \leq R_\odot$.

$$\frac{\partial \rho'}{\partial t} + \Upsilon' = 0 , \tag{3.1}$$

$$\frac{\partial \Upsilon'}{\partial t} + \boldsymbol{\nabla} : (\mathbf{m}'\tilde{\mathbf{u}} + \tilde{\rho}\tilde{\mathbf{u}}\mathbf{u}') = -\nabla^2 (p') - \nabla \cdot (\rho'\tilde{g}_r\hat{\mathbf{r}}) + \nabla \cdot S\hat{\mathbf{r}} , \tag{3.2}$$

$$\frac{\partial p'}{\partial t} = -\frac{\Gamma_1\tilde{p}}{\tilde{\rho}} \left(\nabla \cdot \tilde{\rho}\mathbf{u}' + \rho'\nabla \cdot \tilde{\mathbf{u}} - \frac{p'}{\tilde{p}}\tilde{\mathbf{u}} \cdot \nabla\tilde{\rho} + \tilde{\rho}\mathbf{u}' \cdot \frac{N^2}{g}\hat{\mathbf{r}} \right) . \tag{3.3}$$

Here the acoustic perturbations are denoted by a prime, and the background field terms are denoted by a tilde. Υ is defined as the divergence of the momentum field $\mathbf{m}$ ($\Upsilon = \nabla \cdot \mathbf{m} = \nabla \cdot \rho\mathbf{u}$), computing perturbations in the acoustic (potential) flow field and omitting solenoidal terms in our governing equations. The acoustic oscillations are initiated by a randomized source function (S), modeling the stochastic excitation of acoustic modes at the top boundary of the convection zone. Governing equations (3.1) - (3.3) are solved using a pseudo-spectral computational method through spherical harmonic decomposition ($f = \sum_{lm} aY_{lm}$) of field terms (Υ, ρ, p, $\mathbf{u}$) using the Libsharp spherical harmonic library (Reinecke and Seljebotn 2013). The governing equations (Eqs 3.1 - 3.3) are solved in a vector spherical harmonic (VSH) basis, while the material derivative is solved in its Cauchy conservation form ($\nabla : (\mathbf{m}'\tilde{\mathbf{u}} + \tilde{\rho}\tilde{\mathbf{u}}\mathbf{u}')$) using a tensor spherical harmonic (TSH) basis. This formulation allows for the use of recursion relations to compute derivatives tangent to the surface of the sphere ($\partial/\partial\theta, \partial/\partial\phi$), resulting in a system of 1D equations for radial relations.

We use the GALE code to perform forward modeling of helioseismic data and acoustic travel times for various models of the meridional circulation (Stejko et al. 2021b). Then, by comparing the forward-modeling results with the available observational data, we validate the meridional circulation models. In particular, we performed the forward for four meridional circulation models shown in Figure 2a-d. Models M1 and M2 are the self-consistent mean-field dynamo models (Pipin and Kosovichev 2019) and K1 and K2 are the mean-field hydrodynamics models without magnetic field. The meridional circulation in these models can consist of one or two cells along the radius. The two-cell circulation appears when the radial dependence of the Coriolis number is taken into account (Bekki and Yokoyama 2017), which is due to the radial dependence of the convection turn-over time.

The acoustic oscillations generated for these background flow models were analyzed using the same deep-focusing scheme for selecting the acoustic ray paths (Fig. 2e) and the same cross-covariance fitting procedure, as in the analysis of the solar observations (Zhao et al. 2013). The signal-to-noise ratio for the travel-time measurements was improved by applying specially designed phase-speed filters. The calculated travel times are plotted as a function of the distance between the correlation points of the surface, which corresponds to different depths of the acoustic turning points. The solid curve with error bars shows the measurements obtained from solar observation (Gizon et al. 2020b). Evidently, models M1, M2 and K2 with relatively shallow return flows are consistent with the observations, irrespective of whether the meridional circulation has a single- or double-cell structure. But, model K1 with a deep return flow significantly deviates from the observations.

4. Torsional Oscillations, Extended Solar Cycle and Evidence for Dynamo Waves

Migrating during the solar cycles zonal flows ('torsional oscillations') carry essential information about the global-Sun dynamics and dynamo processes. The torsional oscillations were first discovered in Doppler-shift measurements of the surface rotational velocity as zones of faster and slower solar rotation. Helioseismic observations showed that the torsional oscillation flows are extended into the deep interior. Figure 3 shows the flow evolution during the last two sunspot cycles in the upper half of the convection zone. It is evident that the flow pattern, which consists of a prominent polar-ward branch and a weaker equatorward branch repeats every 11-years (like the sunspot cycle), but the duration of this pattern is about 22 years. This phenomenon is called the 'extended solar cycle'. Both polar and equatorial branches of faster rotation start at about 60

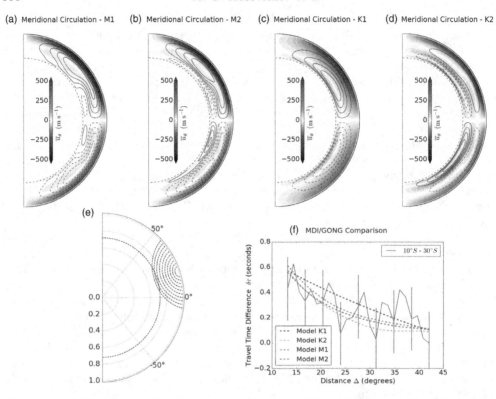

Figure 2. a) The mean-field models of meridional circulation (Pipin and Kosovichev 2018, 2019): a) Single-cell meridional circulation with a shallow return flow at $\sim 0.80 R_\odot$. b) Double-cell meridional circulation with a weak reversal near the bottom of the convection zone. c) Single-cell meridional circulation with a deep return flow near the base of the tachocline. d) Double-cell meridional circulation with a strong reverse flow. Solid and dashed contours represent counterclockwise and clockwise rotation respectively. Meridional circulation speed is increased by a factor of 36 to reduce the duration of computational runs. e) The acoustic ray-paths associated with the selected travel distances for the deep-focus time-distance helioseismology method. f) The North-South travel-time differences ($\delta\tau_{NS}$) as a function of travel distance (Δ) for the models obtained from the helioseismic observations (solid curve) and the computational models (dashed lines).

degrees latitude, approximately at the start of sunspot cycles. The polar branches relatively quickly migrate to the polar regions while the equator-ward branches migrate much slower. They reach the sunspot formation latitudes by the start of the following sunspot cycle and continue through this cycle, forming a 22-year 'extended solar cycle'.

The tracking of the torsional oscillation pattern with depth (Fig. 3) shows that the torsional oscillations originate in the deep convection zone, and thus reflect the internal dynamics associated with the dynamo processes. Zonal acceleration, calculated as the time derivative of zonal velocity (after applying a Gaussian smoothing), (Fig. 4a), and overlaid the magnetic butterfly diagram (Fig. 4b), shows that the zones of magnetic flux in the sunspot zone and the polar regions coincide with the zonal deceleration (blue areas).

This result indicates that tracking the zonal deceleration with depth may reveal the evolution of internal magnetic fields. The snapshots of the distribution of the zonal acceleration with the depth and latitude during three solar minima and two solar maxima, obtained from the SoHO and SDO data are shown in Figure 5. During the solar minima, the zonal deceleration regions are extended from the surface mid-latitudes through the

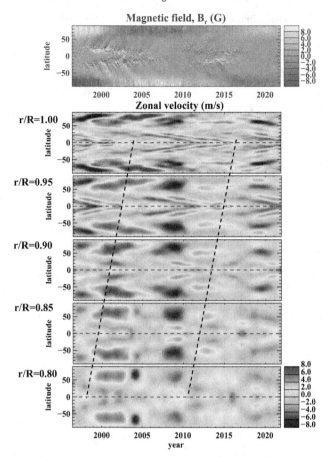

Figure 3. Time-latitude diagrams for the radial magnetic field in 1996-2021, obtained from SoHO/MDI, SDO/HMI and SOLIS data, and the zonal flow velocity calculated from the solar rotation inversion data of SoHO/MDI and SDO/HMI, available from JSOC. The inclined dashed lines track the points of convergence of the fast rotating zonal flows at the equator, indicating the end of the extended Cycles 22 and 23.

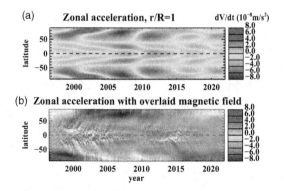

Figure 4. a) Time-latitude diagram of the zonal flow acceleration close to the solar surface from the SoHO/MDI and SDO/HMI data from 1996 to 2022. b) Overlay of the zonal acceleration and the magnetic butterfly diagram for the same period.

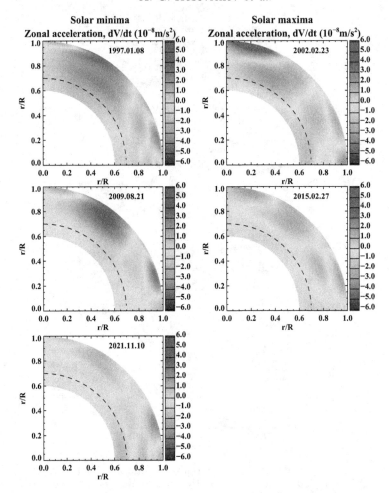

Figure 5. Distributions of the zonal acceleration in the solar interior during the periods corresponding to the solar minima in 1997, 2009, and 2021 (left column) and during the solar maxima in 2002 and 2015 (right columns).

convection zone, and also show near-surface deceleration close to the equator. Evidently, these two regions correspond to the start of a sunspot cycle at mid-latitudes and the end of the previous cycle near the equator. During the solar maxima, the blue deceleration zones in subsurface regions at mid and low latitudes coincide with the sunspot formation zones. In addition, large deceleration regions appear at high latitudes. These regions coincide with the areas of near-polar magnetic field (see Fig. 4b), and their extension to the bottom of the convection zones indicates the polar magnetic field depends on dynamo processes in a high-latitude zone at the bottom of the convection (the tachocline).

The evolution of the zonal deceleration with depth (Fig. 6) shows that both the equatorward and polar-ward branches originate at the bottom of the convection zone at about 60 degrees latitude. The polar branch migrates to the surface in 1-2 years, but it takes about 11 years for the equatorial branch to appear at mid-latitudes in the sunspot formation zone. During the following 11 years, it slowly migrates towards the equator. It is intriguing that the migration slows down at the bottom boundary of the subsurface shear layer, at $r/R_\odot \approx 0.95$.

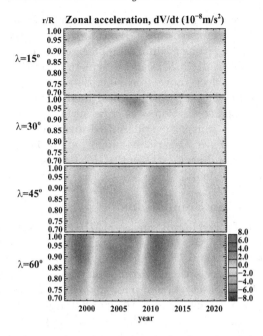

Figure 6. Evolution of the zonal acceleration with depth at four latitudes: 15, 30, 45 and 60 degrees in 1996–2022.

The observed evolution of the subsurface dynamics corresponds to the dynamo wave theory introduced by Parker (1955), which leads to the following interpretation of the helioseismic observations. In this theory, the magnetic field evolution is described as 'dynamo waves', which represent turbulent diffusion of the magnetic field, amplified by helical motions and differential rotation. The cyclic evolution of solar magnetic fields begins at the bottom of the convection zone at about 60 degrees latitudes. In the high-latitude regions, the Coriolis force acting on flows in seed toroidal magnetic flux tubes twists and amplifies the magnetic field, creating a large-scale poloidal magnetic field (by so-called 'the alpha-effect'). The generated poloidal field quickly migrates to the surface in the high-latitude regions, but on its way to the surface at lower latitudes it is stretched and amplified by the differential rotation (so-called 'the Lambda-effect'). It takes about 11-years for the generated toroidal field to reach the surface at about 30 degrees latitude where it emerges in the form of bipolar sunspot groups, initiating the sunspot butterfly diagram.

The helioseismology data (Fig. 6) show that the subsurface shear plays a significant role in the magnetic flux emergence and evolution. The radial gradient of the differential rotation in the near-surface shear layer determines the equator-ward direction of the toroidal field migration in the butterfly diagram (Brandenburg 2005; Pipin and Kosovichev 2011). The data also indicate a downward migrating branch of the zonal deceleration in the near-surface shear layer, which may be related to the submergence of the magnetic field. Observations show that the surface magnetic field rotates with the speed corresponding to the bottom part of the near-surface shear layer. It means that the magnetic field of sunspots is anchored in this layer. The magnetic time-latitude diagram shows that the decaying toroidal magnetic field is transported by meridional flows and turbulent diffusion to high-latitude regions, where it submerges to the bottom of the convection zone, providing a seed field for the next solar cycle.

In the dynamo-wave theory, the duration of solar cycles is primarily determined by three parameters: the strength of cyclonic motions at the bottom of the convection zone, the differential rotation and turbulent diffusion. Thus, the stability of the 11-year cycles ('the solar clock', Russell et al. 2020) can be explained by the long-term stability of the internal dynamics in the deep convection zone. However, the strength of the sunspot cycles changes significantly. The helioseismic measurements show that the amplitude of the zonal acceleration in the convection zone substantially decreased in the past cycle, which is in line with the low sunspot maximum (Kosovichev and Pipin 2019).

In general, the dynamo-wave scenario is supported by helioseismic measurements. However, the underlying non-linear turbulent MHD processes can only be understood through computational modeling.

5. 3D MHD Global-Sun Simulations

The simulations are performed in a 3D anelastic MHD approximation and in a spherical shell, $0 \leq \phi \leq 2\pi$, $0 \leq \theta \leq \pi$, $0.61 \geq r/R_\odot \leq 0.96$. It includes a transition zone between the radiative and convection zones at $0.7R_\odot$ (the tachocline), but does not include the shallow subsurface layers where the anelastic approximation is invalid. The mathematical model is described in terms of the conservation of mass, momentum, energy, and magnetic flux. The energy equation is written for perturbations of potential temperature, Θ, related to the specific entropy: $s = c_p \ln \Theta + \text{const}$) (Ghizaru et al. 2010):

$$\nabla \cdot (\rho_s \mathbf{u}) = 0, \tag{5.1}$$

$$\frac{D\mathbf{u}}{Dt} + 2\Omega \times \mathbf{u} = -\nabla \left(\frac{p'}{\rho_s} \right) + \mathbf{g}\frac{\Theta'}{\Theta_s} + \frac{1}{\mu_0 \rho_s}(\mathbf{B} \cdot \nabla)\mathbf{B} , \tag{5.2}$$

$$\frac{D\Theta'}{Dt} = -\mathbf{u} \cdot \nabla \Theta_e - \frac{\Theta'}{\tau} , \tag{5.3}$$

$$\frac{D\mathbf{B}}{Dt} = (\mathbf{B} \cdot \nabla)\mathbf{u} - \mathbf{B}(\nabla \cdot \mathbf{u}) , \tag{5.4}$$

where $D/Dt = \partial/\partial t + \mathbf{u} \cdot \nabla$ is the total (material) derivative, $\mathbf{u}$ is the velocity in the frame rotating with the angular velocity $\Omega = \Omega_0(\cos\theta, -\sin\theta, 0)$, p' is a pressure perturbation that accounts for both the gas and magnetic pressure, $\mathbf{B}$ is the magnetic field, and Θ' is the perturbation of potential temperature with respect to an ambient state Θ_e (see Sec.3 of Guerrero et al. 2013, for a comprehensive discussion). Furthermore, ρ_s and Θ_s are the density and potential temperature of the reference state; $\mathbf{g} = GM/r^2 \hat{\mathbf{e}}_r$ is the gravity acceleration, and μ_0 is the magnetic permeability.

The term Θ'/τ represents dissipation of the heat flux. For the models including a tachocline, we use the polytropic model with indexes $m_r = 2$ in the radiative zone and $m_{cz} = 1.499978$ in the convection zone (this value is just below the convective instability threshold). The transition between the radiative and convection zone is modeled by a smooth polytropic index variation with width $w_t = 0.015R_\odot$. The relaxation time of the potential temperature perturbations is chosen $\tau = 1.036 \times 10^8$ s (~ 3.3 yr).

The equations are solved numerically using the EULAG-MHD code (Smolarkiewicz and Charbonneau 2013; Guerrero et al. 2013). The time evolution is calculated using a special semi-implicit method based on a high-resolution, non-oscillatory forward-in-time advection scheme MPDATA (Multidimensional Positive Definite Advection Transport Algorithm; Smolarkiewicz 2006). The truncation terms in MPDATA evince viscosity comparable to the explicit sub-grid scale viscosity used in Large-Eddy Simulation (LES) models. Thus, the results of MPDATA are often

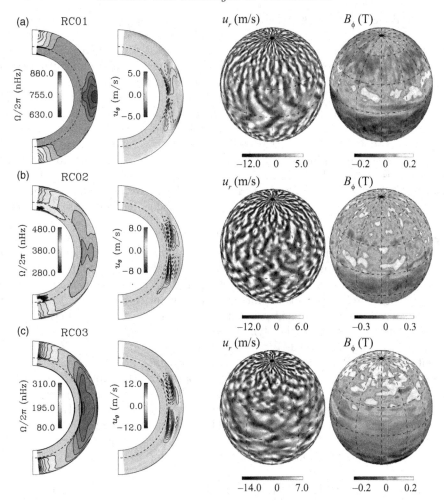

Figure 7. From left to right: mean profiles of the rotation rate, $\Omega/2\pi$, and meridional velocity, v_θ and snapshots at $r = 0.95R_\odot$ of the vertical velocity, u_r, and the toroidal field, B_ϕ, for the simulated models a) RC01: $\Omega = 2\Omega_\odot$, b) RC02: $\Omega = \Omega_\odot$, and c) RC03: $\Omega = 0.5\Omega_\odot$. In the meridional circulation panels, continuous (dashed) lines correspond to clockwise (counterclockwise) circulation. The differential rotation and meridional circulation profiles are calculated from the mean azimuthal values averaged over ~ 3 years.

interpreted as an implicit version of the LES method to account for unresolved turbulent transport. The boundary conditions are impermeable stress-free for the velocity field. The convective heat flux is set to zero at the top, and the flux divergence is zero at the bottom. In addition, it is assumed that the magnetic field is radial at the boundaries.

Figure 7 shows the differential rotation (left column) and meridional circulation (second column) profiles of the computational models with different rotation rate: a) RC01 - $\Omega = 2\Omega_\odot$, b) RC02 - $\Omega = \Omega_\odot$, c) RC03 - $\Omega = 0.5\Omega_\odot$. All these models exhibit the solar-type differential rotation – low-latitude regions rotate faster than high-latitude regions. Model RC01 has the lowest Rossby number; thus, it has the strongest influence of the Coriolis force. In this model, the latitudinal differential rotation is concentrated in the low-latitude regions. In the other two models, it is distributed over latitudes.

The magnetic field is generated by the turbulent dynamo from random initial perturbations. It affects the mean differential rotation (making it solar-like). In addition, the magnetic feedback on the flow generates zonal flows (torsional oscillations). The latitudinal flow structure resembles the observations, but the zonal flow velocity amplitude is several times larger than the observed velocity. The torsional oscillations are driven by the magnetic torque and by a modulation of the latitudinal angular momentum transport mediated by the meridional circulation (Guerrero et al. 2016).

The model reproduces the near-surface rotational shear layer (NSSL). However, it is mainly pronounced at higher latitudes, which contradicts the helioseismic observations showing that the NSSL is almost uniform in latitude. The meridional circulation is multicellular in all models. However, as expected, the meridional velocity in model RC03 (with the lower rotation rate) is higher than the other models. Also, it exhibits a dominant counterclockwise cell in the Northern hemisphere and a clockwise cell in the Southern hemisphere. Noteworthy, a low amplitude poleward flow develops in the upper part of the convection zone at latitudes $> 30°$ in all these models (see in the second column of Fig. 7).

The two right columns of Fig. 7 show snapshots of the radial velocity and magnetic field at the top boundary, $r = 0.96R_\odot$. Evidently, the scale of convection varies with latitude: smaller convective structures are formed at higher latitudes. At low latitudes, the structures are elongated, resembling 'banana' cells. Such convection structuring is not observed on the surface. Therefore, in future work, it is important to investigate the spectrum of subsurface solar convection by helioseismology to check the model predictions.

The simulations show that the dynamo processes strongly depend on the rotation rate. In the fast-rotating model, RC01, the dynamo-generated magnetic field oscillates in amplitude but does not show clear polarity reversals. The field's topology consists of wreaths of the toroidal field of opposite polarity in the Northern and Southern hemispheres. Magnetic cycles with a full period of ~ 30 yr are developed in model RC02. However, in this model, the toroidal magnetic field polarity is the same in both hemispheres, contrary to the observations. The amplitude of the surface magnetic field correlates directly with the rotation rate. The mean field strength reaches 10^4 G in the tachocline because of the strong radial shear.

6. Mean-Field Models of the Solar Dynamics and Dynamo

The mean-field model (Pipin and Kosovichev 2018) describes the evolution of the mean axisymmetric velocity, magnetic field, and entropy. The mean axisymmetric velocity is represented in terms of the angular velocity, Ω, and the meridional circulation velocity $\overline{\mathbf{U}}^m$: $\overline{\mathbf{U}} = \overline{\mathbf{U}}^m + r\sin\theta\Omega\hat{\phi}$, where $\hat{\phi}$ is the azimuthal unit vector. Similar to the 3D simulations, we employ the anelastic approximation. The full MHD system in the anelastic approximation includes the conservation of the angular momentum, the equation for the azimuthal component of vorticity $\overline{\omega} = \left(\boldsymbol{\nabla} \times \overline{\mathbf{U}}^m\right)_\phi$, the heat transport equation in terms of mean entropy $\overline{s}$, and the induction equation for the large-scale azimuthal magnetic field $\overline{\mathbf{B}}$:

$$\frac{\partial}{\partial t}\overline{\rho}r^2\sin^2\theta\Omega = -\boldsymbol{\nabla}\cdot\left(r\sin\theta\overline{\rho}\left(\hat{\mathbf{T}}_\phi + r\sin\theta\Omega\overline{\mathbf{U}}^m\right)\right) + \boldsymbol{\nabla}\cdot\left(r\sin\theta\frac{\overline{\mathbf{B}}\overline{B}_\phi}{4\pi}\right) \quad (6.1)$$

$$\frac{\partial\omega}{\partial t} = r\sin\theta\boldsymbol{\nabla}\cdot\left(\frac{\hat{\phi}\times\boldsymbol{\nabla}\cdot\overline{\rho}\hat{\mathbf{T}}}{r\overline{\rho}\sin\theta} - \frac{\overline{\mathbf{U}}^m\omega}{r\sin\theta}\right) + r\sin\theta\frac{\partial\Omega^2}{\partial z} - \frac{g}{c_p r}\frac{\partial\overline{s}}{\partial\theta} + F_L^{(p)} \quad (6.2)$$

$$\overline{\rho T}\left(\frac{\partial\overline{s}}{\partial t} + \left(\overline{\mathbf{U}}\cdot\boldsymbol{\nabla}\right)\overline{s}\right) = -\boldsymbol{\nabla}\cdot\left(\mathbf{F}^c + \mathbf{F}^r\right) - \hat{T}_{ij}\frac{\partial\overline{U}_i}{\partial r_j} - \boldsymbol{\mathcal{E}}\cdot\left(\boldsymbol{\nabla}\times\overline{\mathbf{B}}\right), \quad (6.3)$$

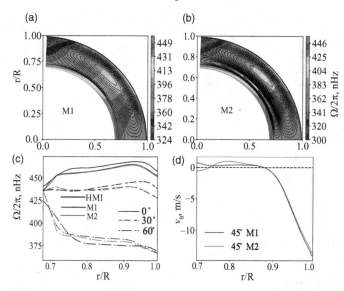

Figure 8. The angular velocity and meridional circulation distributions for model M1, which has a single-cell meridional circulation, b) model M2 with a double-cell meridional circulation c) radial profiles of the angular velocity at 0, 30 and 60 degrees latitudes (blue for model M1, red - M2), and obtained from the SDO/HMI helioseismology data archive (black); d) radial profiles of the meridional circulation velocity at latitude 45°.

$$\frac{\partial \overline{\mathbf{B}}}{\partial t} = \boldsymbol{\nabla} \times \left(\boldsymbol{\mathcal{E}} + \overline{\mathbf{U}} \times \overline{\mathbf{B}} \right), \tag{6.4}$$

where $\partial/\partial z = \cos\theta \partial/\partial r - \sin\theta/r \cdot \partial/\partial\theta$ is the gradient along the axis of rotation, $\hat{\mathbf{T}}$ is the turbulent stress tensor that includes small-scale fluctuations of velocity and magnetic field:

$$\hat{T}_{ij} = \overline{u_i u_j} - \frac{1}{4\pi\overline{\rho}} \left(\overline{b_i b_j} - \frac{1}{2}\delta_{ij}\overline{\mathbf{b}^2} \right), \tag{6.5}$$

$\mathbf{u}$ and $\mathbf{b}$ are the turbulent fluctuating velocity and magnetic field, $\overline{\rho}$ is the mean density, $\overline{T}$ - the mean temperature, $\mathbf{F}^c$ is the eddy convective flux, $\mathbf{F}^r$ is the radiative flux, $F_L^{(p)}$ is the poloidal component of the large-scale Lorentz force, $\boldsymbol{\mathcal{E}}$ is the mean electromotive force in the form:

$$\mathcal{E}_i = (\alpha_{ij} + \gamma_{ij})\,\overline{B}_j - \eta_{ijk}\nabla_j\overline{B}_k. \tag{6.6}$$

where symmetric tensor α_{ij} models generation of the large-scale magnetic field by the α-effect, which includes kinetic and magnetic helicities; anti-symmetric tensor γ_{ij} controls the mean drift of the large-scale magnetic fields in the turbulent medium; the tensor η_{ijk} describes the anisotropic turbulent diffusion (for more details, see Pipin and Kosovichev 2018).

Traditional mean-field dynamo models prescribe the differential rotation and meridional circulation and solve only the induction equation (Eq. 6.4). But, this model solves the full system of the MHD equations, and reproduces the differential rotation, meridional circulation, magnetic field and their variations with the solar cycles. Figure 8 shows the differential rotation and meridional circulation for two models M1 and M2. Their differential rotation profiles are in good agreement with the helioseismic measurements from the SDO/HMI instrument. However, the subsurface shear layer in the models is

346 A. G. Kosovichev *et al.*

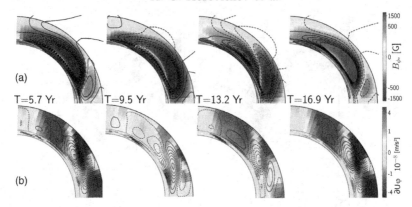

Figure 9. Model M1. Snapshots for a half of the dynamo cycle of: a) the toroidal magnetic field (background image) and streamlines of the poloidal field (contours); b) variations of the zonal acceleration (background image) and the azimuthal force caused by variations of the meridional circulation (contour lines are plotted in the range ± 50 m/s^2).

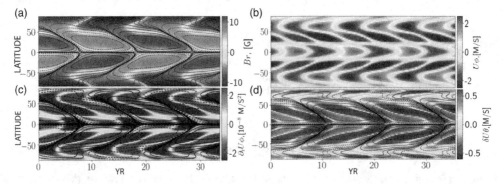

Figure 10. Model M1. a) Time-latitude diagram of the radial magnetic field at the surface (background image) and the toroidal magnetic field at $r_s = 0.9 R_\odot$, contour lines are plotted in range of ± 1kG with an exponential decrease to the low values of ± 4G; b) time-latitude diagram of the torsional oscillations (background image) at the surface; c) time-latitude diagram of the zonal flow acceleration (background image) at the surface and the toroidal magnetic field in the subsurface shear layer (same as panel (a)); d) variations of the meridional circulations at the surface (positive values correspond to the poleward flow).

somewhat broader and less steep compared to the observations (Fig. 8c). Models M1 and M2 are calculated for identical conditions, except that Model M2 takes into account the radial inhomogeneity of the Coriolis number, which depends on the convective turnover time. This effect results in a weak secondary meridional circulation cell at the bottom of the convection zone (Fig. 8b,d; Pipin and Kosovichev 2018).

Figure 9 shows snapshots of the large-scale magnetic field and zonal acceleration in the convection zone for half of the solar cycle for model M1. A new dynamo cycle starts near the bottom of the convection zone at about 60 degrees latitude in the region subjected to zonal deceleration. The acceleration pattern in the upper part of the convection zone drifts toward the equator following the dynamo wave propagation (see, Fig. 9a,b). We see that in the latitude range from $50°$ to $60°$, where the extended dynamo mode is initiated, and the acceleration is provided by the inertia force. The polar branch of the torsional oscillations in model M1 is due to the effects of the Lorentz force and variations of the meridional circulation.

Figure 10 shows the radial magnetic field evolution in model M1 at the surface and the corresponding evolution of dynamo-induced zonal variations of the rotational velocity and zonal acceleration. The large-scale toroidal magnetic field, shown at the bottom of the near-surface shear layer at $r \sim 0.9 R_\odot$, varies with the magnitude of 1.5 kG. In the subsurface shear layer, the dynamo wave of the toroidal magnetic field starts at about 60° latitude, approximately 1-2 years after the end of a previous activity cycle. The toroidal magnetic field strength at this latitude is about 4 G. The wave propagates toward the equator in ~ 22 years. The polar and equatorial branches of the dynamo waves almost completely overlap in time. The evolution of the radial magnetic field also reveals the extended cycle and agrees with the observational results (Stenflo 2012).

The dynamo wave forces variations of the angular velocity and meridional circulation. It is seen in Fig. 10b that the induced zonal acceleration is $\sim 2 - 4 \times 10^{-8}$ m s^{-2}, which is in agreement with the observational results of Kosovichev and Pipin (2019). However, the individual force contributions are stronger than their combined action by more than an order of magnitude. Another interesting finding is that two components of the azimuthal force show the extended 22-year modes. These forces are associated with variations of the meridional circulation, and the inertial force. The polar branch of the torsional oscillations in the models is due to the effects of the Lorentz force and variations of the meridional circulation.

The model shows weak meridional circulation variations in the main part of the convection zone, where its magnitude is about 10–20 cm/s. The surface variations of the meridional circulation in the dynamo cycle are about 1 m/s (Fig. 10d). They correspond to meridional flows converging towards the activity belts, in agreement with results of local helioseismology (Komm et al. 2012; Zhao et al. 2014; Kosovichev and Zhao 2016).

The results show that the large-scale toroidal magnetic field results in a reduction of the convective flux. This phenomenon, called the magnetic shadow effect, was discussed earlier by, e.g., Brandenburg et al. (1992) and Pipin (2004), and is usually considered in the problem of the solar-cycle luminosity variation. In the model, the magnetic shadow effect induces variations of the latitudinal gradient of the mean entropy. This results in perturbation of the Taylor-Proudman balance and variations of the meridional circulation. In agreement with other studies (e.g., Durney 1999; Rempel 2006; Miesch et al. 2011), the variations are concentrated near the boundaries of the convection zone.

Thus, the model shows that the torsional oscillations are driven by a combination of magnetic field effects acting on turbulent angular momentum transport and the large-scale Lorentz force. We find that the 22-year 'extended' cycle of the torsional oscillations results from a combined effect of the overlap of subsequent magnetic cycles and magnetic quenching of the convective heat transport. This quenching results in variations of the meridional circulation. The variations of the meridional circulation together with other drivers of the torsional oscillations maintain their migration to the equator forming the 22-year extended cycle.

7. Using Computational Models for Prediction of Solar Dynamics and Activity

Observations established that the strength of the polar magnetic field during solar minima correlates with the next sunspot maximum (Schatten et al. 1978). The mean-field dynamo model explains this relationship. It also predicts a correlation between the strength of the zonal acceleration at the base of the convection zone at high latitudes (where according to the model and interpretation of the helioseismic measurements, the solar cycle originates) and the amplitude of the following solar cycle (Pipin and Kosovichev 2020). The predicted correlation depends on the mechanism that causes the variations of the solar cycles, which is not yet established. Therefore, two types of dynamo models were considered: models with regular variations of the α-effect, and models with stochastic fluctuations, simulating 'long-memory' and 'short-memory'

A. G. Kosovichev *et al.*

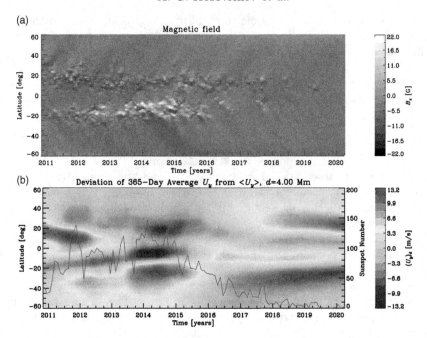

Figure 11. Time-latitude diagrams: a) monthly averaged Bz, b) time-latitude diagrams representing the deviation of the 365-day running average of the meridional velocity, Uy, from its mean over the whole interval at depth 4 Mm.

variations of magnetic activity. It is found that torsional oscillation properties, such as the zonal acceleration, correlate with the magnitude of the subsequent cycles with a time lag of 11–20 yr. The correlation sign and the time-lag depend on the depth and latitude where the torsional oscillations are measured and on the properties of the long-term variations of the dynamo cycles. The strongest correlations with the future cycles are found for the zonal acceleration at the base of the convection zone at high (~ 60 deg) latitudes. The modeling results demonstrate that helioseismic observations of the torsional oscillations can be useful for advanced prediction of the solar cycles, 1-2 sunspot cycles ahead. However, uncertainties remain because the mechanism of the solar cycles variations is not known, and the results for short- and long-memory models differ. The continuing helioseismic observations will allow us to test these predictions.

The extended solar cycle of the migrating meridional flow pattern is an important model prediction. It can be tested using already available local helioseismology data. Previous data analyses found variations of the meridional circulation associated with near-surface flows converging around active regions. However, the model predicts that such variations of the meridional circulation are part of the global dynamo processes. Similarly to the torsional oscillations, they extend through periods of low sunspot activity, forming the extended solar cycle pattern (Fig. 10d).

This prediction is confirmed by analysis of subsurface flow maps obtained by the Time-Distance (Getling et al. 2021) and Ring-Diagram (Komm 2021) techniques. Figure 11 shows the time-latitude diagram of the surface magnetic field and the subsurface South-North velocity variations, obtained after subtracting the mean velocity at 4 Mm depth from May 2010 to September 2020. The velocity variations change the sign across the equator, forming bands of flows converging towards the zones of sunspot formation during Solar Cycle 24, in 2010–16. The flow migration pattern with latitude is similar to the magnetic butterfly diagram. Still, it continues (with a slight latitudinal shift) in 2016–20 when there was no significant surface magnetic activity. Thus, the variations of the meridional circulation form the extended solar-cycle pattern, which is very similar to the extended cycle of the torsional oscillations. Figure 11b reveals the flow velocity increases during periods of high solar activity. These enhancements correspond to the previously discovered converging flows around active regions. Thus, the variations of the meridional circulation consist of two components: flows converging around active regions and a 22-year extended solar-cycle component associated with the solar dynamo. Due to the North-South asymmetry of active regions, the first component is asymmetric (causing apparent cross-equatorial meridional flows). The second component is mostly North-South symmetrical. Such behavior suggests that the global dynamo processes are mostly North-South symmetrical, and that the hemispheric asymmetry of active regions on the surface is associated with the process of magnetic flux emergence.

8. Discussion

In this paper, we presented some examples of computational heliophysics models that help understand the complex turbulent dynamics inside the Sun and the origin of the Sun's magnetic activity and cycles. The processes beneath the visible surface of the Sun can be observed only indirectly by helioseismic techniques. However, the helioseismic inversion of noisy measurements of oscillation frequencies and acoustic travel times are intrinsically ill-posed and cannot provide a unique solution. The computational solar acoustics, presented in Sec. 3, provides an important tool for directly testing theoretical models of the solar dynamics by forward modeling the observational data. In particular, it has shown that the current travel-time measurements are consistent with the single- and double-cell structure of the meridional circulation and that the helioseismic data can establish only an upper constraint on the potentially existing deep secondary cell. This result shifts the focus of the current debates on the meridional circulation structure. Further development of the forward helioseismic modeling will help to determine the strength and spectrum of subsurface convective flows, the structure of the tachocline and its variations with the solar cycle, the evolution of the torsional oscillations and their relationship to internal magnetic fields, and solve other critical problems of solar activity.

Measurements of the internal rotation of the Sun during the last two solar cycles revealed zones of fast and slow rotation ('torsional oscillations'), the patterns of which migrating with latitude and depth form 'extended' 22-cycles. Our analysis showed that the photospheric magnetic field distribution on the time-latitude diagrams coincides with the regions of zonal deceleration near the surface. This suggests that the tracking of the zonal deceleration in the convection zone may reveal the evolution of internal magnetic fields. Indeed this tracking showed migrating patterns of the zonal deceleration resembling dynamo waves predicted by Parker's dynamo theory (Parker 1955). Our results also showed that the near-surface rotational shear layer plays a crucial role in the formation of the magnetic butterfly diagram ('shaping the solar cycle'), as suggested by Brandenburg (2005).

Understanding the observed dynamics can be achieved only through detailed computational models. Unfortunately, the currently available computational resources do not allow us to perform realistic MHD simulations of the whole convection zone from the tachocline to the solar surface. Most global simulations are performed in the anelastic approximation, which neglects the flow compressibility and excludes the surface and near-surface layers. In addition, the resolution of such simulations is not sufficient for describing the whole spectrum of essential turbulent scales. The inability to model the multi-scale turbulent convection is probably the primary reason why these models do not reproduce the observed differential rotation, meridional circulation, and the butterfly diagram. Nevertheless, they provide insight into the complex 3D interaction of turbulent flows and magnetic fields in the highly stratified and rotating convection zone. In particular, we presented the simulation results using a computational model which attempts to describe unresolved turbulent scales using the so-called Implicit Large-Eddy Simulations (ILES) approach. The model reproduces the solar-type differential rotation, tachocline formation, and the near-surface shear layer, albeit only qualitatively.

The quantitative agreement with observations can be obtained in the mean-field MHD approximation, which separates small-scale turbulence from large-scale flows and magnetic fields. In this approximation, the equations for fluctuating turbulent properties are solved, assuming that the large-scale flows and magnetic fields are slowly evolving, and prescribing the turbulence spectrum. The turbulent transport coefficients are then expressed in terms of the large-scale parameters and substituted in the equations describing the large-scale evolution. The mean-field theory has been developed to a high degree of sophistication. In addition to the usual turbulent diffusion transport, it predicts non-diffusive transport of magnetic field and momentum, so-called alpha- and Lambda-effects. The non-diffusive turbulent transport explains the differential rotation and the cyclic evolution of dynamo-generated magnetic fields. With an appropriate choice of free parameters describing the strength and spectrum of the small-scale turbulence, the mean-field models reproduce the observed differential rotation, meridional circulation, and the global magnetic field evolution.

We presented a recently-developed 2D mean-field MHD model, which, in addition, reproduced the evolution of the migrating zonal flows–torsional oscillations and zonal acceleration in close agreement with the observations. It is essential that the model reproduced and explained the 22-year cyclic evolution of the torsional oscillations, which had been a puzzle for many years. The model analysis showed the 'extended' cycle is caused by magnetic quenching of the turbulent heat transport, which affects the meridional circulation and the angular-momentum balance. Remarkably, the model predicted the extended cycle of the meridional circulations, confirmed by the helioseismic observations.

Currently, among the computational models, only the mean-field MHD model can reproduce the basic features of the solar dynamics and activity. However, the limitation of this model is that it is two-dimensional and thus cannot describe magnetic field structures emerging on the solar surface in the form of compact bipolar active regions. The formation and evolution of these structures may significantly affect the global evolution of magnetic fields and subsurface flows. Therefore, initial 3D mean-field MHD models describing emerging-flux effects are being developed (Pipin 2021).

Further progress in our understanding of the global dynamics and dynamo of the Sun will be based on a synergy of helioseismic observations with theoretical and computational models. The computational solar acoustics can improve the accuracy of helioseismic inferences of the differential rotation, meridional circulation, and large-scale subsurface flows and their solar-cycle variations. It also provides a tool for the forward modeling of helioseismic observables and direct testing of MHD models. The 3D MHD modeling will be focused on a better understanding of the multi-scale coupling of the turbulent and large-scale flows and fields, using helioseismic constraints and advantages of the mean-field theory and computational models.

Acknowledgments. The work was partially supported by NASA grants NNX14AB70G, 80NSSC20K1320, and 80NSSC20K0602.

References

Bekki, Y. and Yokoyama, T. 2017, ApJ **835**(1), 9

Böning, V. G. A., Roth, M., Jackiewicz, J., and Kholikov, S. 2017, ApJ **845**(1), 2

Brandenburg, A. 2005, ApJ **625(1)**, 539

Brandenburg, A., Moss, D., and Tuominen, I. 1992, A&A **265**, 328

Charbonneau, P. and Smolarkiewicz, P. K. 2013, *Science* **340(6128)**, 42

Chen, R. and Zhao, J. 2017, ApJ **849(2)**, 144

Chen, R. and Zhao, J. 2018, ApJ **853(2)**, 161

Durney, B. R. 1999, ApJ **511**, 945

Duvall, T. L., J., Jefferies, S. M., Harvey, J. W., and Pomerantz, M. A. 1993, *Nat* **362(6419)**, 430

Getling, A. V., Kosovichev, A. G., and Zhao, J. 2021, ApJ **908(2)**, L50

Ghizaru, M., Charbonneau, P., and Smolarkiewicz, P. K. 2010, ApJ **715(2)**, L133

Gizon, L., Cameron, R. H., Pourabdian, M., Liang, Z.-C., Fournier, D., Birch, A. C., and Hanson, C. S. 2020a, *Science* **368(6498)**, 1469

Gizon, L., Cameron, R. H., Pourabdian, M., Liang, Z.-C., Fournier, D., Birch, A. C., and Hanson, C. S. 2020b, *Science* **368(6498)**, 1469

Gough, D. O. 1969, *Journal of Atmospheric Sciences* **26(3)**, 448

Gough, D. O. and Toomre, J. 1983, Sol. Phys. **82(1-2)**, 401

Guerrero, G., Smolarkiewicz, P. K., de Gouveia Dal Pino, E. M., Kosovichev, A. G., and Mansour, N. N. 2016, ApJ **819(2)**, 104

Guerrero, G., Smolarkiewicz, P. K., Kosovichev, A. G., and Mansour, N. N. 2013, ApJ **779(2)**, 176

Guerrero, G., Zaire, B., Smolarkiewicz, P. K., de Gouveia Dal Pino, E. M., Kosovichev, A. G., and Mansour, N. N. 2019, ApJ **880(1)**, 6

Hartlep, T., Zhao, J., Kosovichev, A. G., and Mansour, N. N. 2013, ApJ **762(2)**, 132

Hartlep, T., Zhao, J., Mansour, N. N., and Kosovichev, A. G. 2008, ApJ **689(2)**, 1373

Hill, F. 1989, ApJ **343**, L69

Jackiewicz, J., Serebryanskiy, A., and Kholikov, S. 2015, ApJ **805(2)**, 133

Kholikov, S., Serebryanskiy, A., and Jackiewicz, J. 2014, ApJ **784(2)**, 145

Khomenko, E., Kosovichev, A., Collados, M., Parchevsky, K., and Olshevsky, V. 2009, ApJ **694(1)**, 411

Kitchatinov, L. L., Pipin, V. V., and Ruediger, G. 1994, *Astronomische Nachrichten* **315**, 157

Kitiashvili, I. N., Kosovichev, A. G., Mansour, N. N., and Wray, A. A. 2015, ApJ **809(1)**, 84

Komm, R. 2021, Sol. Phys. **296(12)**, 174

Komm, R., González Hernández, I., Hill, F., Bogart, R., Rabello-Soares, M. C., and Haber, D. 2012, Sol. Phys. p. 177

Kosovichev, A. G. and Duvall, T. L., J. 1997, in F. P. Pijpers, J. Christensen-Dalsgaard, and C. S. Rosenthal (eds.), *SCORe'96 : Solar Convection and Oscillations and their Relationship*, Vol. 225 of *Astrophysics and Space Science Library*, pp 241–260

Kosovichev, A. G. and Pipin, V. V. 2019, ApJ **871(2)**, L20

Kosovichev, A. G. and Zhao, J. 2016, in J.-P. Rozelot and C. Neiner (eds.), *Lecture Notes in Physics, Berlin Springer Verlag*, Vol. 914, p. 25

Krause, F. and Rädler, K.-H. 1980, *Mean-Field Magnetohydrodynamics and Dynamo Theory*, Berlin: Akademie-Verlag

Lin, C.-H. and Chou, D.-Y. 2018, ApJ **860(1)**, 48

Miesch, M. S., Brown, B. P., Browning, M. K., Brun, A. S., and Toomre, J. 2011, in N. H. Brummell, A. S. Brun, M. S. Miesch, and Y. Ponty (eds.), *IAU Symposium*, Vol. 271 of *IAU Symposium*, pp 261–269

Parchevsky, K. V. and Kosovichev, A. G. 2009, ApJ **694(1)**, 573

Parchevsky, K. V., Zhao, J., Hartlep, T., and Kosovichev, A. G. 2014, ApJ **785(1)**, 40

Parker, E. N. 1955, ApJ **122**, 293

Pipin, V. V. 2004, *Astronomy Reports* **48**, 418

Pipin, V. V. 2021, *arXiv e-prints* p. arXiv:2112.09460

Pipin, V. V. and Kitchatinov, L. L. 2000, *Astronomy Reports* **44**, 771

Pipin, V. V. and Kosovichev, A. G. 2011, ApJ **727(2)**, L45

Pipin, V. V. and Kosovichev, A. G. 2018, ApJ **854(1)**, 67

Pipin, V. V. and Kosovichev, A. G. 2019, ApJ **887(2)**, 215

Pipin, V. V. and Kosovichev, A. G. 2020, ApJ **900(1)**, 26

Reinecke, M. and Seljebotn, D. S. 2013, A&A **554**, A112

Rempel, M. 2006, ApJ **647**, 662

Rempel, M., Schüssler, M., and Knölker, M. 2009, ApJ **691(1)**, 640

Russell, C. T., Luhmann, J. G., and Jian, L. K. 2020, in A. Kosovichev, S. Strassmeier, and M. Jardine (eds.), *Solar and Stellar Magnetic Fields: Origins and Manifestations*, Vol. 354, pp 127–133

Schad, A., Timmer, J., and Roth, M. 2013, ApJ **778(2)**, L38

Schatten, K. H., Scherrer, P. H., Svalgaard, L., and Wilcox, J. M. 1978, Geophys. Res. Lett. **5(5)**, 411

Schou, J., Antia, H. M., Basu, S., Bogart, R. S., Bush, R. I., Chitre, S. M., Christensen-Dalsgaard, J., Di Mauro, M. P., Dziembowski, W. A., Eff-Darwich, A., Gough, D. O., Haber, D. A., Hoeksema, J. T., Howe, R., Korzennik, S. G., Kosovichev, A. G., Larsen, R. M., Pijpers, F. P., Scherrer, P. H., Sekii, T., Tarbell, T. D., Title, A. M., Thompson, M. J., and Toomre, J. 1998, ApJ **505(1)**, 390

Simitev, R. D., Kosovichev, A. G., and Busse, F. H. 2015, ApJ **810(1)**, 80

Smolarkiewicz, P. K. 2006, *International Journal for Numerical Methods in Fluids* **50(10)**, 1123

Smolarkiewicz, P. K. and Charbonneau, P. 2013, *Journal of Computational Physics* **236**, 608

Stein, R. F. and Nordlund, Å. 2012, ApJ **753**, L13

Stejko, A. M., Kosovichev, A. G., and Mansour, N. N. 2021a, ApJS **253(1)**, 9

Stejko, A. M., Kosovichev, A. G., and Pipin, V. V. 2021b, ApJ **911(2)**, 90

Stenflo, J. O. 2012, A&A **547**, A93

Zhao, J., Bogart, R. S., Kosovichev, A. G., Duvall, T. L., J., and Hartlep, T. 2013, ApJ **774(2)**, L29

Zhao, J., Kosovichev, A. G., and Bogart, R. S. 2014, ApJ **789(1)**, L7

The Predictive Power of Computational Astrophysics as a Discovery Tool
Proceedings IAU Symposium No. 362, 2023
D. Bisikalo, D. Wiebe & C. Boily, eds.
doi:10.1017/S1743921322001405

From evolved Long-Period-Variable stars to the evolution of M31

Maryam Torki[1][iD]**, Mahdieh Navabi[1], Atefeh Javadi[1]**[iD]**,**
Elham Saremi[1], Jacco Th. van Loon[2] and Sepideh Ghaziasgar[1]

[1]School of Astronomy, Institute for Research in Fundamental Sciences (IPM), Tehran,

[2]Lennard-Jones Labratories, Keele University, ST5 5BG, UK

Abstract. One of the ways to understand the genesis and evolution of the universe is to know how galaxies have formed and evolved. In this regard, the study of star formation history (SFH) plays an important role in the accurate understanding of galaxies. In this paper, we used long-period variable stars (LPVs) for estimating the SFH in the Andromeda galaxy (M31). These cool stars reach their peak luminosity in the final stage of their evolution also their birth mass is directly related to their luminosity. Therefore, using stellar evolution models, we construct the mass function and hence the star formation history.

Keywords. stars: AGB and post-AGB - stars: luminosity function, mass function - galaxies: evolution - galaxies: formation - galaxies: individual: M31 - galaxies: stellar content - galaxies: structure

---

1. Introduction

Andromeda as the closest spiral giant galaxy offers us a unique opportunity to know how its various components have formed and evolved. M31 is located at 785 ± 25 kpc, $(m - M)_o = 24.47$ mag (McConnachie et al. 2004) and has low foreground reddening E(B-V)=0.06 mag. Our approach to investigating the star formation history (SFH) is based on employing long-period variable stars (LPVs) which we have successfully applied to other galaxies in the Local Group such as M33 (Javadi et al. 2011a,b, 2017), Magellanic Clouds (Rezaeikh et al. 2014), NGC147 and NGC185 (Golshan et al. 2017), IC1613 (Hashemi et al. 2019), Andromeda VII (Navabi et al. 2020, 2021) and Andromeda I (Saremi et al. 2021). Asymptotic giant branch (AGB) and red super giants (RSGs) are basic pillars in our research.

2. DATA and Technique of Star Formation

In this paper, we used the catalogue of LPVs in the M31 from (Mould et al. 2004) to determine star formation (Torki et al. 2019). It should be noted that their results included near-infrared photometry of almost 2000 variables which most of them were AGBs. We used the Padova stellar evolutionary models (Marigo et al. 2017) to obtain SFH and assuming that the metallicity is constant over time, we were able to calculate the mass, age, and pulsation duration of the stars. The SFH is described by the star formation rate (SFR) and conveys the concept of how much gas has become a star each year in the past. In fact, SFR, ξ (in $M_\odot$ yr^{-1}) is a function of time and estimated by:

$$\xi(t) = \frac{\int_{\min}^{\max} f_{\mathrm{IMF}}(m)m \ dm}{\int_{m(t)}^{m(t+dt)} f_{\mathrm{IMF}}(m) \ dm} \frac{dn'(t)}{\delta t}. \tag{2.1}$$

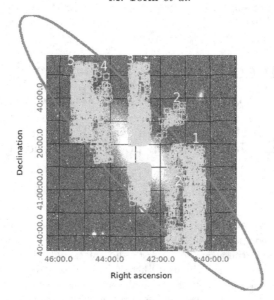

Figure 1. The 2MASS image of the M31 galaxy. The green dots indicate the LPVs from Mould et al. (2004). The magenta ellipse shows the twice half-light radius.

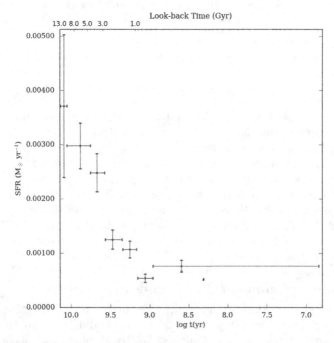

Figure 2. The SFH of the M31 galaxy for constant metallicity of Z = 0.008 based on data in box 1. The horizontal "error bars" show the spread in age within each bin and the statistical errors are presented in the vertical "error bars".

where the m is mass, f_{IMF} is the Initial Mass Function (IMF) (Kroupa 2001). The minimum and maximum mass in Kroupa's IMF are considered 0.02 and 200 M $_{\odot}$ respectively. In the above equation, it is assumed that stars with mass between $m(t)$ and $m(t + dt)$ (t is look back time) are LPVs at the present time, and δt is the duration of the star oscillation which the number of variables observed ($dn'(t)$) in intervals t and $t + dt$ depends on it.

3. Results

The LPVs are scattered in 5 rectangular boxes (Figure 1). In this paper, we estimated the SFH of the M31 galaxy during the broad time interval from 11.8 Gyr ($\log t = 10.07$) to 400 Myr ($\log t = 8.59$) ago in Box 1, which has dimensions of 1.49 kpc×13.09 kpc. The SFR as a function of look-back time in M31 is shown in Figure 2. The horizontal "error bars" correspond to the start and end time for which the SFR was calculated in that bin. We see a peak of star formation at 11.8 Gyr ago, after that, the star formation begins to decrease until the present-day.

References

Golshan, R. H. et al. 2017, MNRAS, 466, 1764
Hashemi, S.A. et al. 2019, MNRAS, 483, 4751
Javadi, A. et al. 2011a, ASPC, 445, 497
Javadi, A. et al. 2011b, MNRAS, 414, 3394
Javadi, A. et al. 2017, MNRAS, 464, 2103
Marigo P. et al., 2017, ApJ, 835, 19
Mould, J., et al. 2004, ApJ, 154, 623
McConnachie, A. W. et al. 2004, MNRAS, 350, 243
Rezaeikh, S. et al. 2014, MNRAS, 445, 2214
Navabi, M. et al. 2020, Proc.conf. Stars and their Variability Observed from Space, 383, 385
Navabi, M. et al. 2021, ApJ, 910, 127
Saremi, E. et al. 2021, ApJ, 923, 164
Torki, M. et al. 2019, Proceedings of IAU Symposium, 343, 512

The Predictive Power of Computational Astrophysics as a Discovery Tool
Proceedings IAU Symposium No. 362, 2023
D. Bisikalo, D. Wiebe & C. Boily, eds.
doi:10.1017/S1743921322001582

Applicability of the Bulirsch-Stoer algorithm in the circular restricted three-body problem

Tatiana Demidova

Crimean Astrophysical Observatory of Russian Academy of Sciences
p. Nauchny, Bakhchisaray, Crimea, Russia, 298409
email: `proxima1@list.ru`

Abstract. The dynamics of massless planetesimals in the gravitational field of a star with a planet in a circular orbit is considered. The invariant of this problem is the Jacobi integral. Preserving the value of the Jacobi integral can be a test for numerical algorithms solving the equation of motion. The invariant changes for particles in the planetary chaotic zone due to numerical errors that occur during close encounters with the planet. The limiting distances from the planet, upon reaching which the value of the Jacobi integral changes, are determined for Bulirsch-Stoer algorithm. It is shown that violation of the Jacobi integral can be used to define the boundaries of the planetary chaotic zone.

Keywords. Bulirsch-Stoer algorithm, circular restricted three-body problem, simulation

---

The size of the planetary chaotic zone depends on the mass of the planet (Wisdom (1980)). The asymmetry of the zone with respect to the planet's orbit was shown in Morrison & Malhotra (2015). The stepped form of the dependence was determined in Demidova & Shevchenko (2020). Over time, the zone is freed from the matter. Therefore, the dependence can be used to interpret observations of debris disks in the presence of a matter free gap and to determine the mass of an invisible planet. Thus, the study of the structure of the planetary chaotic zone remains an urgent task.

Since the debris disks are flat and low-mass structures, it is appropriate to carry out calculations in the approximation of the plane restricted three-body problem. Energy and angular momentum are not conserved in the problem (Murray & Dermott (1999)). But there is an invariant called the Jacobi integral and defined as follows:

$$C_J = 2\left(\frac{GM_*}{R} + \frac{m_p}{r}\right) + 2n(r\dot{y} - y\dot{x}) - \dot{x}^2 - \dot{y}^2, \qquad (0.1)$$

where M_*, m_p are masses of a star and a planet, n is the mean motion of the planet, G is the gravitation constant, R, r are distances from a particle to the star and the planet, x, y and $\dot{x}, \dot{y}$ are coordinates and velocities of the particle. When one simulates the dynamics of a swarm of particles in the gravitational field of a single star with a planet, the value of the Jacobi integral must be constant throughout the calculations.

At the initial moment of time, the star and the planet are located on the same axis with the distance of a_p, and 10^4 particles are equally distributed along the radius within $[-4R_h, 4R_h]$, where $R_h = \left(\frac{m_p}{3M_*}\right)^{1/3} a_p$ is Hill radius, along one line inclined at an angle ϕ to the axis. The particle with coordinates $(x, y) = (r \cdot cos\phi, r \cdot sin\phi)$ has velocity components $(v_x, v_y) = (-[G(M+m)/r]^{1/2} sin\phi, [G(M+m)/r]^{1/2} cos\phi)$. The calculations were performed in a rectangular barocentric non-rotating coordinate system during 10^4 of

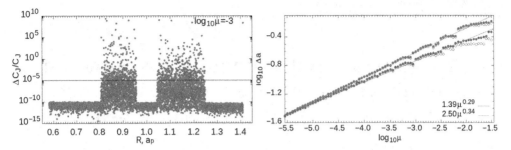

Figure 1. On the left is relative deviation of the value of the Jacobi integral from the initial value after 10^4 periods of the planet depending on the initial position of the particles for the case $\phi = 180°$. On the right the width of the inner (blue) and outer (red) chaotic zone for $\phi = 90°$ (filled circles) and $\phi = 180°$ (open circles), depending on the ratio of the masses of the planet and the star. The approximations of dependence for $\phi = 90°$ are in right bottom corner of the picture.

the planet convolutions. To solve the differential equations, we used Bulirsch-Stoer algorithm (Press *et al.* (1992)) with a fractional truncation error tolerance of $\epsilon = 10^{-14}$. During this time, the major semi-axis of the planet is preserved with an error of less than 10^{-10}.

Calculations have shown that with close approaches of a particle to a planet, the value of the Jacobi integral can change. It can be seen in Fig. 1 that the values of the Jacobi integral vary substantially in the vicinity of the planet's orbit, where, due to the overlap of the first-order mean motion resonances, there is a chaotic zone (Wisdom (1980)). The violations of the Jacobi integral for some particles were also identified in Morrison & Malhotra (2015).

It is interesting that the position of the boundaries of the chaotic zone can be traced from the violations of the Jacobi integral. The boundary positions of the particles were determined if the relative violation of the Jacobi integral was $> 10^{-5}$. The results show that the position of the boundaries of the chaotic zone is weak, but depends on ϕ, especially for the case of large planet masses. The averaged dependences of the size of the inner part of the chaotic zone on the ratio of the masses of the planet and the star (μ) is $\Delta a_{in} = 1.39\mu^{0.29}a_p$, and of the outer one is $\Delta a_{out} = 2.5\mu^{0.34}a_p$ (Fig. 1). They are close to those obtained in Morrison & Malhotra (2015); Demidova & Shevchenko (2020).

Thus, when solving the restricted three-body problem using the Bulirsch-Stoer method, it is necessary to set the accretion radius around the planet. Particles entering this radius should be considered to have left the system, because the Bulirsch-Stoer integrator failed in case of close encounters. The maximum deviation of the Jacobi integral from the initial value depends on the accretion radius. The calculations show if the accretion radius does not exceed $0.01R_h$ the maximum deviation of the integral is $\sim 10^{-10}$. It should be noted that maximum of the distribution of the physical radius of the planets is about $5 \cdot 10^{-2}R_h$ (Morrison & Malhotra (2015)). So, it is good enough for simulation if the accretion radius equals to the planet radius.

References

Wisdom, J. 1980, *AJ*, 85, 1122

Morrison, S., & Malhotra, R. 2015, *ApJ*, 799, 41.

Demidova, T. & Shevchenko, I.I. 2020, *Astron. Lett.*, 46, 774.

Murray, C.D. & Dermott, S.F. *Solar system dynamics* (Cambridge University Press), p. 69.

Press, W.H., Teukolsky, S.A., Vetterling, W.T., & Flannery, B.P. 1992, *Numerical recipes in C. The art of scientific computing* (Cambridge University Press), p. 724

The Predictive Power of Computational Astrophysics as a Discovery Tool
Proceedings IAU Symposium No. 362, 2023
D. Bisikalo, D. Wiebe & C. Boily, eds.
doi:10.1017/S1743921322001429

A radiation hydrodynamics scheme on adaptive meshes using the Variable Eddington Tensor (VET) closure

Shyam H. Menon[1] , **Christoph Federrath**[1,2] , **Mark R. Krumholz**[1,2],
Rolf Kuiper[3], **Benjamin D. Wibking**[1,2] **and Manuel Jung**[4]

[1]Research School of Astronomy and Astrophysics, Australian National University, Canberra, ACT 2611, Australia

[2]ARC Centre of Excellence for Astronomy in Three Dimensions (ASTRO-3D), Canberra, ACT 2611, Australia

[3]Zentrum für Astronomie der Universität Heidelberg, Institut für Theoretische Astrophysik, Albert-Ueberle-Straße 2, 69120 Heidelberg, Germany

[4]Hamburger Sternwarte, Universität Hamburg, Gojenbergsweg 112, 21029 Hamburg, Germany
email: shyam.menon@anu.edu.au

Abstract. We present a new algorithm to solve the equations of radiation hydrodynamics (RHD) in a frequency-integrated, two-moment formulation. Novel features of the algorithm include i) the adoption of a non-local Variable Eddington Tensor (VET) closure for the radiation moment equations, computed with a ray-tracing method, ii) support for adaptive mesh refinement (AMR), iii) use of a time-implicit Godunov method for the hyperbolic transport of radiation, and iv) a fixed-point Picard iteration scheme to accurately handle the stiff nonlinear gas-radiation energy exchange. Tests demonstrate that our scheme works correctly, yields accurate rates of energy and momentum transfer between gas and radiation, and obtains the correct radiation field distribution even in situations where more commonly used – but less accurate – closure relations like the Flux-limited Diffusion and Moment-1 approximations fail. Our scheme presents an important step towards performing RHD simulations with increasing spatial and directional accuracy, effectively improving their predictive capabilities.

Keywords. radiative transfer, methods: numerical, radiation mechanisms: general, hydrodynamics

---

1. Introduction

Radiation hydrodynamics (RHD) plays a crucial role in the evolution of various astrophysical systems from stellar atmospheres (e.g., Mihalas 1978) to cosmological reionization (Gnedin and Ostriker 1997). There has been significant progress in recent years to develop the numerical algorithms that are required to solve the stiff, coupled equations that govern these systems (see Dale 2015; Teyssier and Commerçon 2019 for recent reviews). There are well-known fundamental difficulties associated with numerically solving the RHD equations, one of which is the multidimensional nature of the radiation intensity – a function of space, time, frequency, and angular direction – that is governed by the radiative transfer equation. A common approach to circumvent this difficulty is to integrate the RT equation over all frequencies and angles to obtain the gray radiation moment equations, reducing the dimensionality of the RHD system (Mihalas and Klein 1982; Castor 2004). However, this introduces the need for an extra closure equation to estimate the moments of the radiation intensity, whose evolution is not explicitly computed.

One commonly-used approximate closure is the flux-limited diffusion (FLD) method, which closes the equations at the first moment (the radiation flux), and uses the Eddington approximation, i.e., the assumption that the Eddington tensor is locally isotropic (Levermore and Pomraning 1981). However, this method often suffers from inaccuracies in the optically thin regime, or when a mixture of low- and high-opacity gas is present. A more accurate closure scheme that has recently been adopted widely is the M_1 closure (e.g., Skinner and Ostriker 2013; Wibking and Krumholz 2021), which adopts a local closure relation for the radiation pressure tensor, or equivalently the Eddington tensor, in terms of the local radiation energy density and flux (Levermore 1984). While the M_1 closure can handle transitions in optical depths for a single beam of radiation, it fails for other non-trivial geometrical distributions of radiation sources.

In the algorithm described in this paper, we use the so-called Variable Eddington Tensor (VET) scheme (e.g., Stone et al. 1992; Jiang et al. 2012), a non-local scheme that does not adopt a closure relation or model *a priori*, but rather computes the Eddington tensor self-consistently through a formal solution of the time-independent RT equation along discrete rays using a ray-tracing approach (e.g., Davis et al. 2012). The self-consistently computed closure is combined with the radiation moment equations to solve for the radiation quantities. While more computationally expensive due to the required non-local ray-trace solution and its associated communication overheads, the VET approach does not face the shortcomings of the more approximate closure models discussed above (Jiang et al. 2012). Below, we describe the equations solved by our scheme, which we couple to the **FLASH** (magneto-)hydrodynamics code, outline the brief features of our algorithm, and conclude by demonstrating the novel advantages offered by our scheme.

2. Equations

We solve the equations of non-relativistic gray (frequency-integrated) RHD in conservative form, written in the mixed-frame formulation, i.e., where the moments of the radiation intensity are written in the lab frame, and the opacities are written in the comoving frame (e.g., Mihalas and Klein 1982). We neglect scattering for simplicity, and assume the matter is always in local thermodynamic equilibrium (LTE), and treat the material property coefficients as isotropic in the comoving frame. The equations solved in our scheme are the equations of (magneto-)hydrodynamics along with the radiation moment equations, given by

$$\frac{\partial E_r}{\partial t} + \nabla \cdot \mathbf{F}_r = -cG^0 \tag{2.1}$$

$$\frac{\partial \mathbf{F}_r}{\partial t} + \nabla \cdot (c^2 E_r \mathbb{T}) = -c^2 \mathbf{G}, \tag{2.2}$$

and

$$G^0 = \rho \kappa_E E_r - \rho \kappa_P a_r T^4 + \rho (\kappa_F - 2\kappa_E) \frac{\mathbf{v} \cdot \mathbf{F}_r}{c^2} + \rho (\kappa_E - \kappa_F) \left[\frac{v^2}{c^2} E_r + \frac{\mathbf{vv}}{c^2} : \mathbb{P}_r \right], \tag{2.3}$$

and

$$\mathbf{G} = \rho \kappa_R \frac{\mathbf{F}_r}{c} - \rho \kappa_R E_r \frac{\mathbf{v}}{c} \cdot (\mathbb{I} + \mathbb{T}), \tag{2.4}$$

are the time-like and space-like parts of the specific radiation four-force density for a direction-independent flux spectrum to leading order in all regimes. Source terms of magnitude cG^0 and $\mathbf{G}$ are added to the gas energy and energy momentum equations, respectively, to ensure the conservation of total (radiation + gas) energy and momentum.

In the above equations ρ is the mass density, $\mathbf{v}$ the gas velocity, T the gas temperature, $\mathbb{I}$ the identity matrix, c the speed of light in vacuum, and a_r the radiation constant. E_r is the lab-frame radiation energy density, $\mathbf{F}_r$ the lab-frame radiation momentum density, and $\mathbb{P}_r$ is the lab-frame radiation pressure tensor. These represent the zeroth, first and second (gray) angular moments of the radiation intensity, respectively. The material opacities κ_E and κ_F are the gray energy- and flux-mean opacities in the comoving frame, which we set equal to κ_P, the Planck mean opacity, and κ_R, the Rosseland mean opacity, respectively. The radiation closure relation is used to close the above system of equations, and is of the form

$$\mathbb{P}_r = \mathbb{T}E_r, \tag{2.5}$$

where $\mathbb{T}$ is the Eddington Tensor, given by the relation between the second and zeroth moments of the gray radiation intensity $I_r(\hat{\mathbf{n}}_k)$ travelling in direction $\hat{\mathbf{n}}_k$. I_r is computed independently from a formal solution of the time-independent radiative transfer equation

$$\frac{\partial I_r}{\partial s} = \kappa_P(S - I_r), \tag{2.6}$$

where S is the source function, which, for the purposes of modelling the emission from dust grains, we set equal to the frequency-integrated Planck function $B(T) = ca_R T^4/(4\pi)$. This independent solution is used to compute the angular moments $\mathbb{P}_r$ and E_r, to obtain the corresponding $\mathbb{T}$, which is then used in the radiation moment equations.

3. Numerical Scheme

In our scheme, we operator-split the hyperbolic hydrodynamic subsystem of equations from the radiation moment equations and treat the two separately. To evolve the hyperbolic transport for the gas quantities, we use the existing hydrodynamic solver capabilities in the FLASH code. The hyperbolic transport of E_r and $\mathbf{F}_r$ is done using a piecewise constant (first-order) finite-volume Godunov method using a Harten-Lax-van Leer (HLL)-type Riemann solver, similar to the scheme described in Jiang et al. (2012). The hyperbolic fluxes and source terms are discretised in an implicit backward-Euler fashion, to permit the radiation quantities, which are governed by the light-crossing signal speed, to be evolved at the significantly larger hydrodynamic timescale. We also introduce a new, modified characteristic wavespeed criterion for the Riemann solver on AMR grids. Our criterion extends the condition described in Jiang et al. (2013), which performs an optical depth-dependent correction to obtain accurate solutions in the optically thick regime, to AMR grids. The Eddington tensor $\mathbb{T}$, required to solve the radiation moment equations, is pre-computed at the start of the radiation step using a solution of Equation 2.6, which we obtain using the hybrid characteristics ray-tracer described in Buntemeyer et al. (2016). In addition, our scheme adopts a fixed-point Picard iteration method to couple the stiff radiation-gas nonlinear term. This is achieved by using an estimate for the gas temperature (T_*) in the radiation moment equations, based on the corresponding solution of E_r and $\mathbf{F}_r$ to obtain an updated value of T_* from the gas internal energy equation, and iterating the two stages to convergence. The converged values of E_r and $\mathbf{F}_r$ are used to apply the radiation source terms to the gas energy and momentum densities, respectively. This series of steps summarizes the working of the algorithm in each simulation timestep.

4. Tests

We verified that our scheme works correctly by performing a suite of numerical tests that have analytical/semi-analytical relations to compare with. Due to limited space, we only reproduce two tests here, which demonstrate the novel advantages offered by our scheme - i.e., i) the spatial accuracy – with reduced computational costs – offered by the support for AMR grids, and ii) the accuracy in the radiation field distribution for general problems, offered by the VET closure. All other tests, including those that specifically test the hyperbolic transport in different optical depth regimes, radiation-gas energy exchange, and other standard tests, are outlined in Menon et al. (2022).

4.1. *Radiation Pressure Driven Shell Expansion*

This test simulates the radiation pressure-driven expansion of a thin, dusty, spherical shell as given in Skinner and Ostriker (2013). At $t = 0$, a radial (r) density profile, representing a thin dusty shell of gas, is initialized as

$$\rho_{sh}(r) = \frac{M_{sh}}{4\pi r^2 \sqrt{2\pi R_{sh}^2}} \exp\left(-\frac{r^2}{2R_{sh}^2}\right), \tag{4.1}$$

where M_{sh} is the gas mass in the thin shell, and $R_{sh} \equiv H/(2\sqrt{2\ln 2})$ is the half-width of the shell, where we adopt $H = 1.5\,\mathrm{pc}$. A central radiating source is introduced, given by

$$j_*(r) = \frac{L_*}{(2\pi R_*^2)^{3/2}} \exp\left(-\frac{r^2}{2R_*^2}\right), \tag{4.2}$$

where L_* is the luminosity of the source and R_* the size of the source. We set a value of $L_* = 1.989 \times 10^{42}\,\mathrm{erg\,s^{-1}}$, $R_* = 0.625\,\mathrm{pc}$, and a constant dust opacity of $\kappa_0 = 20\,\mathrm{cm^2\,g^{-1}}$. We use an isothermal equation of state for the thermal pressure, with the sound speed set to $a_0 = 2\,\mathrm{km\,s^{-1}}$, which corresponds to a gas temperature $T \sim 481\,\mathrm{K}$, assuming a mean particle mass $\mu = m_H$. The simulation is performed on a cubic domain with $x = y = z = [-10, 10]\,\mathrm{pc}$, with the source at $x = 0$, with outflow boundary conditions on the gas and radiation. We use AMR for this test, with a base resolution of 32^3 grid cells, and allow up to four levels of refinement, corresponding to an effective resolution of 256^3 grid cells. We also perform simulations on uniform grids of resolution 64^3, 128^3, and 256^3 grid cells, to study the dependence of shell evolution on resolution. We compare the obtained numerical results at these resolutions in Figure 1 with the semi-analytical ODE solution given in Skinner and Ostriker (2013). We see that the numerical solution agrees with the ODE solution quite well, and increasingly so at higher resolutions. We find similar accuracy in the 256^3 effective resolution AMR version and the 256^3 uniform run, however, the AMR run uses about 30% less CPU time than the uniform-grid run.

4.2. *Comparison of closure schemes: Advantage of VET*

Here we compare our new VET approach with the more commonly adopted local closures, Eddington (used by the FLD method) and M_1. We set up a test where we introduce two point-like sources of radiation, modeled with the Gaussian source function $j_*(r)$ given in Equation 4.2, where we use $L_* = 10\,L_\odot$ and $R_* = 54\,\mathrm{AU}$ for both the sources. We place these sources at 90^{deg} with respect to each other, at $(635, 0, 0)\,\mathrm{AU}$ and $(0, 635, 0)\,\mathrm{AU}$ respectively in a $(2000\,\mathrm{AU})^3$ computational domain. We place a dense clump of material at the center of the domain, with radius $267\,\mathrm{AU}$ and density $\rho_c = 3.89 \times 10^{-17}\,\mathrm{g\,cm^{-3}}$, and an optically thin ambient medium with density $\rho_a = 3.89 \times 10^{-20}\,\mathrm{g\,cm^{-3}}$ elsewhere. The gas temperature is spatially uniform with a value of $20\,\mathrm{K}$, and the opacity for the radiation field is set to $\kappa_P = \kappa_R = 100\,\mathrm{cm^2\,g^{-1}}$. We perform three

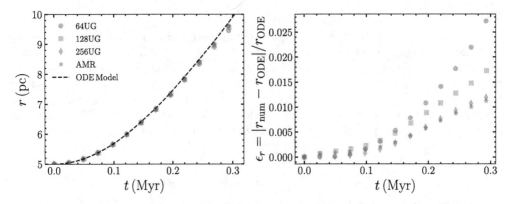

Figure 1. Numerical solution (left) for the radiation-driven thin spherical shell expansion test obtained on uniform grids (UG) with resolutions of 64^3, 128^3, and 256^3 cells, and on an AMR grid with effective resolution of 256^3 cells. Dashed lines indicate the semi-analytical ODE solution for the problem given in Skinner and Ostriker (2013), and the right panel shows the relative errors for the numerical runs. The simulations are converging with increasing resolution. All errors are $< 3\%$ for all relevant times, and the AMR solution is comparable to the 256^3 UG solution, but required 30% less computational time.

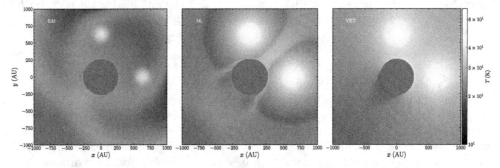

Figure 2. Comparison of the gas temperature fields obtained in the shadow test (Section 4.2) with Eddington (FLD), M_1 and VET closures. The simpler closure schemes produce unphysical solutions for this problem, whereas the VET reproduces the qualitatively correct solution, demonstrating the advantage of VET over other closures in semi-transparent RHD problems.

versions of this test, with only a difference of the adopted closure relation between the versions; corresponding to the Eddington (FLD) closure, the M_1 closure, and the VET closure, respectively. In Figure 2, we see that the Eddington and M_1 closures do not cast qualitatively correct shadows, whereas the VET does. This demonstrates that the VET ensures the consistent propagation of radiation in non-trivial geometrical distributions of diffuse radiation sources.

5. Summary

We have described an algorithm for solving the radiation hydrodynamics (RHD) equations, closed with a non-local Variable Eddington Tensor (VET), and coupled to the FLASH AMR (magneto-)hydrodynamics code (Menon et al. 2022). Using numerical tests, we show that our scheme works in concert with AMR, allowing for very high resolution applications, with adequate accuracy of the radiation field. Unlike the FLD and M_1 closures, our VET method casts shadows as expected for complex geometrical configurations of the gas and radiation field. To our knowledge, our method is the *first* VET closure-based RHD method with AMR support in the literature, and thus represents a step toward, improving the predictive capabilities of numerical RHD simulations.

References

Buntemeyer, L., Banerjee, R., Peters, T., Klassen, M., & Pudritz, R. E. 2016, Radiation hydrodynamics using characteristics on adaptive decomposed domains for massively parallel star formation simulations. *New Astronomy*, 43, 49–69.

Castor, J. I. 2004,. *Radiation Hydrodynamics*. Cambridge University Press.

Dale, J. E. 2015,. The modelling of feedback in star formation simulations.

Davis, S. W., Stone, J. M., & Jiang, Y.-F. 2012, A Radiation Transfer Solver for Athena Using Short Characteristics. *ApJS*, 199(1), 9.

Gnedin, N. Y. & Ostriker, J. P. 1997, Reionization of the Universe and the Early Production of Metals. *ApJ*, 486(2), 581–598.

Jiang, Y. F., Stone, J. M., & Davis, S. W. 2012, A godunov method for multidimensional radiation magnetohydrodynamics based on a variable Eddington tensor. *Astrophysical Journal, Supplement Series*, 199(1).

Jiang, Y.-F., Stone, J. M., & Davis, S. W. 2013, Saturation of the Magneto-rotational Instability in Strongly Radiation-dominated Accretion Disks. *ApJ*, 767(2), 148.

Levermore, C. D. 1984, Relating Eddington factors to flux limiters. *J. Quant. Spectrosc. Radiat. Transfer*, 31(2), 149–160.

Levermore, C. D. & Pomraning, G. C. 1981, A flux-limited diffusion theory. *ApJ*, 248, 321–334.

Menon, S. H., Federrath, Ch., Krumholz, M. R., Kuiper, R., Wibking, B. D., & Jung, M. 2022, VETTAM: a scheme for radiation hydrodynamics with adaptive mesh refinement using the variable Eddington tensor method. *MNRAS*, 512(1), 401.

Mihalas, D. 1978,. *Stellar atmospheres*. W.H. Freeman, San Francisco.

Mihalas, D. & Klein, R. I. 1982, On the solution of the time-dependent inertial-frame equation of radiative transfer in moving media to O(v/c). *Journal of Computational Physics*, 46, 97–137.

Skinner, M. A. & Ostriker, E. C. 2013, A two-moment radiation hydrodynamics module in athena using a time-explicit Godunov method. *Astrophysical Journal, Supplement Series*, 206(2), 21.

Stone, J. M., Mihalas, D., & Norman, M. L. 1992, ZEUS-2D: A Radiation Magnetohydrodynamics Code for Astrophysical Flows in Two Space Dimensions. III. The Radiation Hydrodynamic Algorithms and Tests. *ApJS*, 80, 819.

Teyssier, R. & Commerçon, B. 2019, Numerical Methods for Simulating Star Formation. *Frontiers in Astronomy and Space Sciences*, 6, 51.

Wibking, B. D. & Krumholz, M. R. 2021, Quokka: A code for two-moment AMR radiation hydrodynamics on GPUs. *arXiv e-prints*, arXiv:2110.01792.

Discussion

CHIA-YU HU: The VET is expected to be more computationally expensive than the FLD and M_1 methods, and hence it is possible to probe higher resolution simulations with the latter class of methods. Did that play a role in why earlier methods choose these approaches?

SHYAM: Yes, completely agree; the VET, although more accurate, is certainly more computationally expensive than the other methods, and hence it is crucial to apply it on problems where the added accuracy could possibly matter. These represent alternate approaches to simulations – a less accurate method, cheaper method that allows for higher resolution and further parameter space exploration vs a method with higher accuracy, that can only be used in a smaller parameter range and resolution due to its computational costs.

MIIKKA VÄISÄLÄ: I was wondering whether you could comment on the computational performance of your scheme, and the number of angular rays required for reasonable results?

SHYAM: Thank you for the very good question! Although the VET method is known to be computationally expensive, its not straightforward to provide general performance statistics, since the performance is very much problem-dependent due to the use of implicit matrix inversion algorithms, whose rate convergence depends on the stiffness of the matrix. That being said, the ray-tracer is typically the bottleneck in terms of performance in our scheme, with higher angular ray resolution aggravating the performance. We find in our shadow test, however, that an angular resolution of 48 rays (using HEALPix sampling) is a reasonably good balance between performance and accuracy, and is the fiducial value we use in our scheme.

The Predictive Power of Computational Astrophysics as a Discovery Tool
Proceedings IAU Symposium No. 362, 2023
D. Bisikalo, D. Wiebe & C. Boily, eds.
doi:10.1017/S174392132200165X

Interplay of various particle acceleration processes in astrophysical environment

Sayan Kundu*[iD] and Bhargav Vaidya

Department of Astronomy, Astrophysics and Space Engineering
Indian Institute of Technology, Indore
Madhya Pradesh, India - 452020
*`sayan.astronomy@gmail.com`

Abstract. Astrophysical systems possess various sites of particle acceleration, which gives rise to the observed non-thermal spectra. Diffusive shock acceleration (DSA) and stochastic turbulent acceleration (STA) are the candidates for producing very high energy particles in weakly magnetized regions. While DSA is a systematic acceleration process, STA is a random energization process, usually modelled as a biased random walk in energy space with a Fokker-Planck equation. In astrophysical systems, different acceleration processes work in an integrated manner along with various energy losses.

Here we study the interplay of both STA and DSA in addition to various energy losses, in a simulated RMHD jet cocoon. Further, we consider a phenomenologically motivated STA timescale and discuss its effect on the emission profile of the RMHD jet. A parametric study on the turbulent acceleration timescale is also conducted to showcase the effect of turbulence damping on the emission structure of the simulated jet.

Keywords. Acceleration of particles, Radiation mechanisms: nonthermal, Plasmas, Turbulence

1. Introduction

Particle acceleration is an ubiquitous phenomenon in astrophysical environments. It presents a possible reason for the observed abundance of non-thermal particles and their spatial distribution throughout the source in various extra-galactic systems, thus explaining the observed emission signatures from these sources. The existing literature (Blandford 1994; Marcowith et al. 2020) suggests three main mechanisms to energize charged particles in various astrophysical plasma environments: diffusive shock acceleration (DSA), coherent electric field acceleration due to reconnection, and stochastic turbulent acceleration (STA). Among these, DSA and STA are plausible mechanisms for accelerating charged particles in weakly magnetized medium and magnetic reconnection is more efficient in accelerating particles in magnetically dominated systems. Since Fermi (1949), magneto-hydrodynamic turbulence is known to be an important source for accelerating particles through STA, while DSA requires shocks to be present in the system. Even-though, comparing by the acceleration timescale, DSA is more efficient, STA, however, has been invoked to explain the particle acceleration processes in various astrophysical systems (see for example, Petrosian 2012; Vurm & Poutanen 2009; Ferrand & Marcowith 2010; Schlickeiser & Dermer 2000; Asano & Hayashida 2018; O'Sullivan, Reville, & Taylor 2009; Fan et al. 2008; Donnert & Brunetti 2014).

Incorporating these particle acceleration processes in large-scale numerical simulations is an area of active research (Hanasz, Strong, & Girichidis 2021; Vazza et al. 2021;

Donnert & Brunetti 2014) and recently higher order numerical schemes are being developed in this regard (Kundu, Vaidya, & Mignone 2021; Winner et al. 2019). Considering the length-scale constraint of numerical simulations, multi-scale realistic astrophysical simulations demand various micro-physical quantities, such as viscosity, various acceleration timescale, resistivity, damping rate of turbulence etc., to be calculated apriori and fed into the simulation as an input. This poses a problem, as apriori calculation of the micro-physical quantities in such complex environments is very difficult. So in the present work, we choose a phenomenologically motivated timescale for stochastic acceleration considering the micro-physics of turbulence damping phenomena, observed in real astrophysical scenarios, and show the interplay of STA and DSA on the emission profile of a simulated toy RMHD jet. We also performed a parametric study on the new timescale and analyze its effect on the emission structure.

2. Stochastic Turbulent Acceleration

Due to random scattering, between charged particles and various MHD waves in a magnetized environment the charged particle or cosmic ray distribution follows a Fokker-Planck equation in momentum space (Webb 1989; Vaidya et al. 2018; Kundu, Vaidya, & Mignone 2021),

$$\frac{\partial \chi_p}{\partial \tau} + \frac{\partial}{\partial \gamma}\left[(S + D_A)\chi_p\right] = \frac{\partial}{\partial \gamma}\left(D\frac{\partial \chi_p}{\partial \gamma}\right), \tag{2.1}$$

where, τ is the proper time, $\gamma \approx p/m_0 c$ is the Lorentz factor of the cosmic ray, with m_0 being the mass of the cosmic ray particle and c is the speed of light in vacuum, $\chi_p = N/n$, with $N(p, \tau)$ being the number density of the non-thermal particles with momentum between p and $p + dp$ and n being the number density of the fluid at the position of the macro-particle, S corresponds to various radiative and adiabatic losses; $D_A = 2D/\gamma$ corresponds to the acceleration due to Fermi II order with D being the diffusion coefficient.

2.1. Modelling momentum diffusion coefficient

All the micro-physical processes of the random scattering phenomena are contained in the momentum diffusion coefficient D of Eq. (2.1). Even-though the mathematical form of this diffusion coefficient, due to interactions of cosmic ray with turbulent magnetized medium, have analytically been derived (see, for instance, Schlickeiser 2002; Brunetti & Lazarian 2007; O'Sullivan, Reville, & Taylor 2009) for specific turbulent cases, a general form is still lacking. Due to this reason we choose a parametrized acceleration timescale for our simulations,

$$t_A = \tau_A \exp\{\tau/\tau_d\} \tag{2.2}$$

where, t_A is the acceleration timescale, τ_A, τ_d are some arbitrary parameters and τ is time. The motivation for choosing such kind of acceleration timescale is the finite energy constraint, which implies that, in realistic situations the amount of energy, which could be channelled to the non-thermal particles due to turbulence, is limited and particles can not get accelerated forever. So for a realistic scenario one should consider the effect of damping of the turbulence in the system, which has been taken care of by the presence of the exponential term in t_A. Time-exponential decay in the charged particles' velocity-velocity correlation, for spatial diffusion coefficient, or in the pitch angle auto-correlation has already been reported by various authors (see for example Fraschetti & Giacalone 2012, and the references therein), we have prescribed an extension of such formalism for the momentum diffusion scenario. The momentum diffusion coefficient D, related to the acceleration timescale, therefore becomes,

$$D = \frac{\gamma^q}{2t_A} = \frac{\gamma^q}{2\tau_A} \exp\left\{-\frac{\tau}{\tau_d}\right\}, \tag{2.3}$$

where q is the exponent and for all our simulations we assume $q = 2$.

2.2. *Numerical Algorithm*

In this work, we use a finite volume relativistic magneto-hydrodynamic code PLUTO (Mignone et al. 2007) to do the simulations and also utilize the Lagrangian particle module (Mukherjee et al. 2021; Vaidya et al. 2018) to analyze the emission signatures of the simulated structure. The Lagrangian particle module employs a 2nd order accurate conservative RK-IMEX scheme (Kundu, Vaidya, & Mignone 2021) to solve Eq. (2.1). For this work, we modified the scheme considering the fact that due to various cooling processes, the particle spectrum falls off very rapidly in the higher γ region. Thus, following Winner et al. (2019) we floor the spectrum after a certain threshold χ_{cut} and treat the numerical values below it as zeros. Further note that, the flooring of the particle spectra is done to only those macro-particles which has encountered at least one shock. Moreover, for all our simulations we consider $\chi_{cut} = 10^{-21}$ and the value of CFL number is 0.8 while solving Eq. (2.1).

3. Numerical Setup

We choose to simulate an axisymmetric RMHD jet as a toy model to analyze the interplay of various particle acceleration processes and its effect on the emission signatures. The simulation is performed with a unit length of $\hat{L}_0 = 100\,\mathrm{pc}$, unit density of $\hat{\rho}_0 = 1.67 \times 10^{-24}\,\mathrm{gr\,cm^{-3}}$ and a unit velocity of $\hat{v}_0 = c = 3 \times 10^{10}\,\mathrm{cm/s}$. A unit timescale for the simulation can be computed as $\hat{\tau}_0 = \hat{L}_0/\hat{v}_0 = 326.4\,\mathrm{Myr}$. Note that, all the physical parameters of the simulation are suitably normalized by these scaled units.

We consider a $2D$ cylindrical grid $\{R, Z\} \in \{0, 0\}$ to $\{20\hat{L}_0, 50\hat{L}_0\}$ using 160×400 grid cells as our computational domain. The ambient medium is considered to be static initially with a constant density $\rho_m = 10^3 \hat{\rho}_0$. An under-dense beam of density $\rho_j = \hat{\rho}_0$ is injected into such an ambient medium along the Z direction through a circular nozzle of unit radius $R_j = \hat{L}_0$ from the lower Z boundary. Further, a conserved tracer quantity is injected with the beam and its value is considered to separate different regions in the simulated structure, as described in more details in the following section. The injection velocity of the beam is prescribed using an initial Lorentz factor $\gamma_j = 10$.

The magnetic field is taken to be purely poloidal i.e, $\vec{B} = B_z \hat{e}_z$ and is initially prescribed in the jet nozzle and in the ambient medium following,

$$B_z = \sqrt{2\sigma_z P_j}. \tag{3.1}$$

where, P_j is the jet pressure at $R = R_j$ estimated from the Mach number $M = v_j \sqrt{\rho_j/(\Gamma P_j) + 1/(\Gamma - 1)} = 6$ with an adiabatic index $\Gamma = 5/3$. The value for the magnetization parameter σ_z is taken to be 10^{-4} for all our simulations.

We further inject 25 Lagrangian macro-particles every two time steps so that they sample the entire jet cocoon uniformly. The macro-particles are injected with an initial power-law spectral distribution of index -9 on a γ grid spanning from $\gamma_{\min} = 1$ to $\gamma_{\max} = 10^5$ discretized with 128 bins.

The energy spectrum of the macro-particles are calculated for two different cases: (i) DSA with synchrotron, adiabatic and inverse compton (IC) losses and (ii) considering stochastic acceleration in addition to case (i). For case (i), we solve Eq. (2.1) without the acceleration (D_A) and the diffusion (D) terms. While for case (ii), we include these terms along with the losses in the transport equation. We choose four different acceleration timescales for case (ii) to examine the influence of stochastic acceleration and turbulence damping on the emission features of the simulated jet. To clearly describe these four different scenarios we cast the acceleration timescale in the following way,

$$t_A = \frac{\tau_c(\gamma_{max}, \gamma_{min})}{\alpha} \exp\{(\tau - \tau_{inj})/\tau_d\}; \tag{3.2}$$

where $\tau_c(\gamma_{max}, \gamma_{min}) = \frac{3m_0^2 c^3}{4\sigma_T[U_B(t)+U_{rad}(t)]} \left(\frac{1}{\gamma_{min}} - \frac{1}{\gamma_{max}}\right)$ is the radiative cooling time for a cosmic ray particle from γ_{max} to γ_{min} with σ_T, U_B and U_{rad} being the Thompson cross-section, magnetic and radiation field energy densities respectively. These energy densities can be computed by following $U_B = \frac{B^2}{8\pi}$ and $U_{rad} = a_{rad}T_0^4$, with a_{rad} being the radiation constant and $T_0 = 2.728\,\mathrm{K}$ being the temperature of the Cosmic Microwave Background (CMB) radiation field. $\tau_d = \eta\tau_c(\gamma_{max}, \gamma_{min})/\alpha$, we choose $\eta = 1.1$ and $\alpha = 10^4, 10^5, 10^6, 10^7$ as four different acceleration scenarios. τ_{inj} is the injection time of the Lagrangian particle in a turbulent region. For the particles that have encountered a shock, the τ_{inj} is set to be the shock hitting time, whereas, for the particles that have not crossed any shocks, the τ_{inj} is its initial injection time in the computational domain.

Moreover, for both cases we consider only the synchrotron emissivity which is calculated implicitly in the code for each Lagrangian-particle based on their local spectral distribution and interpolate it on the underlying grid (Vaidya et al. 2018).

4. Results

In this section we proceed to describe the results from our simulations. In Fig. (1), we show the synchrotron emissivity (J_ν) computed from the Lagrangian macro-particles at a frequency $\nu = 43\,GHz$ and at the time $\tau = 200\,\tau_0$ for all the test scenarios. The top left panel shows the profile of J_ν for the case when only DSA is active along with various energy losses, while all the other plots correspond to the scenario when STA is taken into account in addition to DSA. The top right panel depicts the emission profile for the case, when the turbulent acceleration timescale is calculated with $\alpha = 10^4$. For both the plots presented in the top panel, the emission structure could be seen to be very similar except at the regions where strong shocks are present (viz. along the jet spine, head and at the boundary of the cocoon), where the corresponding turbulent acceleration scenario dominates the emission. This difference is expected as the downstream region of the shock is known to be highly turbulent, thus capable of further accelerating the particles by Fermi II process, once they cross the shock. On the contrary, for the case of only DSA, once a particle crosses a shock it starts to lose its energy due to various radiative losses without any further continuous acceleration which could compensate for the energy loss. Moreover for the case of turbulent acceleration, due to the presence of the exponential term in the acceleration timescale, the energization occurs for a finite amount of time after the particle crosses the shock. This clearly imitates the turbulence damping phenomena.

The emission signatures further change as we modulate the value of α. In the lower panel of Fig. (1) the corresponding emission profiles are presented for $\alpha = 10^5, 10^6$ and 10^7. Compared to the $\alpha = 10^4$ run, all the plots in the bottom panel show a significant enhancement in the emission. This is expected, as an increase in α value manifests itself by reducing τ_A (see Eqs. (2.2) and (3.2)). Also note that, when compared to the only shock acceleration scenario (top left plot of Fig. (1)), the emission profiles in the bottom panel show a gradual enhancement in the emission from the back-flow region of the jet cocoon, which is expected to be turbulent in nature (Matthews et al. 2019), as α increases.

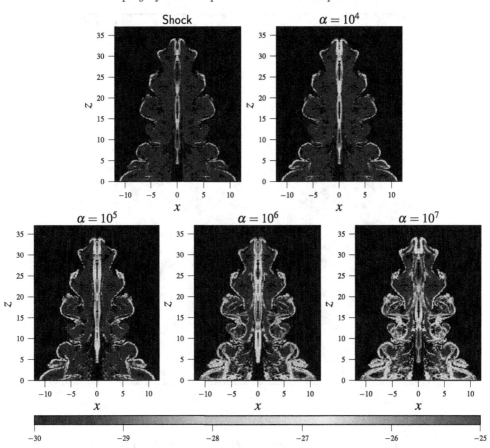

Figure 1. Emission profiles of the simulated RMHD jet, with various acceleration scenarios. *Top left:* Emission structure is shown when only DSA is considered, and all the other profiles correspond to stochastic acceleration with different values of α. The color-bar shown below corresponds to the value of emissivity in units of $erg\ s^{-1}\ cm^{-3}\ Hz^{-1}\ str^{-1}$ in logarithmic scale.

Further, observe that for both the cases with $\alpha = 10^5$ and 10^6, the emission from the jet spine region are more enhanced compared to $\alpha = 10^7$. While on the contrary, in the region inside the cocoon where weak shocks are present, the emission is more dominated for the the latter one. To analyze this phenomena in more detail, we show the magnetic field $B = \sqrt{B_r^2 + B_z^2}$ map for time $\tau = 200\tau_0$ in Fig. (2), where one can observe an order of magnitude higher value of the magnetic field in the jet spine region compared to the jet cocoon. For further quantification, we have calculated the mean magnetic field strength, weighted with the tracer, at the jet spine and in the cocoon region at time $\tau = 200\ \tau_0$. To implement this we consider tracer values > 0.8 and < 0.8 to account for the jet spine and the jet cocoon region respectively (Mukherjee et al. 2021). As expected, we find an order of magnitude higher value of the mean magnetic field in the spine region ($\approx 288\ \mu G$) compared to the cocoon region ($\approx 21\ \mu G$). With the mean magnetic field strength for both the regions, we proceed to calculate the turbulent acceleration timescale. The acceleration timescales are calculated relative to the cooling timescale $\tau_c(\gamma_{max}, \gamma_{min})$, considering the mean $\vec{B}$ field (tracer weighted) from the entire computational domain. Further to show the dependence of the acceleration timescale on different α values, we plot the variation of $\log_{10}(t_A)$, with $\log_{10}(\alpha)$ in Fig. (3) for the two different regions. We observe that,

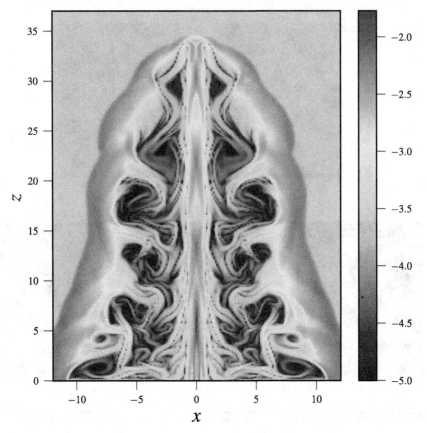

Figure 2. Normalized magnetic field map (B/B_0) of the simulated relativistic AGN jet. The color-bar shows a logarithmic scale of B/B_0. The value of the unit magnetic field being $B_0 = 1.374 \times 10^{-1}$ G.

initially in the jet spine region the timescale decays with increasing α until α reaches a critical point, $\alpha = \alpha_{jc} = 10^6$. Beyond this point the timescale increases exponentially. Due this particular behaviour of the acceleration timescale, from Fig (3) one could observe an order of magnitude increment in t_A as one moves from $\alpha = 10^6$ to $\alpha = 10^7$, in the jet spine region, thereby implying a faster acceleration for $\alpha = 10^6$. A similar functional behaviour of the timescale could be observed in the cocoon region as well. From Fig. (3) it can also be seen that in the cocoon region $\alpha = 10^7$ provides a faster turbulent acceleration compared to $\alpha = 10^6$ and below.

In summary we can say that, as the macro-particles are injected in the computational domain, the particles encounter shocks while moving along the jet spine and proceed to the turbulent downstream region. In the downstream region, due to the presence of higher magnetic field, turbulent acceleration with $\alpha = 10^6$ provides faster acceleration compared to $\alpha = 10^7$. This implies radiative losses are more dominant for $\alpha = 10^7$, in the spine region, therefore a lower emission is expected. Subsequently, as the macro-particles move into the cocoon region, due to the presence of weak shocks, the particles advect into the turbulent downstream and because of comparatively lower magnetic field than the spine region, $\alpha = 10^7$ leads to more efficient acceleration than $\alpha = 10^6$ case.

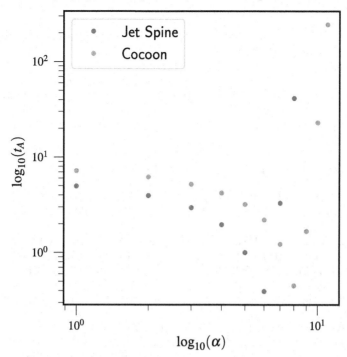

Figure 3. Dependence of (t_A) on various α values, with t_A being the turbulent acceleration timescale for both the regions, jet spine and jet cocoon.

5. Summary

Astrophysical systems provide a playground for various complex physical processes. Among them particle acceleration processes play a fundamental role in shaping the emission features observed in these systems. In this work we focus on studying the interplay of various particle acceleration processes and their effect on the emission structure of astrophysical sources. In particular, we consider a stochastic acceleration timescale, which has the ability to mimic turbulence damping phenomena. Such phenomena could be observed in various realistic turbulent environments. We have demonstrated the effect of this damping on the emission profile of synthetic astrophysical objects by analyzing a RMHD jet simulation as a toy problem. We also present a parametric study on the stochastic acceleration timescale and showed that, due to the presence of exponential damping, the stochastic acceleration could lead to a region based enhancement in the emission.

Acknowledgement

The authors express their gratitude to the referee for the insightful remarks and suggestions on the work. The authors appreciate the financial assistance provided by the Max Planck partner group award at the Indian Institute of Technology, Indore. All the simulations have been carried out using the computing facility at Indian Institute of Technology, Indore and Max Planck Gesellschaft(MPG) super-computing resources. SK would also like to thank Arghyadeep Paul, Sarvesh Mangla, Sriyasriti Acharya, Suchismita Banerjee and Yoshini Bailung for the valuable discussion sessions during the course of this work.

References

Matthews J. H., Bell A. R., Araudo A. T., Blundell K. M., 2019, EPJWC, 210, 04002. doi:10.1051/epjconf/201921004002

Fraschetti F., Giacalone J., 2012, ApJ, 755, 114. doi:10.1088/0004-637X/755/2/114

Hanasz M., Strong A. W., Girichidis P., 2021, LRCA, 7, 2. doi:10.1007/s41115-021-00011-1

Winner G., Pfrommer C., Girichidis P., Pakmor R., 2019, MNRAS, 488, 2235. doi:10.1093/mnras/stz1792

Vazza F., Wittor D., Brunetti G., Brüggen M., 2021, A&A, 653, A23. doi:10.1051/0004-6361/202140513

Petrosian V., 2012, SSRv, 173, 535. doi:10.1007/s11214-012-9900-6

Vurm I., Poutanen J., 2009, ApJ, 698, 293. doi:10.1088/0004-637X/698/1/293

Ferrand G., Marcowith A., 2010, A&A, 510, A101. doi:10.1051/0004-6361/200913520

Schlickeiser R., Dermer C. D., 2000, A&A, 360, 789

O'Sullivan S., Reville B., Taylor A. M., 2009, MNRAS, 400, 248. doi:10.1111/j.1365-2966.2009.15442.x

Fan Z.-H., Liu S., Wang J.-M., Fryer C. L., Li H., 2008, ApJL, 673, L139. doi:10.1086/528372

Donnert J., Brunetti G., 2014, MNRAS, 443, 3564. doi:10.1093/mnras/stu1417

Asano K., Hayashida M., 2018, ApJ, 861, 31. doi:10.3847/1538-4357/aac82a

Blandford R. D., 1994, ApJS, 90, 515. doi:10.1086/191869

Marcowith A., Ferrand G., Grech M., Meliani Z., Plotnikov I., Walder R., 2020, LRCA, 6, 1. doi:10.1007/s41115-020-0007-6

Webb G. M., 1989, ApJ, 340, 1112. doi:10.1086/167462

Fermi E., 1949, PhRv, 75, 1169. doi:10.1103/PhysRev.75.1169

Vaidya B., Mignone A., Bodo G., Rossi P., Massaglia S., 2018, ApJ, 865, 144. doi:10.3847/1538-4357/aadd17

Kundu S., Vaidya B., Mignone A., 2021, ApJ, 921, 74. doi:10.3847/1538-4357/ac1ba5

Mukherjee D., Bodo G., Rossi P., Mignone A., Vaidya B., 2021, MNRAS, 505, 2267. doi:10.1093/mnras/stab1327

Brunetti G., Lazarian A., 2007, MNRAS, 378, 245. doi:10.1111/j.1365-2966.2007.11771.x

Schlickeiser R., 2002, cra.book

Mignone A., Bodo G., Massaglia S., Matsakos T., Tesileanu O., Zanni C., Ferrari A., 2007, ApJS, 170, 228. doi:10.1086/513316

Discussion

TOMOYUKI: Has the coefficient of diffusion for the particle acceleration been calculated from the MHD simulation?

SAYAN: Yes, it has been calculated from the MHD simulation. The simulation gave the value of the magnetic field at each points on the grid and we calculate the diffusion coefficient from that. We also consider the dependence of the diffusion coefficient on the particle Lorentz factor as $D \propto \gamma^2$.

The Predictive Power of Computational Astrophysics as a Discovery Tool
Proceedings IAU Symposium No. 362, 2023
D. Bisikalo, D. Wiebe & C. Boily, eds.
doi:10.1017/S1743921322001363

GP-MOOD: a positivity-preserving high-order finite volume method for hyperbolic conservation laws

Dongwook Lee[1] and Rémi Bourgeois[2]

[1]Department of Applied Mathematics, The University of California, Santa Cruz, CA, United States, email: `dlee79@ucsc.edu`

[2]Maison de la Simulation, CEA Saclay, Université Paris-Saclay, Gif-sur-Yvette, France, email: `remi.bourgeois@cea.fr`

Abstract. We present an *a posteriori* shock-capturing finite volume method algorithm called GP-MOOD. The method solves a compressible hyperbolic conservative system at high-order solution accuracy in multiple spatial dimensions. The core design principle in GP-MOOD is to combine two recent numerical methods, the polynomial-free spatial reconstruction methods of GP (Gaussian Process) and the *a posteriori* detection algorithms of MOOD (Multidimensional Optimal Order Detection). We focus on extending GP's flexible variability of spatial accuracy to an *a posteriori* detection formalism based on the MOOD approach. The resulting GP-MOOD method is a positivity-preserving method that delivers its solutions at high-order accuracy, selectable among three accuracy choices, including third-order, fifth-order, and seventh-order.

Keywords. Gaussian Process modeling, MOOD method, high-order method, finite volume method, a posteriori detection, positivity-preserving method

---

1. Introduction

The modern trend in designing numerical methods for astrophysical flow simulations bears critical design principles driven by the vital need for high-performance computing (HPC). In today's HPC, the hardware progression of the memory capacity per compute core has become gradually saturated. This hardware trend pushes the HPC community to put vast efforts to find more efficient ways to best exercise computing resources in pursuing computer simulations. From the mathematical perspective, one desirable way is developing highly efficient numerical algorithms, which has become an important subject in computational fluid dynamics (CFD) research fields as part of utilizing HPC resources as efficiently as possible.

In this paper, we delve into advancing high-arithmetic-intensity numerical approximation by using the Gaussian Process (GP) modeling. This work has been reported in a series of papers, which have appeared in our past studies by Reyes et al. (2018a, 2019); Reeves et al. (2021); Bourgeois and Lee (2022). Our approach is based on the theory of GP regression that furnishes a polynomial-free method, extended to both finite difference and finite volume methods as a conservative high-order solver. In our former studies, we showed that GP can deliver high-order accuracy at $(2R+1)$th order, where R is an integer value that represents the GP stencil radius in the unit of grid-scale, e.g., Δx. We designed GP as an alternative to the conventional high-order polynomial-based reconstruction algorithms, featuring attractive algorithmic benefits such as GP's selectable order of accuracy *within* a single algorithmic framework.

In designing high-order algorithms for astrophysical simulations, it is necessary to implement stable shock-capturing mechanisms to advance discrete solutions stably in the vicinity of sharp gradients. A majority of well-known shock-capturing methods rely on the so-called *a priori* shock-detection paradigm. This is the classical approach in most of the widely-used polynomial-based reconstruction schemes, in which each method detects the magnitude of local flow gradients in an *a priori* fashion before updating the solution, to evolve it stably while meeting underlying physical principles (e.g., positivity, conservation). The mathematical treatments relevant to *a priori* shock detection mechanisms are truly nonlinear, demanding to calculate cell-by-cell local switches, such as slope limiters, to prevent the evolution of unphysical oscillations near strong gradients. The use of nonlinear limiters accounts for an important portion of the computational cost of the simulation at the price of numerical stability, which is essential for non-smooth flow simulations. Putting this into a different perspective, one gains simulation speed-up by deactivating the nonlinear limiters of a shock-capturing algorithm in smooth flow simulations. For instance, our 2D smooth flow experiments show that one can obtain a factor of two or three gain, revealing the added expense of executing nonlinear switches in modern shock-capturing schemes. Another undesirable consequence of nonlinear limiters is the inherent numerical dissipation. It is the main operational mechanism of nonlinear limiters to gain stability at the cost of adding numerical dissipation by switching to a more stable low-order approximation at shocks and discontinuity. The study in Kent et al. (2014) shows that a large increase in numerical diffusion and dispersion errors are observed at the first time step in limited schemes, significantly reducing the effective grid resolution compared to the corresponding unlimited schemes.

The above discussion leads us to consider a completely different shock-capturing paradigm, called an *a posteriori* scheme. The first *a posteriori* shock-capturing method was proposed in Clain et al. (2011) and further investigated in Diot et al. (2012, 2013); Diot (2012). Referred to as the MOOD (Multidimensional Optimal Order Detection) method, the new paradigm advances shock-dominant discontinuous solutions via the *a posteriori* fashion of "repeat-until-valid." The main focus of this paper is to combine GP's high-order *linear* spatial reconstruction methods for enhanced solution accuracy and the MOOD approach for an *a posteriori* shock-capturing strategy within GP. In what follows, we provide a brief description on how these two methods can be integrated to a new *a posteriori* GP-MOOD algorithm.

2. GP-MOOD: a positivity-preserving high-order method

2.1. *GP linear spatial reconstruction*

By being a polynomial-free, radial kernel-based algorithm, GP is well-suited to reconstruct a pointwise fluid value at any arbitrary location using volume-averaged fluid quantities in a local stencil, called a GP stencil of radius R. A GP stencil is genuinely multi-dimensional. Its shape resembles a blocky-diamond centered at the (i, j) cell in the case of 2D, and a blocky-octahedron centered at the (i, j, k) cell in 3D. See Fig. 1 for two simple examples in 2D. The solution accuracy of the GP reconstruction on these GP stencils in Fig. 1 follows linearly as $(2R+1)$th accurate, which is achieved easily by varying the size of the local GP stencil radius, $R = 1, 2, \ldots$. The $(2R+1)$th order GP reconstructor is primarily derived by formulating a GP kernel. In our former studies, we have used one of the popular GP kernels called the square exponential kernel (or SE), denoted by K and defined by

$$K(\mathbf{x}, \mathbf{y}) = K_{\mathrm{SE}}(\mathbf{x}, \mathbf{y}) = \exp\left[-\frac{(\mathbf{x} - \mathbf{y})^2}{2\ell^2}\right], \qquad (2.1)$$

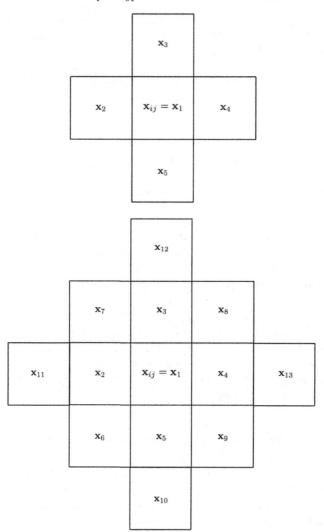

Figure 1. (**top**) The five-point GP stencil of radius $R = 1$ for the 3rd-order GP method. (**bottom**) The 13-point GP stencil of radius $R = 2$ for the 5th-order GP method. In both, the ordered labeling illustrate how to reshape the volume-averaged cell variables at $t = t^n$ into a long one-dimensional array, $\overline{\mathbf{q}}_{ij} = (\mathbf{q}_1, \dots, \mathbf{q}_N)$, where $N = 5$ in (a) and 13 in (b).

where the length hyperparameter, ℓ, controls the characteristic length scale of the functions in the GP function space distributed with a prior mean function $m(\mathbf{x})$ and a prior covariance function $K(\mathbf{x}, \mathbf{y})$. With this SE kernel, one applies the conditioning property of Bayes' theorem to the joint Gaussian distribution on the given grid cell data $\mathbf{q}_{ij}$ to make GP's inference on $f(\mathbf{x}_*)$ given $\mathbf{q}_{ij}$. Denoted as $\mathbf{x}_* \neq \mathbf{x}_i$ is a new point at which GP makes a probabilistic prediction of a function evaluation $f(\mathbf{x}_*)$. For example, $\mathbf{x}_* = \mathbf{x}_{i \pm 1/2, j}$ if one solves Riemann problems at cell face-centers. Here, the choice of functions f is agnostic and is only available probabilistically in GP in the sense of GP regression. Assuming a zero mean, the conditioning property furnishes a new pointwise *posterior mean function* $\tilde{m}(\mathbf{x}_*)$ given by

$$\tilde{m}(\mathbf{x}_*) = \mathbf{k}_*^T \mathbf{K}^{-1} \mathbf{q}_{ij} = \mathbf{z}_*^T \mathbf{q}_{ij}, \tag{2.2}$$

where $[\mathbf{K}]_{mn} \equiv K(\mathbf{x}_m, \mathbf{x}_n)$ and $[\mathbf{k}_*]_m \equiv K(\mathbf{x}_*, \mathbf{x}_m)$. The data-independent vector $\mathbf{z}_*^T = \mathbf{k}_*^T \mathbf{K}^{-1}$ is called the prediction vector, following the same convention in Reyes et al. (2018b, 2019); Reeves et al. (2021).

To use GP for finite volume methods where the input data $\overline{\mathbf{q}}_{ij}$ is given as volume-averaged quantities (rather than pointwise quantities $\mathbf{q}_{ij}$ in finite difference methods), we further integrate the kernels $\mathbf{k}_*$ and $\mathbf{K}$ in Eq. (2.2) (see Reyes et al. (2018a, 2019); Reeves et al. (2021); Bourgeois and Lee (2022) for details). The output is a new GP finite volume reconstructor,

$$\tilde{m}(\mathbf{x}_*) = \mathbf{t}_*^T \mathbf{C}^{-1} \overline{\mathbf{q}}_{ij} = \mathbf{z}_*^T \overline{\mathbf{q}}_{ij}, \qquad (2.3)$$

where $\mathbf{z}_*$ and $\mathbf{C}$ are the integral versions of $\mathbf{k}_*$ and $\mathbf{K}$, respectively. We note here that Eq. (2.3) is a *linear* reconstruction without any limited switches, unlike the conventional non-linear limited polynomial-based reconstructions. The resulting GP reconstruction varies its order of accuracy depending on the size of the GP stencil (see Fig. 1) where the local data $\overline{\mathbf{q}}_{ij}$ is sampled from. The sizes of the GP kernels, $\mathbf{k}_*$ and $\mathbf{K}$, change accordingly to the size of the GP stencil, which are $N \times 1$ and $N \times N$, respectively. In the case of a uniform grid calculation, they need to be computed, transposed, inverted, and saved at the initial step *only once* when a computational grid is configured. They are then reused during simulations. For adaptively varying grid configurations such as adaptive mesh refinements (AMR), one can pre-compute them on each expected AMR grid resolution, save them once-and-for-all and reuse them during simulations. In both, it is shown that GP provides computational efficiency in computing $(2R+1)$th accurate reconstructed pointwise Riemann states at any arbitrary locations, $\mathbf{x}_*$. This paper adopts multiple Gaussian quadrature points along each cell face.

2.2. *The a posteriori MOOD shock-capturing method*

The GP-MOOD method builds upon the existing MOOD methods by Clain et al. (2011); Diot et al. (2012, 2013); Diot (2012). The primary difference between our GP-MOOD and the conventional MOOD approaches are two-fold, including (i) how the multidimensional spatial reconstruction is calculated (i.e., high-order GP vs. high-order polynomials) and (ii) the new addition of "Compressibility-Shock Detection" or CSD criterion to the MOOD iterative loop. See Fig. 2.

As displayed in Fig. 2, GP-MOOD bears the baseline "order-decrement" architecture of the original MOOD loop, which selectively re-calculate pointwise Riemann states $\mathbf{U}_{ij,g_m}^n$ on "failed cells" according to the following three failed-cell detection criteria: (i) PAD (Physical Admissible Detection): positivity preservation on density and pressure variables, (ii) NAD (Numerical Admissibility Detection): numerical validity that monitors CAD (Computer science Admissibility Detection, i.e., NAN & Inf) and DMP (Discrete Maximum Principle) that monitors any excessive numerical oscillations, (iii) CSD (Compressibility-Shock Detection): compressibility and shock strengths, i.e., $\nabla p / p > \sigma_p$ and $\nabla \cdot \mathbf{V} < -\sigma_v$, where $\sigma_p > 0$ and $\sigma_v > 0$ are (heuristically) tunable threshold parameters. The GP-MOOD method uses these detection checks and drops GP's reconstruction order from the highest 7th-order GP-R3, to 5th-order GP-R2 and 3rd-order GP-R1, and down to the safest (also most diffusive and positivity-preserving) first-order. In general, the highest order solution in GP-MOOD can be any arbitrary high-order $(2R+1)$th accurate GP scheme beyond the current choice of 7th-order, which needs to extend the GP stencil size further, e.g., the 41-point blocky-diamond stencil for the 9th-order GP-R4. Such an increase in the GP stencil size will result in higher arithmetic intensity, thereby slowing down the overall time-to-solution. Interested readers are

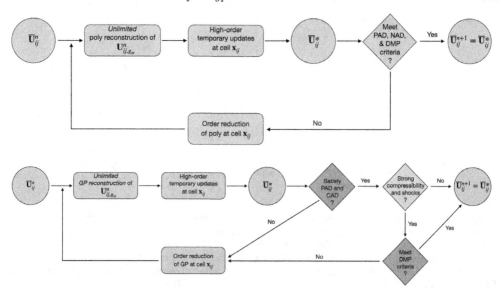

Figure 2. The logical flow of the solution updating procedure in the MOOD loop. (**top**) The flow chart in the existing *a posteriori* polynomial MOOD method. The solution accuracy of a group of unlimited polynomial data reconstruction methods cascades down from high to low until all the MOOD criteria are met in each troubled cell. (**bottom**) The flow chart of the GP-MOOD method. A new CSD condition that checks the strength of flow compressibility (the yellow diamond) is added between the positivity/NAN & Inf check (i.e., PAD and CAD in the top sky blue diamond) and the rest MOOD criteria on DMP (the bottom sky blue diamond). In both, $\mathbf{U}_{ij,g_m}^n$ denotes the pointwise Riemann states at each Gaussian quadrature point g_m; $\overline{\mathbf{U}}_{ij}^*$ denotes a volume-averaged, pre-validated candidate solution using the highest-order method available in each MOOD cascading loop.

encouraged to follow our recent study (Bourgeois and Lee 2022) for further details on GP-MOOD.

3. Results and Conclusion

We display two numerical results in Fig. 3. Shown on the left panel is a grid convergence result, tested on the 2D nonlinear isentropic vortex advection problem Shu (1998). The results of L_1 errors are reported on four different grid resolutions, $N_x = N_y = 50, 100, 200,$ and 400. As shown, the convergence rates of the three GP-MOOD methods on the corresponding diamond GP stencil follow the analytical convergence rates (dotted lines) of $(2R+1)$, showing the expected 3rd-, 5th-, and 7th-order rates.

On the right, we present a 1D shock-tube test called the Shu-Osher problem (Shu and Osher 1989). Simulated results are displayed in a zoomed-in view of the entire profile to focus on the solution comparison over the high-frequency region. Overall, all results produce acceptable density profiles capturing the assumed high-frequency amplitudes reasonably well, conforming with the reference solution. The amplitude closest to the reference profile is achieved by GP-MOOD7, followed by GP-MOOD5, GP-MOOD3, and POL-MOOD3. It is pretty impressive to see how closely the solutions produced by GP-MOOD5 and GP-MOOD7 follow the reference profile, with only on the grid resolution 16 times lower than the reference solution.

Next, we test GP-MOOD on a highly compressible astrophysical problem involving strong shocks and discontinuities to demonstrate our method's robustness and shock-capturing capability. The test problem initializes two Mach 800 jets, one at the top of

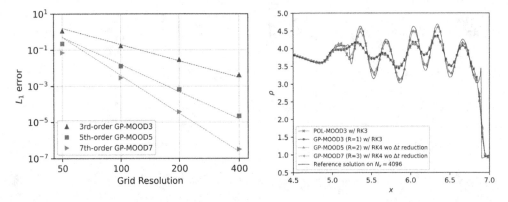

Figure 3. (**left**) Convergence study on the 2D isentropic vortex advection for three different radii, stencils, and reconstructions. (**right**) Result of the Shu-Osher shock tube test at $t = 1.8$ using the 3rd-order GP-MOOD3, 5th-order GP-MOOD5, and 7th-order GP-MOOD7 with CFL $=0.8$, resolved on 256 grid cells. The GP solutions are plotted along with the 3rd-order POL-MOOD3 solution on the same grid resolution. The reference solution is computed using POL-MOOD3 with SSP-RK3 without reducing Δt on $N_x = 4096$.

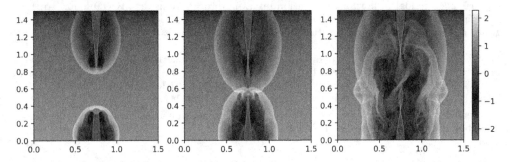

Figure 4. The double Mach 800 jet collision problem is resolved on a 600×600 grid resolution. Density profiles are displayed. The results are computed using GP-MOOD7. From left to right the densities are plotted at $t = 0.002, 0.003$, and 0.005.

the square domain $[0, 1.5] \times [0, 1.5]$, and the other at the bottom. There are two narrow slits, $0.7 \leq x \leq 0.8$, at $y = 0$ and 1.5, through which the Mach 800 jets are injected into the domain via the inflow boundary condition fixed by the jet condition. See Bourgeois and Lee (2022) for more details. The two jets undergo a head-on collision, producing highly turbulent fluid motions that are progressively amplified as the two jets continue to make their ways in the opposite directions for $t \geq 0.003$. The results in Fig. 4 illustrate that GP-MOOD can produce highly accurate turbulent flow dynamics of the two highly compressible jet evolution dynamics closely relevant to astrophysical applications. We remark that the algorithm's accuracy, stability, and positive-preserving are extremely crucial to successfully run this test problem.

We conclude this paper with a summary. We have developed a new GP-MOOD algorithm for a high-order hyperbolic algorithm. GP-MOOD combines (i) the high-order solution property of the GP *linear* reconstruction schemes (thereby furnishing affordable computational approximation cheaper than the conventional nonlinear limited reconstruction methods) and (ii) a new improved MOOD *a posteriori* strategy (thereby reducing numerical dissipation in the MOOD detection criteria). Our GP-MOOD is a strong positivity-preserving method by design, and monitors shocks and discontinuities by detecting a sequence of conditions in the MOOD loop in the *a posteriori* fashion.

A comprehensive study on GP-MOOD, including extensive results and mathematical analysis, is available in Bourgeois and Lee (2022).

4. Acknowledgement

This work was supported in part by the National Science Foundation under grant AST-1908834. We also acknowledge the use of the Lux supercomputer at UC Santa Cruz, funded by NSF MRI grant AST-1828315.

References

A. Reyes, D. Lee, C. Graziani, and P. Tzeferacos. *Journal of Scientific Computing*, 76:443–480, 2018a. ISSN 0885-7474. doi: 10.1007/s10915-017-0625-2

Adam Reyes, Dongwook Lee, Carlo Graziani, and Petros Tzeferacos. *Journal of Computational Physics*, 381:189–217, 2019.

Steve Reeves, Dongwook Lee, Adam Reyes, Carlo Graziani, and Petros Tzeferacos. *arXiv preprint arXiv:2003.08508; Accepted for publication in CAMCoS*, 2021.

Rémi Bourgeois and Dongwook Lee. *Journal of Computational Physics*, 471:111603, 2022.

James Kent, Christiane Jablonowski, Jared P Whitehead, and Richard B Rood. *Journal of Computational Physics*, 278:497–508, 2014.

Stéphane Clain, Steven Diot, and Raphaël Loubère. *Journal of Computational Physics*, 230(10): 4028–4050, 2011.

Steven Diot, Stéphane Clain, and Raphaël Loubère. *Computers & Fluids*, 64:43–63, 2012.

Steven Diot, Raphaël Loubère, and Stephane Clain. *International Journal for Numerical Methods in Fluids*, 73(4):362–392, 2013.

Steven Diot. PhD thesis, Université de Toulouse, Université Toulouse III-Paul Sabatier, 2012.

Adam Reyes, Dongwook Lee, Carlo Graziani, and Petros Tzeferacos. *Journal of Scientific Computing*, 76(1):443–480, 2018b.

Chi-Wang Shu. In *Advanced numerical approximation of nonlinear hyperbolic equations*, pages 325–432. Springer, 1998.

Chi-Wang Shu and Stanley Osher. In *Upwind and High-Resolution Schemes*, pages 328–374. Springer, 1989.

The Predictive Power of Computational Astrophysics as a Discovery Tool
Proceedings IAU Symposium No. 362, 2023
D. Bisikalo, D. Wiebe & C. Boily, eds.
doi:10.1017/S1743921322000345

Converting sink particles to stars in hydrodynamical simulations

Kong You Liow[1], **Steven Rieder[1,2]**, **Clare Dobbs[1]** and **Sarah Jaffa[3,4]**

[1]University of Exeter, Exeter EX4 4QL, UK

[2]RIKEN Center for Computational Science, 650-0047, Hyogo, Japan

[3]University of Hertfordshire, Hertfordshire AL10 9AA, UK

[4]University College London, London WC1H 9NE, UK

Abstract. To form stars in hydrodynamical simulations, we introduce the *grouped* star formation prescription to convert the grouped sink particles into stars that follow the IMF. We show that this method is robust in different physical scales. Such methods to form stars are likely to become more important as galactic or even cosmological scale simulations begin to probe sub-parsec scales.

Keywords. galaxies: ISM – ISM: clouds – stars: formation – galaxies: star clusters: general

1. Introduction

Modelling star formation and resolving individual stars in numerical simulations of molecular clouds and galaxies is highly challenging. Simulations on very small scales can be sufficiently well resolved to consistently follow the formation of individual, stars, whilst on larger scales sinks that have masses sufficient to fully sample the initial mass function (IMF) can be converted into realistic stellar populations. However, as yet, these methods do not work for intermediate scale resolutions whereby sinks are more massive compared to individual stars but do not fully sample the IMF.

We introduce the *grouped* star formation prescription, whereby sinks are first grouped according to their positions, velocities, and ages, then stars are formed by sampling the IMF using the mass of the groups.

2. Method

Our group assignment prescription is simple. For a new sink to join an existing group, the sink must be within:

(*a*) a distance d_g from the group's centre-of-mass (CoM),
(*b*) a speed v_g from the group's CoM speed, and
(*c*) an age τ_g from the group's oldest member.

Else, the sink creates a new group. Then, a population of stars, sampled from the Kroupa IMF (Kroupa 2001), is introduced for each group of sinks. Similar to Wall et al. (2019), the stars are placed within the sinks and the velocities follow the local gas dispersion. Each group is approximately a star-forming region.

Table 1. The physical scales of the models to test the *grouped* star formation method.

Model	Length (pc)	Speed (km/s)	Age (Myr)	Reference
Isolated cloud	< 1	0	< 1	Bate (2012)
Cloud-cloud collision	~ 10	10	~ 1	Liow & Dobbs (2020)
Isolated cloud	~ 10	0	20	Jaffa et al. (in prep)
Spiral arm	~ 1000	0	~ 1	Rieder et al. (2022)

3. Simulation

We use `Ekster` via `AMUSE` (Rieder & Liow 2021; Rieder et al. 2022), a multiphysical code that combines gas hydrodynamics using `Phantom` (Price et al. 2018), high-performance gravitational dynamics using `PeTar` (Wang et al. 2020), and stellar evolution using `SeBa` (Portegies Zwart & Verbunt 1996). The gravitational dynamics between gas and non-gas is coupled using `Bridge` (Fujii et al. 2007).

We tested the *grouped* star formation prescription on the models of varying length, speed and time scales, as shown in Table 1. For each model, we run several sub-models by varying the grouping parameters d_g, v_g and τ_g to find the optimal values in these different scales.

4. Results

In smaller length scale system (< 1 pc), grouped star formation is essential in reproducing the IMF; on the other hand, forming stars using individual sink mass severely undersamples higher mass stars. Even though greater degree of grouping is better in reproducing the IMF, extreme grouping causes violation of local mass conservation. We find that setting $d_g = 1$ pc, $v_g =$ turbulent velocity dispersion, and $\tau_g =$ free-fall time is optimal.

In intermediate scale system (~ 10 pc), grouped star formation is essential too, although to a lesser extent as compared to the smaller scale system. We find that the grouping parameters described above is suitable in this length scale regardless of the speed and time scales of the system.

Finally, in larger scale system (~ 1 kpc), we verify that there is no need to adopt grouped star formation as each sink is massive enough to sample a full population of stars that is consistent to the IMF.

This work is presented in detail in Liow et al. (2022).

References

Bate, M. 2012, *MNRAS*, 419, 3115
Fujii, M., Iwasawa, M., Funato, Y. & Makino, J. 2007, *PASJ*, 59, 1095
Jaffa, S. E., et al., in preparation
Kroupa, P. 2001, *MNRAS*, 322, 231
Liow, K. Y. & Dobbs, C. L. 2020, *MNRAS*, 499, 1099
Liow, K. Y., Rieder, S., Dobbs, C. L. & Jaffa, S. E. 2022, *MNRAS*, 510, 2657
Portegies Zwart, S. & Verbunt, F. 1996, *A&A*, 309, 179
Price, D J., Wurster, J., Tricco, T. S., Nixon, C., Toupin, S., Pettitt, A., Chan, C., Mentiplay, D., Laibe, G., Glover, S., Dobbs, C., Nealon, R., Liptai, D., Worpel, H., Bonnerot, C., and Dipierro, G., Ballabio, G., Ragusa, E., Federrath, C., Iaconi, R., Reichardt, T., Forgan, D., Hutchison, M., Constantino, T., Ayliffe, B., Hirsh, K. & Lodato, G. 2018, *PASA*, 35, 31
Rieder, S., Dobbs, C. L., Bending, T., Liow, K. Y. & Wurster, J. 2022, *MNRAS*, 509, 6155
Rieder, S. & Liow, K. Y. 2021, doi:10.5281/zenodo.5520944
Wall, J. E., McMillan, S. L. W., Mac Low, M.-M., Klessen, R. S., & Portegies Zwart, S. 2019, *ApJ*, 887, 12
Wang, L., Iwasawa, M., Nitadori, K. & Makino, J. 2020, *MNRAS*, 497, 536

The Predictive Power of Computational Astrophysics as a Discovery Tool
Proceedings IAU Symposium No. 362, 2023
D. Bisikalo, D. Wiebe & C. Boily, eds.
doi:10.1017/S1743921322001600

Modelling astrophysical fluids with particles

Stephan Rosswog ⓘ

Astronomy and Oskar Klein Centre, Stockholm University,
AlbaNova, SE-10691 Stockholm, Sweden
email: `stephan.rosswog@astro.su.se`

Abstract. Computational fluid dynamics is a crucial tool to theoretically explore the cosmos. In the last decade, we have seen a substantial methodological diversification with a number of cross-fertilizations between originally different methods. Here we focus on recent developments related to the Smoothed Particle Hydrodynamics (SPH) method. We briefly summarize recent technical improvements in the SPH-approach itself, including smoothing kernels, gradient calculations and dissipation steering. These elements have been implemented in the Newtonian high-accuracy SPH code MAGMA2 and we demonstrate its performance in a number of challenging benchmark tests. Taking it one step further, we have used these new ingredients also in the first particle-based, general-relativistic fluid dynamics code that solves the full set of Einstein equations, `SPHINCS_BSSN`. We present the basic ideas and equations and demonstrate the code performance at examples of relativistic neutron stars that are evolved self-consistently together with the spacetime.

Keywords. hydrodynamics; relativity; stars: neutron; black hole physics; methods: numerical; shock waves

---

1. Introduction

A large fraction of the matter in the Universe can be modelled as fluids, which makes computational gas dynamics a powerful tool in the theoretical exploration of the Cosmos. While also widespread in engineering applications, *astrophysical* gas dynamics comes with its own set of requirements and these sometimes trigger developments in new directions. Contrary to engineering applications, in astrophysics hard boundary conditions rarely play a role and often additional physical processes beyond pure gas dynamics, e.g. magnetic fields, radiation or nuclear reactions, are main drivers of the evolution. Gravity plays a central role in astrophysical gas dynamics. As a long range force, it can easily accelerate gas to velocities that substantially exceed the local sound speed. Therefore, shocks are ubiquitous in astrophysics, but they only occasionally play a role in engineering applications. As a corollary, an astrophysical gas usually cannot –as in many engineering applications– be treated as "incompressible", i.e. obeying a $\nabla \cdot \vec{v} = 0$-condition, and instead the full set of compressible gas dynamics equations needs to be solved. A number of timely astrophysical topics involve gas dynamics in curved spacetime, for example accretion flows around black holes or mergers of neutron stars.

Even if gravity can be accurately treated in the physically rather simple Newtonian approximation, its long-range nature makes it computationally very expensive and the resulting, often filamentary gas structures can pose enormous methodological challenges with respect to geometric adaptivity. To illustrate this in an extreme example, we show in Fig. 1 the tidal disruption of a stellar binary system (67 and 36.8 $M_\odot$) by a supermassive black hole (10^6 $M_\odot$), located at the origin (simulation from Rosswog 2020a). Such binary disruptions where actually both components become shredded have been found to make

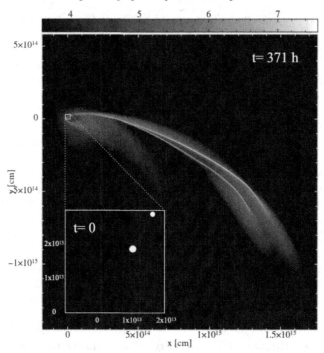

Figure 1. Tidal disruption of a stellar binary system (67 and 36.8 M$_\odot$) by a 10^6 M$_\odot$ black hole (located at the origin; color-coded is the column density). About 10% of all stellar tidal disruption events could be disruptions of binary systems (Mandel & Levin 2015). The simulation has been performed with the code `MAGMA2` (Rosswog 2020a).

up a non-negligible fraction of all tidal disruption events (Mandel & Levin 2015). The initial configuration of this simulation† is shown as inset in the lower left corner. Such a simulation with huge changes in length and density scales, a complicated final geometry with the stars being stretched into extremely thin gas streams that are held together by self-gravity and the majority of the "computational volume" being empty (the initial stars cover $< 10^{-9}$ of the shown volume), are very serious computational challenges. For such applications, particle methods have clear benefits: no additional computational infrastructure (such as an adaptive mesh) is needed, no computational resources are wasted on simulating the vast regions of empty space and the particles simply move where the gas wants to flow. Moreover, the excellent advection properties of particle schemes allow to reliably follow the ejecta out to huge distances.

Methodologically, astrophysical gas dynamics was for a long time split into predominantly Eulerian (mostly Finite Volume) and Lagrangian (mostly Smoothed Particle Hydrodynamics) methods. But in the last decade computational methods have diversified, often combining elements from different methods into "hybrids". One example of such a hybridization are so-called "moving mesh methods" (Springel 2010; Duffel & MacFadyen 2011; Duffel 2016; Ayache et al. 2022) where space is tessellated into Voronoi-cells. Within these cells familiar Finite Volume techniques such as slope-limited reconstructions are applied and at cell interfaces (either exact or approximate) Riemann solvers are used to determine the numerical inter-cell fluxes. The cells themselves are often moved in a (quasi-)Lagrangian way, but, in principle, they can also be kept fixed in space or move with a velocity that is different from the local fluid velocity. In other words,

† The initial conditions for this simulation were kindly provided by I. Mandel and the corresponding stellar profiles by S. Justham.

these are "Adaptive-Lagrangian-Eulerian (ALE)" methods. These methods inherit good shock capturing capabilities and are at the same time highly adaptive and show good (though not perfect) numerical conservation properties.

Such ALE Finite Volume methods, however, are by no means restricted to *non-overlapping* Voronoi cells as basic geometric elements, they can also be applied to freely moving, *overlapping* particles. This has been known in the numerical mathematics community for a long time (see e.g. Ben Moussa, Lanson & Vila 1999; Vila 1999; Hietel, Steiner & Struckmeier 2000; Junk 2003), but has only found its way into astrophysics about a decade ago (e.g. Gaburov & Nitadori 2011; Hopkins 2015; Hubber et al. 2018).

Our main focus here is on the Smoothed Particle Hydrodynamics (SPH) method and its recent developments. We aim at an improved SPH version that keeps the robustness, geometric flexibility and excellent conservation properties of the original method, but is further improved in terms of accuracy. Even more ambitiously, our goal is an accurate particle modelling of a relativistic fluid within a self-consistently evolving, general relativistic spacetime. This goal has recently been reached (Rosswog & Diener 2021) after a string of new elements has been introduced which improve the accuracy of SPH (Rosswog 2010a; Cullen & Dehnen 2010; Rosswog 2010b; Dehnen & Aly 2012; Rosswog 2015b; Frontiere et al. 2017; Rosswog 2020a,b). Most of these new elements are implemented into the Newtonian high-accuracy SPH code MAGMA2 (Rosswog 2020a) which served also as a "test-engine" for many methodological experiments. The new elements have also found their way into the first fully general relativistic, Lagrangian hydrodynamics code SPHINCS_BSSN (Rosswog & Diener 2021).

This paper is organized as fllows: in Sec. 2 we discuss the recent improvements that have been implemented into MAGMA2 and we demonstrate its performance in a number of challenging benchmark tests, in Sec. 3 we discuss the method and implementation of the first general relativistic SPH code that consistently solves the full set of Einstein equations and our results are finally summarized in Sec. 4.

2. Recent improvements of Smoothed Particle Hydrodynamics

Here we briefly summarize frequently used SPH equations to set the stage for further improvements and for a smooth transition to the relativistic case which will be described below. A key ingredient of most SPH formulations is the density estimation at the position of a particle a

$$\rho_a = \sum_b m_b W_{ab}(h_a). \tag{2.1}$$

Here m the particle mass, $W_{ab}(h_a) = W(|\vec{r}_a - \vec{r}_b|, h_a)$ is a smooth kernel function, usually with compact support, and h is the "smoothing length" which determines the support size of W. The usual approach is to keep each particle's mass constant in time so that exact mass conservation is ensured and no continuity equation needs to be solved. One can derive SPH equations elegantly from the SPH-discretized Lagrangian of an ideal fluid (e.g. Monaghan 2005)

$$L = \sum_b m_b \left\{ \frac{1}{2} v_b^2 - u(\rho_b, s_b) - \Phi_b \right\}. \tag{2.2}$$

Here, v_b is the velocity of SPH particle b, u_b its specific internal energy, s_b its specific entropy and Φ_b the gravitational potential. Applying the Euler-Lagrange equations and the adiabatic form of the first law of thermodynamics,

$$\frac{d}{dt}\frac{\partial L}{\partial \vec{v}_a} - \frac{\partial L}{\partial \vec{r}_a} = 0 \quad \text{and} \quad \left(\frac{\partial u}{\partial \rho}\right)_a = \frac{P_a}{\rho_a^2}, \tag{2.3}$$

yields the SPH momentum equation†

$$\frac{d\vec{v}_a}{dt} = -\sum_b m_b \left\{ \frac{P_a}{\rho_a^2}\nabla_a W_{ab}(h_a) + \frac{P_b}{\rho_b^2}\nabla_a W_{ab}(h_b) \right\} + \vec{f}_{g,a} \tag{2.4}$$

with $\vec{f}_g$ being the gravitational acceleration (Price & Monaghan 2007). A consistent energy evolution equation follows in a straight forward way by translating the first law of thermodynamics (see e.g. Rosswog 2009)

$$\frac{du_a}{dt} = \frac{P_a}{\rho_a^2}\sum_b m_b \vec{v}_{ab} \cdot \nabla_a W_{ab}(h_a), \tag{2.5}$$

where $\vec{v}_{ab} = \vec{v}_a - \vec{v}_b$. It is worth mentioning that one has, of course, some freedom in the choice of variables and one can, for example, also evolve the specific thermo-kinetic energy $\hat{e} = u + v^2/2$ according to

$$\frac{d\hat{e}_a}{dt} = -\sum_b m_b \left\{ \frac{P_a \vec{v}_b}{\rho_a^2} \cdot \nabla_a W_{ab}(h_a) + \frac{P_b \vec{v}_a}{\rho_b^2} \cdot \nabla_a W_{ab}(h_b) \right\}. \tag{2.6}$$

As will be seen later, this equation is very similar to the general relativistic evolution equation for the canonical energy per baryon. For practical applications, the energy and momentum equations need to be enhanced by a mechanism to produce entropy in shocks, see Sec. 2.3.

2.1. *Kernel function*

The kernel function W is a core ingredient of any SPH formulation. Traditionally cubic spline kernels have been used (Monaghan 1992), but they are of moderate accuracy in density and gradient estimates (see e.g. Fig. 4 in Rosswog 2015b) and, for large neighbour numbers, they are prone to a "pairing instability", where particles begin to form pairs so that resolution is effectively lost. A necessary condition for stability against pairing is the non-negativity the kernel's Fourier transform (Dehnen & Aly 2012). Wendland (1995) suggested a class of positively definite, radial basis functions of minimal degree and these kernels are immune against the pairing instability.

After exhaustive experiments with various kernels (Rosswog 2015b), we settled for our MAGMA2 code (Rosswog 2020a) on a C^6-smooth Wendland kernel (Schaback & Wendland 2006) which overall delivered the best results. For experiments on *static*, perfect lattices other high-order kernels actually delivered density and gradient estimates of even higher accuracy, but in *dynamic* test cases the Wendland kernel was by far superior. This is because it maintains even in dynamical simulations a very regular particle distribution which is crucial for accurate kernel estimates. This kernel, however, needs a large number of neighbour particles in the SPH summations for accurate density and gradient estimates (see e.g. Figs. 4 and 5 in Rosswog 2015b) and so this improvement comes at some computational cost. As a measure to keep the noise level very low, we choose in MAGMA2 the smoothing length at each time step so that *exactly* 300 neighbour particles contribute in the summations. Technically this is achieved via a very fast tree structure (Gafton & Rosswog 2011), see Rosswog (2020a) for more technical details.

---

† Note that we are neglecting here small corrective terms, usually called "grad-h" terms, see Springel 2002, Monaghan 2002.

2.2. *Accurate gradients via matrix-inversion*

The standard approach in SPH is to represent the $\nabla P/\rho$-term on the RHS of the Euler equations via expressions involving gradients of the kernel functions, as shown in Eqs. (2.4) and (2.5). This gradient representation is anti-symmetric, $\nabla_a W_{ab} = -\nabla_b W_{ab}$, and therefore allows for a straight forward enforcement of exact numerical conservation‡. While individual gradient estimates can be of moderate accuracy only, the overall simulation may still have a high degree of accuracy since it strictly obeys Nature's conservation laws. It is important, though, that any potential improvement of gradient accuracy does not sacrifice one of SPH's most salient features, its excellent conservation properties.

Such an improvement is actually possible and one way to achieve it is by enforcing the exact reproduction of linear functions via a matrix inversion (Garcia-Senz et al. 2012). In the resulting gradient expression one term (that vanishes for an ideal particle distribution) can be dropped and this omission guarantees the desired anti-symmetry of the gradient expression with respect to the exchanging $a \leftrightarrow b$. This gradient prescription delivers gradient estimates that are several orders of magnitude more accurate than the standard SPH approach, see Fig. 1 in Rosswog (2015b). The new set of SPH equations uses the standard density calculation (2.1), but has momentum and energy equations modified according to†

$$\frac{d\vec{v}_a}{dt} = -\sum_b m_b \left\{ \frac{P_a}{\rho_a^2} \vec{G}_a + \frac{P_b}{\rho_b^2} \vec{G}_b \right\}, \tag{2.7}$$

$$\left(\frac{du_a}{dt}\right) = \frac{P_a}{\rho_a^2} \sum_b m_b \vec{v}_{ab} \cdot \vec{G}_a, \tag{2.8}$$

where the gradient functions read

$$\left(\vec{G}_a\right)^k = \sum_{d=1}^{3} C^{kd}(\vec{r}_a, h_a)(\vec{r}_b - \vec{r}_a)^d W_{ab}(h_a) \ \& \ \left(\vec{G}_b\right)^k = \sum_{d=1}^{3} C^{kd}(\vec{r}_b, h_b)(\vec{r}_b - \vec{r}_a)^d W_{ab}(h_b) \tag{2.9}$$

and the "correction matrix"

$$\left(C^{ki}(\vec{r}, h)\right) = \left(\sum_b \frac{m_b}{\rho_b} (\vec{r}_b - \vec{r})^k (\vec{r}_b - \vec{r})^i W(|\vec{r} - \vec{r}_b|, h) \right)^{-1} \tag{2.10}$$

accounts for the local particle distribution. This equation set is based on *much* more accurate gradients, but equally good at numerically conserving physically conserved quantities. This follows directly from the anti-symmetry of the gradient functions with respect to the exchange $a \leftrightarrow b$ and is also confirmed practically in a simulation of the violent collision between two main sequence stars (see Sec. 3.7.3 in Rosswog (2020a)) which is usually considered a worst-case scenario for energy conservation (Hernquist 1993). In this test the conservation accuracy of the matrix inversion approach is on par with the standard SPH formulation.

2.3. *Slope-limited reconstruction in the dissipative terms*

SPH has a reputation of being overly dissipative, but the SPH equations as derived from the above Lagrangian do not involve any dissipation at all. Therefore, the applied

‡ Since one uses radial kernels, their gradients point in the directions of the line joining two particles, this ensure angular momentum conservation. See e.g. Sec. 2.4 in Rosswog 2009b for a detailed discussion of conservation in SPH.

† In the `MAGMA2` code paper we also explore an additional SPH formulation that also uses this accurate gradient prescription.

dissipation is the responsibility of the code developer. One way to add the dissipation that is needed in shocks is, as in typical Finite Volume methods, via Riemann solvers. This approach has occasionally been followed in SPH (Initsuka 2002; Cha & Whitworth 2003; Cha, Inutsuka & Nayakshin 2010; Puri & Ramachandran 2014), but more common is the use of artificial viscosity. While artificial viscosity is one of the oldest concepts in computational fluid dynamics (von Neumann & Richtmyer 1950), its modern forms are actually not that different from approximate Riemann solvers (e.g. Monaghan 1997).

One way to add artificial viscosity (that is actually very close to the original suggestion of von Neumann & Richtmyer (1950) is to simply enhance the physical pressure P by a viscous contribution Q, i.e. one replaces everywhere $P \to P + Q$, where the viscous pressure is given by (Monaghan & Gingold 1983; Frontiere et al. 2017)

$$Q_a = \rho_a \left(-\alpha c_{s,a} \mu_a + \beta \mu_a^2 \right), \tag{2.11}$$

and

$$\mu_a = \min \left(0, \frac{\tilde{v}_{ab} \cdot \tilde{\eta}_a}{\eta_a^2 + \epsilon^2} \right), \tag{2.12}$$

$\vec{v}_{ab} = \vec{v}_a - \vec{v}_b$, $\tilde{\eta}_a$ is the separation vector between particles a and b, de-dimensionalized with particle a's smoothing length, $\tilde{\eta}_a = (\vec{r}_a - \vec{r}_b)/h_a$. The min-function ensures that the artificial pressure is only applied between approaching particles. The quantity μ_a is a measure of the "velocity jump" between the particles a and b. While this prescription works well in strong shocks, it can be more dissipative than actually needed, especially when there is no shock at all.

This unwanted dissipation can be reduced in a similar way as in Finite Volume methods: rather than applying the velocity jump calculated as the difference of the particle velocities $\vec{v}_a - \vec{v}_b$ (in Finite Volume language: applying a zeroth-order reconstruction between the particles), one can perform a slope-limited reconstruction of the particle velocities from both the a- and the b-side to the mid-point between the two particles and use the jump between these reconstructed velocities at the midpoint when calculating the quantity μ. This different calculation of μ is the only change that is required, otherwise the same equation (2.11) can be used. In `MAGMA2` we use a quadratic reconstruction together with a van Leer slope limiter (van Leer 1974; Frontiere et al. 2017), for the technical details we refer to our original code paper (Rosswog 2020a). This reconstruction dramatically reduces unwanted dissipation even if the dissipation parameters α and β are kept at constant, large values. We found this effect particularly pronounced when simulating weakly triggered Kelvin-Helmholtz instabilities: without reconstruction the instability growth was effectively suppressed (as seen in more traditional SPH approaches) while *with* reconstruction the instability grows even at low resolution and with large, constant dissipation parameters α and β at a rate very close to the expected one (see Fig. 20 in Rosswog 2020a).

2.4. *Steering dissipation by entropy monitoring*

One can go even one step further in reducing dissipation: in addition to the above described slope-limited reconstruction, one can also make the dissipation parameters α and β in Eq. (2.11) time-dependent (Morris & Monaghan 1997; Rosswog et al. 2000; Cullen & Dehnen 2010; Rosswog 2015b). Following Cullen & Dehnen (2010), we calculate in `MAGMA2` at each time step for each particle a desired value α^{des} for the dissipation parameter α (we use $\beta = 2\alpha$). If the current value at a particle a, α_a, is larger than α_a^{des}, we let it decay exponentially according to

$$\frac{d\alpha_a}{dt} = -\frac{\alpha_a - \alpha_0}{30 \tau_a}, \tag{2.13}$$

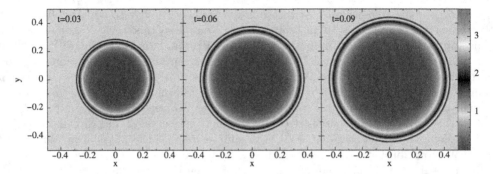

Figure 2. Density evolution in a Sedov-Taylor blast wave. The outermost black ring at the leading edge of the blast is the over-plotted exact solution. Simulation performed with code `MAGMA2`.

where $\tau_a = h_a/c_{s,a}$ is the particle's dynamical time scale and α_0 is a low floor value (which can be zero). Otherwise, if $\alpha_a^{\text{des}} > \alpha_a$, the value of α_a is instantaneously raised to α_a^{des}.

The novel part in our prescription (Rosswog 2020b) is how we determine α^{des} or, in other words, how we determine the exact amount of needed dissipation. The main idea is that we are simulating an ideal fluid which should conserve entropy exactly. In our approach exact entropy conservation is not enforced, so we can monitor it at each particle and use its degree of non-conservation to steer dissipation. A non-conservation of entropy can be the result of a passing shock, or to a much lower extent, it can result from particles becoming noisy for purely numerical reasons. In both cases some amount of dissipation should be applied. For now, we use $s = P/\rho^\Gamma$ as an entropy measure and we monitor over each time step Δt the relative entropy violation $\dot{\xi}_a \equiv (\Delta s_a/\Delta t)/(s_a/\tau_a)$. By numerical experiments we determine a relative entropy violation that can be tolerated without need for dissipation, $\dot{\xi}_0$, and a value $\dot{\xi}_1$, where full dissipation should be applied, in between we smoothly increase the dissipation values, see Rosswog (2020b) for the technical details. This way of steering the dissipation parameter has been found to work very well, it robustly switches on in shocks, but only leads to very low dissipation values otherwise. As an example, we show a Rayleigh-Taylor instability in Fig. 3, where the density (left panel) evolves in very close agreement with literature results (e.g. Frontiere et al. 2017). Non-negligible amounts of dissipation are only triggered in the direct interface between the initial high- and low-density fluid (right panel), elsewhere the dissipation is essentially zero.

2.5. MAGMA2 results

Here we only show a few tests: a Sedov-Taylor explosion as an example for a shock, a Rayleigh-Taylor instability as an instability example and two Schulz-Rinne tests as examples of complex shock-vortex interactions. For a Kelvin-Helmholtz test (density and triggered dissipation) we refer to a movie on the author's website. For more tests and the technical details of a number of benchmark tests we refer to Rosswog (2020a,b).

Sedov blast wave

A classic, multi-dimensional shock problem is the Sedov-Taylor explosion test where a strong, initially point-like blast expands into a low density environment (Sedov 1959; Taylor 1950). For a given explosion energy E, an ambient medium density ρ and polytropic $\Gamma = 5/3$, the blast wave radius propagates according to $r(t) \approx 1.15[(Et^2)/\rho]^{1/5}$ and the density jumps by the strong-explosion limit factor $(\Gamma + 1)/(\Gamma - 1) = 4$. Behind the shock the density drops quickly and finally vanishes at the centre of the explosion. We show in Fig. 2 a cut through the 3D density as a function of time. Also shown

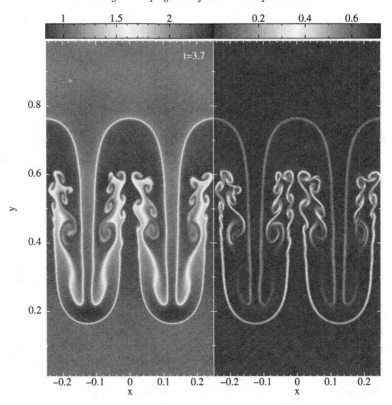

Figure 3. Shown are the density (left) and the applied dissipation parameter α ($\beta = 2\alpha$) as steered via monitoring the local entropy conservation (right) in a Rayleigh-Taylor instability. Note that dissipation is nearly entirely absent throughout most of the flow. Simulation performed with code `MAGMA2`.

(as leading black circle), but hardly visible, is the exact solution, which demonstrates the accurate agreement between our numerical and the exact solution. No deviation from perfect spherical symmetry is visible and also the particle values are (practically noise-free) lying on top of the exact solution, see Fig. 10 in Rosswog (2020a).

Rayleigh-Taylor test
The Rayleigh-Taylor instability is a standard probe of the subsonic growth of a small perturbation. In its simplest form, a layer of higher density rests on top of a layer with lower density in a constant acceleration field, e.g. due to gravity. While the denser fluid sinks down, it develops a characteristic, "mushroom-like" pattern. Simulations with traditional SPH implementations have shown only retarded growth or even a complete suppression of the instability (Abel 2011; Saitoh & Makino 2013). We set up this test case as Frontiere et al. (2017) and our `MAGMA2` results show a healthy growth of the instability, see Fig. 3, left panel. Note in particular, that our entropy steering triggers dissipation in only a very limited region of space, while the bulk of the simulated volume has essentially no dissipation (right panel).

Schultz-Rinne tests
Schulz-Rinne (1993) designed a particularly challenging set of tests in which initially four constant states meet in one corner and the initial values are chosen so that one elementary wave, either a shock, a rarefaction or a contact discontinuity appears at each

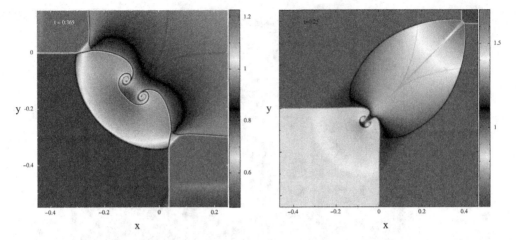

Figure 4. Schulz-Rinne tests. In these challenging tests initially four constant states meet in one corner of the xy-plane. In the subsequent evolution complex shocks and vortex structures form. Our `MAGMA2` results are in close agreement with those found in Eulerian simulations.

interface. During the evolution, complex flow patterns emerge, involving shocks and vorticity, for which no exact solutions are known. These tests are considered challenging benchmarks for multi-dimensional hydrodynamics codes (Schulz-Rinne 1993; Lax & Liu 1998; Kurganov & Tadmor 2002; Liska & Wendroff 2003) and we are only aware of one study that tries to tackle these tests with (a Riemann solver version of) SPH (Puri & Ramachandran 2014), with mixed success. We show in Fig. 4 two such tests (produced with the 3D code; 10 particle layers in z- and $660{\times}660$ particles in xy-direction; polytropic $\Gamma = 1.4$), for further examples we refer to the `MAGMA2` code paper (Rosswog 2020a). Fig. 4 shows crisp and noise-free mushroom-like structures that are in very good agreement with the Eulerian results that can be found in the literature (e.g. Lax & Liu 1998; Liska & Wendroff 2003).

3. Smoothed Particle Hydrodynamics In Curved Spacetime: the SPHINCS_BSSN code

Motivated by the splendid prospects of multi-messenger astrophysics (Rosswog 2015a; Abbott et al. 2017; Barak et al. 2019; Kalogera et al. 2021) our ultimate goal is to develop a general relativistic hydrodynamics code that consistently solves for the evolution of spacetime, but models the fluid with particles. We expect that a particle method has clear benefits (compared to the current Eulerian approaches) in following the small amounts of ejecta, $\sim 1\%$ of the binary mass, that are responsible for the entire electromagnetic display of a compact binary merger.

Since the general relativistic evolution of spacetime is a hyperbolic problem, we need to integrate to spacetime geometry forward in time, while in the Newtonian approach (with an infinite propagation speed of gravity) we solve an elliptic problem where the gravitational forces are calculated from the instantaneous matter state. The methods to evolve spacetime have substantially matured in the last two decades and can be found in recent textbooks on Numerical Relativity (Alcubierre 2008; Baumgarte & Shapiro 2010; Rezzolla & Zanotti 2013; Shibata 2016; Baumgarte & Shapiro 2021), we decided to follow the well-established BSSN-approach for evolving the spacetime on a computational mesh, very similar to what is done in Eulerian approaches, but to evolve matter via Lagrangian particles. This strategy is implemented in our newly developed Numerical Relativity code, SPHINCS_BSSN (Rosswog & Diener 2021).

3.1. *General-relativistic hydrodynamics*

General relativistic SPH equations can be derived similarly to the Newtonian approach (e.g. Monaghan & Price 2001), see Rosswog (2009), Sec. 4.2 for a step-by-step derivation of the equations that we will use. Instead of discretizing the gas into particles of constant mass, one now assigns to each particle a baryon number ν_b that remains a constant-in-time property of each particle. One chooses a "computing frame" in which the simulations are performed and calculates a computing frame baryon number density at the position of a particle a according to

$$N_a = \sum_b \nu_b \, W(|\vec{r_a} - \vec{r_b}|, h_a), \tag{3.1}$$

where W is an SPH smoothing kernel (we use the same C^6-smooth Wendland kernel as in MAGMA2). In other words, we calculate the density N just as the mass density in Newtonian SPH, Eq. (2.1), but with particle masses being replaced with baryon numbers. The number density in the local rest frame density of a particle, n, is related to N via

$$N = \sqrt{-g}\,\Theta n, \tag{3.2}$$

where g is the determinant of the spacetime metric and the generalized Lorentz factor is given by

$$\Theta \equiv \frac{1}{\sqrt{-g_{\mu\nu}v^\mu v^\nu}} \quad \text{with} \quad v^\alpha = \frac{dx^\alpha}{dt}. \tag{3.3}$$

Similar to the Newtonian approach, one can start from a discretized Lagrangian of an ideal fluid

$$L = -\sum_b \nu_b \left(\frac{1+u}{\Theta}\right)_b. \tag{3.4}$$

Note that we have followed here the convention that we measure all energies in units of the baryon rest mass $m_0 c^2$. We base our numerical evolution variables on the canonical momentum per baryon and the canonical energy per baryon as they follow from the above Lagrangian. The canonical momentum per baryon reads

$$(S_i)_a = (\Theta \mathcal{E} v_i)_a, \tag{3.5}$$

where $\mathcal{E} = 1 + u + P/n$ is the relativistic enthalpy per baryon and the canonical energy per baryon

$$e_a = \left(S_i v^i + \frac{1+u}{\Theta}\right)_a = \left(\Theta \mathcal{E} v_i v^i + \frac{1+u}{\Theta}\right)_a. \tag{3.6}$$

These quantities are evolved in time according to

$$\frac{d(S_i)_a}{dt} = -\sum_b \nu_b \left\{ \frac{P_a \sqrt{-g_a}}{N_a^2} \frac{\partial W_{ab}(h_a)}{\partial x_a^i} + \frac{P_b \sqrt{-g_b}}{N_b^2} \frac{\partial W_{ab}(h_b)}{\partial x_a^i} \right\} + \left(\frac{\sqrt{-g}}{2N} T^{\mu\nu} \frac{\partial g_{\mu\nu}}{\partial x^i} \right)_a \tag{3.7}$$

and

$$\frac{de_a}{dt} = -\sum_b \nu_b \left\{ \frac{P_a \sqrt{-g_a}}{N_a^2} v_b^i \frac{\partial W_{ab}(h_a)}{\partial x_a^i} + \frac{P_b \sqrt{-g_b}}{N_b^2} v_a^i \frac{\partial W_{ab}(h_b)}{\partial x_a^i} \right\} - \left(\frac{\sqrt{-g}}{2N} T^{\mu\nu} \frac{\partial g_{\mu\nu}}{\partial t} \right)_a. \tag{3.8}$$

Note that our equations for the conservation of baryon number, Eq. (3.1), momentum, Eq. (3.7), and Eq. (3.8) (compare to Eq. (2.6)), have a very "Newtonian look and feel". But while they are very convenient for the numerical evolution, they are actually not the physical variables that we are really interested in, these are n, v^i and u. This means that we have to recover the physical variables n, v^i, u at every time step from the numerical variables N, S_i and e. But this is a price that also Eulerian approaches have to pay, and we recover the physical variables with very similar methods, see Sec. 2.2.4 in Rosswog & Diener (2021) for the technical details. We also need to add dissipative terms in SPHINCS_BSSN and we follow a strategy similar to the one used in MAGMA2: a) we apply a slope limited reconstruction in the dissipative terms and b) we steer the dissipation by monitoring the entropy change at every particle and time step. The details can be found in Sec. 2.2.3 of Rosswog & Diener (2021).

3.2. *Evolving the spacetime via the BSSN formulation*

To robustly evolve the spacetime, we have implemented two frequently used variants of the BSSN equations in SPHINCS_BSSN, the "Φ-method" (Nakamura, Oohara & Kojima (1987); Shibata & Nakamura (1995); Baumgarte & Shapiro (1999)) and the "W-method" (Tichy & Marronetti (2007); Marronetti et al. (2008)). The complete set of BSSN equations is very lengthy and will therefore not be reproduced here. It is described in detail in a number of Numerical Relativity textbooks (Alcubierre 2008; Baumgarte & Shapiro 2010; Rezzolla & Zanotti 2013; Shibata 2016; Baumgarte & Shapiro 2021) and can also be found in Rosswog & Diener (2021). For all the tests presented here we use the "Φ-method".

3.3. *Coupling between fluid and spacetime*

As can be seen from the hydrodynamic equations Eqs.(3.7) and (3.8), the fluid needs the derivatives of the metric (known on the mesh) at each particle position. The evolution of the metric (evolved on a mesh), in turn, is governed by the energy momentum tensor $T_{\mu\nu}$ that is known at the particle positions. We therefore need to continuously map $T_{\mu\nu}$ from the particles to the mesh ("P2M-step") and $\partial_\lambda g_{\mu\nu}$ from the mesh to the particles ("M2P-step").

In the P2M-step we have experimented with methods that are frequently used in particle-mesh methods (Hochney & Eastwood 1988) and with common SPH kernels. But a general relativistic self-gravitating system is numerically very delicate and we did not find these methods accurate enough for our purposes. For example, an initial neutron star setup according to a Tolman-Oppenheimer-Volkoff solution, did not stay close to its equilibrium solution. We found *much* better solutions when using kernels that are frequently used in the context of "vortex methods" (Cottet & Koumoutsakos 2000). These kernels are very accurate for close to uniform particle distributions, but they are not positive definite and they require the cancellation of positive and negative contributions. If applied naively everywhere, this can lead to Gibbs-phenomena-like spurious oscillations near the surface of the star. Therefore we have implemented a hierarchy of kernels of decreasing order with only the least accurate, "parachute" kernel being strictly positive definite. Applying this hierarchy of kernels led to very good results, for details of this approach, we refer to Sec. 2.4 in Rosswog & Diener (2021).

The M2P-step turned out to be less delicate, here we use a quintic Hermite polynomial in generalization of the procedure described in Timmes & Swesty (2000) to ensure that the interpolated values are C^2 when a particle passes from one grid cell to another.

3.4. *SPHINCS_BSSN results*

Our full evolution code has been scrutinized in a number of standard test cases such as relativistic shock tubes (to test special relativistic hydrodynamics), oscillations of neutron stars in a frozen spacetime ("Cowling approximation"; to test general relativistic hydrodynamics), oscillations of neutron stars when the spacetime is dynamically evolved

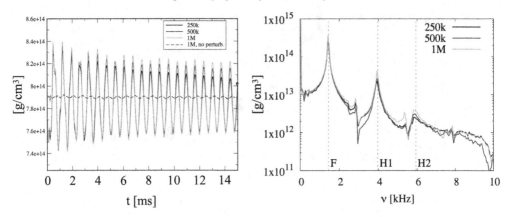

Figure 5. Oscillations of a perturbed neutron star ($\Gamma = 2.0$). Left: central density as function of time. Right: Fourier spectrum of the central density oscillations, the red dashed vertical lines are the fundamental normal mode frequency (F) and the next two higher mode frequencies (H1, H2) as determined by Font et al. (2002). Simulations performed with code SPHINCS_BSSN.

(to test the combined hydrodynamic-plus-spacetime evolution) and last, but not least, the challenging "migration test". In this test, a neutron star is prepared on the unstable branch and migrates, depending on the type of perturbation, either via violent oscillations onto the stable branch, or collapses into a black hole. In the following, we will only describe the fully relativistic, oscillating neutron star and the migration test. We use units in which $G = c = 1$ and masses are measured in solar units. For the other tests and more details we refer to the original paper (Rosswog & Diener 2021), a first set of neutron star merger simulations with SPHINCS_BSSN can be found in Diener, Rosswog & Torsello (2022).

Oscillating neutron star in a dynamical spacetime

In this test, we set up a 1.40 M$_\odot$ (gravitational) neutron star, modelled with a polytropic equation of state ($P = Kn^\Gamma$; $K = 100$ and $\Gamma = 2.0$; keep in mind our convention of measuring energies in $m_0 c^2$), according to the corresponding Tolman-Oppenheimer-Volkoff (TOV) solution. Subsequently, the star receives a small, radial velocity perturbation and is evolved in its dynamical spacetime. The resulting central density evolution is shown in the left panel of Fig. 5 (for 250 000, 500 000 and 1 million SPH particles). The unperturbed stars stay close to but slightly oscillate (due to truncation error) around the TOV solution (black line). We perturb the stars and measure their oscillation frequencies. The fundamental normal mode (F: 2.696 kHz) and the first two overtones (H1: 4.534 kHz, H2: 6.346 kHz) as determined by Font et al. (2002) via a 3D Eulerian high resolution shock capturing code are shown as the red dashed lines in the right panel. We find excellent agreement of the oscillation frequencies at the $\sim 1\%$ level.

Migration of an unstable neutron star to the stable branch

A more complex test case involves an unstable initial configuration of a neutron star (Font et al. 2002; Cordero-Carrillon et al. 2009; Bernuzzi & Hilditch 2010). According to Eulerian studies, the evolution depends delicately on the star's initial perturbation: if just evolved, the truncation error alone drives the star to violent oscillations (with $v > 0.5c$) and it finally settles on the stable branch. If, on the other hand, a small radial inward velocity perturbation of only $\delta v_r = -0.005c \sin{(\pi r/R)}$ is applied, the star collapses and forms a black hole. Can we confirm these results with SPHINCS_BSSN? Yes, we find again very close agreement with the Eulerian studies. Our results for the first case is shown in Fig. 6. The upper panel row shows different stages of the violent oscillation, the lower one

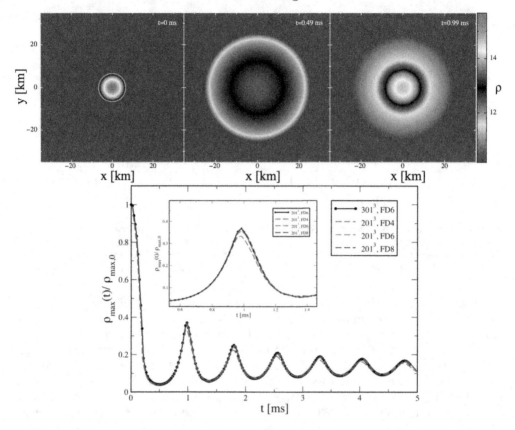

Figure 6. Migration of a neutron star from the unstable to the stable branch. The star evolves, triggered by only truncation error, via violent oscillations ($v > 0.5c$) towards the stable branch. The bottom panel shows the evolution of the central density for different spatial resolutions and Finite Differencing (FD) orders. Simulations performed with code SPHINCS_BSSN.

shows the evolution of the central stellar density (for different numerical resolutions and Finite Difference (FD) orders). When the small inward velocity perturbation is applied, the same star collapses to a black hole, see Fig. 7.

4. Summary and conclusions

In this paper we have described some of the recent developments related to SPH. Our focus was on further improving SPH's accuracy without sacrificing its excellent conservation properties. The new elements include high-order Wendland functions as SPH kernels, accurate gradients that require the inversion of a small matrix and new measures to steer dissipation in SPH. The first of these measures is based on transferring Finite Volume techniques to SPH. More specifically, we perform slope-limited reconstructions between particle pairs and use these reconstructed values in the artificial dissipation terms which massively reduces unnecessary dissipation even if the dissipation parameters are kept at large, constant values. The results can be further improved by additionally making the dissipation parameters time dependent and steer them based on monitoring the exact conservation of entropy.

These new elements have been implemented into two codes that were developed from scratch: the Newtonian code MAGMA2 (Rosswog 2020a) and the fully general relativistic

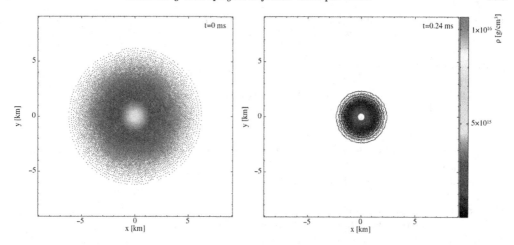

Figure 7. Same star as in the previous figure, but this time a small inward velocity perturbation was applied. This is enough to trigger the collapse to a black hole. Shown are particle positions within $|z| < 0.7$ km. Simulations performed with code SPHINCS_BSSN.

code SPHINCS_BSSN (Rosswog & Diener 2021). Both codes have delivered results of very high accuracy and will be used in our future studies of astrophysical gas dynamics.

Acknowledgements

The author has been supported by the Swedish Research Council (VR) under grant number 2016_03657, by the Swedish National Space Board under grant number Dnr. 107/16, by the research environment grant "Gravitational Radiation and Electromagnetic Astrophysical Transients (GREAT)" funded by the Swedish Research council (VR) under Dnr 2016_06012 and by the Knut and Alice Wallenberg Foundation under grant Dnr. KAW 2019.0112. It is a great pleasure to acknowledge the very productive collaboration with P. Diener, co-developer of SPHINCS_BSSN.

References

Abbott et al. 2017, *ApJL*, 848, L12
Abel, T. 2011, *MNRAS*, 413, 271
Alcubierre, M. 2008, *Introduction to 3+1 Numerical Relativity*, Oxford University Press
Ayache, E., vanEerten H.J. & Eardly, R., 2022, *MNRAS*, 519, 1315
Barack et al., 2019, *Classical and Quantum Gravity*, 36, 143001
Baumgarte, T. & Shapiro, S.L., 1999, *Phys. rev. D*, 59, 024007
Baumgarte, T. & Shapiro, S.L., 2010, *Numerical Relativity: Solving Einstein's Equations on the Computer*, Cambridge University Press
Baumgarte, T. & Shapiro, S.L., 2021, *Numerical Relativity: Starting from Scratch*, Cambridge University Press
Ben Moussa, B., Lanson, N. & Vila, J.P., 1999, *International Series of Numerical Mathematics*, 29, 31
Bernuzzi, S. & Hilditch, D., 2010, *Phys. Rev. D*, 81, 084003
Cha, S.H. & Witworth, A.P., 2003 *MNRAS*, 340, 73
Cha, S.H., Inutsuka, S.I. & Nayakshin, S., 2010 *MNRAS*, 403, 1165
Cottet, G.H. & Koumoutsakos, P.D., 2000, *Vortex Methods*, Cambridge University Press, Cambridge
Cordero-Carrillon, I., et al., 2009, *Phys. Rev. D*, 79, 024017
Cullen, L. & Dehnen, W., 2010, *MNRAS*, 408, 669
Dehnen, W. & Aly, H., 2012, *MNRAS*, 425, 1068

Diener, Rosswog & Torsello, 2022, eprint arXiv:2203.06478

Duffel, P. & MacFadyen, A., 2011, *ApJS*, 197, 15

Duffel, P., 2016, *ApJS*, 226, 2

Font, T. et al., 2002, *Phys. Rev. D*, 65, 084024

Frontiere, N., Raskin, C & Owen, J.M., 2017, *Journal of Computational Physics*, 332, 160

Gaburov, E. & Nitadori, K., 2011, *MNRAS*, 414, 129

Gafton, E. & Rosswog, S., 2011, *MNRAS*, 418, 770

Garcia-Senz, D., Cabezon, R. & Escartin, J.A., 2012, *A & A*, 538, 9

Hernquist, L., 1993, *ApJ*, 404, 717

Hietel, D., Steiner, K. & Struckmeier, J., 2000, *Mathematical Models and Methods in Applied Sciences*, 10, 1363

Hockney, R.W. & Eastwood, J.W., 1988, *Computer Simulation Using Particles, McGraw-Hill, New York*

Hopkins, P., 2015, *MNRAS*, 450, 53

Hubber, P., Rosotti, G.P. & Booth, R.A., 2018, *MNRAS*, 473, 1603

Inutsuka, S.I., 2002 *Journal of Computational Physics*, 179, 238

Junk, M., 2003, *In: Griebel M., Schweitzer M.A. (eds) Meshfree Methods for Partial Differential Equations. Lecture Notes in Computational Science and Engineering, vol 26. Springer, Berlin, Heidelberg*, 26, 223

Kalogera et al., 2022, *The Next Generation Global Gravitational Wave Observatory: The Science Book*, arXiv:2111.06990

Kurganov, A. & Tadmor, E., 2002 *Numerical Methods for Partial Differential Equations*, 18, 584

Lax, P. & Liu, X.D., 1998 *SIAM J. Sci. Comput.*, 19, 319

Liska, R. & Wendroff, B., 2003 *SIAM J. Sci. Comput.*, 25, 995

Mandel, I., & Levin, Y., 2015, *ApJL*, 805, L4

Marronetti et al., 2008, *Phys. Rev. D*, 77, 064010

Monaghan, J.J., 1992, *Ann. Rev. Astron. Astrophys*, 30, 543

Monaghan, J.J., 1997, *Journal of Computational Physics*, 136, 298

Monaghan, J.J. & Price, D.J., 2001, *MNRAS*, 328, 381

Monaghan, J.J., 2002, *MNRAS*, 335, 843

Monaghan, J.J., 2005, *Reports on Progress in Physics*, 68, 1703

Morris, J. & Monaghan, J.J., 1997, *Journal of Computational Physics*, 136, 41

Nakamura, T., Oohara, K, & Kojima, Y., 1987, *Prog. Theor. Phys. Suppl.*, 90, 1

Price, D.J. & Monaghan, J.J., 2007, *MNRAS*, 374, 1347

Monaghan, J.J. & Gingold, R.A., 1983, *Journal of Computational Physics*, 149, 135

Puri, K. & Ramachandran, P., 2014, *Journal of Computational Physics*, 270, 432

Rezzolla, L. & Zanotti, O., 2013, *Relativistic Hydrodynamics, Oxford University Press*

Rosswog, S., et al. 2000, *A&A*, 360, 171

Rosswog, S., 2009, *New Astronomy Reviews*, 53, 78

Rosswog, S., 2010a, *Classical and Quantum Gravity*, 27, 114108

Rosswog, S., 2010b, *Journal of Computational Physics*, 229, 8591

Rosswog, S., 2015a, *International Journal of Modern Physics D*, 24, 1530012

Rosswog, S., 2015b, *MNRAS*, 448, 3628

Rosswog, S., 2015c, *Living Reviews of Computational Astrophysics*, 1, 1

Rosswog, S., 2020a, *MNRAS*, 498, 4230

Rosswog, S., 2020b, *ApJ*, 898, 60

Rosswog, S. & Diener, P., 2021, *Classical and Quantum Gravity*, 38, 115002

Saitoh, T.R. & Makino, J., 2013, *ApJ*, 768, 44

Schaback, R. & Wendland, H., 2006, *Acta Numerica*, 15, 543

Schulz-Rinne, C.W., 1993, *SIAM Journal of Mathematical Analysis*, 24, 76

Sedov, L.I., 1959, *Proceedings of the Royal Society of London Series A*, New York: Academic Press, 1959

Shibata, M. & Nakamura, T. 1995, *Phys. Rev. D*, 52, 5428

Shibata, M., 2016, *Numerical Relativity, World Scientific*

Springel, V., & Hernquist, L., 2002, *MNRAS*, 333, 649

Springel, V., 2010, *MNRAS*, 401, 791

Taylor, G., 1950, *Proceedings of the Royal Society of London Series A*, 201, 159

Tichy, W. & Marronetti, P., 2007, *Phys. Rev. D*, 76, 061502

Timmes, F. & Swesty, D., 2000, *ApJS*, 126, 501

van Leer, B., 1974, *Journal of Computational Physics*, 14, 361

Vila, J.P., 1999, *Mathematical Models and Methods in Applied Science*, 02, 161

von Neumann, J. & Richtmyer, R.D. 1950, *Journal of Applied Physics*, 21, 232

Wendland, H., 1995, *Advances in Computational Mathematics*, 4, 389

The Predictive Power of Computational Astrophysics as a Discovery Tool
Proceedings IAU Symposium No. 362, 2023
D. Bisikalo, D. Wiebe & C. Boily, eds.
doi:10.1017/S1743921322001454

Detecting vortices in fluid dynamics simulations using computer vision

Thomas Rometsch ⓘ

Universität Tübingen
Auf der Morgenstelle 10, 72076 Tübingen, Germany
email: `thomas.rometsch@uni-tuebingen.de`

Abstract. Vortices are patches of fluid revolving around a central axis. They are ubiquitous in fluid dynamics. To the human eye, detecting vortices is a trivial task thanks to our inherent ability to identify patterns. To solve this task automatically, we developed the Vortector pipeline which was used to identify and characterize vortices in around one million snapshots of planet-disk interaction simulations in the context of planet formation. From the emergence of two regimes of vortex lifetime, one of which shows very long-lived vortices, we conclude that future resolved disk observations will predominantly detect vortices in the outer parts of protoplanetary disks.

Keywords. protoplanetary disks – planet–disk interaction – fluid dynamics – methods: numerical

---

1. Introduction

Protoplanetary disks (PPDs) are flat rotating, mostly gaseous objects that form together with their host stars and are the birthplaces of planets.

Goodman et al. (1987) showed that vortices exist in PPDs by providing an analytical vortex solution of the fluid dynamics equations. Later, vortices were also observed in numerical fluid dynamics models, including models in which the vortex was caused by an embedded giant planet (Godon & Livio 1999; De Val-Borro et al. 2007). Since the advent of high-resolution radio astronomy observations (e.g., with ALMA), non-axisymmetric structures have been observed in protoplanetary disks and interpreted as vortices (van der Marel et al. 2013; Bae et al. 2016; Pérez et al. 2018).

Early numerical modelling relied on the assumption that the temperature in the disk is not changing with time. This assumption of an isothermal disk has the advantage of reducing simulation runtime and reducing physical complexity. In recent years, however, it has become evident that the isothermal assumption results in incorrect features in the disks under certain conditions (Ziampras et al. 2020; Miranda & Rafikov 2020), which is especially relevant for the comparison to resolved disk observations and the interpretation of observed substructures.

To improve the models, the radiative effects in the disk can be modeled using the β-cooling prescription (Gammie 2001). In Rometsch et al. (2021) we used this prescription to model vortices in protoplanetary disks that emerge as a consequence of gap opening by an embedded giant planet.

One challenge in numerical studies of vortices is that vortices are not represented as simulation objects but emerge as a pattern in the velocity field. For a small number of simulations and output snapshots, the detection and characterization might be performed manually. In our study, we needed to detect vortices in more than $1\,000\,000$ simulation snapshots, which had to be done using a computer algorithm.

In this document, we highlight the approach we took to identify and characterize vortices in our recent study. We also present a prediction about the expected location of vortices as they will be observed with upcoming high-resolution disk observations.

2. Overview

Vortex models.

The motion of a vortex can fundamentally be described by the curl of the velocity of the flow, $\vec{v}$, which is called vorticity

$$\vec{\omega} = \nabla \times \vec{v}. \tag{2.1}$$

In the case of vortices in protoplanetary disks, the disk motion is approximately Keplerian, meaning the velocity at each radius r away from the star is given by the Keplerian velocity $v \approx v_\mathrm{K} = r\Omega_\mathrm{K}$, with the Keplerian angular velocity Ω_K. This results in a background vorticity of the disk $\omega_\mathrm{K} = \Omega_\mathrm{K}/2$. Vortices being formed at the outer edge of a gap carved by an embedded giant planet usually rotate anticyclonically compared to the background disk. Inside an anticyclonic vortex, the vorticity of the fluid is lowered — being anticyclonic motion, it must have the opposite sign compared to the background disk. Associated with an anticyclonic vortex is an increase in mass density.

Considering a vertically integrated disk model (2D) in the $r - \Phi$ plane of cylindrical coordinates centered on the star, vortices have an elliptical shape (a crescent shape in a top-down view on the disk) and the surface density Σ is approximately a Gaussian bell curve in the vortex region, with a different extent in each direction (for an illustrative explanation see Sect 3.1 in Lin & Pierens (2018)),

$$\Sigma(r, \Phi) = \Sigma_\mathrm{p} \exp\left(-\frac{(r - r_0)^2}{2\sigma_r^2} - \frac{(\Phi - \Phi_0)^2}{2\sigma_\Phi^2}\right). \tag{2.2}$$

Here, Σ_p is the peak density inside the vortex, r_0 and Φ_0 are the center coordinates of the vortex, and σ_r and σ_Φ are a measure for the extent of the vortex in each direction.

Vortex detection algorithm.

These three characteristics of anticyclonic vortices in disks,

- the local minimum of vorticity,
- the elliptical shape,
- and the Gaussian density peak

are used as the basis for our detection algorithm.

For the detection of vortices in the velocity field, we first compute ω and then divide by Σ to get the vortensity $\varpi = \omega_z / \Sigma$. The additional division by Σ (which peaks inside of the vortex) enhances the contrast between the background disk and the vortex interior, where ω_z has a minimum. To be able to use the algorithm in different scenarios, the vortensity is also normalized to the background vorticity ϖ_K, usually resulting in vortensity values between -1 and 1. This normalization has the benefit of removing the radial variation due to the background profile of Σ and Ω_K.

The first step of the algorithm is the identification of vortex candidates. We use the fact that the vortensity has a minimum inside of the vortex. Thus, iso-value lines of vortensity must be closed around the vortex center. Hence, closed iso-value lines are candidates for a vortex. The iso-vortensity lines are extracted using the computer vision library `OpenCV`.

The second step is to identify possible vortices by their elliptical shape. To compare the contours to ellipses, a fit of the ellipse equation to the contour lines is performed, and the difference between the area of the fitted ellipse and the area enclosed by the closed contour is computed. If the relative difference between the two areas is smaller than a

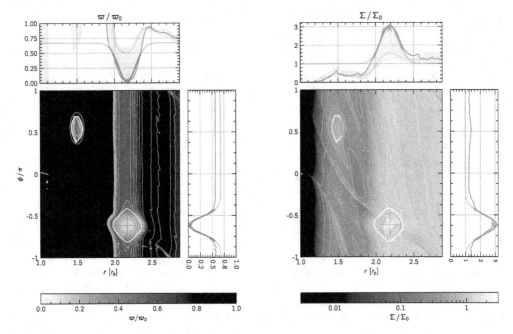

Figure 1. Vortex detection in a planet-disk simulation based on iso-vortensity lines using the `Vortector` utility. Normalized vortensity ϖ/ϖ_K and surface density Σ are shown in the left and right panels, respectively, with iso-value lines in grey. The detected vortices are indicated by the white contour line and the ellipses (corresponding to the FWHM extent). The side panels show cuts through the center of the vortex marked by the crosshair and Gaussian bell curves fitted to the 2D data. Adapted from Fig. B.2 in Rometsch et al. (2021).

threshold (we used 0.122), we consider the contour as a candidate for a vortex. Because multiple iso-value lines might surround a minimum, we only keep the outermost contours that enclose all other contours around the same center.

The final step is to fit the model density peak from Eq. (2.2) to the data inside the candidate contour. An example of the resulting vortex identification for a model from Rometsch et al. (2021) is shown in Fig. 1. Having a fit to the data allows for defining the vortex independent of detection parameters. We defined the vortex region as the area being contained within an ellipse in the $r - \Phi$ plane with semi-axes of $\sqrt{2\ln(2)}\sigma_r$ and $\sqrt{2\ln(2)}\sigma_\Phi$, thus corresponding to the full width at half maximum (FWHM) of the Gaussian fit.

This procedure was repeated for all simulation snapshots, resulting in time series data of vortex properties such as the minimum ϖ inside the vortex, the total mass enclosed in the vortex region, and the vortex location.

The algorithm is implemented in an optimized version in the `Vortector` Python package, available to the community on `github`. The identification of vortices in a single simulation snapshot with 1058x2048 grid cells took around 50-100 ms and the fitting of the Bell curves took another 150-300 ms on modern hardware, enabling vortex detection in well over the targeted 10^6 simulation snapshots within hours of computation on a compute cluster.

Astrophysical application.

In Rometsch et al. (2021), we simulated a grid of 2D fluid dynamics models of a PPD orbiting around a solar mass star and being host to an embedded Jupiter mass planet. We varied the Shakura & Sunyaev (1973) α-viscosity parameter between 10^{-6} and 10^{-3}

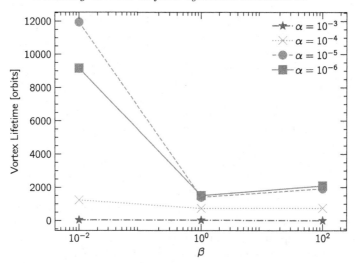

Figure 2. Dependence of vortex lifetimes on the cooling timescale β and the α viscosity parameter. The figure illustrates the existence of a long-lived regime for $\beta = 0.01$ and low viscosity $\alpha \leqslant 10^{-5}$. Lines are only shown to guide the eye. A simplified version of Fig. 4 in Rometsch et al. (2021) showing only the averages of the lifetimes obtained from the PLUTO and FARGO simulations.

and the β parameter between 10^{-2} and 10^{2} to investigate the combined effect of the cooling timescale and viscosity on the lifetime of vortices.

The mass of the planet was increased over a period of 100 to 1000 planetary orbits to mimic giant planet formation. The growing giant planet opens a gap in the disk, and vortices appear at the edges of the gap due to the Rossby-wave instability (Lovelace et al. 1999) — in this study, we focused on the vortices located at the outer gap edge. First, multiple small vortices emerge, which usually merge into one large vortex. This large vortex then dissipates due to viscous spreading such that the lifetime increases for lower viscosities. One inquiry in the study was the lifetime of these planet-induced vortices at low values of α.

An overview of the resulting vortex lifetimes for the chosen set of parameters is shown in Fig. 2. Each model was simulated with the PLUTO (Mignone et al. 2007) and the FARGO (Masset 2000) codes, which use different numerical schemes to solve the fluid dynamics equations, to increase confidence in the validity of the numerical results.

One of the studies' main results is the apparent existence of two regimes of vortex life at low values of viscosity. For low viscosities, $\alpha \leqslant 10^{-5}$, vortices in a disk with a short cooling timescale ($\beta = 0.01$) can live longer for nearly one order of magnitude compared to disks with a higher cooling timescale ($\beta \geqslant 1$), in agreement with earlier studies (Les & Lin 2015; Tarczay-Nehéz et al. 2020). While several possible vortex decay mechanisms other than viscous spreading exist, e.g., the elliptical instability (Lesur & Papaloizou 2009), the results hint at the existence of another vortex dispersion mechanism that depends on the thermodynamics and radiative processes of the disk and which only becomes dominant at low viscosities. The precise mechanism of vortex dissipation needs further investigation in future studies.

3. Implications

A reference for vortex characterization in simulations.

With the development of the Vortector, we introduced a well-defined and repeatable vortex detection and characterization algorithm. We hope that this tool provides a reproducible way for the identification and characterization of vortices in grid-based fluid dynamics simulations of PPDs and that it facilitates the comparison of vortex properties across different studies.

Prediction for the distance at which vortices will be observed.
The cooling timescale is expected to vary in PPD, decreasing with distance from the central star from $\beta = 1 - 10$ at ∼5 au to $\beta = 0.1 - 0.01$ at ∼50 au (Ziampras et al. 2020). Assuming low values of viscosity and combining these expected cooling timescales with the results in Rometsch et al. (2021), planet-induced vortices can be expected to live for several ten kyrs at 5 au (in the short-lived regime for $\beta \geqslant 1$), several hundred kyrs at 50 au, and several Myrs at 100 au (in the long-lived regime for $\beta < 1$). Based on lifetime consideration alone, the results of our study, therefore, suggest that vortices in PPDs are more likely to be observed at large distances (≥ 50 au) from the star.

The author acknowledges funding by the Deutsche Forschungsgemeinschaft (DFG, German Research Foundation) – 325594231 and support by the High Performance and Cloud Computing Group at the Zentrum für Datenverarbeitung of the University of Tübingen, the state of Baden-Württemberg through bwHPC and the German Research Foundation (DFG) through grant INST 37/935-1 FUGG.

References

Bae, J., Zhu, Z., & Hartmann, L. 2016, *ApJ*, 819, 134

De Val-Borro, M., Artymowicz, P., D'Angelo, G. & Peplinski, A. 2007, *A&A*, 471, 1043

Gammie, C. F. 2001, *ApJ*, 553, 174

Godon, P. & Livio, M. 1999, *ApJ*, 523, 350

Goodman, J., Narayan, R., & Goldreich, P. 1987 *MNRAS* 225, 695–711

Les, R., & Lin, M.-K. 2015, *MNRAS*, 450, 1503

Lesur, G., & Papaloizou, J. C. B. 2009, *A&A*, 498, 1

Lin, M.-K., &, Pierens, A. 2018, *MNRAS*, 478, 575-591

Lovelace, R. V. E., Li, H., Colgate, S. A., & Nelson, A. F. 1999, *ApJ*, 513, 805

Masset, F. 2000, *A&AS*, 141, 165

Mignone, A., Bodo, G., Massaglia, S., et al. 2007, *ApJS*, 170, 228

Miranda, R., & Rafikov, R. R. 2020, *ApJ*, 904, 121

Pérez, L. M., Benisty, M., Andrews, S. M., et al. 2018, 2018 *ApJ*, 869, L50

Rometsch, T., Ziampras, A., Kley, W. & Béthune, W. 2021, *A&A*, 656, A130

Shakura, N. I., & Sunyaev, R. A. 1973, *A&A*, 500, 33

Tarczay-Nehéz, D., Regály, Z., & Vorobyov, E. 2020, *MNRAS*, 493, 3014

Marel, N. v. d., Dishoeck, E. F. v., Bruderer, S. et al. 2013, *Science*, 340, 1199

Ziampras, A., Kley, W., & Dullemond, C. P. 2020, *A&A*, 637, A50

Discussion

SIMON PORTEGIES ZWART: The vortices seem to grow very quickly, very massive, then stay for a while and then sort of die out, but the quick appearance also suggests that there might be some influence by the initial conditions. How do you explain the sudden appearance of these huge vortices?

ROMETSCH: The vortex grows as the planet is introduced into the simulation, and its mass is gradually increased to mimic the gas accretion of the giant planets. The timescale is on the lower side of timescales expected for runaway accretion. We also checked for a planet growth timescale of 1000 orbits instead of 100 orbits, and the same effect appears. As soon as the giant planet opens a gap, the Rossby-wave instability kicks in, and the vortices form very quickly.

SIMON PORTEGIES ZWART: But it's a bit of a chicken and the egg problem, right? Who is first, the vortex or the planet?

ROMETSCH: Yes, it is. But you can also form vortices just by viscosity transitions alone, e.g., at a MRI dead zone. The fast vortex growth is a feature of the Rossby-wave instability rather than the planet growth process.

SHYAM MENON: You mentioned that vortices are also locations of dust traps, so you have a density peak there. Naively, one would expect that you would scale with surface density rather than its inverse when you are looking for vortices.

ROMETSCH: The background Keplerian disk has a positive vorticity contribution. The anticyclonic vortices have a negative contribution, thus they lower the vorticity. Then dividing by the surface density enhances the contrast between the background disk and vortices.

The Predictive Power of Computational Astrophysics as a Discovery Tool
Proceedings IAU Symposium No. 362, 2023
D. Bisikalo, D. Wiebe & C. Boily, eds.
doi:10.1017/S1743921322001818

Modeling gravitational few-body problems with tsunami and okinami

Alessandro A. Trani[1,2] and Mario Spera[3,4,5]

[1]Department of Earth Science and Astronomy, College of Arts and Sciences, The University of Tokyo, 3-8-1 Komaba, Meguro-ku, Tokyo 153-8902, Japan

[2]Okinawa Institute of Science and Technology, 1919-1 Tancha, Onna-son, Okinawa 904-0495, Japan

[3]SISSA, Via Bonomea 265, I-34136, Trieste, Italy

[4]INFN, Sezione di Trieste, I-34127 Trieste, Italy

[5]INFN, Sezione di Padova, Via Marzolo 8, I-35131, Padova, Italy
email: aatrani@gmail.com

Abstract. In recent years, an increasing amount of attention is being paid to the gravitational few-body problem and its applications to astrophysical scenarios. Among the main reasons for this renewed interest there is large number of newly discovered exoplanets and the detection of gravitational waves. Here, we present two numerical codes to model three- and few-body systems, called TSUNAMI and OKINAMI. The TSUNAMI code is a direct few-body code with algorithmic regularization, tidal forces and post-Newtonian corrections. OKINAMI is a secular, double-averaged code for stable hierarchical triples. We describe the main methods implemented in our codes, and review our recent results and applications to gravitational-wave astronomy, planetary science and statistical escape theories.

Keywords. stars: kinematics and dynamics, methods: numerical, gravitational waves, gravitation

---

1. Introduction

The gravitational three-body problem has a 300 year old history, dating back to Newton, Poincaré and many others. Rather than just being a didactic tool for the mathematical physicist, the three-body problem has numerous applications to modern astrophysical conundrums. Thanks to the recent advancement in observational astronomy, the three-body problem (and more generally, the few-body problem) is experiencing a renewed interest. Such interest is driven by the detection of gravitational waves in 2015, and the subsequent birth of gravitational-wave astronomy (The LIGO Scientific Collaboration et al. 2021). In fact, three-body interactions between compact objects have been proposed as one of the key formation mechanisms of gravitational-wave sources.

Another area of interest for three-body problems is exoplanet formation and evolution. This was made possible thanks to the rapid increase in exoplanet detections from transit surveys (K2, TESS, Howell et al. 2014; Ricker et al. 2015), and the characterization of numerous exotic planetary systems (i.e. hot Jupiters, ultra-short period planets, compact resonant chains). The formation of such exotic systems can be explained with gravitational few-body interactions between planets or passing stars. In addition, the recent reports on exomoon candidates have opened up questions on how extrasolar moons form and evolve.

Modeling few-body gravitational interactions is not an easy task. One issue arises from the nature of the gravitational force, which scales as $\propto r^{-2}$, where r is the separation between two particles. When two particles get very close, $r \to 0$ and the acceleration increases dramatically. Using traditional integrators, like the Runge-Kutta or Hermite methods, as the acceleration increases, the timestep needs to be reduced accordingly, in order to time-resolve the trajectory of the particles with sufficient accuracy. This can possibly lead to the halt of the integration, or to the faster accumulation of integration errors due to the increased number of timesteps.

We have developed two codes, named TSUNAMI and OKINAMI that employ different techniques in order to accurately model few-body gravitational interactions. Here, we describe the main numerical methods that we implemented, along with their applications to astrophysical scenarios.

2. Overview of the codes

TSUNAMI and OKINAMI implement different methods and therefore have slightly different scopes. The main difference is that while TSUNAMI can simulate systems of hundreds of particles in arbitrary configurations, OKINAMI can model only hierarchical stable triples. Both codes can be interfaced through a dedicated Python library and come with several example scripts.

2.1. *The* TSUNAMI *code*

TSUNAMI is based on the following techniques: regularization of the equations of motion, chain coordinates to reduce round-off errors and Bulirsch–Stoer extrapolation. The first technique (regularization) takes care of the singularity of the gravitational potential for $r \to 0$. The second technique (chain coordinates) helps reducing the round-off errors in hierarchical systems, which arise with the center-of-mass coordinates, without the need to include numerically expensive techniques of compensated summation. The third method (Bulirsch–Stoer extrapolation) increases the accuracy of the integration and makes it adaptable over a wide dynamical range.

TSUNAMI solves the Newtonian equations of motion derived from a modified, extended Hamiltonian (Mikkola and Tanikawa 1999a,b). As a consequence, time is another variable that is integrated along positions and velocities of the particles. Along one timestep of fictitious time ΔS, the physical time ΔT is advanced by:

$$\Delta T = \frac{\Delta S}{\alpha U + \beta \Omega + \gamma} \tag{2.1}$$

where U is the potential energy, Ω is a function of positions, and α, β and γ are arbitrary coefficients. Setting the values for (α, β, γ) effectively changes the regularization algorithm. $(\alpha, \beta, \gamma) = (1, 0, 0)$ corresponds to the logarithmic Hamiltonian algorithm, $(\alpha, \beta, \gamma) = (0, 1, 0)$ is equivalent to the time-transformed leapfrog scheme, and for $(\alpha, \beta, \gamma) = (0, 0, 1)$ the integration scheme reduces to the non-regularized leapfrog.

Positions and velocities are integrated in a chain coordinate system, rather than in the center-of-mass coordinates. This has the effect of reducing by 1 the number of equations to be integrated, and more importantly it reduces round-off errors when calculating distances between close particles far from the center of mass of the system (Mikkola and Aarseth 1993). These errors can quickly arise if the inter-particle separation is very small compared to the distance from the center of mass, due to the limits of floating-point arithmetic, which can happen, for example, in case of close binaries far

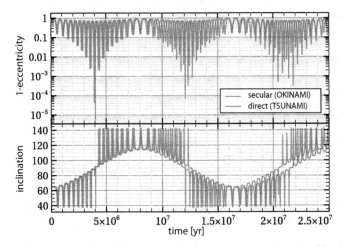

Figure 1. Comparison between TSUNAMI (red, direct integration) and OKINAMI (blue, secular average) in evolving a Jupiter-Sun-brown dwarf system. The top panel shows the eccentricity of the Jupiter's orbit around the Sun-like star, while the bottom panel shows the mutual inclination between the Jupiter's orbit and the brown dwarf's orbit. As the Jupiter is perturbed by the outer brown dwarf, the Jupiter exhibits von Zeipel-Kozai-Lidov oscillations, with the typical flip of the orbit associated to the octupole-level secular interaction. This figure reproduces fig. 3 from (Naoz et al. 2013).

from a massive black hole. The chain of inter-particle vectors is formed so that all particles are included in the chain. The first segment of the chain is chosen to be the shortest inter-particle distance in the system. The next segment is included so that it connects the particle closest to one of the ends of the current chain. This process is repeated until all particles are included. As the system evolves, care is taken to update the chain so that any chained vector is always shorter than adjacent non-chained vectors. It is possible to directly transform the old coordinates into new chain coordinates without passing through the center-of-mass coordinates.

Finally, a simple leapfrog integration might not be accurate enough for some applications. Therefore, the accuracy of the integration can be improved with the Bulirsch-Stoer extrapolation. The idea behind Bulirsch-Stoer extrapolation is to consider the results of a numerical integration as being an analytic function of the stepsize h. The solution of a given time interval ΔS is computed for smaller and smaller substeps $h = \Delta S/N_{\text{steps}}$ and then it is extrapolated to $h \to 0$, using rational or polynomial functions.

The above integration scheme works well for Newtonian gravity. However, this is not enough to model some systems like binary black holes or planets, which require additional physics. TSUNAMI implements additional forces, like equilibrium tides (Hut 1981), dynamical tides (Samsing et al. 2018) and post-Newtonians corrections of order 1, 2 and 2.5, using the midpoint step described in Mikkola and Merritt (2008).

2.2. *The* OKINAMI *code*

Unlike TSUNAMI, OKINAMI is limited to stable hierarchical triples, that is, a binary whose center of mass forms another binary with a tertiary body. At its core, OKINAMI integrates the equations of motion derived from a three-body Hamiltonian, expanded at the octupole-level interaction and averaged over the mean anomalies of the inner and outer orbits. The double-average has the advantage of considerably speeding up the integration, because it avoids the integration of the "fast angles". On the other hand, this has two consequences: the information about the individual positions of the bodies along their orbit is lost, and the equations cannot describe the evolution of the system on timescales shorter than the inner and outer orbital periods. After the double average, what we obtain is a set of ordinary differential equations for the inner and outer eccentricities, (e_1, e_2), arguments of pericenter (ω_1, ω_2), longitudes of the ascending nodes (Ω_1, Ω_2) and orbital inclinations (i_1, i_2). OKINAMI integrates these equations using and adaptive Runge-Kutta-Fehlberg of order 7. In addition, OKINAMI implements also equilibrium tides and post-Newtonian terms of orders 1, 2 and 2.5 for the inner orbit.

As an example, Figure 1 shows the evolution of inclination and eccentricity of the orbit of a Jupiter-sized planet around a Sun-like star, orbited by a distant brown dwarf. The integration with TSUNAMI takes about 5 minutes, while the one with OKINAMI takes less than a second.

3. Applications

3.1. *Exoplanets and exomoons*

In Trani et al. (2020) we investigated the fate of exomoons around migrating hot Jupiters. We considered the scenario in which hot Jupiters experience high-eccentricity, tidally driven migration, due to the gravitational perturbation from a distant companion star. Physically, the system is a 4-body problem composed of two nested hierarchical triple systems: one triple is composed of the primary star, companion star and the Jupiter, the second triple is constituted by the moon, its host Jupiter and the main star.

This kind of system is an ideal test bench for TSUNAMI. Our code can accurately model the tidal forces required for the migration of the Jupiter and general relativity precession that can alter the long-term dynamics of the Jupiter and its moon. We found that exomoons are unlikely to survive the migration process of the host Jupiter. Massive moons can prevent the migration process entirely, by suppressing the eccentricity excitation induced by the secondary star. If the moon cannot shield the planet from perturbations, the Jupiter's orbit becomes increasingly eccentric, triggering the dynamical instability of the moon. Subsequently, most exomoons end up being ejected from the system or colliding with the primary star and the host planet. Only a few escaped exomoons can become stable planets after the Jupiter has migrated, or by tidally migrating themselves.

Even though close-in giants are ideal candidates for exomoon detections, our results suggest that it is unlikely for exomoons to be discovered around them, at least for planets migrated via high-eccentricity tidal circularization. Nonetheless, tidally disruptions or collisions of exomoons can still leave observational signatures, such as debris disks or chemically altered stellar atmospheres.

Besides exotic scenarios like exomoons and hot Jupiters, TSUNAMI is an excellent tool to assess the stability and long-term evolution of planetary systems (e.g. Livingston et al. 2019).

3.2. *Gravitational-wave radiation sources*

The astrophysical origin of gravitational-wave events from coalescing black holes binaries is still debated, though many of the proposed formation scenarios involve some kind of few-body gravitational interaction. Specifically, chaotic three-body interactions between compact-object binaries and single black holes can happen frequently in dense stellar systems. These interactions can alter the spin-orbit orientation of black hole binaries, which can then be inferred from gravitational-wave observations.

In Trani et al. (2021b), we estimated the spin parameter distributions of merging black-hole binaries, comparing them with the currently available data. Here, we introduced a new formation scenario that combines elements from both the isolated and the dynamical formation scenarios. We ran an extensive set of highly-accurate simulations with

TSUNAMI, and we used the results to estimate the intrinsic merger rates of black-hole binaries in combination with a semi-analytic model.

Assuming low natal black-hole spins ($\chi < 0.2$), our scenario reproduces the distributions of $\chi_{\rm eff}$ and $\chi_{\rm p}$ inferred from current observations. In particular, this model can explain the peak at positive $\chi_{\rm eff}$ with a tail at negative $\chi_{\rm eff}$, and the broad peak at $\chi_{\rm p} \sim 0.2$. This is in sharp contrast with the predictions of the isolated and the dynamical scenarios: the first fails to produce negative $\chi_{\rm eff}$, while the second predicts a symmetric distribution around $\chi_{\rm eff} \sim 0$.

In Trani et al. (2021a) we examined merging compact-object binaries in hierarchical triple systems. We first obtained a large sample of triples formed in low mass clusters through dynamical interactions, simulated using direct N-body methods (Rastello et al. 2021). Because we selected only stable triples, we evolved them using OKINAMI. We obtained the merger properties of binary black holes, black hole–neutron stars, and black hole–white dwarfs. The rates for binary black holes, black hole–neutron stars are about 100 times lower than those of binary mergers from the same clusters. This is caused by the lower merger efficiency of triple systems, which is about 100 times lower than that of binaries. Nonetheless, compact objects merging from triples have unique properties that can be used to discriminate them from other formation channels.

Compared to binary black-hole mergers from open clusters, mergers from triples have more massive primaries, with a mass distribution peaking at around $30\,{\rm M}_\odot$ rather than $10\,{\rm M}_\odot$. The mass ratio also peaks at smaller values of 0.3, in contrast to the cluster binaries pathway, which favors equal-mass binaries. This is caused by the von Zeipel-Kozai-Lidov mechanism, whose eccentricity-pumping effect is enhanced at low mass ratios.

TSUNAMI is also ideal to model compact objects and stars in proximity to massive black holes. We investigated the impact of three-body encounters around massive black holes on binary black hole coalescence in Trani et al. (2019b, see also Trani et al. 2019a and Trani 2020)

3.3. *Statistical solutions to the three-body problem*

The gravitational three-body problem is chaotic and has no general analytic solution, and only partial statistical solutions have been achieved so far (see Stone and Leigh 2019, and references therein). The main idea behind these statistical escape theories is to leverage chaos to predict the evolution of a three-body problem only in a statistical sense, using the assumption of thermodynamical ergodicity. Recently, Kol (2021) introduced a novel statistical theory based on the flux of phase space, rather than on phase-space volume like all the previous theories.

In a series of papers, we have been testing the statistical theories by simulating a large ensembles of three-body systems with TSUNAMI and comparing the final outcome distributions to the theoretical predictions. Our results in Manwadkar et al. (2020) and Manwadkar et al. (2021) show that the flux-based theory is in tighter agreement with the outcome of three-body simulations, compared to the previous statistical theories. We are now in the process of further testing the potential of this theory with the aim of providing a complete, accurate statistical description of the three-body problem.

References

Howell, S. B., Sobeck, C., Haas, M., Still, M., Barclay, T., Mullally, F., Troeltzsch, J., Aigrain, S., Bryson, S. T., Caldwell, D., Chaplin, W. J., Cochran, W. D., Huber, D., Marcy, G. W., Miglio, A., Najita, J. R., Smith, M., Twicken, J. D., and Fortney, J. J.: 2014, *PASP* **126(938)**, 398

Hut, P.: 1981, *A&A* **99**, 126

Kol, B.: 2021, *Celestial Mechanics and Dynamical Astronomy* **133(4)**, 17

Livingston, J. H., Dai, F., Hirano, T., Gandolfi, D., Trani, A. A., Nowak, G., Cochran, W. D., Endl, M., Albrecht, S., Barragan, O., Cabrera, J., Csizmadia, S., de Leon, J. P., Deeg, H., Eigmüller, P., Erikson, A., Fridlund, M., Fukui, A., Grziwa, S., Guenther, E. W., Hatzes, A. P., Korth, J., Kuzuhara, M., Monta nes, P., Narita, N., Nespral, D., Palle, E., Pätzold, M., Persson, C. M., Prieto-Arranz, J., Rauer, H., Tamura, M., Van Eylen, V., and Winn, J. N.: 2019, *MNRAS* **484(1)**, 8

Manwadkar, V., Kol, B., Trani, A. A., and Leigh, N. W. C.: 2021, *MNRAS* **506(1)**, 692

Manwadkar, V., Trani, A. A., and Leigh, N. W. C.: 2020, *MNRAS* **497(3)**, 3694

Mikkola, S. and Aarseth, S. J.: 1993, *Celestial Mechanics and Dynamical Astronomy* **57**, 439

Mikkola, S. and Merritt, D.: 2008, *AJ* **135**, 2398

Mikkola, S. and Tanikawa, K.: 1999a, *MNRAS* **310**, 745

Mikkola, S. and Tanikawa, K.: 1999b, *Celestial Mechanics and Dynamical Astronomy* **74**, 287

Naoz, S., Farr, W. M., Lithwick, Y., Rasio, F. A., and Teyssandier, J.: 2013, *MNRAS* **431(3)**, 2155

Rastello, S., Mapelli, M., di Carlo, U. N., Iorio, G., Ballone, A., Giacobbo, N., Santoliquido, F., and Torniamenti, S.: 2021, *arXiv e-prints* p. arXiv:2105.01669

Ricker, G. R., Winn, J. N., Vanderspek, R., Latham, D. W., Bakos, G. Á., Bean, J. L., Berta-Thompson, Z. K., Brown, T. M., Buchhave, L., Butler, N. R., Butler, R. P., Chaplin, W. J., Charbonneau, D., Christensen-Dalsgaard, J., Clampin, M., Deming, D., Doty, J., De Lee, N., Dressing, C., Dunham, E. W., Endl, M., Fressin, F., Ge, J., Henning, T., Holman, M. J., Howard, A. W., Ida, S., Jenkins, J. M., Jernigan, G., Johnson, J. A., Kaltenegger, L., Kawai, N., Kjeldsen, H., Laughlin, G., Levine, A. M., Lin, D., Lissauer, J. J., MacQueen, P., Marcy, G., McCullough, P. R., Morton, T. D., Narita, N., Paegert, M., Palle, E., Pepe, F., Pepper, J., Quirrenbach, A., Rinehart, S. A., Sasselov, D., Sato, B., Seager, S., Sozzetti, A., Stassun, K. G., Sullivan, P., Szentgyorgyi, A., Torres, G., Udry, S., and Villasenor, J.: 2015, *Journal of Astronomical Telescopes, Instruments, and Systems* **1**, 014003

Samsing, J., Leigh, N. W. C., and Trani, A. A.: 2018, *ArXiv e-prints*

Stone, N. C. and Leigh, N. W. C.: 2019, *Nature* **576(7787)**, 406

The LIGO Scientific Collaboration, the Virgo Collaboration, the KAGRA Collaboration, Abbott, R., Abbott, T. D., Acernese, F., Ackley, K., Adams, C., Adhikari, N., Adhikari, R. X., Adya, V. B., Affeldt, C., Agarwal, D., Agathos, M., Agatsuma, K., Aggarwal, N., Aguiar, O. D., Aiello, L., Ain, A., Ajith, P., Akcay, S., Akutsu, T., Albanesi, S., and Allocca, A. a.: 2021, *arXiv e-prints* p. arXiv:2111.03606

Trani, A. A.: 2020, in A. Bragaglia, M. Davies, A. Sills, and E. Vesperini (eds.), Star Clusters: From the Milky Way to the Early Universe, Vol. 351, pp 174–177

Trani, A. A., Fujii, M. S., and Spera, M.: 2019a, *ApJ* **875(1)**, 42

Trani, A. A., Hamers, A. S., Geller, A., and Spera, M.: 2020, *MNRAS* **499(3)**, 4195

Trani, A. A., Rastello, S., Di Carlo, U. N., Santoliquido, F., Tanikawa, A., and Mapelli, M.: 2021a, *arXiv e-prints* p. arXiv:2111.06388

Trani, A. A., Spera, M., Leigh, N. W. C., and Fujii, M. S.: 2019b, *ApJ* **885(2)**, 135

Trani, A. A., Tanikawa, A., Fujii, M. S., Leigh, N. W. C., and Kumamoto, J.: 2021b, *MNRAS* **504(1)**, 910

The Predictive Power of Computational Astrophysics as a Discovery Tool
Proceedings IAU Symposium No. 362, 2023
D. Bisikalo, D. Wiebe & C. Boily, eds.
doi:10.1017/S1743921322001272

Plasmoid Dominated Magnetic Reconnection and Particle Acceleration

Arghyadeep Paul[1] , Sirsha Nandy[1] and Bhargav Vaidya[1]

Dept. of Astronomy Astrophysics and Space Engineering
Indian Institute of Technology Indore,
Khandwa Road, Simrol, Indore 453552, India
email: `arghyadeepp@gmail.com`

Abstract. The effect of a parallel velocity shear on the explosive phase of magnetic reconnection in a double tearing mode is investigated within the 2D resistive magneto-hydrodynamic framework. All the systems follow a three phase evolution pattern with the phases delayed in time for an increasing shear speed. We find that the theoretical dependence of the reconnection rate with shear remains true in more general scenarios such as that of a plasmoid dominated double current sheet system. We also find that the power-law distribution of plasmoid sizes become steeper with an increasing sub-Alfvénic shear. We further demonstrate the effect of a velocity shear on acceleration of test particles pertaining to the modification in the energy spectrum.

Keywords. Explosive reconnection, Plasmoids, Particle acceleration, Double Tearing Mode

1. Introduction

The conversion of magnetic energy to kinetic and thermal energy through the process of magnetic reconnection is ubiquitous in astrophysical and laboratory plasmas e.g, solar flares, tokamaks etc. (Mann et al. (2009), Krasheninnikov (1999)). The well known Sweet-Parker model of magnetic reconnection predicts a dependence of the reconnection rate on the Lundquist number (S) of the system as $S^{-1/2}$. This rate, however, is much slower compared to the observations from the heliophysics domain with typical high Lundquist number values. Though Petschek type reconnection improves upon the Sweet-Parker model with a weaker dependence on S, sustaining it requires localized resistivity enhancement near current sheets (Huang & Bhattacharjee (2010)). A secondary tearing instability called "plasmoid instability" occurs in a high Lundquist number ($> 10^4$) regime, which can fragment an elongated current sheet into multiple X points, thereby facilitating a much faster magnetic reconnection with a weak dependence on Lundquist number (Daughton et al. (2006)). Multiple current sheet systems are quite prevalent in the reconnection regions (Crooker et al. (1993)) and a simplification of the same is the double tearing mode or DTM that exhibits a fast structure-driven non-linear growth phase (Janvier et al. (2011)). The dynamic nature of typical reconnection regions naturally suggests the presence of a velocity shear near the current sheets. A velocity shear flow parallel to the current sheets generally tends to throttle the growth of the tearing instability, thereby suppressing the reconnection rate. Though the relation of reconnection rate and shear speed in tearing instability (Cassak & Otto (2011)) is known to be applicable for plasmoid instability in a single current sheet (Hosseinpour et al. (2018)), the resonant flux-feedback effect of two current sheets in DTMs can affect the overall scaling of the reconnection rate with shear. Studies of particle acceleration to suprathermal energies near the reconnection regions reveal that the energy spectrum of the particles generally follows a power law of the form $N(E) \sim E^{-1.5}$ (Drake et al. (2013)). Drake et al. (2013) have also shown that this power law is the signature of a multi-island magnetic reconnection system where the acceleration time is shorter than the loss rate of the particles, albeit the dynamics in their study were treated as non-relativistic. Akramov & Baty (2017) have also performed test particle simulations in the non-relativistic regime to study the phenomena of particle acceleration in explosive DTM reconnection with particle integration being carried out in a static fluid background. However for the study of particle acceleration in explosive reconnection phases with fast moving transient features, it is preferable to have a fluid background that evolves with the particle. In this article, we investigate the dynamics of a double current sheet system in the presence of a parallel shear flow using resistive magneto-hydrodynamic simulations in a slab geometry with emphasis on the reconnection rate and the plasmoid size distribution. We also explore the phenomena of particle acceleration in such rapidly evolving systems by introducing test particles in a moving fluid background in order to study the effect of such a shear flow from the particle energization perspective. The article is organized as follows: Section 2 describes the numerical model setup with initial conditions. Section 3 elaborates on the results of the current study and Section 4 contains the summary and some relevant additional discussions.

2. Numerical Model

Fluid setup

The numerical setup solves a set of resistive MHD equations in a 2D slab geometry using the PLUTO code (Mignone et al. (2007)). The computational domain is confined within $-L_x \leq x \leq L_x$ and $-L_y \leq y \leq L_y$ where $L_x = 64.0$ and $L_y = 96.0$ and is resolved by an uniform Cartesian grid having 1280 and 1920 grid cells along the x and y directions respectively. Since the MHD equations are scale invariant, we omit any relation of the length scales in our setup with any physical length units at this stage. Periodic boundary conditions are imposed along the x-boundaries and the y-boundaries are set to be reflective. The system is considered to be isothermal and the initial magnetic field configuration is that of a double Harris-sheet equilibrium with current sheets along $y = \pm16$, with an asymptotic field strength $B_0 = \sqrt{2}$ and a magnetic field shear width of unity. The asymptotic Alfvén speed of the system, v_A, is defined by $B_0/\sqrt{\rho_\infty}$, and has the value $\sqrt{2}$ for our setup. Here, ρ_∞ is the density far away from the current sheet. An initial equilibrium was achieved by asserting a corresponding variation in density. An explicit resistivity was prescribed with a value of $\eta = 2 \times 10^{-5}$, which defines the corresponding Lundquist number to be, $S \sim 9 \times 10^6$. The time scale is normalised to the Alfvén transit time across the current sheet which is defined as $\tau_A = w_B/v_A$, where w_B is the magnetic field shear width. The in-plane velocity shear was prescribed to be of the same mathematical form as the magnetic field and the strength of the shear speed was varied from $0.0v_A$ (no shear) to $1.0v_A$ (Alfvénic shear) in steps of $0.25v_A$ for this study. A small divergence-free perturbation was applied to initiate the reconnection process.

Particle Setup

To study the impact of the shear flow on the particle acceleration mechanisms in the system, we inject a total of 4 million test particles in a subset of the domain bounded by $-64 \leq x \leq 64$ and $-40 \leq y \leq 40$. To incorporate the three degrees of freedom of the particles, the domain was also extended along the z direction between $-10 \leq z \leq 10$, along which the fluid quantities are considered to be invariant. The test particles used in our setup correspond to protons with a dimensionless charge to mass ratio of unity in code units. The particles were evolved using a relativistic Boris pusher algorithm

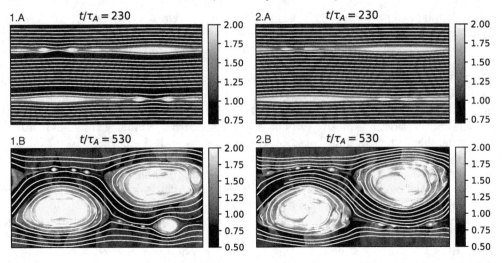

Figure 1. Figure shows the plasma density in code units over plotted by magnetic field streamlines. Figure numbers 1 and 2 (columns) correspond to a shear speed of $0.0v_A$ and $0.50v_A$ respectively and the labels A and B (rows) correspond to different times ($t/\tau_A \sim 230$ and $t/\tau_A \sim 530$).

(Mignone et al. (2018)). With the introduction of particles, the length scales are now represented in terms of c/ω_{pi}, which is the plasma skin depth with ω_{pi} being the ion plasma frequency. The time is given in the units of the inverse cyclotron frequency $\Omega^{-1} = c/(\omega_{pi}v_A)$. The particles were initialised with a gaussian distribution of velocities having a mean of zero and a standard deviation of $0.1v_A$. An extended description of the particle setup can be found in Paul & Vaidya (2021).

3. Results

General Evolution

All the systems, irrespective of the presence of a shear flow show a similar evolution pattern with three distinct phases: (a) A slow initial growth phase, (b) A fast plasmoid dominated explosive phase and (c) A turbulent gradual relaxation phase. These phases however, are delayed in time by a certain amount with an increase in shear speed. We thus describe below the general evolution characteristics for the system with no shear as a benchmark.

During the slow growth phase ($t/\tau_A \sim 0$ to $t/\tau_A \sim 170$), two large islands initially form and grow. The growth rate during this phase is consistent with the wave number of the applied initial perturbation. Subsequently, the portion of the current sheets outside of the two monster islands (the secondary current sheets) start to gradually thin due to their tendency to approach a steady state Sweet-Parker inverse aspect ratio given as $\delta_{SP}/L \sim S^{-0.5}$. This gradual thinning eventually fragments the current sheet into multiple small plasmoids and the system enters into the explosive plasmoid dominated fast reconnection phase. The smaller plasmoids merge into the monster plasmoids thereby increasing their size as seen from panels 1.B and 2.B of Figure 1. This increase in size enhances the coupling between the two current layers and has an effect similar to that of a coupled double tearing mode highlighted by Janvier et al. (2011). The growth rate normalised to the Alfvén transit time was calculated to be $\gamma\tau_A \sim 4.3 \times 10^{-2}$, by fitting an exponential curve to the time evolution of the maximum value of the B_y component ($|B_y|_{max}$) during the early phases ($t/\tau_A \sim 180$ to $t/\tau_A \sim 220$) of the plasmoid instability.

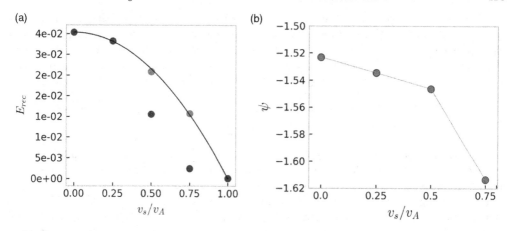

Figure 2. Panel 2(a) shows the variation of the reconnection rate with shear. The solid black line corresponds to the theoretical reconnection rate. The blue circles correspond to the reconnection rates measured during the same time interval for all the shear speeds ($t/\tau_A \sim 283$ to $t/\tau_A \sim 293$). The red circles correspond to the reconnection rates measured at different times ($t/\tau_A \sim 311$ for $v_s/v_A \sim 0.5$ and $t/\tau_A \sim 336$ for $v_s/v_A \sim 0.75$), when the island widths of all the systems are similar. Figure 2b shows the variation of the median power law indices (ψ) of the plasmoid size distribution with shear v_s/v_A.

Reconnection rate

To measure the effects of the velocity shear on the reconnection rate (E_{rec}) of the system, we measure the combined reconnection rates of the two current sheets for each system following the method given in Paul & Vaidya (2021). The reconnection rate, E_{rec} is given as the sum of the average values of the magnitude of the out of plane electric field E_z in two thin strips ($\Delta y = 2.0$) along the two current sheets. This is normalized to $B_0 v_A$ and its mathematical formulation is given below (Paul & Vaidya (2021)):

$$E_{\text{rec}} = \frac{1}{B_0 v_A} \left(\left\langle \left| \frac{\iint_{y_1}^{y_2} E_z dx dy}{\iint_{y_1}^{y_2} dx dy} \right| \right\rangle + \left\langle \left| \frac{\iint_{-y_1}^{-y_2} E_z dx dy}{\iint_{-y_1}^{-y_2} dx dy} \right| \right\rangle \right).$$

Here $y_1 = 15$ and $y_2 = 17$ enclosing the current sheets. Following Cassak & Otto (2011), one expects the reconnection rate to decrease with an increase in shear following the relation $E = E_0 \left(1 - \left(v_s^2/v_A^2 \right) \right)$, where E_0 is the reconnection rate in the absence of a shear, v_s is the strength of the shear flow and v_A is the Alfvén speed of the system. We find that the system follows the theoretical scaling of the reconnection rate quite well during the early phases of the evolution before the plasmoid instability has set in (figure not shown). However, during the explosive reconnection phase (E_{rec} measured between $t/\tau_A \sim 283$ to $t/\tau_A \sim 293$), the systems with higher shear speeds deviate significantly from the theoretical scaling as seen from the blue circles in Figure 2a. The reason for this is, in a double current sheet system, the monster plasmoid of either current sheet has a feedback effect on the other wherein they tend to push magnetic flux towards the X-points of the other current sheet (Janvier et al. (2011)). Thereby, the systems with larger sized primary magnetic islands, which are the systems with a lower shear speed, experience an enhancement in the reconnection rate as seen in Figure 2a for the two systems with a shear speed of zero and $0.25v_A$. We ascertain the above statement by measuring the reconnection rate of the systems with a shear of $0.5v_A$ and $0.75v_A$ when the width of the primary islands in those setups are similar to that of the setup with zero shear (the setups with zero and $0.25v_A$ shear have very similar island widths). This happens at $t/\tau_A \sim 311$ for $v_s/v_A \sim 0.5$ and $t/\tau_A \sim 336$ for $v_s/v_A \sim 0.75$ and the same has been plotted using red circles on Figure 2a which coincide well with the theoretical curve. This ascertains that the deviation is primarily due to the difference in the island sizes between setups with different shears which has a significant effect on the reconnection rate.

Plasmoid Size Distribution

We use the method of granulometry on the density distribution data to find the variation of the size distribution of plasmoids with an increasing shear flow. Granulometry is the application of the method of binary opening on a binary field with the mask size increasing gradually over each iteration. Before applying binary opening, the 2-dimensional plasma density data obtained from each data file saved at an interval of $1.4\tau_A$ is made binary by making use of the Otsu filter available in python skimage library. For our case, the mask shape is circular which calculates the equivalent radii of the plasmoids. We find that the obtained plasmoid size variation fits well with a power-law distribution $N(r) \sim r^\psi$, where 'r' is the equivalent radius of plasmoid. This power law distribution is consistent with simulations by Petropoulou et al. (2018) as well as observations (Patel et al. (2020)). We first fit the size distribution data for each time step to obtain the power-law index for that step. We then calculate the median of the power-law index (ψ) during a period when the variation of the index with time is nearly stationary. We repeat the whole process mentioned above for each of the four different sub-Alfvénic shear values excluding the Alfvénic shear. Figure 2b shows a decreasing trend in the values of the median power-law indices with increasing shear, indicating that the plasmoid size distribution gets steeper with an increasing shear. This suggests that as the shear speed increases, smaller plasmoids tend to be more prevalent in the system.

Particle Energy Spectrum

To investigate the energy distribution of the particles accelerated during the explosive plasmoid dominated phase, we inject test particles in a sub-domain of two setups, one with zero shear and other with a significant shear of $0.75\ v_A$. Since the explosive phase starts at different times for the two setups, the injection time is different. Therefore the energy spectrum is plotted with respect to $\Omega\Delta t'$ which denotes the duration for which the particles were integrated. We note here that the quantity $\Omega t'$ is equivalent to the fluid evolution time scale t/τ_A.

The particle energy E_{tot} is given by $0.5v_p^2$, where v_p is the particle velocity. The particle spectrum gradually evolves to form a time invariant power law distribution with a spectral index $\alpha = -1.24$, as seen from Figure 3 which shows the particle energy spectrum at three different times, namely, the early evolution phase ($\Omega\Delta t' = 68$), an intermediate stage ($\Omega\Delta t' = 109$) and the late evolution phase ($\Omega\Delta t' = 328$). The red and blue solid lines in Figure 3 correspond to the late evolution phase exhibiting the time invariant power law index. We highlight here that the presence of a shear has negligible effect on the power law index of the spectrum as seen from Figure 3 and the difference in the final time invariant spectrum of the two systems lie in the high energy tail which is truncated at lesser energies for the setup with a shear of $0.75v_A$. An inspection of the trajectories of some of the highly energized particles show that the particles accelerate in the in-plane direction as well as the out of plane direction where the in-plane energization occurs due to the interaction of the particles with fast moving transient magnetic structures in the domain, whereas, the out of plane energization is provided by the electric field which has

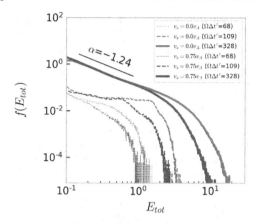

Figure 3. The normalised particle energy spectrum for the setup with zero shear (blue) and the setup with a shear of $0.75v_a$ (red) for various integration time durations. The dotted, dashed and solid lines represent time integration duration of $\Omega \Delta t' = 68$, 109 and 328 respectively.

a convective $(\mathbf{v} \times \mathbf{B})$ and a resistive $(\eta \mathbf{J})$ component. The former in-plane energization process was captured due to the fact that the particles were integrated in an evolving fluid background.

4. Summary and Discussions

We use resistive magneto-hydrodynamic simulations to examine the evolution of a high Lundquist number double current sheet system that exhibits an explosive plasmoid dominated reconnection phase. We also introduce test particles in a dynamic fluid background to investigate the phenomena of particle acceleration and the effect of the shear flow on the same. The principal highlights of the study are:

• All the systems show a three phase evolution pattern with an initial slow growth phase followed by a fast explosive reconnection phase which terminates into a gradual turbulent relaxation phase. An in-plane velocity shear generally works to suppress the tearing instability and thus, the second and the third phases are delayed in time by a certain amount with an increase in the shear flow magnitude.

• The reconnection rate obtained during the early phases is well in agreement with the theoretical values. During the later stages however, the reconnection rate agrees with the theoretical values only when the rates are measured at different times when the size of the primary magnetic islands (monster plasmoids) are similar. This shows that the variation of size of the primary magnetic islands must be taken into account while determining the scaling of the reconnection rate.

• The plasmoid size distribution is obtained using the granulometry technique. We found that the plasmoid size distribution fits well with a power law distribution. We also found that the median power law index ψ for the setups with different sub-Alfvénic shears lies within a range of -1.61 to -1.52 and has a decreasing trend with an increase in the magnitude of shear. This indicates that smaller plasmoids are more prevalent in setups with higher shear speed.

• The energy spectrum of the accelerated particles also eventually attains a time invariant power law distribution with a spectral index of -1.24. The presence of a shear flow has negligible effect on the spectral index of the particles. However, the presence of a shear flow does make the overall acceleration process less efficient (see Paul & Vaidya (2021) for details), which results in a truncation of the high energy portion of the spectrum at a lower value.

416 A. Paul, S. Nandy & B. Vaidya

In this study, however, we have not considered the effect of a guide field which can have significant effects on the particle acceleration due to ion heating. This study also considers the ions as test particles meaning that they do not provide any feedback to the fluid. As a natural extension of this study, the effect of the particle feedback, wherein, the motion of the particles deposits fields back to the grid, would be an interesting future prospect.

References

Paul, A. & Vaidya, B. 2021, Physics of Plasmas, 28, 082903. doi:10.1063/5.0054501
Mignone, A., Bodo, G., Massaglia, S., et al. 2007, apjs, 170, 228. doi:10.1086/513316
Mignone, A., Bodo, G., Vaidya, B., et al. 2018, apj, 859, 13. doi:10.3847/1538-4357/aabccd
Janvier, M., Kishimoto, Y., & Li, J. Q. 2011, prl, 107, 195001. doi:10.1103/PhysRevLett.107.195001
Cassak, P. A. & Otto, A. 2011, Physics of Plasmas, 18, 074501. doi:10.1063/1.3609771
Patel, R., Pant, V., Chandrashekhar, K., et al. 2020, aap, 644, A158. doi:10.1051/0004-6361/202039000
Petropoulou, M., Christie, I. M., Sironi, L., et al. 2018, mnras, 475, 3797. doi:10.1093/mnras/sty033
Hosseinpour, M., Chen, Y., & Zenitani, S. 2018, Physics of Plasmas, 25, 102117. doi:10.1063/1.5061818
Drake, J. F., Swisdak, M., & Fermo, R. 2013, apjl, 763, L5. doi:10.1088/2041-8205/763/1/L5
Akramov, T. & Baty, H. 2017, Physics of Plasmas, 24, 082116. doi:10.1063/1.5000273
Huang, Y.-M. & Bhattacharjee, A. 2010, Physics of Plasmas, 17, 062104. doi:10.1063/1.3420208
Daughton, W., Scudder, J., & Karimabadi, H. 2006, Physics of Plasmas, 13, 072101. doi:10.1063/1.2218817
Crooker, N. U., Siscoe, G. L., Shodhan, S., et al. 1993, jgr, 98, 9371. doi:10.1029/93JA00636
Mann, G., Warmuth, A., & Aurass, H. 2009, aap, 494, 669. doi:10.1051/0004-6361:200810099
Krasheninnikov, S. I. 1999, APS Division of Plasma Physics Meeting Abstracts

5. Questions and Answers

Q1. There have been studies of particle acceleration in the presence of velocity gradients without considering reconnection where the particles accelerate in some kind of Fermi mechanisms. Do you see any of that in your setup?

Ans: In our setup, the acceleration was actually due to a combined effect of electric fields and Fermi mechanisms. It is difficult to isolate the particle energization that occurs solely due to the velocity gradient.

Q2. How complex would it be to set up your code in different geometries? You gave some examples of Cartesian meshing but are you able to adapt the geometry of your code to the problem that you want to treat?

Ans: The simulations that we did were solved using the MHD code PLUTO which has this hybrid MHD-PIC framework. All the simulations that I discussed are indeed done in Cartesian coordinates because the particle module in PLUTO code currently only supports the Cartesian geometry for now. Maybe in the future we will be able to address different geometries.

Author index

Arca Sedda, M. – 203
Armijos-Abendaño, J. – 250
Asano, T. – 116
Avtaeva, A. A. – 164

Baba, J. – 116
Babyk, I. V. – 100
Banda-Barragán, W. E. – 56, 250
Barbieri, L. – 134
Bédorf, J. – 116
Bending, T. – 105
Benitez-Llambay, A. – 15
Berczik, P. – 128, 203
Binoy, J. – 175
Bisikalo, D. – 228
Bisikalo, D. V. – 152, 167
Bourgeois, R. – 373
Brighenti, F. – 56
Brose, R. – 262
Brüggen, M. – 56
Bufferand, H. – 134

Candlish, G. N. – 26
Casavecchia, B. – 56
Casetti, L. – 134
Celeste, M. – 262
Cernic, V. – 148
Cerutti, B. – 184
Chan, T. K. – 15
Chattopadhyay, T. – 122
Chiaki, G. – 45
Ciraolo, G. – 134
Cournoyer-Cloutier, C. – 300
Crinquand, B. – 184

Danehkar, A. – 64
Dattathri, S. – 51
Demidova, T. – 356
Demidova, T. V. – 279
Dénes, H. – 250
di Carlo, U. N. – 134
Di Cintio, P. – 134
Dmitriev, D. V. – 279
Dobbs, C. – 105, 380
Dobrycheva, D. – 111
Doğan, S. – 177
Dudorov, A. – 273

Eapen, P. E. – 175
El Mellah, I. – 184

Federrath, C. – 358
Fischer, M. S. – 8
Frenk, C. – 15
Fujibayashi, S. – 190
Fujii, M. – 209
Fujii, M. S. – 116, 258

Garaldi, E. – 1
Getling, A. – 333
Ghaziasgar, S. – 353
Ghendrih, P. – 134
Giersz, M. – 203
Giri, G. – 70
Girish, T. E. – 175, 330
Gladysheva, Y. G. – 167
Gómez de Castro, A. I. – 234
Gopinath, S. – 330
Granovsky, S. – 169, 324
Gray, W. J. – 64
Green, S. – 262
Grigoryev, V. V. – 279
Guerrero, G. – 333
Gunn, J. P. – 134
Gupta, S. – 134

Hanawa, T. – 288
Haworth, T. J. – 262
Hayashi, K. – 190
Hernquist, L. – 1
Hirai, Y. – 209, 258
Hirashima, K. – 209

Igarashi, T. – 76
Irodotou, D. – 33
Ishchenko, M. – 128
Izviekova, I. O. – 100

Jaffa, S. – 380
Javadi, A. – 353
Johansson, P. H. – 33
Jung, M. – 358

Kalinicheva, E. S. – 173
Kang, H. – 87
Kannan, R. – 1
Kargaltseva, N. – 273
Kato, Y. – 76
Kavanagh, R. D. – 262
Khaibrakhmanov, S. – 273
Kharchenko, N. – 128

Khramtsov, V. – 111
Kim, H. – 134
Kimura, K. – 246
King, A. R. – 177
Kirilova, D. – 21
Kirsanova, M. S. – 268
Kitiashvili, I. – 324
Kitiashvili, I. N. – 169
Kiuchi, K. – 190
Kochina, O. V. – 282
Komaki, A. – 294
Kompaniiets, O. V. – 100
Kosovichev, A. – 333
Kovaleva, D. – 150
Krumholz, M. R. – 358
Kuiper, R. – 358
Kumar, V. M. – 175
Kundu, S. – 365

Lake, W. – 45
Lee, D. – 373
Lepri, S. – 134
Lewis, S. – 300
Liao, S. – 33
Liow, K. Y. – 105, 380
Livi, R. – 134
Longhinos, B. – 175

Mac Low, M.-M. – 300
Mackey, J. – 262
Makino, J. – 209
Maksimova, L. – 306
Maliszewski, K. – 203
Mannerkoski, M. – 33
Matsumoto, R. – 76
Matsumoto, Y. – 76
McMillan, S. L. W. – 300
Menon, S. H. – 358
Mezzacappa, A. – 215
Mitani, H. – 158
Mondal, D. – 122
Moriwaki, K. – 209
Moutzouri, M. – 262
Muru, M. M. – 54

Nakatani, R. – 158, 294
Nakazato, Y. – 45
Nandy, S. – 410
Naoz, S. – 45
Navabi, M. – 353
Nixon, C. J. – 177

O'Rourke, C. – 262
Oey, M. S. – 64
Ohsuga, K. – 76

Pakmor, R. – 1
Panayotova, M. – 21
Parfrey, K. – 184
Pasquato, M. – 134
Paul, A. – 410
Pavlyuchenkov, Y. – 306
Pavlyuchenkov, Y. N. – 268
Peña, G. A. – 26
Pipin, V. – 333
Pogorelov, N. V. – 309
Polak, B. – 300
Portegies Zwart, S. – 116, 300
Price, D. – 177
Pringle, J. E. – 177

Qiu, Y. – 82

Rantala, A. – 33
Rawlings, A. – 33
Rieder, S. – 105, 380
Rizzuto, F. – 33
Rometsch, T. – 398
Rosswog, S. – 382
Ryu, D. – 87

Saitoh, T. – 209
Saitoh, T. R. – 258
Sakai, N. – 288
Sapozhnikov, S. – 150
Saremi, E. – 353
Sathyan, T. N. – 175
Schmidt, W. – 94
Selg, S. – 94
Sellentin, E. – 116
Seo, J. – 87
Sharma, M. – 39
Sharma, P. – 51
Shematovich, V. I. – 152, 164, 173
Shibata, M. – 190
Shimajiri, Y. – 258
Simon Petit, A. – 134
Smith, A. – 1
Sobolenko, M. – 203
Sobolev, A. – 228
Spera, M. – 404
Springel, V. – 1
Spurzem, R. – 203
Stejko, A. – 333

Takahashi, H. R. – 76
Theuns, T. – 15
Torki, M. – 353
Torniamenti, S. – 141
Tran, A. – 300
Trani, A. A. – 134, 404

Vaidya, B. – 70, 365, 410
van Loon, J. T. – 353
Van Looveren, G. – 282
Vasylenko, A. A. – 100
Vasylenko, M. – 111
Vavilova, I. – 111
Vavilova, I. B. – 100
Vinod Kumar, R. – 330
Vipindas, V. – 330
Vogelsberger, M. – 1

Wanajo, S. – 190
Wang, L. – 258
Wibking, B. D. – 358
Wiebe, D. S. – 282

Wilhelm, M. J. C. – 300
Wray, A. – 324
Wray, A. A. – 169
Wurster, J. – 105

Yamamoto, S. – 288
Yeou, C. – 45
Yoon, S.-J. – 134
Yoshida, N. – 45, 158, 294

Zamozdra, S. – 273
Zargaryan, D. – 262
Zhilkin, A. – 228, 273
Zhilkin, A. G. – 167

Printed in the United States
by Baker & Taylor Publisher Services